Lecture Notes in Computer S

Commenced Publication in 1973
Founding and Former Series Editors:
Gerhard Goos, Juris Hartmanis, and Jan van Leeuwen

Steven D. Galbraith   Kenneth G. Paterson (Eds.)

Pairing-Based Cryptography – Pairing 2008

Second International Conference
London, UK, September 1-3, 2008
Proceedings

Volume Editors

Steven D. Galbraith
University of London
Royal Holloway
Department of Mathematics
Egham, Surrey, TW20 0EX, UK
E-mail: steven.galbraith@rhul.ac.uk

Kenneth G. Paterson
University of London
Royal Holloway
Information Security Group
Egham, Surrey, TW20 0EX, UK
E-mail: Kenny.Paterson@rhul.ac.uk

Library of Congress Control Number: Applied for

CR Subject Classification (1998): E.3, D.4.6, F.2.2, G.2, K.6.5

LNCS Sublibrary: SL 4 – Security and Cryptology

ISSN 0302-9743
ISBN-10 3-540-85503-3 Springer Berlin Heidelberg New York
ISBN-13 978-3-540-85503-3 Springer Berlin Heidelberg New York

Springer is a part of Springer Science+Business Media

springer.com

Printed in Germany

Typesetting: Camera-ready by author, data conversion by Scientific Publishing Services, Chennai, India
Printed on acid-free paper SPIN: 12454020 06/3180 5 4 3 2 1 0

Preface

The Pairing 2008 Conference was held at Royal Holloway, University of London during September 1–3, 2008. This conference followed on from the Pairing in Cryptography workshop (held in Dublin, Ireland on June 12–15, 2005) and the Pairing 2007 conference (held in Tokyo, Japan on July 2–4, 2007, with proceedings published in Springer's LNCS 4575). The aim of this series of conferences is to bring together leading researchers and practitioners from academia and industry, all concerned with problems related to pairing-based cryptography.

The programme consisted of 3 invited talks and 20 contributed papers. The invited speakers were Xavier Boyen (Voltage Security, USA), Florian Hess (Technical University Berlin, Germany) and Nigel Smart (University of Bristol, UK). Special thanks are due to these speakers, all three of whom provided papers which are included in these proceedings.

The contributed talks were selected from fifty submissions. The accepted papers cover a range of topics in mathematics and computer science, including hardware and software implementation of pairings, cryptographic protocols, and mathematical aspects and applications of pairings.

We would like to thank all the people who helped with the conference programme and organisation. First, we thank the Steering Committee for their guidance and suggestions. We also heartily thank the Programme Committee and the sub-reviewers listed on the following pages for their thoroughness during the review process. Each paper was reviewed by at least three people and there was significant online discussion about a number of papers.

The submission and review process was greatly simplified by the ichair software developed by Thomas Baignères and Matthieu Finiasz. Thanks also to Jon Hart for running the submissions webserver and Takeshi Okamoto for designing and maintaining the conference webpage.

Thanks go to the authors of all submitted papers for supporting the conference. Authors of accepted papers are thanked again for revising their papers according to the suggestions by the referees and for returning their latex source files in good time. The revised versions were not checked by the Programme Committee, so authors bear full responsibility for their contents. We thank the staff at Springer for their help with producing the proceedings.

Finally, we thank the London Mathematical Society, Microsoft Research and Voltage Security for their generous sponsorship of this event.

June 2008

Steven Galbraith
Kenny Paterson

Organisation

Programme Chairs

Steven Galbraith	Royal Holloway, University of London, UK
Kenny Paterson	Royal Holloway, University of London, UK

General Chairs

Steven Galbraith	Royal Holloway, University of London, UK
Takeshi Okamoto	Tsukuba University of Technology, Japan
Kenny Paterson	Royal Holloway, University of London, UK

Steering Committee

Dan Boneh	Stanford University, USA
Steven Galbraith	Royal Holloway, University of London, UK
Tanja Lange	Technische Universiteit Eindhoven, The Netherlands
Alfred Menezes	University of Waterloo, Canada
Victor Miller	CCR, USA
Eiji Okamoto	University of Tsukuba, Japan (chair)
Takeshi Okamoto	Tsukuba University of Technology, Japan
Tatsuaki Okamoto	NTT, Japan
Takakazu Satoh	Tokyo Institute of Technology, Japan
Michael Scott	Dublin City University, Ireland
Tsuyoshi Takagi	Future University Hakodate, Japan

Sponsors

London Mathematical Society
Microsoft Research
Voltage Security

Programme Committee

Paulo Barreto	University of São Paulo, Brazil
Jan Camenisch	IBM Zurich Research Laboratory, Switzerland
Liqun Chen	Hewlett-Packard Labs, UK
Andreas Enge	INRIA Saclay–Île-de-France and École Polytechnique, France

Steven Galbraith	Royal Holloway, University of London, UK (Co-chair)
David Galindo	University of Malaga, Spain
Marc Joye	Thomson R&D, France
Eike Kiltz	CWI, The Netherlands
Soonhak Kwon	Sungkyunkwan University, Korea
Tanja Lange	Technische Universiteit Eindhoven, The Netherlands
Kristin Lauter	Microsoft Research, USA
Alfred Menezes	University of Waterloo, Canada
Eiji Okamoto	University of Tsukuba, Japan
Tatsuaki Okamoto	NTT, Japan
Dan Page	University of Bristol, UK
Kenny Paterson	Royal Holloway, University of London, UK (Co-chair)
Takakazu Satoh	Tokyo Institute of Technology, Japan
Michael Scott	Dublin City University, Ireland
Hovav Shacham	UCSD, USA
Igor Shparlinski	Macquarie University, Australia
Tsuyoshi Takagi	Future University Hakodate, Japan
Frederik Vercauteren	Katholieke Universiteit Leuven, Belgium

Sub-reviewers

Michel Abdalla
Miho Aoki
Gaëtan Bisson
John Boxall
Reinier Bröker
Sanjit Chatterjee
Jung Hee Cheon
Benoît Chevallier-Mames
Alexander Dent
Kirsten Eisenträger
David Freeman
María Isabel González Vasco
Dong-Guk Han
Javier Herranz
Florian Hess
David Jao
Koray Karabina
Mitsuru Kawazoe
Taekyoung Kwon
Hyang-Sook Lee
Fagen Li
François Morain
Michael Naehrig
Miyako Ohkubo
Cheol-Min Park
Young-Ho Park
Alice Silverberg
Ning Shang
Masaaki Shirase
Chang Shu
Benjamin Smith
Katherine Stange
Kobayashi Tetsutaro
Mark Watkins
Claire Whelan
Go Yamamoto

Table of Contents

Invited Talks

Cryptography I

Mathematics

Constructing Pairing Friendly Curves

Implementation of Pairings

Hardware Implementation

Cryptography II

Pairings in Trusted Computing*

Liqun Chen[1], Paul Morrissey[2], and Nigel P. Smart[2]

[1] Hewlett-Packard Laboratories,
Filton Road,
Stoke Gifford,
Bristol, BS34 8QZ,
United Kingdom
liqun.chen@hp.com

[2] Computer Science Department,
Woodland Road,
University of Bristol,
Bristol, BS8 1UB,
United Kingdom
{paulm,nigel}@cs.bris.ac.uk

Abstract. Pairings have now been used for constructive applications in cryptography for around eight years. In that time the range of applications has grown from a relatively narrow one of identity based encryption and signatures, through to more advanced protocols. In addition implementors have realised that pairing protocols once presented can often be greatly simplified or expanded using the mathematical structures of different types of pairings. In this paper we consider another advanced application of pairings, namely to the Direct Anonymous Attestation (DAA) schemes as found in the Trusted Computing Group standards. We show that a recent DAA proposal can be further optimized by transferring the underlying pairing groups from the symmetric to the asymmetric settings. This provides a more efficient and scalable solution than the existing RSA and pairing based DAA schemes.

1 Introduction

The growth of pairing based cryptography and the growth of elliptic curve cryptography from an implementation perspective closely follow the same path. Originally elliptic curve systems were defined over supersingular elliptic curves. This was mainly due to the difficulty in constructing suitable curves with a known group order of the required size, and also due to perceived performance advantages which accrue from using supersingular curves. With the discovery of the MOV attack [20], implementors moved over to using non-supersingular (or ordinary) elliptic curves. This move was supported by the research conducted into the Schoof algorithm and its variants, [2]. The Schoof algorithm enabled ordinary

* The second and third author would like to thank EPSRC for partially supporting the work in this paper.

S.D. Galbraith and K.G. Paterson (Eds.): Pairing 2008, LNCS 5209, pp. 1–17, 2008.

elliptic curves to be constructed with a known number of group elements, thus enabling standard elliptic curve cryptography to be performed using ordinary elliptic curves.

Pairing based cryptography also started by using supersingular elliptic curves, via the use of symmetric pairings. Again this was mainly because such curves enabled one to compute the number of points very efficiently, and because it appeared that symmetric pairings could be implemented much more efficiently than standard pairings. However, a major drawback of symmetric pairings is that their security properties scale badly. This poor scaling is due to the embedding degree being bounded by six. Thus with the wider acceptance of AES style security levels it has been necessary for pairing protocols to also move to the setting of ordinary elliptic curves, where asymmetric pairings are required. This security concern, which has prompted the move to asymmetric pairings, has been supported by a large body of research into optimizing pairings in the ordinary elliptic curve setting, and in generating the required parameters. Probably, at the time of writing the best choice for parameters is to choose a Barreto-Naehrig curve [1], implement the Ate-pairing [18], and use sextic twists to reduce the complexity of the group operations.

However, whilst protocols in the standard elliptic curve cryptography setting move seamlessly from the supersingular case to the ordinary case, this is not true in the pairing based cryptography setting. For example, issues arise with respect to hashing onto various groups, or from mappings between the two groups in the domain, see [15]. Thus in pairing based cryptography various initial protocol suggestions often needed to be revisited as asymmetric pairings became more accepted as the default implementation choice. For example the original Boneh-Franklin encryption scheme [6] was originally presented for symmetric pairings and in this setting is highly efficient. However, at high security levels it is less attractive than some of the more modern approaches such as the Boneh-Boyen scheme [3] (which is preferred by those who worry about exact security) and the SK-KEM scheme [13] (which has a better performance than the Boneh-Boyen scheme, but worse exact security). Another example of the need to fully evaluate pairings in the asymmetric setting can be found in ID-based key agreement, for which [14] provides a good summary of the issues involved.

Over the years various more advanced protocols have been proposed which use pairings; for example encryption with keyword search [5], group signatures [4], traitor tracing [19]. In this paper we consider another advanced application of pairings, namely to Direct Anonymous Attestation (DAA) schemes as found in the Trusted Computing Group standards [21]. We show that a recent DAA proposal [8] can be further optimized, and hence we provide a more efficient and scalable solution than the existing RSA and pairing based DAA schemes.

The original DAA scheme [7] was based on a signature scheme of Camenisch and Lysyanskaya [10] whose security was based on the strong-RSA assumption and the decisional Diffie–Hellman assumption in a finite field. In [16] another DAA scheme was presented, based on the Camenisch and Michels signature scheme [12], this again results in a DAA scheme which is secure under the strong-RSA and the decisional Diffie–Hellman assumption in a finite field.

Recently, Brickell et. al. [8] have presented a DAA scheme based on symmetric pairings. This DAA protocol is based on another signature scheme of Camenisch and Lysyanskaya [11] which makes use of symmetric pairing groups. This results in a scheme which is secure under the DBDH assumption and the LRSW assumption. This latter assumption is a non-standard assumption which underlies the Camenisch and Lysyanskaya signature scheme. The LRSW assumption was introduced in [17], where it was shown that it holds in the generic group model.

As a starting point we take the pairing based DAA scheme of Brickell et. al. and provide some efficiency improvements. In addition we present the scheme in the asymmetric setting, which requires, as is usual in this situation, some minor modifications to the original scheme. We show that the new scheme is particularly suited to the environment in which the DAA scheme is meant to run. This is because our new scheme places a smaller computational requirement on the TPM, which is a small hardware device which sits on the motherboard of the trusted platform. In fact the TPM has only to perform a single basic elliptic curve point multiplication in the signing protocol, and in all parts of the protocol the TPM requires no operation to be performed in large finite fields nor any pairing calculations. In addition we reduce the number of pairings computed by the Host during a signing operation from three to one.

In the full version of this paper we shall show that our optimized asymmetric pairing based protocol is secure in a security model based on real/ideal world simulations. This is closer to the original security model of [7], than the security model used in [8]. Thus our protocol is not only more efficient than that in [8], but it also enjoys enhanced security properties.

2 Introduction to Direct Anonymous Attestation

In order to give an intuitive explanation of what direct anonymous attestation (DAA) is, its importance and its impact, we use the following scenario. Consider a user, Alice, who owns a laptop computer. Alice uses this computer for online shopping with a given retailer Charlie and to remotely log on to network server Bob in order to work from home as part of her day job. Both Bob and Charlie want some assurance that Alice is using a laptop which contains some combination of hardware and software from some specific set. This is to protect themselves from malicious users who may try to compromise their systems. In other words they want some assurance that Alice's platform can be trusted. On the other hand Alice does not want either Bob or Charlie to know exactly which laptop she owns, what hardware it contains, or what software it runs: just that it can in fact be trusted. Furthermore, Alice wants Bob to be able to link transactions made with her laptop to each other (without giving Bob any more information than that the transactions were made using the same platform), but does not want any other transactions to be linked. Informally, a DAA scheme is a mechanism for achieving each of these seemingly contradictory goals.

We assume that each trusted platform has a certain module, known as a trusted platform module (TPM), embedded into it at the time of manufacture. Each TPM will have a unique endorsement key pair (EK) which is also chosen at the time of manufacture and which is hidden inside the TPM. Usually the TPM is a small chip embedded onto a computer's motherboard and as a result the TPM has only limited computational and storage resources. As such the TPM is a potential bottleneck within a DAA scheme. To make a DAA scheme as efficient as possible the main goal is to minimize the amount of computation that a given TPM will have to perform. We refer to the platform into which the TPM is embedded as the Host and the combination of TPM and Host as a *user*.

In order to convince a verifier that a platform contains such a TPM, and can hence be trusted, the user has to first obtain a credential from some credential issuer. It does this by having its TPM compute a commitment to a secret internal value f that is unique for each issuer/TPM pair. This commitment is then used as evidence to the issuer that the user does in fact have a valid TPM embedded within it. Note that a given user can obtain many credentials from a given issuer.

In order to convince a given verifier that a user owns such a credential the user computes a "signature of knowledge" of such a credential and the associated value f corresponding to this credential. This signature of knowledge is then sent to the verifier. Then, since the credential was issued by a specific issuer, the verifier is convinced that a given user contains a valid TPM but does not know which TPM this is. If a user wants a verifier to be able to link transactions then that user simply computes the signature of knowledge in a certain specified way: by using a given verifier basename.

One last consideration is what happens if a given TPM is compromised and its secret internal values published? In this case we use rogue tagging. Each issuer and verifier maintains their own list of rogue values. When a given value of f is published they then decide whether to add it to their list or not. Then we require, when a user computes a signature of knowledge of a given credential this includes some information that allows for the signature to be recognised as produced by a compromised TPM secret value.

Since the verifier is only given a signature of knowledge of a credential, and not the credential itself, if the issuer and the verifier that computed the credential collude, then they should not be able to identify transactions made by a specific TPM. Yet a given verifier will be assured that any transaction that does take place was made by a platform that contains such a TPM and that this TPM has not been compromised.

2.1 The DAA Players

We refer to each of the entities in a DAA scheme as *players*. We first describe the types of players we consider in our model. This set of players is the same set as in [9] and is intended to represent a DAA scheme in which a given TPM wishes to remotely and anonymously authenticate itself to a given verifier. Intuitively, the set of players will consist of a set of users, each comprising a Host and a TPM, a set of issuers, and a set of verifiers to which users want to authenticate their TPM.

We now give a formal description of each of the DAA players. A general DAA scheme has a set of players that consists of the following.

- A set of users $\mathcal{U}$ where each $U_i \in \mathcal{U}$ consists of
  - A TPM m_i from some set of TPMs $\mathcal{M}$ with an endorsement key ek_i and seed $\mathsf{DaaSeed}_i$;
  - A Host h_i from some set of hosts $\mathcal{H}$ which will have a counter value cnt_i, a set of commitments $\{\mathsf{comm}\}_i$ and a set of credentials $\{\mathsf{cre}\}_i$.
- A set of issuers $\mathcal{I}$ where each $I_k \in \mathcal{I}$ has a public and private key pair $(\mathsf{ipk}_k, \mathsf{isk}_k)$ and long term value K_k (for example a long term public key of the issuer). Each $I_k \in \mathcal{I}$ also maintains a list of rogue TPM internal values, we denote this list by $\mathsf{RogueList}(I_k)$.
- A set of verifiers $\mathcal{V}$. Each verifier $V_j \in \mathcal{V}$ maintains a set of base names $\{\mathsf{bsn}\}_j$ and a list of rogue TPM internal values $\mathsf{RogueList}(V_j)$. Each V_j may optionally maintain a list of message and signature pairs received (this can be used to trade memory for computation in linking).

We assume that initially the sets $\{\mathsf{comm}\}_i$, $\{\mathsf{cre}\}_i$ are empty for all $U_i \in \mathcal{U}$. In addition we assume that the list $\mathsf{RogueList}(I_k)$ are empty for all $I_k \in \mathcal{I}$, and that the list $\mathsf{RogueList}(V_j)$ and the sets $\{\mathsf{bsn}\}_j$ are empty for all $V_j \in \mathcal{V}$.

It is worth describing the various player parameters and how they relate to each other. Generally, at the time of manufacture, each TPM will have a single endorsement key ek_i embedded into the TPM chip. In addition, each TPM generates a TPM-specific secret $\mathsf{DaaSeed}_i$ and stores it in nonvolatile memory, this value will never be disclosed or changed by the TPM. We do not consider choosing and assigning the values ek_i and $\mathsf{DaaSeed}_i$ in the setup algorithm, since the setup algorithm is run only by an issuer. The $\mathsf{DaaSeed}_i$ is generally a 20–byte constant that, together with a given issuer value K_k, allows for the generation and regeneration of a given value of an internal secret key f. Each TPM can have multiple possible values for f (at least one per issuer and possible more if a given issuer has more than one value of K_k). We refer to the set of possible values of f for a given user i as $\{f\}_i$ Since the TPM has limited storage requirements it does not store the current value for f, it regenerates it as required from $\mathsf{DaaSeed}_i$. For each value of f the TPM will be able to compute a single commitment on f. The value cnt_i that a given Host maintains can be thought of as an index for a particular f and commitment pair.

For each commitment, as we will see later, a given issuer could issue multiple credentials. We assume the Host only stores one credential for a given f/commitment pair, and hence the value of cnt_i will also refer to the *current* value of the corresponding credential.

The set $\{\mathsf{bsn}\}_j$ is used to achieve user controlled linkability of signatures.

2.2 Formal Definition of a DAA Scheme

Informally a DAA scheme consists of a system setup algorithm, a protocol for users to obtain credentials, a signing protocol, algorithms for verifying and linking signatures and an algorithm for tagging rogue TPM values. Our definition is similar to that given in [8] but with some modifications. Specifically, we give a single protocol for the joining functionality as opposed to multiple protocols, and our signature functionality is given as a protocol as opposed to an algorithm. Also we have an additional rogue tagging algorithm.

Definition 1 (Daa **Scheme**)**.** Formally, we define a Daa scheme to be a tuple of protocols and algorithms Daa = (Setup, Join, Sign, Verify, Link, RogueTag) where:

- Setup(1^t) is a p.p.t. system setup algorithm. On input 1^t, where t is a security parameter, this outputs a set of system parameters par which contains all of the issuer public keys ipk_k and the various parameter spaces. This algorithm also serves to setup and securely distribute each of the issuer secret keys isk_k.
- Join(U_i, I_k) is a 3 party protocol run between a TPM, a Host and an issuer. In a correct initial run of the protocol with honest players the Host should obtain an additional valid commitment and an additional valid credential. In correct subsequent runs one valid credential should be replaced with another.
- Sign(U_i, msg) is a 2 party protocol run between a TPM and a Host used to generate a signature of knowledge on some message msg. In a correct run of the protocol with honest players the signature of knowledge will be constructed according to some basename for some specified verifier that may or may not allow the signature to be linked to other signatures with this same verifier.
- Verify(σ, msg) is a deterministic polynomial time (d.p.t.) verification algorithm that allows a given verifier to verify a signature of knowledge σ of a credential on a message msg intended for a given verifier with a specific basename. The verification process will involve checking the signature against the list RogueList(V_j). This algorithm returns either *accept* or *reject*.
- Link(σ_0, σ_1) is a d.p.t. linking algorithm that returns either *linked*, *unlinked* or $\perp$. The algorithm should return $\perp$ if either signature was produced with a rogue key, return *linked* if both are valid signatures and the user who produced them wanted these to be linkable to each other, and return *unlinked* otherwise.
- RogueTag(f, σ) is a d.p.t. rogue tagging algorithm that returns true if σ is a valid signature produced using the TPM secret value f and returns false otherwise.

For correctness we require that if

- a user $U_i \in \mathcal{U}$ engages in a run of Join with I_k, resulting in U_i obtaining a commitment comm on a TPM secret value f and a credential cre corresponding to f,
- the user U_i then creates two signatures σ_b on two messages msg_b for $b \in \{0, 1\}$ intended for verifier $V_j \in \mathcal{V}$ with basename bsn (which could be $\perp$),
- and the secret TPM value used to compute these f is not in RogueList.

Then

$$\mathsf{Verify}(\sigma_0, \mathsf{msg}_0) = \mathsf{Verify}(\sigma_1, \mathsf{msg}_1) = accept$$

and if bsn $\neq \perp$ then Link(σ_0, σ_1) = *linked*.

3 The Camensich-Lysyanskaya Signature Scheme

Before proceeding it is worth pausing to present the pairing based Camensich-Lysyanskaya signature scheme which is at the heart of not only our DAA scheme, but also the scheme of [8]. We let $\hat{t} : \mathbb{G}_1 \times \mathbb{G}_2 \rightarrow \mathbb{G}_T$ denote a pairing between three groups of prime order q. We let the generator of $\mathbb{G}_1$ (resp. $\mathbb{G}_2$) be denoted by P_1 (resp. P_2).

- **KeyGeneration:** The private key is a pair $(x, y) \in \mathbb{Z}_q \times \mathbb{Z}_q$, the public key is given by the pair $(X, Y) \in \mathbb{G}_2 \times \mathbb{G}_2$ where $X = xP_2$ and $Y = yP_2$.
- **Signing:** On input of a message $m \in \mathbb{Z}_q$ the signer generates $A \in \mathbb{G}_1$ at random and outputs the signature $(A, B, C) \in \mathbb{G}_1 \times \mathbb{G}_1 \times \mathbb{G}_1$, where $B = yA$ and $C = [x + mxy]A$.
- **Verification:** To verify a signature on a message the verifier checks whether $\hat{t}(A, Y) = \hat{t}(B, P_2)$ and $\hat{t}(A, X) \cdot \hat{t}(mB, X) = \hat{t}(C, P_2)$.

The original signature scheme is given in the symmetric pairing setting (i.e. where $\mathbb{G}_1 = \mathbb{G}_2$), we have chosen the above asymmetric version to reduce the size of the signatures and to have the fastest signing algorithm possible. The key property of this signature scheme is that signatures are re-randomizable without knowledge of the secret key: given (A, B, C) one can re-randomize it by computing (rA, rB, rC) for a random element $r \in \mathbb{Z}_q$.

There is an interesting difference between this signature scheme in the symmetric and the asymmetric settings. In the symmetric setting the signer, on being given two valid signatures (A, B, C) and (A', B', C'), is able to tell that they correspond to a randomization of a previous signature, without knowing what that message is. He can do this by verifying that $A' = rA, B' = rB$ and $C' = rC$, for some value r, by performing the following steps:

$$\hat{t}(A', B) = \hat{t}(A, B') \text{ and } \hat{t}(A', C) = \hat{t}(A, C').$$

This makes use of the fact that the DDH problem is easy in $\mathbb{G}_1$ in the symmetric setting.

In the asymmetric setting a signer is unable to determine if two signatures correspond to the same message, since in this setting the DDH problem is believed to be hard in $\mathbb{G}_1$. Indeed one can show that an adversary who can tell whether (A', B', C') is a randomization of (A, B, C), even if the adversary knows x and y, is able to solve DDH in $\mathbb{G}_1$. This difference provides one of the main optimizations of our scheme below.

4 Previous DAA Schemes

In this section we present prior work on DAA schemes, and we analyse their performance.

4.1 Factoring Based Schemes

The original DAA scheme from [7] makes use of the Camenisch-Lysyanskaya signature scheme [10], and hence is based on the strong-RSA assumption. In particular it makes use of a strong-RSA modulus $N = p \cdot q$, i.e. where $p = 2 \cdot p' + 1$ and $q = 2 \cdot q' + 1$ for primes p' and q'. In addition it uses a finite field of prime order Γ. The difficulty of discrete logarithms in $\mathbb{F}_\Gamma$ and of factoring N should be roughly equivalent, so Γ and N are chosen to be roughly the same size.

As in all systems the Setup procedure is rather involved. However, this is only run once and the resulting parameters are only verified once by each party so we ignore the cost of the Setup algorithm and its verification.

In Table 1 table we present the computational cost for all the other algorithms, with respect to each player. An entry of the form

$$1 \cdot \mathbb{G}_N + 2 \cdot \mathbb{G}_\Gamma + 3 \cdot \mathbb{G}_N^2$$

implies that the cost is about one exponentiation modulo N, two modulo Γ and three multiexponentiations with two exponents modulo N, i.e. three operations of the form $g^a \cdot h^b \pmod{N}$. Note, that a multiexponentiation with m exponents can often be performed significantly faster than m separate exponentiations.

In the table we let P_c denote the cost of generating a prime number of the required size and P_v the cost of verifying that a given number of the required size is prime. We let n denote the number of keys in the verifier's rogue secret key list. We do not specify the time for the linking algorithm, as it is closely related to that of the verification algorithm, and we give the additional time for the RogueTag algorithm over and above the verification algorithm time (which needs to be carried out).

Note that the exponents involved in many of the operations, especially the verification operation, are not of full length. Hence, the above table grossly overestimates the required computational resources. However, one can see that the constrained computing device, namely the TPM is having to perform a considerable number of RSA-length operations.

In [16] a different variant of the DAA protocol is given which tries to reduce the computational cost of the Host, thus allowing trusted computing technologies which use the DAA protocol to be deployed in small devices such as mobile

Table 1. Cost of the DAA protocol from [7]

Operation	Party	Cost
Join	TPM	$3 \cdot \mathbb{G}_\Gamma + 2 \cdot \mathbb{G}_N^3$
	Issuer	$n \cdot \mathbb{G}_\Gamma + 2 \cdot \mathbb{G}_N + 1 \cdot \mathbb{G}_N^4 + 1 \cdot \mathbb{G}_\Gamma^2 + P_c$
	Host	$1 \cdot \mathbb{G}_\Gamma + 1 \cdot \mathbb{G}_N^2 + P_v$
Sign	TPM	$3 \cdot \mathbb{G}_\Gamma + 1 \cdot \mathbb{G}_N^3$
	Host	$1 \cdot \mathbb{G}_\Gamma + 1 \cdot \mathbb{G}_N + 1 \cdot \mathbb{G}_N^2 + 2 \cdot \mathbb{G}_N^3 + 1 \cdot \mathbb{G}_N^4$
Verify	Verifier	$4 \cdot \mathbb{G}_\Gamma^2 + 2 \cdot \mathbb{G}_N^4 + 1 \cdot \mathbb{G}_N^6 + n\mathbb{G}_\Gamma$
RogueTag	Verifier	$1 \cdot \mathbb{G}_N^4$

phones. We do not analyse, for reasons of space, the performance of this protocol, but it is also based on factoring assumptions and so all parties need to compute with large integers.

4.2 Symmetric Pairing Based Schemes

Given the increase in RSA key lengths as required by moving to AES key levels, since AES-128 is equivalent to roughly 3000 bits of RSA security, the above two protocols are not going to be suitable in the long term. This led to Brickell et. al. [8] to propose an elliptic curve variant, which reduced the load of the TPM at the expense of requiring pairings to be computed by the other parties.

The Brickell et. al. protocol uses symmetric pairings $\hat{t} : \mathbb{G}_1 \times \mathbb{G}_1 \longrightarrow \mathbb{G}_T$. As above we let $\mathbb{G}_1^m$ etc denote the cost of a multiexponentiation of m values in the group $\mathbb{G}_1$. We also let P denote the cost of a pairing computation. The associated costs are then given by Table 2.

Table 2. Cost of the DAA protocol from [8]

Operation	Party	Cost
Join	TPM	$3 \cdot \mathbb{G}_1$
	Issuer	$(2+n) \cdot \mathbb{G}_1 + 2 \cdot \mathbb{G}_1^2$
	Host	$6 \cdot P$
Sign	TPM	$4 \cdot \mathbb{G}_T$
	Host	$3 \cdot \mathbb{G}_1 + 2 \cdot \mathbb{G}_T + 3 \cdot P$
Verify	Verifier	$(n+1) \cdot \mathbb{G}_T + 1 \cdot \mathbb{G}_T^2 + 1 \cdot \mathbb{G}_T^3 + 5 \cdot P$
RogueTag	Verifier	$1 \cdot \mathbb{G}_T$

To get some idea of the comparison between the factoring based scheme and the pairing based scheme, consider that the groups $\mathbb{G}_T$ and $\mathbb{G}_N$ (or $\mathbb{G}_\Gamma$) are represented by bit strings of roughly the same size. In addition operations in $\mathbb{G}_T$ can be made slightly more efficient than those in $\mathbb{G}_N$, as in $\mathbb{G}_T$ we can make use of various torus-like representations and tricks, which are not available in standard RSA groups. Finally, the operations in $\mathbb{G}_1$ are about $1/4$ the cost of operations in $\mathbb{G}_T$[1].

In the next section we present a variant of the Brickell et. al. pairing based protocol which uses asymmetric pairings. By using asymmetric pairings and Barreto-Naehrig curves, we are able to obtain, for the same size of $\mathbb{G}_T$, operations in $\mathbb{G}_1$ which are around $144/10 \approx 14$ times more efficient than those in $\mathbb{G}_T$, as opposed to 4 times as above. This is because now $\mathbb{G}_T$ is a subgroup of $\mathbb{F}_{q^{12}}$.

---

[1] This is a rough estimate derived as follows: $\mathbb{G}_T$ is a subgroup of $\mathbb{F}_{q^6}$ and operations in $\mathbb{F}_q$ will be $36 = 6^2$ times more efficient generally than operations in $\mathbb{G}_T$, $\mathbb{G}_1$ is an elliptic curve over $\mathbb{F}_q$ and so will have operations which take around 10 $\mathbb{F}_q$ operations, and $10/36 \approx 1/4$.

5 The Optimized Pairing Based DAA Scheme

We now give a detailed description of the our new DAA scheme based on asymmetric bilinear maps, as opposed to symmetric ones.

5.1 The Setup Algorithm

To set the system up we need to select parameters for each protocol and algorithm used within the DAA scheme well as the long term parameters for each Issuer. On input of the security parameter 1^t the algorithm executes the following:

1. *Generate the Commitment Parameters* $\mathsf{par}_{\mathrm{C}}$. For this three groups, $\mathbb{G}_1, \mathbb{G}_2$ and $\mathbb{G}_T$, of sufficiently large prime order q are selected. Two random generators are selected such that $\mathbb{G}_1 = \langle P_1 \rangle$ and $\mathbb{G}_2 = \langle P_2 \rangle$ along with a pairing $\hat{t} : \mathbb{G}_1 \times \mathbb{G}_2 \mapsto \mathbb{G}_T$. Next a hash function $H_1 : \{0,1\}^* \mapsto \mathbb{Z}_q$ is selected and $\mathsf{par}_{\mathrm{C}}$ is set to be $(\mathbb{G}_1, \mathbb{G}_2, \mathbb{G}_T, \hat{t}, P_1, P_2, q, H_1)$.
2. *Generate the Rogue List Parameters* $\mathsf{par}_{\mathrm{R}}$. A hash function $H_2 : \{0,1\}^* \mapsto \mathbb{Z}_q$ is selected. The rogue list parameters $\mathsf{par}_{\mathrm{R}}$ are then set to be (H_2).
3. *Generate Signature and Verification Parameters* $\mathsf{par}_{\mathrm{S}}$. Two additional hash functions are selected: $H_3 : \{0,1\}^* \mapsto \mathbb{Z}_q$, and $H_4 : \{0,1\}^* \mapsto \mathbb{Z}_q$. We set $\mathsf{par}_{\mathrm{S}}$ to be (H_3, H_4).
4. *Generate the Issuer Parameters* $\mathsf{par}_{\mathrm{I}}$. For each $I_k \in \mathcal{I}$ the following is performed. Two integers are selected $x, y \leftarrow \mathbb{Z}_q$ and the issuer secret key isk_k is assigned to be (x, y). Then the values $X = x \cdot P_2 \in \mathbb{G}_2$ and $Y = y \cdot P_2 \in \mathbb{G}_2$ are computed and the issuer public key ipk_k is assigned to be (X, Y).
   Then an issuer value K_k is computed according to the issuer public values in some predefined manner (we leave the specific details of how this is done as an implementation detail).
   Finally, $\mathsf{par}_{\mathrm{I}}$ is set to be $(\{\mathsf{ipk}_k, \mathrm{K}_k\})$ for each issuer $I_k \in \mathcal{I}$.
5. *Publish Public Parameters.* Finally, the system public parameters par are set to be $(\mathsf{par}_{\mathrm{C}}, \mathsf{par}_{\mathrm{R}}, \mathsf{par}_{\mathrm{S}}, \mathsf{par}_{\mathrm{I}})$ and are published.

The grouping of system parameters is according to usage. For example the set $\mathsf{par}_{\mathrm{C}}$ contains all system parameters necessary for computing commitments and the set $\mathsf{par}_{\mathrm{R}}$ contains those for any rogue checking computations (and also linking).

The group order q is selected so that solving the decisional Diffie–Hellman problem in $\mathbb{G}_1, \mathbb{G}_2$ and $\mathbb{G}_T$ takes time 2^t, as does solving the appropriate bilinear Diffie–Hellman problem with respect to the pairing $\hat{t}$.

An additional optional check of issuer public key values can be added by having each issuer compute $X' = x \cdot P_1$ and $Y' = y \cdot P_1$ and publishing these as part of par. Then to check that both X and Y are correctly formed one simply checks that $\hat{t}(P_1, X) = \hat{t}(X', P_2)$ and $\hat{t}(P_1, Y) = \hat{t}(Y', P_2)$.

5.2 The Join Protocol

This is a protocol between a given TPM $m \in \mathcal{M}$, the corresponding Host $h \in \mathcal{H}$, and an Issuer $I \in \mathcal{I}$. We first give an overview of how a general Join protocol proceeds. There are 3 main stages to a Join protocol. First the TPM m generates some secret message f using the value K_k provided by the issuer and its internal seed DaaSeed. The TPM then computes a commitment on this value and passes this to its Host who adds this to the list of commitments for that user and forwards it to the Issuer. In the second stage the issuer performs some checks on the commitment it receives and, if these correctly verify, computes a credential such that the correctness of this credential can be check by the TPM and Host working together. This credential is passed to the Host in an authenticated manner (using ek). The final stage of a Join protocol involves the Host and TPM working together to verify the correctness of the credential. In our case the Host first performs some computations and stores some values related to these before passing part of the credential on to the TPM prior to verifying the correctness of the credential and then adding this to the list of credentials for that user.

Our protocol proceeds as shown in Figure 1. The following notes should be born in mind when examining this protocol.

- If the points P_1, P_2, X, Y are not formed correctly then this could leak information about the value of a given f, for example due to small subgroup attacks. To

TPM (m)		Host (h)		Initiator (I)
				$n_I \leftarrow \{0,1\}^t$
	$\xleftarrow{\mathsf{comm}_{\mathrm{req}}}$		$\xleftarrow{\mathsf{comm}_{\mathrm{req}}}$	$\mathsf{comm}_{\mathrm{req}} \leftarrow n_I$
$\mathsf{str} \leftarrow 1 \Vert X \Vert Y \Vert n_I$				$\mathsf{str} \leftarrow 1 \Vert X \Vert Y \Vert n_I$
$f \leftarrow H_1(0 \Vert \mathsf{DaaSeed} \Vert \mathrm{K}_k)$				
$u \leftarrow \mathbb{Z}_q$				
$U \leftarrow u \cdot P_1; F \leftarrow f \cdot P_1$				
$c \leftarrow H_1(\mathsf{str} \Vert F \Vert U)$				
$s \leftarrow u + c \cdot f \pmod q$				
$\mathsf{comm} \leftarrow (F, c, s)$	$\xrightarrow{\mathsf{comm}}$		$\xrightarrow{\mathsf{comm}}$	$U' \leftarrow sP_1 - cF$
				If $F = f \cdot P_1$ for some f on the rogue list, or $c \neq H_1(\mathsf{str} \Vert F \Vert U')$ then **abort**
				$r \leftarrow \mathbb{Z}_q$
				$A \leftarrow r \cdot P_1; B \leftarrow y \cdot A$
				$C \leftarrow (x \cdot A + rxy \cdot F)$
				$\mathsf{cre} \leftarrow (A, B, C)$
	$\xleftarrow{\mathcal{E}}$		$\xleftarrow{\mathcal{E}}$	$\mathcal{E} \leftarrow E_{\mathsf{ek}}(\mathsf{cre})$
$\mathsf{cre} \leftarrow E_{\mathsf{ek}}^{-1}(\mathcal{E}); E \leftarrow f \cdot B$	$\xrightarrow{\mathsf{cre}, E}$	$\rho_a \leftarrow \hat{t}(A, X)$		
		$\rho_b \leftarrow \hat{t}(B, X)$		
		$\rho_c \leftarrow \hat{t}(C, P_2)$		
		If $\hat{t}(A, Y) \neq \hat{t}(B, P_2)$ or $\hat{t}(A + E, X) \neq \rho_c$ then **abort**		

Fig. 1. The Join Protocol

prevent this from happening each TPM needs to verify that P_1 generates $\mathbb{G}_1$, P_2 generates $\mathbb{G}_2$ and that $X, Y \in \mathbb{G}_2$. This need be done once for each TPM so we do not give this as part of the Join protocol. Algorithms for checking whether points are elements of particular pairing groups are given in [14].

- The value of cre is not sent in the clear and hence only the intended user can obtain the complete credential. This is done by encrypting the value cre under a public key corresponding to the TPM endorsement key ek. Again, we do not consider these calculations in the performance analysis of the scheme.
- In contrast with the RSA-based DAA schemes we do not require a relatively complicated proof of knowledge of the correctness of a given commitment. Instead, the proof of knowledge is provided by a very efficient Schnorr signature, on the value F computed using the secret key f.
- Once a credential is issued from I, the TPM and the Host verify that this credential is correctly formed. This is to avoid performing computations with a credential that is incorrectly formed since this could lead to leaking information about the value f held by the TPM. The last part of the protocol therefore performs the verification algorithm from the Camenisch-Lysyanskaya signature scheme. In addition the TPM should check that the value of B it receives in the credential is correctly formed. Since $B \in \mathbb{G}_1$ this can be performed very efficiently and so we ignore its cost when computing the cost of running the protocol.
- We note that the Host does not perform any verification on values that are provided by the TPM. Since we assume that it is harder to compromise a TPM than a Host, we do not model the case of a corrupt TPM inside an honest Host and hence the Host will always trust the correctness of values provided by its TPM.
- The values ρ_a, ρ_b, ρ_c and E are stored for later use by the Host in the signing algorithm. This improves the performance by avoiding recomputation of various pairing values.

5.3 The Sign Protocol

This is a protocol run between a given TPM $m \in \mathcal{M}$ and Host $h \in \mathcal{H}$. The objective of the sign protocol is for m and h to work together to produce a signature of knowledge on some message. The signature should prove knowledge of a discrete logarithm f, knowledge of a valid credential and that this credential was computed for the same value f. We note that the Host will know a lot of the values needed in the computation and will be able to take on a lot of the computational workload. However, if the TPM has not had its internal value of f published (i.e. it is not a rogue module) then the Host will not know f and will be unable to compute the whole signature without the aid of the TPM.

We again assume that we could have an adversarially controlled Host and honest TPM and, as a result, the TPM will have to do a number of checks on the data passed to it from its Host. We let msg denote the message to be signed and bsn denote the base name of the verifier. The protocol then proceeds as in Figure 2, so as to produce the signature σ.

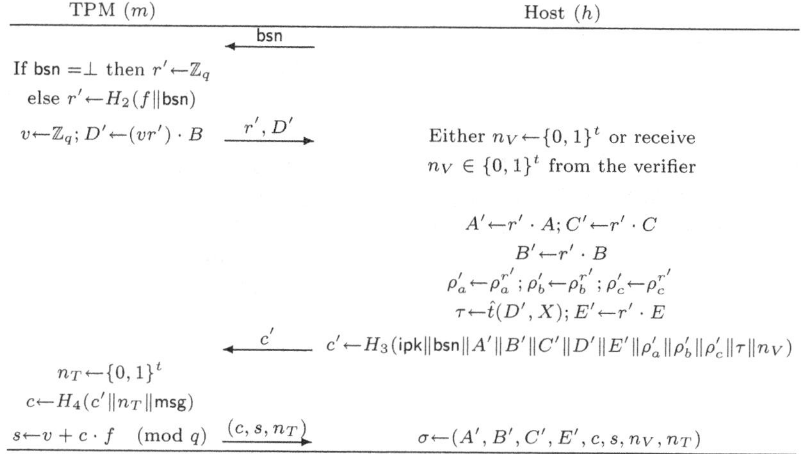

Fig. 2. The Sign Protocol

Again we provide some notes as to the rational behind some of the steps:

- In most applications of the Sign protocol, the signature is generated as a request from the verifier, and the verifier supplies its own value of n_V, to protect against replays of previously requested signatures. If a signature is produced in an offline manner we allow the Host to generate its own value of n_V.
- Prior to running the protocol the Host decides if it wants σ to be linkable to other signatures produced for the same verifier. If it does not want the signature to be linkable to any existing or future signatures then it chooses $\mathsf{bsn} = \perp$. If it decides that it wants the signature to be linked to some previously generated signatures with this verifier then it sets bsn to be the same as that used for the signature it wants to link to. Otherwise, if the Host decides it may want other signatures future signatures to be able to be link to this one then it chooses a verifier bsn that it has not used before.
- The use of E' allows the verifier to identify if the signature was produced by a rogue TPM by computing $f_i \cdot B'$ for all f_i values on the rogue list and comparing these to E'. This is performed during the verification algorithm. Without E' the rogue test algorithm can be performed by using elements in $\mathbb{G}_T$, however in practice this is much less efficient than using E'.
- During the run of the signature protocol two nonces are used: one from the verifier n_V and one from the TPM n_T. These are used to ensure each signature is different from previous signatures and to ensure that no adversarially controlled TPM and Host pair or no honest TPM and adversarially controlled Host can predict or force the value of a given signature.
- The value r' is used to mask the signature created from the other players in the scheme including the issuer. Without using r' the credential on which the signature is computed would be sent in the clear and hence other parties

would be able to link signatures. That the issuer cannot link the signatures follows from the earlier mentioned property of the Camenisch-Lysyanskaya signature scheme in the asymmetric setting. Thus it provides two types of linking resistance: it stops any issuer from being able to link a given signature to a given signer (since issuers know the values of r used to compute a credential and without r' the credential is sent in the clear), and it stops any player in the system from being able to tell if any two signatures were produced by the same signer (if different bsn are used).
Note, that the Host is trusted to keep anonymity because it is assumed that the Host has the motivation to protect privacy and also because the host can always disclose the platform identity anyway. However, the Host is not trusted to be honest for not trying to forge a DAA signature without the aid of the TPM.

5.4 The Verification Algorithm

This is an algorithm run by a verifier V. Intuitively the verifier checks that a signature provided proves knowledge of a discrete logarithm f, checks that it proves knowledge of a valid credential issued on the same value of f and that this value of f is not on the list of rogue values.

We now describe the details of our Verify algorithm. On input a signature σ of the form $\sigma = (A', B', C', E', c, s, n_V, n_T)$ this algorithm performs the following steps:

1. *Check Against Rogue List.* If $E' = f_i \cdot B'$ for any f_i in the set of rogue secret keys then return *reject.*
2. *Check Correctness of A' and B'.* If $\hat{t}(A', Y) \neq \hat{t}(B', P_2)$ then return *reject.*
3. *Verify Correctness of Proofs.* This is done by performing the following sets of computations:
   - $\rho_a^\dagger \leftarrow \hat{t}(A', X)$, $\rho_b^\dagger \leftarrow \hat{t}(B', X)$ and $\rho_c^\dagger \leftarrow \hat{t}(C', P_2)$.
   - $\tau^\dagger \leftarrow (\rho_b^\dagger)^s \cdot (\rho_c^\dagger / \rho_a^\dagger)^{-c}$
   - $D^\dagger \leftarrow sB' - cE'$.
   - $c^\dagger \leftarrow H_3(\mathsf{ipk} \| \mathsf{bsn} \| A' \| B' \| C' \| D^\dagger \| E' \| \rho_a^\dagger \| \rho_b^\dagger \| \rho_c^\dagger \| \tau^\dagger \| n_V)$.

   Finally if $c \neq H_4\left(c^\dagger \| n_T \| \mathsf{msg}\right)$ return *reject* and otherwise return *accept.*

5.5 The Linking Algorithm

This is an algorithm run by a given verifier $V_j \in \mathcal{V}$ which has a set of basenames $\{\mathsf{bsn}\}_j$ in order to determine if a pair of signatures were produced by the same TPM. Signatures can only be linked if they were produced by the same TPM and the user wanted them to be able to be linked together.

Formally, on input a pair of signatures σ_b for $b \in \{0, 1\}$ each having the form $\sigma_b = (A'_b, B'_b, C'_b, E'_b, c_b, s_b, n_{V,b}, n_{T,b})$ the algorithm performs the following steps:

1. *Verify Both Signatures.* For each signature σ_b the verifier runs $\mathsf{Verify}(\sigma_b)$ and if either of these returns *reject* then the value $\perp$ is returned.

2. *Compare Signatures and Basenames.* If the two basenames which verify the signatures are equal, and if $A'_0 = A'_1$ then return *linked*, else return *unlinked*.

It may be the case that one or both signatures input to the the Link algorithm have previously been received and verified by the verifier. Regardless of this we insist that the verifier re-verify these as part of the Link algorithm since the list of rogue TPM values may have been updated since the initial verification.

Note, our linking algorithm works due to the way that r' is computed in the signing algorithm. Also note that anyone who knows bsn can link the two signatures, but they cannot link the signatures with the signers.

5.6 The Rogue Tagging Algorithm

The purpose of the rogue tagging algorithm is to ensure that an adversary is not able to tag a given value of TPM internal secret as rogue if the TPM that owns that particular value is not corrupted.

On input a value of f and a signature σ intended for a given verifier V_j the algorithm then proceeds as follows:

1. *Verify the Signature.* If $\mathsf{Verify}(\sigma) = \mathit{reject}$ then the value $\perp$ is returned.
2. *Check Value of* f. If $E' \neq f \cdot B'$ then return $\perp$ and otherwise add an entry f to $\mathsf{RogueList}(V_j)$.

We note that, since the credential computed for a given user is sent using a secure channel, the only way that an adversary can produce a valid signature would be if it knew the value of the credential and hence had corrupted that user to some extent. This prevents the adversary from adding arbitrary values of f to $\mathsf{RogueList}(V_j)$.

5.7 Efficiency Comparison

Table 3 presents the performance analysis of our optimized version of the pairing based DAA protocol. We use a similar notation for computational cost as in our previous tables. The main advantages of our version can be listed as follows:

- Due to DDH being hard in $\mathbb{G}_1$ we can remove a number of the checks and masks in the original pairing based DAA protocol.
- In addition we move the computation of τ in the signature from the TPM to the Host. This removes the need for the TPM to perform any $\mathbb{G}_T$ operations at all.
- The Host precomputes some pairing values at the Join stage so as to remove the need to perform these at the signing stage. This comes at the expense of a couple more $\mathbb{G}_T$ operations. But an exponentiation in $\mathbb{G}_T$ is cheaper than a pairing.

The main point to note is that the TPM is only required to perform operations in $\mathbb{G}_1$, which can be an elliptic curve over a relatively small finite field. Thus the TPM does not have to perform any expensive operations at all.

Table 3. Cost of our DAA protocol

Operation	Party	Cost
Join	TPM	$3 \cdot \mathbb{G}_1$
	Issuer	$(2+n) \cdot \mathbb{G}_1 + 2 \cdot \mathbb{G}_1^2$
	Host	$6 \cdot P$
Sign	TPM	$1 \cdot \mathbb{G}_1$
	Host	$4 \cdot \mathbb{G}_1 + 3 \cdot \mathbb{G}_T + 1 \cdot P$
Verify	Verifier	$n\mathbb{G}_1 + 1 \cdot \mathbb{G}_1^2 + 1 \cdot \mathbb{G}_T^2 + 5 \cdot P$
RogueTag	Verifier	$1 \cdot \mathbb{G}_1$

In conclusion we have presented a DAA protocol based on pairings for which the TPM, i.e. the constrained device in the system, needs very little computational resources in comparison to other variants of the DAA protocol. This efficiency has been achieved by moving to the asymmetric pairings setting and by various precomputations. Our protocol can be proved secure in the random oracle model under the strongest of the two security notions in the literature for DAA schemes. The security proof will be in the full version of the paper.

References

1. Barreto, P.S.L.M., Naehrig, M.: Pairing-friendly elliptic curves of prime order. In: Preneel, B., Tavares, S. (eds.) SAC 2005. LNCS, vol. 3897, pp. 319–331. Springer, Heidelberg (2006)
2. Blake, I.F., Seroussi, G., Smart, N.P.: Elliptic curves and cryptography. Cambridge University Press, Cambridge (1999)
3. Boneh, D., Boyen, X.: Efficient selective-ID secure identity-based encryption without random oracles. In: Cachin, C., Camenisch, J.L. (eds.) EUROCRYPT 2004. LNCS, vol. 3027, pp. 223–238. Springer, Heidelberg (2004)
4. Boneh, D., Boyen, X., Shacham, H.: Short group signatures. In: Franklin, M. (ed.) CRYPTO 2004. LNCS, vol. 3152, pp. 41–55. Springer, Heidelberg (2004)
5. Boneh, D., Di Crescenzo, G., Ostrovsky, R., Persiano, G.: Public key encryption with keyword search. In: Cachin, C., Camenisch, J.L. (eds.) EUROCRYPT 2004. LNCS, vol. 3027, pp. 506–522. Springer, Heidelberg (2004)
6. Boneh, D., Franklin, M.: Identity-based encryption from the Weil pairing. In: Kilian, J. (ed.) CRYPTO 2001. LNCS, vol. 2139, pp. 213–229. Springer, Heidelberg (2001)
7. Brickell, E., Camenisch, J., Chen, L.: Direct anonymous attestation. In: Proceedings of the 11th ACM Conference on Computer and Communications Security, pp. 132–145. ACM Press, New York (2004)
8. Brickell, E., Chen, L., Li, J.: Simplified security notions for direct anonymous attestation and a concrete scheme from pairings. Cryptology ePrint Archive. Report 2008/104, `http://eprint.iacr.org/2008/104`
9. Brickell, E., Chen, L., Li, J.: A new direct anonymous attestation scheme from bilinear maps. In: Proceedings Trust 2008. LNCS, vol. 4968, pp. 166–178. Springer, Heidelberg (to appear, 2008)
10. Camenisch, J., Lysyanskaya, A.: A signature scheme with efficient protocols. In: Cimato, S., Galdi, C., Persiano, G. (eds.) SCN 2002. LNCS, vol. 2576, pp. 268–289. Springer, Heidelberg (2003)

11. Camenisch, J., Lysyanskaya, A.: Signature schemes and anonymous credentials from bilinear maps. In: Franklin, M. (ed.) CRYPTO 2004. LNCS, vol. 3152, pp. 56–72. Springer, Heidelberg (2004)
12. Camenisch, J., Michels, M.: A group signature scheme based on an RSA-variant. Technical Report RS-98-27, BRICS, University of Aarhus (1998)
13. Chen, L., Cheng, Z., Malone-Lee, J., Smart, N.P.: An efficient ID-KEM based on the Sakai-Kasahara key construction. IEE Proceedings, Information Security 153, 19–26 (2006)
14. Chen, L., Cheng, Z., Smart, N.P.: Identity-based key agreement protocols from pairings. Int. Journal of Information Security 6, 213–242 (2007)
15. Galbraith, S., Paterson, K., Smart, N.P.: Pairings for cryptographers (to appear, 2008)
16. Ge, H., Tate, S.R.: A Direct Anonymous Attestation Scheme for Embedded Devices. In: Okamoto, T., Wang, X. (eds.) PKC 2007. LNCS, vol. 4450. Springer, Heidelberg (2007)
17. Lysyanskaya, A., Rivest, R., Sahai, A., Wolf, S.: Pseudonym systems. In: Heys, H.M., Adams, C.M. (eds.) SAC 1999. LNCS, vol. 1758, pp. 184–199. Springer, Heidelberg (2000)
18. Hess, F., Smart, N.P., Vercauteren, F.: The Eta pairing revisited. IEEE Transactions on Information Theory 52, 4595–4602 (2006)
19. Mitsunari, S., Sakai, R., Kasahara, M.: A new traitor tracing. IEICE Transactions on Fundamentals E85-A(2), 481–484 (2002)
20. Menezes, A.J., Okamoto, T., Vanstone, S.A.: Reducing elliptic curve logarithms to logarithms in a finite field. IEEE Trans. Inf. Theory 39, 1639–1646 (1993)
21. Trusted Computing Group, `http://www.trustedcomputinggroup.org`

Pairing Lattices

Florian Hess

Technische Universität Berlin, Germany
hess@math.tu-berlin.de

Abstract. We provide a convenient mathematical framework that essentially encompasses all known pairing functions based on the Tate pairing and also applies to the Weil pairing. We prove non-degeneracy and bounds on the lowest possible degree of these pairing functions and show how endomorphisms can be used to achieve a further degree reduction.

1 Introduction

The cryptographic importance of efficiently computable, bilinear and non-degenerate pairings that are hard to invert in various ways has been amply demonstrated. The currently only known instantiations of pairings suitable for cryptography are the Weil and Tate pairings on elliptic curves or on Jacobians of more general algebraic curves. In view of the applications, efficient algorithms for computing these pairings are of great importance.

Let us take a look at the problem of defining efficiently computable pairings on elliptic curves starting from a general point of view.

Let E be an elliptic curve over $\mathbb{F}_q$ and let G_1, G_2 be two subgroups of $E(\mathbb{F}_q)$ of prime order r satisfying $r \mid (q-1)$. Let μ_r be the subgroup of r-th roots of unity of $\mathbb{F}_q^\times$. We are interested in bilinear pairings $e : G_1 \times G_2 \to \mu_r$. Such a pairing can in principle be defined by taking any generator of μ_r as the pairing value of a generator of G_1 and a generator of G_2 and by extending via linearity. Since the computation of pairing values would then require taking discrete logarithms, this is not a practical approach.

A different approach avoiding the problem with the discrete logarithms would be to use an algebraic representation of e such that pairing values are obtained by substituting the coordinates of the input points with respect to a short Weierstrass form of E into an algebraic expression. This can in principle generally be achieved by using polynomial interpolation and would for example lead to a representation

$$e(P, Q) = f(x_P, y_P, x_Q, y_Q)$$

where $P = (x_P, y_P) \in G_1$, $Q = (x_Q, y_Q) \in G_2$ and $f \in \mathbb{F}_{q^k}[x_1, y_1, x_2, y_2]$ is a fixed polynomial of total degree about r^2 (or r if viewed in x_1, y_1 and x_2, y_2 separately). However, this approach will also be impractical unless some efficient, i.e. at least polynomial time in $\log(r)$, way of storing and evaluating f is found.

The approach currently employed is to use specific rational functions f_P and f_Q on E depending on P and Q instead of interpolation polynomials such that the pairing values are obtained by a function evaluation of the form

S.D. Galbraith and K.G. Paterson (Eds.): Pairing 2008, LNCS 5209, pp. 18–38, 2008.

$$e(P,Q) = f_P(Q)^{(q-1)/r} \tag{1}$$

or

$$e(P,Q) = f_P(Q)/f_Q(P). \tag{2}$$

The functions f_P and f_Q are defined by means of principal divisors with large coefficients but small support. One then essentially applies the Riemann-Roch theorem in form of Miller's algorithm to find a polynomial-in-log(r)-sized representation of f_P and f_Q, consisting of a short product of quotients of linear polynomials in x and y with large exponents, which enables the efficient evaluation of $f_P(Q)$ and $f_Q(P)$.

The Tate pairing is based on (1) and the Weil pairing is based on (2). The function $(P,Q) \mapsto f_P(Q)$ alone is in general not bilinear and does not take values in μ_r. The effect of raising $f_P(Q)$ to the power of $(q^k-1)/r$ or of dividing $f_P(Q)$ by $f_Q(P)$ is to force the resulting functions to be bilinear and to take values in μ_r. We may refer to pairings of the form (1) as pairings defined by the Tate pairing methodology and to pairings of the form (2) as pairings defined by the Weil pairing methodology.

The Ate pairing of [2] and the pairings of [5,10] are pairings defined by the Tate methodology whose pairing functions have reduced degree in comparison with the Tate pairing. Products of the Tate pairing and these pairings with the goal of a further degree reduction have been considered in [4]. This idea has been much extended in [9]. In the case of the Weil pairing methodology considerably less work has been done. In [11] the reduction idea of [2] is applied to the Weil pairing.

The objective of this paper is to present a unified and extended treatment of the idea to find new pairing functions of small degree by using products of existing pairing functions. We provide a convenient mathematical framework that allows to formulate a much clearer non-degeneracy condition and relation with the Tate pairing in comparison to [2,5,10,4,9]. We also show that our framework applies to the Weil pairing, based on an improvement and extension of [11], and prove (or give heuristic arguments) for the optimality and exhaustiveness of our results for ordinary elliptic curves.

While we strive to find suitable pairing functions of smallest degree, the objective of the paper is not to give the most efficiently evaluated pairing functions. This is illustrated best with the following example. The polynomial $f(x) = (x-a)^n \in \mathbb{F}_q[t]$ can have very large degree but still has efficient representation and can be efficiently evaluated at elements of $\mathbb{F}_q$. On the other hand, $g(x) = \prod_{i=1}^{m}(x-a_i) \in \mathbb{F}_q[x]$ may have much smaller degree than f while the cost of representing and evaluating g can be much higher. On the other hand, if there are suitable relations between the a_i, the cost might also be smaller. In this paper we will go from pairing functions of a form analogous to f to pairing functions of a form analogous to g, but with rather small m. It is open whether our pairing functions will lead to more efficiently evaluated pairing functions. Some positive examples are given in [9]. Our intention is to provide a good overview over (all) possible pairing functions and we hope that this will prove useful for finding new efficiently evaluated pairing functions.

We give a brief guideline to the paper. The main results are Theorem 1, Theorem 2, Theorem 3 and Theorem 5. Theorem 1 is just a special, but arguably the most important case of Theorem 3. Theorem 3 is based on Theorem 2 , which provides a direct generalisation (and improvement) of [2,5,10,11] that makes use of endomorphisms. Theorem 5 is an independent add on to the other theorems and shows how the pairings from these theorems can be used in parametric families of elliptic curves. The reader who wants to get a quick overview of the results of this paper is advised to read Section 2.1, Section 3 and Theorem 5, then continue with Theorem 3 and the rest of the paper.

2 Preliminaries

2.1 Notation

In this paper we will consider ordinary elliptic curves only, although the general logic behind the construction can be applied to supersingular curves and higher genus curves as well. Let us first briefly define the standard notation and setting for pairings on such elliptic curves.

Let E be an ordinary elliptic curve over a finite field $\mathbb{F}_q$. Let $r \geq 5$ be a prime factor of $\#E(\mathbb{F}_q)$ with embedding degree $k \geq 2$ such that $k \mid (r-1)$. Then $E(\mathbb{F}_{q^k})[r] \cong \mathbb{Z}/r\mathbb{Z} \times \mathbb{Z}/r\mathbb{Z}$ and there exists a basis P, Q of $E(\mathbb{F}_{q^k})[r]$ satisfying $\pi(P) = P$ and $\pi(Q) = qQ$, where π is the q-power Frobenius endomorphism on E. We define $G_1 = \langle P \rangle$ and $G_2 = \langle Q \rangle$. Note that $G_1 \cap G_2 = \{\mathcal{O}\}$.

Let $\mathcal{O}$ be the point at infinity and $z \in \mathbb{F}_q(E)$ a fixed local uniformiser at $\mathcal{O}$. We say that $f \in \mathbb{F}_{q^k}(E)$ is monic if $(fz^{-v})(\mathcal{O}) = 1$ where v is the order of f at $\mathcal{O}$. In other words this says that the Laurent series expansion of f in terms of z is of the form $f = z^v + O(z^{v+1})$. We will consider monic functions f throughout the paper without further mentioning.

If $f \in E(\mathbb{F}_{q^k})^\times$ then the degree of f is defined as the sum of the positive coefficients of the divisor (f) of f, which is equal to sum of the negative coefficients.

For $s \in \mathbb{Z}$ and $R \in E(\mathbb{F}_{q^k})$ we let $f_{s,R} \in \mathbb{F}_{q^k}(E)$ be the uniquely determined monic function with divisor $(f_{s,R}) = ((sR) - (\mathcal{O})) - s((R) - (\mathcal{O}))$ where (R) is the prime divisor corresponding to the point R (note that our definition is just the inverse of the standard definition $(f_{s,R}) = s((R) - (\mathcal{O})) - ((sR) - (\mathcal{O}))$). Miller's algorithm expresses $f_{s,R}$ as a product of about $\log_2(|s|)$ quotients of monic linear functions with exponents of bitlength up to about $\log_2(|s|)$. Note that for $R \in E(\mathbb{F}_q)$ we have $f_{s,R} \in \mathbb{F}_q(E)$.

The r-th roots of unity in $\mathbb{F}_{q^k}$ are denoted by μ_r. The n-th cyclotomic polynomial is denoted by Φ_n, and its degree by $\varphi(n)$.

2.2 Tate, Ate and Weil Pairings

Recall that the reduced Tate pairing and ate pairings are bilinear pairings $G_2 \times G_1 \to \mu_r$ and are given as follows. The reduced Tate pairing is

$$t : G_2 \times G_1 \to \mu_r, \quad (Q, P) \mapsto f_{r,Q}(P)^{(q^k-1)/r}.$$

It is in fact defined on all $E(\mathbb{F}_{q^k})[r] \times E(\mathbb{F}_{q^k})[r]$ and is non-degenerate on $G_2 \times G_1$.

Let s be an arbitrary integer such that $s \equiv q \bmod r$. Let $N = \gcd(s^k - 1, q^k - 1)$, $L = (s^k - 1)/N$ and $c = \sum_{j=0}^{k-1} s^{k-1-j} q^j \bmod N$. The ate pairing with respect to s is given by

$$a_s : G_2 \times G_1 \to \mu_r, \quad (Q, P) \mapsto f_{s,Q}(P)^{c(q^k-1)/N}.$$

The relation with the Tate pairing is $a_s(Q, P) = t(Q, P)^L$. It is thus non-degenerate if and only if $r \nmid L$ (see [5]).

For $k \mid \#\mathrm{Aut}(E)$ the twisted ate pairing with respect to s is given by

$$a_s^{\mathrm{twist}} : G_1 \times G_2 \to \mu_r, \quad (P, Q) \mapsto f_{s,P}(Q)^{c(q^k-1)/N}.$$

The relation with the Tate pairing is $a_s^{\mathrm{twist}}(P, Q) = t(P, Q)^L$. It is thus non-degenerate if and only if $r \nmid L$ (see [5]).

It is possible to have the same final exponent in the ate and twisted ate pairing as in the Tate pairing. Consider the modified ate pairing

$$a_s : G_2 \times G_1 \to \mu_r, \quad (Q, P) \mapsto f_{s,Q}(P)^{(q^k-1)/r}$$

and the modified twisted ate pairing

$$a_s : G_1 \times G_2 \to \mu_r, \quad (P, Q) \mapsto f_{s,P}(Q)^{(q^k-1)/r}.$$

Since $r \mid N$ and $r \nmid c$ these are always bilinear, and using the relation with the Tate pairing it is not difficult to show that they are non-degenerate if and only if $s^k \not\equiv 1 \bmod r^2$ (see also Theorem 2 and its proof).

The Weil pairing (see [6]) is

$$e : G_1 \times G_2 \to \mu_r, \quad (P, Q) \mapsto (-1)^r f_{r,P}(Q)/f_{r,Q}(P).$$

It is in fact defined on all $E(\mathbb{F}_{q^k})[r] \times E(\mathbb{F}_{q^k})[r]$ and is non-degenerate on $G_1 \times G_2$. Since r is an odd prime we always have $(-1)^r = -1$. For $k \mid \#\mathrm{Aut}(E)$ and $s \equiv q \bmod r$ the Weil pairing with ate reduction[1] with respect to s is given by

$$e_s : G_1 \times G_2 \to \mu_r, \quad (P, Q) \mapsto -w f_{s,P}(Q)/f_{s,Q}(P)$$

for some suitable k-th root of unity $w \in \mathbb{F}_q$. A variant of this pairing, but with final exponentiation, is considered in [11]. For our version see Theorem 2.

It is in general not true that the ate pairing, twisted ate pairing or Weil pairing with ate reduction can be extended to a bilinear pairing on the full r-torsion $E(\mathbb{F}_{q^k})[r]$. Moreover, the twisted ate pairing and the Weil pairing with ate reduction will in general not be bilinear for $k \nmid \#\mathrm{Aut}(E)$.

[1] Following the naming analogy of the Tate and ate pairing we might call this pairing also the eil pairing. Note that eil is the german word for hurry. For a suitable choice of s the eil pairing can indeed be computed faster than the Weil pairing.

3 Pairing Functions of Lowest Degree

Let s be an integer. For $h = \sum_{i=0}^{d} h_i x^i \in \mathbb{Z}[x]$ with $h(s) \equiv 0 \bmod r$ let $f_{s,h,R} \in \mathbb{F}_{q^k}(E)$ for $R \in E(\mathbb{F}_{q^k})[r]$ be the uniquely defined monic function satisfying

$$(f_{s,h,R}) = \sum_{i=0}^{d} h_i((s^i R) - (\mathcal{O})).$$

Furthermore, define

$$\|h\|_1 = \sum_{i=0}^{d} |h_i|.$$

A relation of $\|h\|_1$ with $\deg(f_{s,h,R})$ is given in Lemma 1 below.

Theorem 1. *Assume that s is a primitive k-th root of unity modulo r^2.*
Let $h \in \mathbb{Z}[x]$ with $h(s) \equiv 0 \bmod r$. Then

$$a_{s,h} : G_2 \times G_1 \to \mu_r, \quad (Q, P) \mapsto f_{s,h,Q}(P)^{(q^k-1)/r}$$

defines a bilinear pairing. If $k \mid \#\mathrm{Aut}(E)$ then

$$a_{s,h}^{\mathrm{twist}} : G_1 \times G_2 \to \mu_r, \quad (P, Q) \mapsto f_{s,h,P}(Q)^{(q^k-1)/r}$$

and

$$e_s : G_1 \times G_2 \to \mu_r,$$
$$(P, Q) \mapsto \Big((-1)^{h(1)} f_{s,h,P}(Q)/f_{s,h,Q}(P)\Big)^{\gcd(k,q-1)}$$

define bilinear pairings. The pairings $a_{s,h}$, $a_{s,h}^{\mathrm{twist}}$ and $e_{s,h}$ are non-degenerate if and only if $h(s) \not\equiv 0 \bmod r^2$ holds.

The relation with the Tate and Weil pairing is

$$a_{s,h}(Q, P) = t(Q, P)^{h(s)/r}, \quad a_{s,h}^{\mathrm{twist}}(P, Q) = t(P, Q)^{h(s)/r},$$
$$e_{s,h}(P, Q) = e(P, Q)^{h(s)/r}.$$

There exists an efficiently computable $h \in \mathbb{Z}[x]$ with $h(s) \equiv 0 \bmod r$, $\deg(h) \le \varphi(k) - 1$ and $\|h\|_1 = O(r^{1/\varphi(k)})$ such that the above pairings are non-degenerate. The O-constant depends only on k.

Any $h \in \mathbb{Z}[x]$ with $h(s) \equiv 0 \bmod r$ such that the above pairings are non-degenerate satisfies $\|h\|_1 \ge r^{1/\varphi(k)}$.

Proof. Theorem 1 is a special case of Theorem 3 with s a primitive k-th root of unity modulo r and $d \ge 0$ such that $s = q^d \bmod r$, thus $e = 1$. □

Some remarks on the theorem are in order.

Choice of s. Suppose that s is an integer with $s^k \equiv 1 \bmod r$. Since k is coprime to r we can find i such that $(s+ir)^k \equiv 1 \bmod r^2$. Replacing s by $s+ir$ we can thus assume that $s^k \equiv 1 \bmod r^2$ without loss of generality.

The pairings a_s, a_s^{twist} and e_s depend only on the value of s modulo r^2, as is directly seen from the relations with the Tate and Weil pairing. Since there are no further congruence conditions on s, the value of s can be freely changed modulo r^2 without affecting a_s, a_s^{twist} and e_s.

Computation of h. The polynomial h of Theorem 1 can be determined as follows. Let m be an integer with $\phi(n) \le m \le n$ and consider the $m \times m$ integer matrix

$$M = \begin{pmatrix} r & 0 & & \dots & & 0 \\ -s & 1 & 0 & & \dots & 0 \\ -s^2 & 0 & 1 & 0 & \dots & 0 \\ & & \vdots & & & \\ -s^{m-1} & 0 & & \dots & 0 & 1 \end{pmatrix}.$$

Suppose $m = \phi(n)$ and $w = (w_0, w_1, \dots, w_{m-1})$ is a shortest $\mathbb{Z}$-linear combination of the rows of M, then we can take $h = \sum_{i=0}^{m-1} w_i x^i$. An (approximation of) w can be computed using the first LLL reduced basis element obtained by the LLL algorithm when applied to the rows of M.

As a variation, it is also possible to choose m such that $\phi(n) < m \le n$. We apply the LLL algorithm in the same manner and take w as the smallest LLL reduced basis element satisfying $\|w\|_1 \ge r^{1/\varphi(n)}$.

Exponent. The final exponent satisfies $\gcd(k, q-1) \in \{1, 2, 3, 4, 6\}$. If it is one or even (or q is even) then the term $(-1)^{h(1)}$ can of course be discarded.

Completeness. The construction of pairings of the form $a_{s,h}$ and $a_{s,h}^{\text{twist}}$ of Theorem 1 is complete in the following sense: Consider the case of a_s and let $f_Q \in E(\mathbb{F}_{q^k})^\times$ be any function supported on $Z = \{\pi^i(Q) \,|\, 0 \le i \le k-1\}$ such that $S \mapsto f_Q(S)^{(q^k-1)/r}$ defines a homomorphism $G_1 \to \mu_r$. Then there are $w, h_i \in \mathbb{Z}$ such that $(f_Q) = \sum_{i=0}^{k-1} h_i(\pi^i(Q)) - w(\mathcal{O})$. Then $\sum_{i=0}^{k-1} h_i q^i \equiv 0 \bmod r$ and $\sum_{i=0}^{k-1} h_i(\pi^i(T)) - w(\mathcal{O})$ is a principal divisor for every $T \in G_2$. Let $f_T \in \mathbb{F}_{q^k}(E)^\times$ be monic such that $(f_T) = \sum_{i=0}^{k-1} h_i(\pi^i(T)) - w(\mathcal{O})$ for every $T \in G_2$. Then $(T, S) \mapsto f_T(S)^{(q^k-1)/r}$ defines a bilinear pairing equal to $a_{s,h}$ for $h = \sum_{i=0}^{k-1} h_i x^i \in I^{(1)}$ by Theorem 1. Hence the homomorphism defined by f_Q is obtained by a pairing $a_{s,h}$ from Theorem 1 with fixed first argument Q.

The promised relation of $\|h\|_1$ with $\deg(f_{s,h,R})$ is given by the following lemma.

Lemma 1. *Assume that $s \not\equiv 0 \bmod r$, d is less than the order of s modulo r and $R \neq \mathcal{O}$. We then have*

$$\|h\|_1/2 \le \deg(f_{s,h,R}) \le \|h\|_1.$$

Proof. Let $(f_{s,h,R}) = \sum_{j=-1}^{n} \lambda_j(P_j)$ with pairwise distinct P_j and $P_{-1} = \mathcal{O}$. We have $\sum_j \lambda_j = 0$ and hence $\deg(f_{s,h,R}) = \sum_{\lambda_i > 0} |\lambda_i| = \sum_{\lambda_i < 0} |\lambda_i|$. We may thus assume $\lambda_{-1} \leq 0$. This implies $\sum_{\lambda_j>0} |\lambda_j| \leq \sum_{j\geq 0} |\lambda_j|$. If $j \geq 0$, every λ_j is a sum of some h_i and every h_i occurs in at most one of the λ_j, hence $\sum_{j\geq 0} |\lambda_j| \leq \sum_{i=0}^{d} |h_i| = \|h\|_1$, which proves the upper degree bound without using the assumption on s, d, R.

For the proof of the lower degree bound observe that $\deg(f_{s,h,R}) = \sum_j |\lambda_j|/2$ since $\sum_{\lambda_j>0} |\lambda_j| = \sum_{\lambda_j<0} |\lambda_j|$, again using $\sum_j \lambda_j = 0$. Also note that the assumption on s, d, R implies that the $s^i R$ are pairwise distinct for $0 \leq i \leq d$, hence we can assume $P_j = s^j R$, $\lambda_j = h_j$ for $0 \leq j \leq n$ and $n = d$. Then $\sum_j |\lambda_j|/2 \geq \sum_{j\geq 0} |\lambda_j|/2 = \|h\|_1/2$, which proves the lower degree bound. □

4 Extended Pairings

The next theorem extends the ate pairing, twisted ate pairing and Weil pairing with ate reduction with respect to s to a possibly slightly larger set of admissible values of s. We will then apply this to extend Theorem 1 in order to make use of automorphisms of E. We let $v_r(m)$ denote the maximal exponent of r in m.

Theorem 2. *Let s be any primitive n-th root of unity modulo r with $n \mid \mathrm{lcm}(k, \#\mathrm{Aut}(E))$. Let $u = sq^{-d} \bmod r$ be some primitive e-th root of unity modulo r with $e \mid \gcd(n, \#\mathrm{Aut}(E))$ and $d \geq 0$. Define $v = s^{-1}q^d = u^{-1} \bmod r$. Let $\alpha \in \mathrm{Aut}(E)$ of order e with $\alpha(Q) = uQ$.*

Then

$$a_s : G_2 \times G_1 \to \mu_r, \quad (Q, P) \mapsto \left(\prod_{j=0}^{e-1} f_{s,Q}(\alpha^{-j}(P))^{v^j} \right)^{(q^k-1)/r}$$

defines a bilinear pairing. If $n \mid \#\mathrm{Aut}(E)$ then

$$a_s^{\mathrm{twist}} : G_1 \times G_2 \to \mu_r, \quad (P, Q) \mapsto \left(\prod_{j=0}^{e-1} f_{s,P}(\alpha^{j}(Q))^{v^j} \right)^{(q^k-1)/r}$$

defines a bilinear pairing. The pairings a_s and a_s^{twist} are non-degenerate if and only if $s^n \not\equiv 1 \bmod r^2$ holds.

Suppose $n \mid \#\mathrm{Aut}(E)$ and let $\nu = \min(2, v_r(q^k - 1)) \geq 1$. With e, d as above let $v = s^{-1}q^d \bmod r^\nu$. Then there is an n-th root of unity $w \in \mathbb{F}_q$ such that

$$e_s : G_1 \times G_2 \to \mu_r,$$
$$(P, Q) \mapsto \prod_{j=0}^{e-1} \left(-w f_{s,P}(\alpha^j(Q)) / f_{s,\alpha^j(Q)}(P) \right)^{v^j}$$

defines a bilinear pairing. The pairing e_s is non-degenerate if and only if $s^n \not\equiv 1 \bmod r^2$ holds.

We refer to these pairings as extended ate, extended twisted ate pairing and extended Weil pairing with ate reduction (or simply extended eil pairing).

Some remarks on the theorem are in order.

Special inputs. If $P = \mathcal{O}$ or $Q = \mathcal{O}$ then the pairing values are defined to be equal to 1.

Existence of primitive n-th roots modulo r. Let $m = \mathrm{lcm}(k, \#\mathrm{Aut}(E))$. The proof of Theorem 2 will show that $m \mid (r-1)$ and that $\mathbb{F}_r$ contains all m-th roots of unity.

Choice of d, e and s. An easy calculation with cyclic groups shows that it is always possible to choose $d \geq 0$ such that $u = sq^{-d} \bmod r$ has order e modulo r for some $e \mid \#\mathrm{Aut}(E)$. The value of s can be changed modulo r^2 without changing the pairings a_s, a_s^{twist} and e_s.

Possible cases. Since the automorphism group of an ordinary elliptic curve can only be cyclic of order 2, 4 or 6 there are only few new cases in which Theorem 2 can be applied. On the other hand, there is some freedom of choice regarding the parameters e, d. If $s = q$ then $e = d = 1$ is possible and we recover the non-extended versions of the pairings.

Point multiples. The proof of Theorem 2 will show the existence of $\alpha, \beta \in \mathrm{Aut}(E)$ such that $sQ = (\alpha\pi^d)(Q)$ and $sP = \beta(P)$ (the latter only if $n \mid \#\mathrm{Aut}(E)$).

Computation of w. There is only very few possibilties for $w \in \mu_n \cap \mathbb{F}_q$, and it is probably easiest to try these cases in turn and check for which choice of w the condition $e_s(2P, Q) = e_s(P, Q)^2$ holds.

Another approach is as follows. Let e_s^{raw} denote the function obtained from the definition of e_s using $w = -1$. Then there is a $2n$-th root of unity $w_s \in \mathbb{F}_q$ such that $e_s(S, T) = w_s e_s^{\mathrm{raw}}(S, T)$ for all $S \in G_1$ and $T \in G_2$. The element w_s can be computed from the failing bilinearity of e_s^{raw}: We have $w_s = e_s^{\mathrm{raw}}(2P, Q)/e_s^{\mathrm{raw}}(P, Q)^2$.

Proof (of Theorem 2). We first show the general reduction equation (6). Suppose that $T, S \in E(\mathbb{F}_{q^k})[r]$ and ψ is a purely inseparable $\mathbb{F}_q$-rational isogeny of degree q^d with $\psi(T) = sT$ and $\psi(S) = s^{-1}q^d S = vS$, where the order of s modulo r is equal to n and the order of $s^{-1}q^d$ modulo r is equal to e. We compute

$$f_{r,T}^{(s^n-1)/r} = f_{s^n-1,T} = f_{s^n,T}, \tag{3}$$

where the second equality holds because $s^n \equiv 1 \bmod r$. Lemma 2 of [1] yields

$$f_{s^n,T} = f_{s,T}^{s^{n-1}} f_{s,sT}^{s^{n-2}} \cdots f_{s,s^{n-1}T}. \tag{4}$$

Since ψ is purely inseparable of degree q^d and $\mathbb{F}_q$-rational, we obtain from Lemma 4 in [2]

$$f_{s,\psi^i(T)} \circ \psi^i = w_{s,\psi} f_{s,T}^{q^{id}} \tag{5}$$

for some n-th root of unity $w_{s,\psi} \in \mathbb{F}_q$ (recall that all functions are assumed to be monic). We have $\psi^i(T) = s^i T$ and $\psi^{ie}(S) = S$. Let $k' = n/e$. Combining this with (3), (4), (5) and a short calculation collecting functions that are evaluated at the same points gives

$$\begin{aligned} f_{r,T}(S)^{(s^n-1)/r} &= w \prod_{m=0}^{n-1} f_{s,T}(\psi^{-m}(S))^{s^{n-1-m}q^{dm}} \\ &= w \left(\prod_{j=0}^{e-1} f_{s,T}(\psi^{-j}(S))^{s^{e-1-j}q^{dj}} \right)^{\sum_{i=0}^{k'-1}(s^e)^{k'-1-i}(q^{ed})^i} \end{aligned} \tag{6}$$

for some n-th root of unity $w \in \mathbb{F}_q$. Thus raising $f_{r,T}(S)$ to the power $(s^n-1)/r$ yields a reduced expression. In the following we will choose T, S as Q, P or P, Q. The choice of ψ requires a closer look at the automorphism group of E and its operation on G_1 and G_2.

Automorphisms of additive cyclic groups operate by non-zero integer multiplication. We thus get isomorphisms $\mathrm{Aut}(G_1) \cong \mathrm{Aut}(G_2) \cong \mathbb{F}_r^\times$. Because E is ordinary, $\mathrm{Aut}(E)$ is a cyclic group (of order 2, 4 or 6) and operates faithfully on G_2 and G_1. The Frobenius endomorphism π operates faithfully on G_2 with order k. Since $\mathrm{Aut}(G_2)$ is cyclic, $\mathrm{Aut}(E)$ and π generate a cyclic subgroup H of $\mathrm{Aut}(G_2)$ of order $n = \mathrm{lcm}(k, \#\mathrm{Aut}(E))$. The image of H in $\mathbb{F}_r^\times$ is the group of n-th roots of unity, which shows that s can be written as $s \equiv uq^d \bmod r$ with u of order e modulo r and $e \mid \#\mathrm{Aut}(E)$.

In the ate pairing case, since $u^e \equiv 1 \bmod r$ and $e \mid \#\mathrm{Aut}(E)$, there is $\alpha \in \mathrm{Aut}(E)$ corresponding to the multiplication-by-u automorphism of G_2 such that $(\alpha\pi^d)(Q) = uq^dQ = sQ$. Define $T = Q$, $S = P$ and $\psi_\alpha = \alpha\pi^d$. Then $\psi_\alpha(P) = \alpha(P) = (s^{-1}q^d)P = vP$ and (6) holds with these definitions, giving

$$f_{r,Q}(P)^{(s^n-1)/r} = w \left(\prod_{j=0}^{e-1} f_{s,Q}(\alpha^{-j}(P))^{s^{e-1-j}q^{dj}} \right)^{\sum_{i=0}^{k'-1}(s^e)^{k'-1-i}(q^{ed})^i} \tag{7}$$

for some n-th root of unity $w \in \mathbb{F}_q$.

In the twisted ate pairing case, since $s^{\#\mathrm{Aut}(E)} \equiv 1 \bmod r$, there is $\beta \in \mathrm{Aut}(E)$ corresponding to the multiplication-by-s automorphism of G_2 such that $\beta(P) = sP$. Define $T = P$, $S = Q$ and $\psi_\beta = \beta\pi^d$. Then $\psi_\beta(Q) = (s^{-1}q^d)Q = vQ$ and (6) holds with these definitions. Note that $\alpha(Q) = uQ$ and $\psi_\beta(Q) = vQ = \alpha^{-1}(Q)$, so we obtain

$$f_{r,P}(Q)^{(s^n-1)/r} = w \left(\prod_{j=0}^{e-1} f_{s,P}(\alpha^{j}(Q))^{s^{e-1-j}q^{dj}} \right)^{\sum_{i=0}^{k'-1}(s^e)^{k'-1-i}(q^{ed})^i} \tag{8}$$

for some n-th root of unity $w \in \mathbb{F}_q$.

In order to conclude the proof for the ate and twisted ate pairing we raise (7) and (8) to the power $(q^k-1)/r$, observing $w^{(q^k-1)/r} = 1$. The left hand sides then become $t(Q,P)^{(s^n-1)/r}$ and $t(P,Q)^{(s^n-1)/r}$ respectively, so the right hand sides define bilinear pairings that are non-degenerate if and only if $s^n \not\equiv 1 \bmod r^2$. We then consider the exponents occuring in (7) and (8) modulo r. We have $s^e \equiv (uq^d)^e \equiv q^{ed} \bmod r$, so $c = \sum_{i=0}^{k'-1}(s^e)^{k'-1-i}(q^{ed})^i \equiv k'q^{ed(k'-1)} \not\equiv 0 \bmod r$. Hence the outer exponent c can be omitted without affecting bilinearity or non-degeneracy. Finally, $s^{e-1-j}q^{dj} = s^{e-1}(q^d s^{-1})^j \equiv s^{e-1}v^j \bmod r$. By omitting s^{e-1} for the same reason we arrive at the pairings of the assertion.

For the Weil pairing we apply both cases simultaneously. By means of the chinese remainder theorem we make some additional assumptions on s without changing $s \bmod r^2$. We assume $\nu = v_r(q^k-1)$, $s \equiv 0 \bmod (q^k-1)/r^\nu$ and that s is even. Also also assume that $u = sq^{-d} \bmod r^\nu$ and $v = u^{-1} \bmod r^\nu$ for this new ν. These new assumptions will be removed at the end of the proof. Dividing (8) and (7) gives

$$\begin{aligned} e(P,Q)^{(s^n-1)/r} &= (-1)^{s^n-1} f_{r,P}(Q)^{(s^n-1)/r} / f_{r,Q}(P)^{(s^n-1)/r} \\ &= -w' \left(\prod_{j=0}^{e-1} \left(f_{s,P}(\alpha^j(Q))/f_{s,Q}(\alpha^{-j}(P))\right)^{s^{e-1-j}q^{dj}} \right)^c \\ &= -w'' \left(\prod_{j=0}^{e-1} \left(f_{s,P}(\alpha^j(Q))/f_{s,\alpha^j(Q)}(P)\right)^{s^{e-1-j}q^{dj}} \right)^c \qquad (9) \end{aligned}$$

with $c = \sum_{i=0}^{k'-1}(s^e)^{k'-1-i}(q^{ed})^i \not\equiv 0 \bmod r$ as above, where the last equation holds because α is an automorphism with $\alpha(\mathcal{O}) = \mathcal{O}$. The elements $w', w'' \in \mathbb{F}_q$ are again n-th roots of unity. Since $s \equiv 0 \bmod (q^k-1)/r^\nu$ we get $s \equiv 0 \bmod r'$ for all prime numbers $r' \neq r$ dividing $q^k - 1$. Then $c \equiv q^{ed(k'-1)} \not\equiv 0 \bmod r'$ and $\gcd(c, q^k-1) = 1$, so c can be omitted from the final exponentiation. Let $\bar{c}c \equiv 1 \bmod q^k - 1$. Since s is even we have that q is even or precisely one of the exponents $s^{e-1-j}q^{dj}$ is odd. Also $w^s = 1$ and $w^q = w$ for any n-th root of unity $w \in \mathbb{F}_q$. We can thus write

$$\begin{aligned} e_{s,r}(P,Q)^{\bar{c}(s^n-1)/r} &= -w \prod_{j=0}^{e-1} \left(f_{s,P}(\alpha^j(Q))/f_{s,\alpha^j(Q)}(P)\right)^{s^{e-1-j}q^{dj}} \\ &= \prod_{j=0}^{e-1} \left(-w f_{s,P}(\alpha^j(Q))/f_{s,\alpha^j(Q)}(P)\right)^{s^{e-1-j}q^{dj}} \qquad (10) \end{aligned}$$

for some n-th root of unity $w \in \mathbb{F}_q$. We know that (10) defines an element in μ_r. Since $s \equiv 0 \bmod (q^k-1)/r^\nu$ the factors of the product in (10) are elements in μ_{r^ν} for $0 \leq j < e-1$. We obtain that the factor for $j = e-1$ is an element of μ_{r^ν} as well. Since its exponent $q^{(e-1)d}$ is coprime to r^ν and since $\alpha^j(Q)$ runs through all points of G_2 we get that

$$-w f_{s,P}(\alpha^j(Q))/f_{s,\alpha^j(Q)}(P) \in \mu_{r^\nu} \qquad (11)$$

for all $0 \leq j \leq e-1$. Now $s^{e-1-j}q^{dj} \equiv s^{e-1}v^j \bmod r^\nu$ by assumption. Let $\bar{s}s \equiv 0 \bmod r$. We replace the exponents $s^{e-1-j}q^{dj}$ by $s^{e-1}v^j$ and raise (10) to the power $\bar{s}^{e-1}$. This gives

$$e_{s,r}(P,Q)^{\bar{s}\bar{c}(s^n-1)/r} = \prod_{j=0}^{e-1} \left(-w f_{s,P}(\alpha^j(Q))/f_{s,\alpha^j(Q)}(P)\right)^{v^j}, \tag{12}$$

and the left hand side of this equation shows that the right hand side defines a bilinear pairing that is non-degenerate if and only if the condition $s^n \not\equiv 1 \bmod r^2$ holds. Now

$$f_{r^2,P}(\alpha^j(Q))/f_{r^2,\alpha^j(Q)}(P) = e(P,\alpha^j(Q))^r = 1. \tag{13}$$

Multiplying the right hand side of (12) with the left hand side of (13) to the power λv^j for $0 \le j \le e-1$ gives

$$e_{s,r}(P,Q)^{\bar{s}\bar{c}(s^n-1)/r} = \prod_{j=0}^{e-1} \left(-w f_{s+\lambda r^2,P}(\alpha^j(Q))/f_{s+\lambda r^2,\alpha^j(Q)}(P)\right)^{v^j}. \tag{14}$$

This finally shows that the right hand side of (12) depends only on the value of s modulo r^2 and thus also only on the value of v modulo r^2. So we can replace the additional assumptions on ν, s, u, v made in the proof before (9) by $\nu = \min(2, v_r(q^k-1))$ and $v = s^{-1}q^d \bmod r^\nu$. This finishes the proof. □

5 Extended Pairing Functions of Lowest Degree

With the extended pairings we obtain an extended version of Theorem 1.

Theorem 3. *We use the notation and assumptions from the beginning of section 3 and from Theorem 2. We additionally assume* $s^n \equiv 1 \bmod r^2$.

Let $h \in \mathbb{Z}[x]$ *with* $h(s) \equiv 0 \bmod r$. *Then*

$$a_{s,h} : G_2 \times G_1 \to \mu_r, \quad (Q,P) \mapsto \left(\prod_{j=0}^{e-1} f_{s,h,Q}(\alpha^{-j}(P))^{v^j}\right)^{(q^k-1)/r},$$

$$a_{s,h}^{\text{twist}} : G_1 \times G_2 \to \mu_r, \quad (P,Q) \mapsto \left(\prod_{j=0}^{e-1} f_{s,h,P}(\alpha^j(Q))^{v^j}\right)^{(q^k-1)/r},$$

$$e_{s,h} : G_1 \times G_2 \to \mu_r,$$

$$(P,Q) \mapsto \left(\prod_{j=0}^{e-1} \left((-1)^{h(1)} f_{s,h,P}(\alpha^j(Q))/f_{s,h,\alpha^j(Q)}(P)\right)^{v^j}\right)^{\gcd(n,q-1)}$$

define bilinear pairings whenever the respective assumptions for a_s, a_s^{twist} *and* e_s *of Theorem 2 are met.*

Each pairing $a_{s,h}$, $a_{s,h}^{\text{twist}}$ *and* $e_{s,h}$ *is non-degenerate if and only if* $h(s) \not\equiv 0 \bmod r^2$. *The relation with the Tate and Weil pairing is*

$$a_{s,h}(Q,P) = t(Q,P)^{eh(s)/r}, \quad a_{s,h}^{\text{twist}}(P,Q) = t(P,Q)^{eh(s)/r},$$
$$e_{s,h}(P,Q) = e(P,Q)^{eh(s)/r}.$$

There exists an efficiently computable $h \in \mathbb{Z}[x]$ *with* $h(s) \equiv 0 \bmod r$, $\deg(h) \leq \varphi(n) - 1$ *and* $\|h\|_1 = O(r^{1/\varphi(n)})$ *such that the above pairings are non-degenerate. The* O*-constant depends only on* n.

Any $h \in \mathbb{Z}[x]$ *with* $h(s) \equiv 0 \bmod r$ *such that the above pairings are non-degenerate satisfies* $\|h\|_1 \geq r^{1/\varphi(n)}$.

Proof. The theorem is an instantiation of the generic Theorem 6 for the three different pairing functions. In the following proof we will thus use the notation from Theorem 6.

Let $h, g \in \mathbb{Z}[x]$ and $R \in E(\mathbb{F}_{q^k})[r]$. Since $f_{s,g(x)(x^n-1)+h(x),R} = f_{s,h(x),R}$ we can consider $f_{s,h,R}$ also for $h \in I^{(1)}$ in a natural way. Note that $f_{s,x-s,R}$ is equal to $f_{s,R}$ using the previous notation. Observing $h(s) \equiv 0 \bmod r$ it is then clear that we have three functions

$$a_s, a_s^{\text{twist}}, e_s : I^{(1)} \to W,$$

where $h \in I^{(1)}$ is mapped to $a_{s,h}, a_{s,h}^{\text{twist}}$ and $e_{s,h}$ respectively. Theorem 1 follows directly from Theorem 6 if we prove the three properties of Theorem 6 for a_s, a_s^{twist} and e_s.

Property 1 is clear for a_s, a_s^{twist} and e_s, since

$$f_{s,h+g,R} = f_{s,h,R} f_{s,g,R} \text{ and } (-1)^{(h+g)(1)} = (-1)^{h(1)}(-1)^{g(1)}$$

for any $h, g \in I^{(1)}$ and $R \in E(\mathbb{F}_{q^k})[r]$.

To show property 2 observe that

$$f_{s,hx,R} = f_{s,h,sR} \text{ and } (-1)^{(hx)(1)} = (-1)^{h(1)}$$

for any $h \in I^{(1)}$ and $R \in E(\mathbb{F}_{q^k})[r]$. Let b denote a_s or a_s^{twist}. Let T, S be admissible input points of b_h and assume $b_h \in W^{\text{bilin}}$. Then

$$b_{hx}(T,S) = b_h(sT,S) = b_h(T,S)^s,$$

as was to be shown. The case of e_s is a little more complicated. Consider $\beta \in \text{Aut}(E)$ from the proof of Theorem 2 with $\beta(P) = sP$. Then $\beta(\alpha^j(sQ)) = \alpha^j(Q)$ and $f_{s,h,\alpha^j(sQ)}(P) = w f_{s,h,\alpha^j(Q)}(sP)$ for some n-th root of unity $w \in \mathbb{F}_q$, where application of β to the left hand side of the equation yields the right hand side. Assuming $e_{s,h} \in W^{\mathrm{bilin}}$ we get

$$\begin{aligned}
e_{s,hx}(P,Q) &= \\
&= \left(\prod_{j=0}^{e-1} \left((-1)^{(hx)(1)} f_{s,hx,P}(\alpha^j(Q)) / f_{s,hx,\alpha^j(Q)}(P) \right)^{v^j} \right)^{\gcd(n,q-1)} \\
&= \left(\prod_{j=0}^{e-1} \left((-1)^{h(1)} f_{s,h,sP}(\alpha^j(Q)) / f_{s,h,\alpha^j(sQ)}(P) \right)^{v^j} \right)^{\gcd(n,q-1)} \\
&= \left(w \prod_{j=0}^{e-1} \left((-1)^{h(1)} f_{s,h,sP}(\alpha^j(Q)) / f_{s,h,\alpha^j(Q)}(sP) \right)^{v^j} \right)^{\gcd(n,q-1)} \\
&= e_{s,h}(sP,Q) = e_{s,h}(P,Q)^s.
\end{aligned}$$

Finally we prove property 3. Consider $\alpha \in \mathrm{Aut}(E)$ from Theorem 2 with $\alpha(Q) = uQ$ and thus $\alpha^{-1}(P) = uP$. Then

$$\begin{aligned}
a_{s,r}(Q,P) &= \left(\prod_{j=0}^{e-1} f_{s,r,Q}(\alpha^{-j}(P))^{v^j} \right)^{(q^k-1)/r} = \prod_{j=0}^{e-1} t(Q, \alpha^{-j}(P))^{v^j} \\
&= \prod_{j=0}^{e-1} t(Q, u^j(P))^{v^j} = t(Q,P)^e
\end{aligned}$$

and similarly

$$\begin{aligned}
a_{s,r}^{\mathrm{twist}}(P,Q) &= \left(\prod_{j=0}^{e-1} f_{s,r,P}(\alpha^j(Q))^{v^j} \right)^{(q^k-1)/r} = \prod_{j=0}^{e-1} t(P, \alpha^j(Q))^{v^j} \\
&= \prod_{j=0}^{e-1} t(P, u^j(Q))^{v^j} = t(P,Q)^e.
\end{aligned}$$

Furthermore,

$$\begin{aligned}
e_{s,r}(P,Q) &= \prod_{j=0}^{e-1} \left((-1)^r f_{s,r,P}(\alpha^j(Q)) / f_{s,r,\alpha^j(Q)}(P) \right)^{v^j} \\
&= \prod_{j=0}^{e-1} e_s(P, \alpha^j(Q))^{v^j} = \prod_{j=0}^{e-1} e_s(P, u^j(Q))^{v^j} = e_s(P,Q)^e.
\end{aligned}$$

The functions $a_{s,x-s}$, $a_{s,x-s}^{\mathrm{twist}}$ and $e_{s,x-s}$ are equal to the respective pairings a_s, a_s^{twist} and e_s from Theorem 2. Because $s^n \equiv 1 \bmod r^2$ they are all degenerate. This concludes the proof of Theorem 3. □

We remark that the comments after Theorem 1 apply to Theorem 3 as well.

The equation for $e_{s,hx}$ in the proof shows that the exponent $\gcd(n, q-1)$ cannot be omitted in general. Otherwise $e_{s,hx}(P,Q) = we_{s,h}(sP,Q)$ should be bilinear, which it is not, if $e_{s,h}(P,Q)$ is bilinear.

Since the automorphism group of ordinary elliptic curves is rather small the best improvement we can get in Theorem 3 is for $\varphi(n) = 2\varphi(k)$. This happens precisely when

1. k is odd and $\#\mathrm{Aut}(E) = 4$, or equivalently $D = -4$,
2. k is not divisible by 3 and $\#\mathrm{Aut}(E) = 6$, or equivalently $D = -3$,

where D denotes the discriminant of the endomorphism ring. In all other cases, $\varphi(n) = \varphi(k)$.

It is interesting to look for further extensions. The key point with the ate pairing reduction is equation (5). But every purely inseparable function of degree q^i is of the form $\gamma\pi^i$ with $\gamma \in \mathrm{Aut}(E)$. Thus we cannot do better than Theorem 3.

On the other hand, we could choose to not use (5). Based on solely (4) it is indeed possible to define non-degenerate bilinear pairings. The following theorem states this for the ate pairing case, the twisted ate pairing and Weil pairing cases are left to the reader. We continue to use the notation from the beginning of section 3.

Theorem 4. *Let n be any divisor of $r-1$ and s a primitive n-th root of unity modulo r^2.*

Let $h \in \mathbb{Z}[x]$ with $h(s) \equiv 0 \bmod r$. Then

$$a_{s,h} : G_2 \times G_1 \to \mu_r, \quad (Q,P) \mapsto \left(\prod_{j=0}^{n-1} f_{s,h,s^jQ}(P)^{s^{n-1-j}} \right)^{(q^k-1)/r}$$

is a bilinear pairing that is non-degenerate if and only if $h(s) \not\equiv 0 \bmod r^2$. The relation with the Tate pairing is

$$a_{s,h}(Q,P) = t(Q,P)^{ns^{n-1}h(s)/r}.$$

There exists an efficiently computable $h \in \mathbb{Z}[x]$ with $h(s) \equiv 0 \bmod r$, $\deg(h) \le \varphi(n)-1$ and $\|h\|_1 = O(r^{1/\varphi(n)})$ such that $a_{s,h}$ is non-degenerate. The O-constant depends only on n.

Any $h \in \mathbb{Z}[x]$ with $h(s) \equiv 0 \bmod r$ such that $a_{s,h}$ is a non-degenerate bilinear pairing satisfies $\|h\|_1 \ge r^{1/\varphi(n)}$.

Proof. Using equation (4) with $T = Q, S = P$ to the power of $(q^k-1)/r$ we find that $a_{s,x-s}$ defines a bilinear pairing that is degenerate. Also $a_{s,r} = t^{ns^{n-1}}$ is quite directly seen. From here the proof is the same as that of Theorem 1 and can be left to the reader. □

Note that the product in the definition of $a_{s,h}$ runs over n function evaluations, as opposed to e function evaluations in Theorem 3. This is precisely the effect of the missing ate pairing reduction. While the product over n function evaluations is a big disadvantage it might be outweighed by using h with very small norm and efficient endomorphisms γ such that $\gamma(Q) = sQ$. An example for a similar construction, which does give a fast pairing, are the NSS curves from [8]. See also [9], where these pairings are called superoptimal pairings.

Of course it would be nice to have $n > k$ and still use a pairing as in Theorem 1, that is only one function evaluation instead of more function evaluations. We have tried some examples of elliptic curves with the computer and n with $k \mid n$ and determined all functions in $\mathbb{F}_{q^k}(E)$ supported in $Z_s = \{s^i Q \mid 0 \le i \le n-1\}$ that would define a bilinear (non-degenerate) pairing. Except for the already known functions supported on $Z = \{q^i Q \mid 0 \le i \le k-1\} \subseteq Z_s$ we did not find any new functions. This suggests that on G_1 and G_2, at least generically, all functions defining pairings are in fact of the form like in Theorem 1.

6 Parametric Families

For parametric families of pairing friendly elliptic curves we get the following theorem. We continue to use the notation from the beginning of section 3. A non-zero polynomial $f \in \mathbb{Z}[t]$ is called primitive if the greatest common divisor of its coefficients is equal to 1.

Theorem 5. *Assume that $n, k \ge 2$ are integers and q, s, r are non-constant polynomials in $\mathbb{Z}[t]$, such that s is a primitive n-th root of unity modulo r^2 and r is a primitive polynomial. Assume furthermore that for all $t_0 \in J$ with J a suitable unbounded subset of $\mathbb{Z}$ there is an elliptic curve E over $\mathbb{F}_{q(t_0)}$ with parameters $n, r(t_0), s(t_0)$ as in Theorem 1 (here $k = n$), Theorem 3 or Theorem 4.*

Then there is $h \in \mathbb{Z}[t][x]$ with $\deg(h) \le \varphi(n) - 1$ and $\deg_t(h) = 1/\varphi(n)\deg(r)$ such that

$$a_{s(t_0),h(t_0,x)} : G_2 \times G_1 \to \mu_r$$

from said theorem is a non-degenerate bilinear pairing for all sufficiently large $t_0 \in J$. The polynomial h can be efficiently computed.

Any $h \in \mathbb{Z}[t][x]$ such that $a_{s(t_0),h(t_0,x)}$ is non-degenerate for all sufficiently large $t_0 \in J$ satisfies $\deg_t(h) \ge 1/\varphi(n)\deg(r)$.

Proof. Follows immediately from Theorem 7.

A consequence of the Theorem is that in parametric families $\deg(r)$ must be divisible by $\varphi(n)$.

The polynomial h can be computed in the same way as the polynomial h from Theorem 1, using the function field LLL (see e.g. [7] and the discussion before Lemma 6) instead of the standard LLL algorithm.

We refer to [9] for examples of this construction.

7 Generic Results

This last section of the paper contains some technical lemmas dealing with the ring A and its ideals $I^{(i)}$ that occured in the proofs of Theorems 1, 3, 4 and 5.

In the following we will work with $R = \mathbb{Z}$ and $R = \mathbb{Q}[t]$. It is hence convenient to deal with these cases simultaneously for a moment. The following notation and assumptions will however be in place for the rest of this section.

Let R be a (principal ideal) domain and let $r, s \in R$ such that $r \neq 0$ is not a unit and s has order $n \geq 2$ in $(R/rR)^\times$. In other words, s is a primitive n-th root of unity modulo r. Define the R-algebra and its ideals

$$A = R[x]/(x^n - 1)R[x],$$
$$I^{(i)} = \{h + (x^n - 1)R[x] \mid h(s) \equiv 0 \bmod r^i R\},$$

for $i \geq 0$ such that $s^n \equiv 1 \bmod r^i R$. In the following we will identify elements of A with their representing polynomials of degree $\leq n-1$. We also define the R-modules

$$I^{(i),m} = \{h \in I^{(i)} \mid \deg(h) \leq m - 1\}.$$

Note $I^{(i),m} \subseteq I^{(j),w}$ for $m \leq w$ and $j \leq i$. Also $I^{(i),n} = I^{(i)}$.

7.1 Ideal Structure

Lemma 2. *The $I^{(i)}$ and $I^{(i),m}$ have the following properties:*

1. $I^{(i)} = r^i A + (x - s)A$.
2. $I^{(i),m}$ *is free of rank m and a basis is* $r^i, x - s, x^2 - s^2, \ldots, x^{m-1} - s^{m-1}$.
3. *If $m \geq \varphi(n)$ then $I^{(i),m} = M \oplus I^{(i),\varphi(n)}$ with $M = \{h \in I^{(i),m} \mid h \equiv 0 \bmod \Phi_n\}$.*

Proof. From the definition of $I^{(i)}$ it is clear that $r^i A + (x-s)A \subseteq I^{(i)}$. Conversely, let $h \in I^{(i)}$. Polynomial division by $x-s$ with remainder shows $h = g\cdot(x-s)+h(s)$ with $g \in A$ and $h(s) \in R$. By definition of $I^{(i)}$ we have $h(s) \in r^i R$. Thus $h = h(s) + g \cdot (x - s) \in r^i A + (x - s)A$. This proves the first assertion.

The second assertion follows easily from the first assertion and a short Hermite normal form calculation applied to the basis $r^i, x - s, x(x - s), \ldots, x^{m-2}(x - s)$ of $I^{(i),m}$.

The third assertion follows using polynomial division by Φ_n with remainder: The projection $I^{(i),m} \to I^{(i),\varphi(n)}$, $h \mapsto h \bmod \Phi_n$ is split by the inclusion $I^{(i),\varphi(n)} \to I^{(i),m}$. Here $h \bmod \Phi_n \in I^{(i),\varphi(n)}$ since $\Phi_n(s) \equiv 0 \bmod r^i$. Note that M is a free R-module with basis $\Phi_n, \ldots, x^{m-\varphi(n)-1}\Phi_n$. □

We remark that in addition to Lemma 2 one can show $I^{(i)} = (I^{(1)})^i$ if $R = nR + rR$ (for example $R = \mathbb{Z}$ and r a prime). Since the ideals $I^{(i)}$ are closed under multiplication by x we see that they are closed under rotation of the coefficients of $h \in I^{(i)}$.

7.2 Lattice Arguments for $R = \mathbb{Z}$

We keep the notation and assumptions from the beginning of Section 7 for $R = \mathbb{Z}$ and $r \geq 2$. For $h = \sum_{i=0}^{d} h_i x^i \in \mathbb{Z}[x]$ define

$$\|h\|_1 = \sum_{i=0}^{d} |h_i| \text{ and } \|h\|_2 = \left(\sum_{i=0}^{d} |h_i|^2\right)^{1/2}.$$

Extend this definition to A by using class representatives of degree $\leq n-1$. This makes $I^{(i)}$ into a lattice. We have $\|\cdot\|_1 = \Theta(\|\cdot\|_2)$ on $I^{(i)}$ where the constants depend only on n.

Lemma 3. *Assume $i \geq 1$ satisfies $s^n \equiv 1 \bmod r^i$ and let $h \in \mathbb{Z}[x]$ such that $h(s) \equiv 0 \bmod r^i$. If $h \not\equiv 0 \bmod \Phi_n$ then*

$$\|h\|_1 \geq r^{i/\varphi(n)}.$$

Proof. Let ζ be a primitive n-th root of unity in $\bar{\mathbb{Q}}$ and $B = \mathbb{Z}[\zeta]$ the ring of integers of the n-th cyclotomic number field $K/\mathbb{Q}$. Let $\mathfrak{a} = r^i B + (\zeta - s)B$. Then $\mathfrak{a}$ is an ideal of B of norm $N_{K/\mathbb{Q}}(\mathfrak{a}) = r^i$, by assumption on s. We have $\zeta \equiv s \bmod \mathfrak{a}$. Thus $h(\zeta) \in \mathfrak{a}\backslash\{0\}$ by assumption on h and therefore

$$|N_{K/\mathbb{Q}}(h(\zeta))| \geq N_{K/\mathbb{Q}}(\mathfrak{a}) = r^i.$$

On the other hand, the $\varphi(n)$ complex conjugates $\zeta^{(j)}$ of ζ satisfy $|\zeta^{(j)}| = 1$. Hence $|h(\zeta^{(j)})| \leq \|h\|_1$ and

$$|N_{K/\mathbb{Q}}(h(\zeta))| = \Big|\prod_{j=1}^{\varphi(n)} h(\zeta^{(j)})\Big| \leq \|h\|_1^{\varphi(n)}.$$

Combining the two inequalities proves the first assertion. □

Lemma 4. *Assume $s^n \equiv 1 \bmod r^2$. Let $m \geq \varphi(n)$ and $w = m - \varphi(n)$. Any length ordered LLL-reduced basis $v_1, \ldots, v_m$ of $I^{(1),m}$ satisfies*

$$\|v_i\|_1 = O(1) \text{ and } v_i \in I^{(2)} \text{ for } 1 \leq i \leq w,$$
$$\|v_i\|_1 = \Theta(r^{1/\varphi(n)}) \text{ and } v_i \notin I^{(2)} \text{ for } w < i \leq m.$$

The O- and Θ-constants depend only on n and the element relations hold for r sufficiently large in comparison to n.

Proof. By Lemma 2 the determinant of $I^{(1),m}$ is r and its dimension is m. We also have $I^{(1),m} = M \oplus I^{(1),\varphi(n)}$ with $M = \{h \in I^{(1),m} \mid h \equiv 0 \bmod \Phi_n\}$. Thus there are at least $\varphi(n)$ basis vectors v_i of $I^{(1),m}$ whose projection onto $I^{(1),\varphi(n)}$ is not zero. By Lemma 3 these v_i satisfy $\|v_i\|_2 = \Omega(r^{1/\varphi(n)})$. On the other hand, the LLL-property shows $\prod_{i=1}^{m} \|v_i\|_2 = O(r)$. Thus there are precisely $\varphi(n)$ basis vectors v_i of size $\Theta(r^{1/\varphi(n)})$ whose projection onto $I^{(1),\varphi(n)}$ is not zero. The other basis vectors v_i are in M and satisfy $\|v_i\|_2 = O(1)$. Since the v_i are assumed to be ordered by length the assertion on the norms follows.

Now $\Phi_n(s) \equiv 0 \bmod r^2$ by assumption on s. Hence $v \in I^{(2)}$ for every $v \in M$. This shows $v_i \in I^{(2)}$ for $1 \leq i \leq w$. On the other hand, if $v \in I^{(1),m}\backslash M$ and $v \in I^{(2)}$, then $v \not\equiv 0 \bmod \Phi_n$ and $v(s) \equiv 0 \bmod r^2$. Then $\|v\|_2 = \Omega(r^{2/\varphi(n)})$ by Lemma 3, which is a contradiction. This finally shows $v_i \notin I^{(2)}$ for $w < i \leq m$. □

The true constants of the O-terms and Θ-terms cannot easily be given, only worst case bounds are available that are usually much too large. Since r will in practice be much larger than n the contribution of these terms is small and can essentially be neglected. In this case the element relations will hold. Note that, unconditionally, any (LLL-reduced) basis of $I^{(1),m}$ must contain at least one basis element that is not in $I^{(2)}$.

7.3 Lattice Arguments for $R = \mathbb{Q}[t]$

The results of this section are needed for the proof of Theorem 5. We keep the notation and assumptions from the beginning of Section 7 for $R = \mathbb{Q}[t]$ and $\deg(r) \geq 1$. For $h = \sum_{i=0}^{d} h_i x^i \in \mathbb{Q}[t, x]$ with $h_i \in \mathbb{Q}[t]$ define

$$\deg_t h = \max_{0 \leq i \leq d} \deg(h_i).$$

Extend this definition to A by using class representatives of degree $\leq n-1$ in x. This makes $I^{(i)}$ into a lattice[2] with respect to deg.

Lemma 5. *Suppose $i \geq 1$ satisfies $s^n \equiv 1 \bmod r^i\mathbb{Q}[t]$ and let $h \in \mathbb{Q}[t, x]$ such that $h(s) \equiv 0 \bmod r^i\mathbb{Q}[t]$. If $h \not\equiv 0 \bmod \Phi_n(x)\mathbb{Q}[t, x]$ then*

$$\deg_t(h) \geq i/\varphi(n) \deg(r).$$

Proof. Let ζ be a primitive n-th root of unity in $\bar{\mathbb{Q}}$ and $B = \mathbb{Q}[t, \zeta]$ the integral closure of $\mathbb{Q}[t]$ in the function field $K = \mathbb{Q}(t, \zeta)/\mathbb{Q}$. Let $\mathfrak{a} = r^i B + (\zeta - s)B$. Then $\mathfrak{a}$ is an ideal of B of norm $N_{K/\mathbb{Q}(t)}(\mathfrak{a}) = r^i$, by assumption on s. We have $\zeta \equiv s \bmod \mathfrak{a}$. Thus $h(\zeta) \in \mathfrak{a}$ by assumption on h and

$$\deg(N_{K/\mathbb{Q}(t)}(h(\zeta))) \geq \deg(N_{K/\mathbb{Q}(t)}(\mathfrak{a})) = i \deg(r).$$

On the other hand, the $\varphi(n)$ Puiseux series expansions of ζ with respect to the degree valuation of $\mathbb{Q}(t)$ are just the constant (i.e. without non-zero powers of t) complex conjugates $\zeta^{(j)}$ of ζ and thus satisfy $\deg(\zeta^{(j)}) = 0$. Hence $\deg(h(\zeta^{(j)})) \leq \deg_t(h)$ and

$$\begin{aligned}\deg(N_{K/\mathbb{Q}(t)}(h(\zeta))) &= \deg\left(\prod_{j=1}^{\varphi(n)} h(\zeta^{(j)})\right) \\ &= \sum_{j=1}^{\varphi(n)} \deg(h(\zeta^{(j)})) \leq \varphi(n) \deg_t(h).\end{aligned}$$

Combining the two inequalities proves the assertion. □

[2] This means that $I^{(i)}$ is a free $\mathbb{Q}[t]$-module of finite rank such that subsets of bounded deg-value are finite dimensional $\mathbb{Q}$-vector spaces.

The following lemma uses the function field LLL (e.g. [7]). On input of $M \in \mathbb{Q}[t]^{n\times n}$ with $\det(M) \neq 0$ the function field LLL outputs $N, T \in \mathbb{Q}[t]^{n\times n}$ such that $N = MT$, $\det(T) = 1$ and the sum of the maximal degrees occuring in each column equals the degree of $\det(M)$. The columns of N are then by definition independent LLL-reduced elements of $\mathbb{Q}[t]^n$.

Lemma 6. *Assume $s^n \equiv 1 \bmod r^2\mathbb{Q}[t]$. Let $m \geq \varphi(n)$ and $w = m - \varphi(n)$. Any length ordered LLL-reduced basis $v_1, \ldots, v_m$ of $I^{(1),m}$ satisfies*

$$\deg_t v_i = 0 \text{ and } v_i \in I^{(2)} \text{ for } 1 \leq i \leq w,$$
$$\deg_t(v_i) = 1/\varphi(n)\deg(r) \text{ and } v_i \notin I^{(2)} \text{ for } w < i \leq m.$$

Proof. The assertion and proof are exactly analogous to Lemma 4 (using the analogy $\deg_t = \log(\|\cdot\|_2)$). □

7.4 Pairing Lattices

Let W denote the multiplicative group of functions $G_1 \times G_2 \to \mu_r$ on two cyclic groups G_1 and G_2 of prime order r. Let W^{bilin} denote the subgroup of bilinear functions.

We can finally wrap up and state our main generic theorems.

Theorem 6. *Assume that r is a prime number and s is a primitive n-th root of unity modulo r^2. Let*

$$a_s : I^{(1)} \to W, \quad h \mapsto a_{s,h}$$

be a map with the following properties:

1. *$a_{s,g+h} = a_{s,g}a_{s,h}$ for all $g, h \in I^{(1)}$,*
2. *$a_{s,hx} = a_{s,h}^s$ for all $h \in I^{(1)}$ with $a_{s,h} \in W^{\text{bilin}}$,*
3. *$a_{s,r} \in W^{\text{bilin}}\backslash\{1\}$ and $a_{s,t-s} = 1$.*

Then $\text{im}(a_s) = W^{\text{bilin}}$ and $\ker(a_s) = I^{(2)}$. More precisely,

$$a_{s,h} = a_{s,r}^{h(s)/r}$$

for all $h \in I^{(1)}$.

There exists an efficiently computable $h \in I^{(1),\varphi(n)}$ with $\|h\|_1 = O(r^{1/\varphi(n)})$ and $a_{s,h} \neq 1$. The O-constant depends only on n.

Any $h \in I^{(1)}$ with $a_{s,h} \neq 1$ satisfies $\|h\|_1 \geq r^{1/\varphi(n)}$.

Proof. From properties 1 and 2 we see

$$a_{s,hg} = a_{s,h}^{g(s)}$$

for all $h \in I^{(1)}$ with $a_{s,h} \in W^{\text{bilin}}$ and $g \in A$. We have $I^{(1)} = rA + (x-s)A$ by Lemma 2, so every $h \in I^{(1)}$ is of the form $h = g_1 r + g_2(x-s)$ with $g_1, g_2 \in A$. Then, using property 3,

$$a_{s,h} = a_{s,g_1 r + g_2(x-s)} = a_{s,r}^{g_1(s)} a_{s,x-s}^{g_2(s)} = a_{s,r}^{g_1(s)} \in W_{\text{bilin}} \tag{15}$$

and thus $\text{im}(a_s) \subseteq W_{\text{bilin}}$. Since $a_{s,r} \neq 1$ and r is prime, we have $\text{im}(a_s) = W_{\text{bilin}}$.

The properties of a_s shown so far can be conveniently summarised as follows. We make W_{bilin} into an A-module via $f^g = f^{g(s)}$ for $f \in W_{\text{bilin}}$ and $g \in A$. Then a_s is an epimorphism of the A-modules $I^{(1)}$ and W_{bilin}.

The kernel of a_s is an A-submodule of $I^{(1)}$ and hence an ideal of A contained in $I^{(1)}$. Since a_s is surjective, the index satisfies

$$(I^{(1)} : \ker(a_s)) = \#W_{\text{bilin}} = r.$$

But $r^2, x-s \in \ker(a_s)$ so $I^{(2)} = r^2 A + (x-s)A \subseteq \ker(a_s)$ by Lemma 2. Again by Lemma 2 we have $(I^{(1)} : I^{(2)}) = r$, so $\ker(a_s) = I^{(2)}$ follows.

Looking at (15) we see that $g_1(s) = h(s)/r \bmod r$ and thus

$$a_{s,h} = a_{s,r}^{h(s)/r},$$

which shows the relation of $a_{s,h}$ with the generator $a_{s,r}$ of W^{bilin}.

Using $\ker(a_s) = I^{(2)}$, the rest of the theorem follows directly from Lemma 4 with $m = \phi(n)$, the LLL algorithm and Lemma 3. □

The ideal $I^{(1)}$ together with the map $a_s : I^{(1)} \to W$ satisfying the properties stated in Theorem 6 is called a pairing lattice with pairing lattice function a_s.

Theorem 7. *Assume that $n \geq 2$ and r, s are non-constant polynomials in $\mathbb{Z}[t]$ such that s is a primitive n-th root of unity modulo r^2 and r is a primitive polynomial. Assume furthermore that there is a pairing lattice function*

$$a_{s(t_0)} : I^{(1)}_{r(t_0),s(t_0)} \to W^{\text{bilin}}_{r(t_0)}$$

for all $t_0 \in J$ with J a suitable unbounded subset of $\mathbb{Z}$.

Then there is $h \in \mathbb{Z}[t][x]$ with $\deg(h) \leq \varphi(n) - 1$ and $\deg_t(h) = 1/\varphi(n)\deg(r)$ such that

$$a_{s(t_0),h(t_0,x)} \neq 1$$

for all sufficiently large $t_0 \in J$. The polynomial h can be efficiently computed.

Any $h \in \mathbb{Z}[t][x]$ such that $a_{s(t_0),h(t_0,x)} \neq 1$ for all sufficiently large $t_0 \in J$ satisfies $\deg_t(h) \geq 1/\varphi(n)\deg(r)$.

Proof. There are only finitely many $t_0 \in J$ such that $s(t_0)$ has order less than n modulo r^2, because these t_0 must be zeros of $s^m - 1 \bmod r$ for $m < n$. Since t_0 is to be chosen large enough we may assume that $s(t_0)$ is a primitive n-th root of unity modulo r^2.

We define $A, I^{(1)}, I^{(2)}$ for r, s and $R = \mathbb{Q}[t]$ as at the beginning of section 7. From Lemma 6 with $m = \phi(n)$ and the function field LLL we see that there is $v_i \in \mathbb{Q}[t][x]$ with $v_i(s) \equiv 0 \bmod r\mathbb{Q}[t]$, $\deg(v_i) \le \phi(n) - 1$ and $\deg_t(v_i) = 1/\varphi(n)\deg(r)$ for $1 \le i \le \phi(n)$. Let $h \in \mathbb{Z}[t][x]$ be the product of v_i with the least common multiple of all denominators of all $\mathbb{Q}$-coefficients of v_i. Then $h(s) \in \mathbb{Z}[t]$ and $h(s) \equiv 0 \bmod r\mathbb{Z}[t]$ by the lemma of Gauss [3, p. 181], since r was assumed to be primitive.

Substituting t_0 for t in this congruence we get $h(t_0, s(t_0)) \equiv 0 \bmod r(t_0)$. From $\deg_t(h) = 1/\varphi(n)\deg(r)$ we see $\|h(t_0, x)\|_1 = O(r(t_0)^{1/\varphi(n)})$. Lemma 3 implies $h(t_0, s(t_0)) \not\equiv 0 \bmod r(t_0)^2$. We conclude that $a_{s(t_0),h(t_0,x)}$ defines a non-degenerate pairing by Theorem 1.

The last statement on the degrees follows since $\|h(t_0, x)\|_1 \ge r(t_0)^{1/\varphi(n)}$ by Lemma 3 for t_0 tending to infinity. □

Acknowledgement

The author thanks Steven Galbraith for a careful reading of a prior version of the paper.

References

1. Barreto, P.S.L.M., Galbraith, S., O'hEigeartaigh, C., Scott, M.: Efficient pairing computation on supersingular abelian varieties. Designs, Codes and Cryptography 42(3), 239–271 (2007)
2. Hess, F., Smart, N.P., Vercauteren, F.: The Eta Pairing Revisited. IEEE Transaction on Information Theory 52(10), 4595–4602 (2006)
3. Lang, S.: Algebra. GTM 211. Springer, Heidelberg (2002)
4. Lee, E., Lee, H.-S., Park, C.-M.: Efficient and Generalized Pairing Computation on Abelian Varieties, Cryptology ePrint Archive, Report 2008/040 (2008), `http://eprint.iacr.org/2008/0040`
5. Matsuda, S., Kanayama, N., Hess, F., Okamoto, E.: Optimised Versions of the Ate and Twisted Ate Pairings. In: Galbraith, S.D. (ed.) Cryptography and Coding 2007. LNCS, vol. 4887, pp. 302–312. Springer, Heidelberg (2007)
6. Miller, V.S.: The Weil pairing, and its efficient calculation. J. Cryptology 17(4), 235–261 (2004)
7. Paulus, S.: Lattice basis reduction in function fields. In: Buhler, J.P. (ed.) ANTS 1998. LNCS, vol. 1423, pp. 567–575. Springer, Heidelberg (1998)
8. Scott, M.: Faster Pairings Using an Elliptic Curve with an Efficient Endomorphism. In: Maitra, S., Veni Madhavan, C.E., Venkatesan, R. (eds.) INDOCRYPT 2005. LNCS, vol. 3797, pp. 258–269. Springer, Heidelberg (2005)
9. Vercauteren, F.: Optimal Pairings, Cryptology ePrint Archive, Report, 2008/096 (2008), `http://eprint.iacr.org/2008/096`
10. Zhao, C.-A., Zhang, F., Huang, J.: A Note on the Ate Pairing, Cryptology ePrint Archive, Report 2007/247 (2007), `http://eprint.iacr.org/2007/247`
11. Zhao, C.-A., Zhang, F.: Reducing the Complexity of the Weil Pairing Computation, Cryptology ePrint Archive, Report 2008/212 (2008), `http://eprint.iacr.org/2008/212`

The Uber-Assumption Family
A Unified Complexity Framework for Bilinear Groups

Xavier Boyen

Voltage, Inc.
xb@boyen.org

Abstract. We offer an exposition of Boneh, Boyen, and Goh's "uber-assumption" family for analyzing the validity and strength of pairing assumptions in the generic-group model, and augment the original BBG framework with a few simple but useful extensions.

1 Introduction

It is no secret that the rapid development of pairing-based cryptography has of late been supported, in no small part, by a dizzying array of tailor-made complexity assumptions. These assumptions, known by acronyms or abbreviations such as BDH or D-Linear, sometimes annotated with dimensional parameters such as ℓ-BDHI or (ℓ, m)-PolySDH, always found their original motivation in the same necessity for compromise: the impetus to strike a balance between the operational objective of conjuring a new useful cryptographic scheme, and the aesthetic appeal of resting the security of one's contraption on the firmest possible grounds. Whereas weaker and established assumptions should always and unequivocally be preferred whenever possible, there are times when, despite our best efforts, we seem to require the power of a novel or stronger assumption to reach an elusive goal.

The newcomer to this particular branch of cryptography will therefore most likely be astonished by the sheer number, and sometimes creativity, of those assumptions. The contrast with the more traditional branches of algebraic cryptography is quite stark indeed. The two most venerable branches of public-key cryptography, "Factoring" and "Discrete-Log", used to rest on a handful of useful assumptions at most: QR and RSA on one side, and DL and DH on the other. Yet, the much younger "Pairing" branch, which strictly speaking is merely an offshoot from discrete-log, is already teeming with dozens of plausible assumptions, whose distinctive features make them uniquely and narrowly suited to specific types of constructions and security reductions.

Far from being a collective whim, this haphazard state of affair stems from the very power of the bilinear pairing, and the tremendous complexity that it brings, in comparison to the admittedly quite simpler algebraic structures of twentieth-century public-key cryptography. Since the new "bilinear" groups offer a much richer palette of cryptographically useful trapdoors than their "unidimensional" counterparts, there are also many more ways to arrange and exploit

S.D. Galbraith and K.G. Paterson (Eds.): Pairing 2008, LNCS 5209, pp. 39–56, 2008.

those trapdoors and make assumptions on those arrangements. As desirable as it might be, the notion of a one-size-fits-all pairing assumption — strong enough to enable a large array of useful schemes, and weak enough to leave room for insightful security reductions — appears to be wishful thinking. Accordingly, it is no surprise that many of the innovative pairing-based cryptographic schemes proposed in recent years were based on original, then-unseen assumptions.

This is not to say, however, that all pairing assumptions are created equal, and that a hefty dose of skepticism and prudence is advisable when presented with a new one. When faced with a new assumption, the key questions are thus:

Is the assumption natural? — per one's subjective sense of plausibility
How strong is the assumption? — for a quantifiable measure of strength

Of course, no cryptographically useful assumption about an NP language can be fully trusted without proving that its defining problem has no polynomial-time solution — a famous problem whose solution is not within reach —, hence the need to make *assumptions* in the first place. What we can do, however, is vet our assumptions in an idealized model in which they can be evaluated, and hope that the model realistically captures the limited powers of all foreseeable attackers for the planned lifetime of the cryptographic designs that rely on them.

2 Generic Bilinear Groups

At the very least, any assumption one is willing to make about a particular family of algebraic groups should be true in the most basic presentation of such family. This is the idea behind Nechaev and Shoup's notion of generic groups, originally used in [29,31] to study the hardness of the Discrete Log (DL) and Diffie-Hellman (DH) problems in prime-order cyclic groups devoid of any other exploitable structural feature. The first recorded use of the generic-group model to justify a pairing assumption was made by Boneh and Boyen, who in [7] extend the generic-group model to capture the case of (pairs of) groups equipped with a pairing in order to provide some justification for their new SDH assumption. More recently, Maurer [26] considers an extension of the generic model that allows random sampling in addition to the deterministic group operations.

A generic group is a fictitious idealized entity that precisely captures the group structure and any additional feature such as a pairing needed for the task at hand, but none of the extraneous structure that an actual presentation of the group might display. Withholding the extraneous structure from an attacker naturally reduces the possibilities of breaking a scheme or its supporting assumptions; however the important point is that the generic-group model is just powerful enough to support the nominal functionality, and that taking away anything from it will render it useless. In that sense, bilinear generic groups epitomize the bare minimal environment in which one can hope to implement a scheme based on bilinear discrete-logarithm problems. Any scheme requiring only a cyclic group structure and a bilinear pairing should be realizable in the generic model; and conversely, if the scheme can be broken in the generic model, then there is no hope that it will be secure under any actual presentation of the bilinear groups.

What makes generic groups very useful is that, in this model, it is relatively easy to prove strong (exponential) lower bounds on the hardness of solving many problems of cryptographic interest, which in turn can be used to justify or "gain confidence in" a particular assumption. The first use of the generic model to lend credence to a bilinear assumption was made by Boneh and Boyen in [7] to support their new (and back then highly non-standard) SDH complexity assumption. Without the support of a bilinear generic-group analysis, the authors of [7] would have never dared make such an unusual assumption.

An important caveat is that the very act of placing oneself in the generic-group model amounts to making an extremely strong assumption. Proving the security of a cryptographic scheme directly in the generic-group model is therefore usually very easy (unless the scheme is, literally, inherently broken), and brings little insight as to its real-world security. The generic-group model shares many of the controversial attributes of the random-oracle model, but is in a sense even stronger and less appealing. After all, random oracles are used to model the destruction of any exploitable structure by "bit-mashing" hash contraptions that are designed with this particular goal as the primary objective: make the number of rounds large enough and you are almost guaranteed to fulfill that requirement, whatever the design of the round function. Generic groups, on the other hand, formalize the belief that the particular presentation of a group will expose some aspects of its structure while hiding others: one's choice of mathematical implement is thus much more constrained, as it has to fulfill the two conflicting goals of structure removal and structure preservation. Once we have chosen a particular type of presentation for our bilinear groups (*e.g.*, supersingular elliptic curves over prime finite fields), the structure-hiding properties are mostly bound by this choice, with little room for subsequent adjustment.

Of course, both the random-oracle and generic-group models are objectionable because they funnel a non-interactive functionality through an interactive interface, thereby giving the simulator unfair "mind-reading" and "challenge-concoction" capabilities in its interaction with the opponent. Such is the nature of idealized models that enforce artificial restrictions on mathematical objects; and we mention that, in some circumstances, the two models exhibit rather similar weaknesses [18]. However, the point is that, whereas a broken random-oracle implement might well be salvaged by increasing a complexity parameter, mathematic weaknesses in generic-group implements are much more likely to be terminal.

The generic-group model should thus be viewed as a meta-assumption, useful not for proving the security of actual schemes, but to assess the plausibility of specific, weaker assumptions on which actual schemes are shown to rest.

Even though the generic-group model itself involves interactive oracles, we emphasize that we use it to analyze the non-interactive BBG uber-assumption family and its members — and if there is one temptation that we shall vigorously resist, it is to bring interactive assumptions into that family.

3 Bilinear Pairings

We write $\mathbb{F}_p$ for the finite field of prime order p. We write $\mathbb{Z}_n$ for the ring of integers modulo a composite $n = p_1 p_2 \ldots$, whose prime factors are the p_i.

We consider cyclic bilinear groups of prime order p or composite order n. For a security parameter κ, we shall respectively suppose that $\lceil \log_2 p \rceil = \Theta(\kappa)$ or that $\lceil \log_2 p_i \rceil = \Theta(\kappa)$ for all prime factors $p_i \mid n$.

In either case, let $\mathbb{G}$, $\hat{\mathbb{G}}$, and $\mathbb{G}_t$ be three cyclic groups of order p or n, written multiplicatively, and whose elements have polynomial-size representations in κ. A bilinear map or pairing in $\langle \mathbb{G}, \hat{\mathbb{G}} \rangle$ is an efficiently computable, non-degenerate function $\mathbf{e} : \mathbb{G} \times \hat{\mathbb{G}} \to \mathbb{G}_t$ that satisfies the bilinearity property $\mathbf{e}(g^r, \hat{g}^s) = \mathbf{e}(g, \hat{g})^{r\,s}$ for all elements $g \in \mathbb{G}$ and $\hat{g} \in \hat{\mathbb{G}}$ and all exponents $r, s \in \mathbb{Z}$.

We assume the existence of an efficient randomized generation procedure $\mathcal{G}$ that, on input the security parameter κ represented in unary as 1^κ, outputs a randomly generated *bilinear instance* $\mathbf{G} = \langle (p \text{ or } n), \mathbb{G}, \hat{\mathbb{G}}, \mathbb{G}_t, g, \hat{g}, \mathbf{e} \rangle \xleftarrow{\$} \mathcal{G}(\kappa)$, such that $\lceil \log_2 p \rceil$ or $\lceil \log_2 n \rceil = \Theta(\kappa)$ as the case may be — the groups' primality or lack thereof will always be clear from context.

In practice, bilinear instances may be realized on certain algebraic varieties or curves over finite fields [4,3,19], by computing the Weil, Tate, or a related pairing using Miller's efficient algorithm [27] or variants thereof [2,25]. We refer the reader to [5,17] for additional information.

It is convenient to typify the bilinear groups depending on whether the group isomorphism $\psi : \hat{\mathbb{G}} \to \mathbb{G}$ and its inverse $\psi^{-1} : \mathbb{G} \to \hat{\mathbb{G}}$ are efficiently computable. Using the terminology from [21], we say that $\langle \mathbb{G}, \hat{\mathbb{G}} \rangle$ is of:

"type 1" — if both ψ and ψ^{-1} are efficiently computable (this includes the case where $\mathbb{G} = \hat{\mathbb{G}}$);

"type 2" — if ψ is efficiently computable, but not ψ^{-1} (if the converse holds, we swap $\mathbb{G}$ and $\hat{\mathbb{G}}$);

"type 3" — if neither ψ nor ψ^{-1} is efficiently computable.

Remark that "not efficiently computable" does not necessarily mean "infeasible to compute". Since such infeasibility can have its usefulness too, several authors have floated the idea of making it an explicit assumption, albeit with differing degrees of prudence [9,1]. The accepted moniker for such infeasibility assumptions is eXtended Diffie-Hellman, leading to the (suffix) acronym XDH.

For clarity, in any bilinear context $\mathbf{G} = \langle p \text{ or } n, \mathbb{G}, \hat{\mathbb{G}}, \mathbb{G}_t, g, \hat{g}, \mathbf{e} \rangle$, we use the "hat-notation" ($\hat{g}$ *vs.* g) to indicate that an element belongs to $\hat{\mathbb{G}}$ rather than $\mathbb{G}$.

4 Examples of Bilinear Assumptions

A hallmark of pairing-based cryptography, that some would qualify as unfortunate, is the plethora of complexity assumptions that come with it. The reason for this abundance is that the pairing is quite powerful and flexible in comparison to earlier cryptographic tools, and accordingly there are many flexible ways in which this power can be exploited.

Rather than make a blanket assumption that the pairing will be "safe" in *all* conceivable ways it *could* be used, the sensible approach is to make the weakest assumption that one possibly can for the intended application. Indeed, should even a minor cryptographic weakness be discovered down the road, it would surely contradict a blanket assumption, but might just leave a tailored assumption unscathed, if sufficiently weak.

This is the same as in a regular finite-field or elliptic-curve cryptographic group $\mathbb{G}$, where one may just make the minimal assumption that there is no feasible discrete-logarithm extraction map $\text{DL} : \mathbb{G} \times \mathbb{G} \to \mathbb{F}_p$ s.t. $g^{\text{DL}(g,h)} = h$; or one may wish to assume something stronger, such as it is hard to compute the Diffie-Hellman function $\text{CDH} : \mathbb{G} \times \mathbb{G} \times \mathbb{G} \to \mathbb{G} : \langle g, g^x, g^y \rangle \mapsto g^{x\,y}$, or even to decide the predicate $\text{DDH} : \mathbb{G} \times \mathbb{G} \times \mathbb{G} \times \mathbb{G} \to \{0,1\} : \langle g, g^x, g^y, g^z \rangle \mapsto [x\,y \stackrel{?}{=} z]$ — where, by "hard", we mean that for any polynomial growth function poly$(\cdot)$, the cost increases as $\omega(\text{poly}(\log_2 p))$ when the prime group order $p \to \infty$.

With the pairing, there are may more ways to arrange "givens" and "goals" when stating a hardness assumption, whether computational or decisional. Each arrangement yields a specific assumption, sometimes equivalent to others, sometimes weaker, stronger, or incomparable. Here are examples from the literature.

Gap Diffie-Hellman. Gap problems are computational problems stated relative to a decisional oracle [30]. Gap-DH is perhaps the simplest possible assumption related to pairing groups: it assumes the hardness of CDH given a pairing that is used only as a DDH predicate "oracle". (Notice that the implementation of the predicate can only be done using symmetric or type-1 pairings in $\mathbb{G}$, or with type-2 pairings if Gap-DH is defined in $\hat{\mathbb{G}}$.) The assumption is:

Gap-DH — with an oracle implied by any (symmetric) pairing [23]
Instance: $g_0 = g,\ g_1 = g^\alpha,\ g_2 = g^\beta \in \mathbb{G}$
Oracle: $\text{DDH}(X, Y, Z, W) := [\mathbf{e}(Y, Z) \stackrel{?}{=} \mathbf{e}(X, W)]$
Goal: $\text{CDH}(g_0, g_1, g_2) = g^{\alpha\beta} \in \mathbb{G}$

Bilinear Diffie-Hellman. Now a classic, this assumption modifies CDH and DDH to account for the pairing. Since a (symmetric) pairing can efficiently perform "trans-group" Diffie-Hellman operations from $\mathbb{G} \times \mathbb{G}$ into $\mathbb{G}_t$, the pairing assumption shall posit that the same cannot be done from $\mathbb{G} \times \mathbb{G} \times \mathbb{G}$ into $\mathbb{G}_t$.

The computational and decisional versions of BDH are as follows, as originally stated for symmetric pairings [22], then as restated for all pairing groups in two different ways (the last one being weaker in type-3 groups).

(D-)BDH — as originally stated for symmetric pairings [22,10]
Instance: $g,\ g^\alpha,\ g^\beta,\ g^\gamma \in \mathbb{G}$
Goal (Computational): $\mathbf{e}(g, g)^{\alpha\beta\gamma} \in \mathbb{G}_t$
Goal (Decisional): $\mathbf{e}(g, g)^{\alpha\beta\gamma} \stackrel{?}{=} v$ for a test value $v \in \mathbb{G}_t$

(D-)BDH' — as restated for all known pairing types [6]
Instance: $g,\ g^\alpha,\ g^\gamma \in \mathbb{G}$ and $\hat{g},\ \hat{g}^\alpha,\ \hat{g}^\beta \in \hat{\mathbb{G}}$

Goal (Computational): $\mathbf{e}(g, \hat{g})^{\alpha\beta\gamma} \in \mathbb{G}_t$
Goal (Decisional): $\mathbf{e}(g, \hat{g})^{\alpha\beta\gamma} \stackrel{?}{=} v$ for a test value $v \in \mathbb{G}_t$

(D-)BDH" — weaker statement for type-3 pairings [20]
Instance: $g^\gamma \in \mathbb{G}$ and $\hat{g}$, $\hat{g}^\alpha$, $\hat{g}^\beta \in \hat{\mathbb{G}}$
Goal (Computational): $\mathbf{e}(g, \hat{g})^{\alpha\beta\gamma} \in \mathbb{G}_t$
Goal (Decisional): $\mathbf{e}(g, \hat{g})^{\alpha\beta\gamma} \stackrel{?}{=} v$ for a test value $v \in \mathbb{G}_t$

Linear. The Linear assumption is another proposal to cope with the limitation that DDH cannot possibly hold in (type-1) bilinear groups. The idea is to restore the hardness of deciding a product equality "in the exponent", by splitting each of the two factors of the product using a linear combination. What makes Linear more useful than BDH in some reductions, is that all the elements in the instance and the goal live in the bilinear group(s) $\mathbb{G}$ and $\hat{\mathbb{G}}$; that is, "at the ground level", and not in the target group.

Again, we have a computational and a decisional version of the assumption, as originally stated in symmetric groups, and twice restated in asymmetric ones.

(D-)Linear — as originally stated for a single group, optionally bilinear [9]
Instance: g, g^α, g^β, $g^{\alpha\gamma}$, $g^{\beta\delta} \in \mathbb{G}$
Goal (Computational): $g^{\gamma+\delta} \in \mathbb{G}$
Goal (Decisional): $g^{\gamma+\delta} \stackrel{?}{=} v$ for a test value $v \in \mathbb{G}$

(D-)Linear' — as restated for all known pairing types [13]
Instance: g, g^α, g^β, $g^{\alpha\gamma}$, $g^{\beta\delta} \in \mathbb{G}$ and $\hat{g}$, $\hat{g}^\alpha$, $\hat{g}^\beta \in \hat{\mathbb{G}}$
Goal (Computational): $g^{\gamma+\delta} \in \mathbb{G}$
Goal (Decisional): $g^{\gamma+\delta} \stackrel{?}{=} v$ for a test value $v \in \mathbb{G}$

(D-)Linear" — weaker statement for type-3 pairings [20]
Instance: g, $g^{\alpha\gamma}$, $g^{\beta\delta} \in \mathbb{G}$ and $\hat{g}$, $\hat{g}^\alpha$, $\hat{g}^\beta \in \hat{\mathbb{G}}$
Goal (Computational): $g^{\gamma+\delta} \in \mathbb{G}$
Goal (Decisional): $g^{\gamma+\delta} \stackrel{?}{=} v$ for a test value $v \in \mathbb{G}$

(Weak) (Bilinear) Diffie-Hellman Inversion. Whereas the instances in the previous assumptions had a small fixed number of elements, one may consider giving the adversary a possibly much greater data set from which to proceed. Accordingly, this makes for stronger assumptions, but ones from which security theorems of actual constructions may be easier to prove. The family of ℓ-BDHI assumptions falls in this category, where the parameter ℓ indicates the size of the problem instance.

A closely related assumption family is the ℓ-"weak Diffie-Hellman" from [28], later renamed ℓ-DHI in [6] for it is significantly stronger than DH for large ℓ. DHI does not require bilinearity; and, like Decision DH, the decisional D-DHI is generically false given a type-1 pairing (or just a DDH oracle).

The (bilinear) ℓ-BDHI assumption itself was stated for the case of symmetric-pairing groups in [6], and subsequently extended for the general case and slightly weakened in [8], where it became known as ℓ-wBDHI for "weak BDHI".

These three families of assumptions are as follows.

ℓ-(D-)DHI — without pairing, a.k.a. "weak Diffie-Hellman" [28]
Instance: $g,\ g^{\alpha},\ g^{\alpha^2},\ \ldots,\ g^{\alpha^\ell} \in \mathbb{G}$ for a fixed parameter $\ell \in \mathbb{N}$
Goal (Computational): $g^{1/\alpha} \in \mathbb{G}$
Goal (Decisional): $g^{1/\alpha} \stackrel{?}{=} v$ for a test value $v \in \mathbb{G}$

ℓ-(D-)BDHI — as stated for symmetric pairings [7]
Instance: $g,\ g^{\alpha},\ g^{\alpha^2},\ \ldots,\ g^{\alpha^\ell} \in \mathbb{G}$ for a fixed parameter $\ell \in \mathbb{N}$
Goal (Computational): $\mathbf{e}(g,g)^{1/\alpha} \in \mathbb{G}_t$
Goal (Decisional): $\mathbf{e}(g,g)^{1/\alpha} \stackrel{?}{=} v$ for a test value $v \in \mathbb{G}_t$

ℓ-(D-)wBDHI — weaker version for all pairing types [8]
Instance: $g,\ g^{\alpha},\ g^{\alpha^2},\ \ldots,\ g^{\alpha^\ell} \in \mathbb{G}$ and $\hat{g},\ \hat{g}^{\beta} \in \hat{\mathbb{G}}$ for a fixed $\ell \in \mathbb{N}$
Goal (Computational): $\mathbf{e}(g,\hat{g})^{\alpha^{\ell+1}\beta} \in \mathbb{G}_t$
Goal (Decisional): $\mathbf{e}(g,g)^{\alpha^{\ell+1}\beta} \stackrel{?}{=} v$ for a test value $v \in \mathbb{G}_t$

Bilinear Diffie-Hellman Exponent. Another parametric assumption very similar to ℓ-wBDHI, this one ℓ-BDHE asks to compute or decide a target-group element whose exponent lies, not at the end of a supplied sequence of other elements (such as α^{-1} or $\alpha^{\ell+1}$ in relation to $\alpha^0,\ldots,\alpha^\ell$), but in the middle of a sequence that contains a hole or missing element (such as α^ℓ in relation to $\alpha^0,\ldots,\alpha^{\ell-1},\alpha^{\ell+1},\ldots,\alpha^{2\ell}$).

ℓ-(D-)BDHE — stated here for all known pairing types [8]
Instance: $g,\ g^{\alpha},\ \ldots,\ g^{\alpha^{\ell-1}},\ g^{\alpha^{\ell+1}},\ \ldots,\ g^{\alpha^{2\ell}} \in \mathbb{G}$ and $\hat{g},\ \hat{g}^{\beta} \in \hat{\mathbb{G}}$ for $\ell \in \mathbb{N}$
Goal (Computational): $\mathbf{e}(g,\hat{g})^{\alpha^\ell\beta} \in \mathbb{G}_t$
Goal (Decisional): $\mathbf{e}(g,\hat{g})^{\alpha^\ell\beta} \stackrel{?}{=} v$ for a test value $v \in \mathbb{G}_t$

Strong Diffie-Hellman. Yet another ℓ-parametric assumption, this one is special in the sense that it has, not one, but exponentially many non-trivially different solutions that are all equally acceptable and cannot feasibly turned into one another. Hence, this assumption inherently computational and has no natural decisional counterpart.

Though originally stated with type-2 groups in mind, where $\hat{\mathbb{G}}$ can be mapped to $\mathbb{G}$ at will, SDH has been later restated to accommodate type-3 pairings, by giving the power sequence directly in the group $\mathbb{G}$ where the solution resides. Incidentally, this makes the restated assumption weaker in type-2 bilinear groups, but makes no difference in type-1 groups.

ℓ-SDH — as originally stated for type-1 and type-2 pairings [7]
Instance: $g \in \mathbb{G}$ and $\hat{g},\ \hat{g}^{\alpha},\ \hat{g}^{\alpha^2},\ \ldots,\ \hat{g}^{\alpha^\ell} \in \hat{\mathbb{G}}$ for a fixed parameter $\ell \in \mathbb{N}$
Goal: any pair $\langle w,\ g^{1/(\alpha+w)}\rangle \in \mathbb{F}_p \times \mathbb{G}$

ℓ-SDH' — as restated for all pairings, weaker in type-2 groups [7, full article]
Instance: $g,\ g^{\alpha},\ g^{\alpha^2},\ \ldots,\ g^{\alpha^\ell} \in \mathbb{G}$ and $\hat{g},\ \hat{g}^{\alpha} \in \hat{\mathbb{G}}$ for fixed parameter $\ell \in \mathbb{N}$
Goal: any pair $\langle w,\ g^{1/(\alpha+w)}\rangle \in \mathbb{F}_p \times \mathbb{G}$

Modified SDH and Hidden SDH. A less elegant but actually weaker statement of the SDH assumption is to give in the instance, not g raised to a sequence of $\ell + 1$ powers of α, but a list of ℓ pairs $\langle w_j,\ g^{1/(\alpha+w_j)}\rangle$ for randomly chosen $w_j \in \mathbb{F}_p$. The resulting assumption is called, Modified SDH.

Another variation on this theme is not to give the w_j in the clear, but hide them in an exponent, as in g^{w_j}. To preclude trivial attacks, we need to hide w_j in *two* exponents, g^{w_j} and h^{w_j}. Problem instances with this hiding mechanism thus consist of triples of the form $\langle g^{w_j},\ g^{1/(\alpha+w_j)},\ h^{w_j}\rangle$. The resulting assumption is called, Hidden SDH.

Modified SDH and Hidden SDH are formally stated as follows.

Modified ℓ-SDH — as originally stated for type-1 pairings [14]
Instance: $g,\ g^{\alpha} \in \mathbb{G}$ and $\ell - 1$ pairs $\langle w_j,\ g^{1/(\alpha+w_j)}\rangle \in \mathbb{F}_p \times \mathbb{G}$ for a fixed parameter $\ell \in \mathbb{N}$
Goal: another pair $\langle w,\ g^{1/(\alpha+w)}\rangle \in \mathbb{F}_p \times \mathbb{G}$

ℓ-HSDH — as originally stated for type-1 pairings [14]
Instance: $g,\ g^{\alpha},\ g^{\beta} \in \mathbb{G}$ and $\ell - 1$ triples $\langle g^{w_j},\ g^{1/(\alpha+w_j)},\ g^{\beta w_j}\rangle \in \mathbb{G}^3$ for fixed parameter $\ell \in \mathbb{N}$
Goal: another triple $\langle g^{w},\ g^{1/(\alpha+w)},\ g^{\beta w}\rangle \in \mathbb{G}^3$

Poly-SDH. Yet another variation of SDH is the Poly-SDH assumption from [12]. It is based on the same notion as SDH, but, in a sense, makes much better use of the pairing without strengthening the assumption "too much".

Recall that valid SDH solutions can be verified just with a DDH (or cross-group DDH) test, which can of course be realized using the pairing but does not fully exploit it. Poly-SDH remedies this shortcoming by giving more leeway in the form of the solutions, which now require an actual (product of) pairings for their verification. This is fine, since the pairing is already available.

The technical idea behind Poly-SDH is to allow the solver to output not one but m pairs $\langle w_i,\ g^{c_i/(\alpha_i+w_i)}\rangle$, under the joint constraint that $\sum_{i=1}^{m} c_i = 1 \in \mathbb{F}_p$. For $m = 1$, it is the same as (Modified) SDH; whereas for $m > 1$, it extends it by allowing convex combinations of m independent ordinary SDH solutions.

(ℓ, m)-Poly-SDH — as originally stated for all pairing types [12]
Instance: $g,\ g^{\alpha_1},\ \ldots,\ g^{\alpha_m} \in \mathbb{G}$ and ℓm pairs $\langle w_{i,j},\ g^{1/(\alpha_i+w_{i,j})}\rangle$ such that $1 \le i \le m$ and $1 \le j \le \ell$ for fixed parameters $\ell, m \in \mathbb{N}$
Goal: any m pairs $\langle w_i,\ g^{c_i/(\alpha_i+w_i)}\rangle$ such that all $w_i \notin \{w_{i,j} : 1 \le j \le \ell\}$ and $\sum_{i=1}^{m} c_i = 1 \pmod{p}$ for undisclosed c_i.

5 The Classic Uber-Assumption

Since many of the useful pairing assumptions can be proven in the generic model using a similar argument, it would be nice to express them all as particular instances of a common template, and justify the template itself in the generic-group model. This idea was originally studied by Boneh, Boyen, and Goh [8], whose provided a "master theorem" for such a template. The new name "uber-assumption" [1] is intended to evoke both a notion of generality and one of power, as a warning that the uber-assumption is in general way too strong to be used by itself, and should thus be reserved as a convenient framework for the analysis of weaker assumptions.

In one sentence, the BBG uber-assumption family from [8] is defined for prime-order bilinear groups with a symmetric pairing $\mathbf{e} : \mathbb{G} \times \mathbb{G} \to \mathbb{G}_t$, and emcompasses the decision problems whose instance elements and challenge are all polynomial powers of the generators $g \in \mathbb{G}$ and $\mathbf{e}(g, g) \in \mathbb{G}_t$.

A precise statement is given next, adapted from [8] and restated for general pairings $\mathbf{e} : \mathbb{G} \times \hat{\mathbb{G}} \to \mathbb{G}_t$.

5.1 The BBG Problem Statement

Let p be some large prime, and let r, s, t, and c be four positive integers. Consider $R \in \mathbb{F}_p[X_1, \ldots, X_c]^r$, $S \in \mathbb{F}_p[X_1, \ldots, X_c]^s$, and $T \in \mathbb{F}_p[X_1, \ldots, X_c]^t$, three tuples of multivariate polynomials over the field $\mathbb{F}_p$, and respectively containing r, s, and t polynomials in the same d variables $X_1, \ldots, X_c$. We write $R = \langle r_1, r_2, \ldots, r_r \rangle$, $S = \langle s_1, s_2, \ldots, s_s \rangle$ and $T = \langle t_1, t_2, \ldots, t_t \rangle$. In the classic BBG template with polynomial exponents, the first components of R, S, and T ought to be the constant polynomial 1; that is, $r_1 = s_1 = t_1 = 1$.

For a set Ω, a function $f : \mathbb{F}_p \to \Omega$, and a vector $\langle x_1, \ldots, x_c \rangle \in \mathbb{F}_p^d$, we use the notation $f(R)$ to denote the application of f to each element of R, namely,

$$f(R(x_1, \ldots, x_c)) = \langle f(r_1(x_1, \ldots, x_c)), \ldots, f(r_r(x_1, \ldots, x_c)) \rangle \in \Omega^r \ ;$$

and use a similar notation for applying f to the s-tuple S and the t-tuple T.

Let then $\mathbb{G}$, $\hat{\mathbb{G}}$, and $\mathbb{G}_t$ be groups of order p, and $\mathbf{e} : \mathbb{G} \times \hat{\mathbb{G}} \to \mathbb{G}_t$ be a non-degenerate bilinear map. Suppose that $g \in \mathbb{G}$ and $\hat{g} \in \hat{\mathbb{G}}$ respectively generate the groups to which they belong, and set $g_t = \mathbf{e}(g, \hat{g}) \in \mathbb{G}_t$ thus generating $\mathbb{G}_t$. Together, these form the bilinear context $\mathbf{G} = \langle p, \mathbb{G}, \hat{\mathbb{G}}, \mathbb{G}_t, g, \hat{g}, \mathbf{e} \rangle$.

Boneh, Boyen, and Goh define the (R, S, T, f)-Diffie-Hellman problem in $\mathbf{G}$ as follows. Given the input vector,

$$U(x_1, \ldots, x_c) = \left\langle g^{R(x_1, \ldots, x_c)}, \hat{g}^{S(x_1, \ldots, x_c)}, g_t^{T(x_1, \ldots, x_c)} \right\rangle \in \mathbb{G}^r \times \hat{\mathbb{G}}^s \times \mathbb{G}_t^t \ ,$$

secretly created from random $\langle x_1, \ldots, x_c \rangle \in_\$ \mathbb{F}_p^d$, compute the output value,

$$V(x_1, \ldots, x_c) = g_t^{f(x_1, \ldots, x_c)} \in \mathbb{G}_t \ .$$

Decisional Problem (Strict). The corresponding decisional problem is then defined in the obvious way: given a vector $U(x_1, \ldots, x_c)$ and a test value v that is

[1] The spelling of "uber" without the *umlaut* is meant to indicate that its usage is less that of the German word *über*, than the English-language colloquialism for "super". The first recorded use of the phrase "uber-assumption" is attributed to Dan Boneh.

either $V(x_1, \ldots, x_c)$ or some $V(x'_1, \ldots, x'_c)$ for independent $\langle x'_1, \ldots, x'_c \rangle \in_\$ \mathbb{F}_p^d$, decide which is the case. The advantage of an algorithm $\mathcal{A}$ that outputs $b \in \{0,1\}$ in solving the Decision (R, S, T, f)-Diffie-Hellman problem in $\mathbf{G}$ is defined as,

$$\mathbf{Adv}_\mathcal{A} = \left| \begin{array}{l} \Pr\left[\mathcal{A}(U(x_1, \ldots, x_c), V(x_1, \ldots, x_c)) = 0\right] \\ - \Pr\left[\mathcal{A}(U(x_1, \ldots, x_c), V(x'_1, \ldots, x'_c)) = 0\right] \end{array} \right| ,$$

where the probability is over the random choice of generators $g \in \mathbb{G}$ and $\hat{g} \in \hat{\mathbb{G}}$, the random choice of secret inputs $\langle x_1, \ldots, x_c \rangle \in \mathbb{F}_p^d$ and $\langle x'_1, \ldots, x'_c \rangle \in \mathbb{F}_p^d$, and the random bits consumed by $\mathcal{A}$.

Decisional Problem (General). An alternative definition of the decisional problem substitutes for $V(x'_1, \ldots, x'_c)$ a uniformly sampled random element $V' \in \mathbb{G}_t$ in the above equation. The alternate advantage definition is given by,

$$\mathbf{Adv}_\mathcal{A} = \left| \begin{array}{l} \Pr\left[\mathcal{A}(U(x_1, \ldots, x_c), V(x_1, \ldots, x_c)) = 0\right] \\ - \Pr\left[\mathcal{A}(U(x_1, \ldots, x_c), V') = 0\right] \end{array} \right| ,$$

where the probability is over the random choice of generators $g \in \mathbb{G}$ and $\hat{g} \in \hat{\mathbb{G}}$, the random choice of secret inputs $\langle x_1, \ldots, x_c \rangle \in \mathbb{F}_p^d$, the uniform random choice of $V' \in \mathbb{G}_t$, and the random bits consumed by $\mathcal{A}$.

Often, it makes no difference to use either definition, unless the function $V(\ldots)$ does not have uniform cover, in which case the second definition will be more demanding (it being easier for the adversary $\mathcal{A}$ to make the distinction). The more general definition should be used if the range one's function V is either not uniform or does not cover the entire codomain, as is often the case in bilinear groups of composite order, as we shall see in Section 6.3.

5.2 The BBG Generic Lower Bound

In order to state a generic lower bound on the complexity of the above problem, we need one more definition.

Independence. Let $R = \langle r_1, \ldots, r_r \rangle \in \mathbb{F}_p[X_1, \ldots, X_c]^r$, $S = \langle s_1, \ldots, s_s \rangle \in \mathbb{F}_p[X_1, \ldots, X_c]^s$, and $T = \langle t_1, \ldots, t_t \rangle \in \mathbb{F}_p[X_1, \ldots, X_c]^t$ as previously defined.

We say that a polynomial $f \in \mathbb{F}_p[X_1, \ldots, X_c]$ is dependent on the triple $\langle R, S, T \rangle$ if there exist $r\,s + t$ constants $\{a_{i,j}\}$ and $\{b_k\}$, and possibly $r^2 + s^2$ additional constants $\{c_{i',i''}\}$ and $\{d_{j',j''}\}$, for $1 \le i, i', i'' \le r$, $1 \le j, j', j'' \le s$, and $1 \le k \le t$, such that,

$$f = \sum_{i=1}^{r} \sum_{j=1}^{s} a_{i,j}\, r_i\, s_j + \sum_{i'=1}^{r} \sum_{i''=1}^{r} c_{i',i''}\, r_{i'}\, r_{i''} + \sum_{j'=1}^{s} \sum_{j''=1}^{s} d_{j',j''}\, s_{j'}\, s_{j''} + \sum_{k=1}^{t} b_k\, t_k \ .$$

The constants $d_{j',j''}$ exist to model the computable isomorphism $\psi : \hat{\mathbb{G}} \to \mathbb{G}$, while the constants $c_{j',j''}$ exist to model the inverse isomorphism $\psi^{-1} : \mathbb{G} \to \hat{\mathbb{G}}$. Failing this, the constants $c_{j',j''}$ and/or $d_{j',j''}$ must be set to zero.

We say that f is independent of $\langle R, S, T \rangle$ if f is not dependent on $\langle R, S, T \rangle$.

Clearly, independence is easier to achieve without any computable isomorphism; and this observation can thus be exploited to express assumptions such as XDH [1] that rely on the hardness of isomorphism(s), in the generic model.

Degrees. For a polynomial $f \in \mathbb{F}_p[X_1, \ldots, X_c]$, we let d_f denote the total degree of f in all the indeterminate variables. For a set $R \subseteq \mathbb{F}_p[X_1, \ldots, X_c]^r$, we let $d_R = \max_{r_i \in R}\{d_{r_i}\}$, the maximum total degree of all polynomials in the set.

Generic-Group Model. In the generic-group model of [31], as extended to the bilinear case in [7], elements are represented as arbitrary bit strings in $\{0,1\}^*$, and are manipulated by making oracle calls to the generic-group functions (one for the group operation in each group, one for the pairing, and possibly one or two for the isomorphism and its inverse). Beyond the initial supply of elements given to $\mathcal{A}$ in the problem instance, $\mathcal{A}$ can obtain new elements only as the result of such oracle calls.

The idea behind the generic-group argument, is that the simulator that interacts with $\mathcal{A}$ will treat as uninstantiated symbolic random variables all the secret scalars that $\mathcal{A}$ is not supposed to see, and simulate the oracle functions by deriving symbolic indices for all the group elements shown to $\mathcal{A}$. At the end, one then argues that $\mathcal{A}$ was unable to answer its challenge, even though with high probability it was given an accurate simulation. The probability that the simulation is accurate is the probability that no two distinct representations end up referring to the same in the group element upon a random assignment of the secret exponents.

In order to reason about the string representation of generic-group elements available to $\mathcal{A}$, we make use of three abstract maps $\xi_{\mathbb{G}}$, $\xi_{\hat{\mathbb{G}}}$, $\xi_{\mathbb{G}_t} : \mathbb{F}_p \to \{0,1\}^*$ such that $\xi_{\mathbb{G}}(x)$, $\xi_{\hat{\mathbb{G}}}(y)$, $\xi_{\mathbb{G}_t}(z)$ represent $g^x \in \mathbb{G}$, $\hat{g}^y \in \hat{\mathbb{G}}$, $g_t^z \in \mathbb{G}_t$, respectively. The simulator internally maintains the indices x, y, z as symbolic expressions of all the uninstantiated secret scalars (*e.g.*, secret exponents α, *etc.*).

Complexity Bounds. We are now in a position to state the BBG generic lower bound on the complexity of the Decision (R, S, T, f)-Diffie-Hellman problem in an idealized bilinear context $\mathbf{G}$. We emphasize, however, that the bound applies only in generic-group idealizations of $\mathbf{G}$, not in actual presentations of $\mathbf{G}$.

Theorem 1 ("Master Theorem" [8])
Let $R = \langle r_1, \ldots, r_r \rangle \in \mathbb{F}_p[X_1, \ldots, X_c]^r$, $S = \langle s_1, \ldots, s_s \rangle \in \mathbb{F}_p[X_1, \ldots, X_c]^s$, and $T = \langle t_1, \ldots, t_t \rangle \in \mathbb{F}_p[X_1, \ldots, X_c]^t$. Let also $f \in \mathbb{F}_p[X_1, \ldots, X_c]$. Let then,

$$d = \begin{cases} \max\{2\,d_R, 2\,d_S, d_R + d_S, d_T, d_f\} & \textit{for type-1 bilinear contexts} \\ \max\{2\,d_S, d_R + d_S, d_T, d_f\} & \textit{for type-2 bilinear contexts} \\ \max\{d_R + d_S, d_T, d_f\} & \textit{for type-3 bilinear contexts}\ . \end{cases}$$

Let $\xi_{\mathbb{G}}$, $\xi_{\hat{\mathbb{G}}}$, $\xi_{\mathbb{G}_t} : \mathbb{F}_p \to \{0,1\}^*$ be three arbitrary external string-encoding maps for the three respective groups $\mathbb{G}$, $\hat{\mathbb{G}}$, $\mathbb{G}_t$, as previously defined.

If f is independent of $\langle R, S, T\rangle$, then, for any algorithm $\mathcal{A}$ that makes a total of at most q queries to (1) the group operation oracles in $\mathbb{G}$, $\hat{\mathbb{G}}$, and $\mathbb{G}_t$, (2) the bilinear pairing oracle $\mathbf{e} : \mathbb{G} \times \hat{\mathbb{G}} \to \mathbb{G}_t$, and (3), if allowed, the homomorphism oracle $\psi : \hat{\mathbb{G}} \to \mathbb{G}$ and its inverse ψ^{-1}, we have:

$$\mathbf{Adv}_{\mathcal{A}} = \left| \Pr \left[\mathcal{A} \begin{pmatrix} p, \\ \xi_{\mathbb{G}}(R(x_1, \ldots, x_c)), \\ \xi_{\hat{\mathbb{G}}}(S(x_1, \ldots, x_c)), \\ \xi_{\mathbb{G}_t}(T(x_1, \ldots, x_c)), \\ \xi_{\mathbb{G}_t}(v_0), \\ \xi_{\mathbb{G}_t}(v_1) \end{pmatrix} = b \ : \ \begin{matrix} x_1, \ldots, x_c, y \in_{\$} \mathbb{F}_p, \\ b \in_{\$} \{0, 1\}, \\ v_b \leftarrow f(x_1, \ldots, x_c), \\ v_{1-b} \leftarrow y \end{matrix} \right] - \frac{1}{2} \right|$$
$$\leq \frac{(q + 2c + 2)^2 \cdot d}{2p} .$$

Not only can this theorem serve to ascertain that a proposed assumption is not tautologically false, it also provides a quantitative basis for comparing the concrete efficiency or tightness of multiple assumptions: Let the assumption that minimizes $\mathbf{Adv}_{\mathcal{A}}$ for some standard number of oracle queries, win! Of course, such benchmark this should be balanced with the usefulness of the contenders, so perhaps Theorem 1 should be used to compare assumptions from which comparably functional cryptosystems can be constructed.

We leave all such other applications of Theorem 1 to the imagination of the reader, and merely mention as an alternative the following corollary which can provide a quick qualitative sanity check.

Corollary 1 (Asymptotic Lower Bound [8])
Let $R \in \mathbb{F}_p[X_1, \ldots, X_c]^r$, $S \in \mathbb{F}_p[X_1, \ldots, X_c]^s$, and $T \in \mathbb{F}_p[X_1, \ldots, X_c]^t$, be tuples of c-variate polynomials over $\mathbb{F}_p$. Let $f \in \mathbb{F}_p[X_1, \ldots, X_c]$ be one more such polynomial. Let d be the sum-total degree defined as in Theorem 1, or $d = \max\{2d_R, 2d_S, d_R + d_S, d_T, d_f\}$ in the worst case.

If f is independent of $\langle R, S, T\rangle$, then any randomized algorithm $\mathcal{A}$ that solves the Decision (R, S, T, f)-Diffie-Hellman problem in the generic-group model with constant advantage $\Omega(1)$, must take time at least $\Omega(\sqrt{p/d} - c)$, asymptotically as the security parameter $\kappa \to \infty$.

5.3 The BBG Bound in Action

Theorem 1 and Corollary 1 are useful — *not* to assert the uber-assumption as a reduction basis for proving the security of pairing-based cryptographic schemes — it is way too strong to be used for that purpose, and furthermore it is interactive as it relies on the generic-group model — but as a template from which the generic validity of specific weaker assumptions can be proved.

To show this, we briefly mention how many standard decisional assumptions follow from the classic BBG uber-assumption, using Corollary 1.

Decision DH in target group $\mathbb{G}_t$:
Set $R = S = \langle 1 \rangle$, $T = \langle 1, \alpha, \beta \rangle$, $f = \alpha\beta$.
Decision BDH in symmetric bilinear group $\mathbb{G}$:
Set $R = S = \langle 1, \alpha, \beta, \gamma \rangle$, $T = \langle 1 \rangle$, $f = \alpha\beta\gamma$.
Decision BDH in asymmetric bilinear groups $\langle \mathbb{G}, \hat{\mathbb{G}} \rangle$:
Set $R = \langle 1, \alpha, \beta \rangle$, $S = \langle 1, \alpha, \gamma \rangle$, $T = \langle 1 \rangle$, $f = \alpha\beta\gamma$.
Decision ℓ-BDHI in $\mathbb{G}$:
Set $R = S = \langle 1, \alpha, \alpha^2, \ldots, \alpha^\ell \rangle$, $T = \langle 1 \rangle$, $f = \alpha^{\ell+1}$.
Decision ℓ-BDHE in $\langle \mathbb{G}, \hat{\mathbb{G}} \rangle$:
Set $R = \langle 1, \alpha, \ldots, \alpha^{\ell-1}, \alpha^{\ell+1}, \ldots, \alpha^{2\ell} \rangle$, $S = \langle 1, \beta \rangle$, $T = \langle 1 \rangle$, $f = \alpha^\ell \beta$.

5.4 The Computational Case

Although the theorem above was stated only for the decisional uber-assumption, it is easy to obtain a similar result for computational instances. The decisional bound gives us the strongest result, but it has a few limitations, *e.g.*, one cannot use it to prove CDH is hard in $\mathbb{G}$, since DDH is easy in $\mathbb{G}$ for type-1 pairings.

In order to derive a generic bound for a computational uber-assumption, one must bound the probability that an adversary $\mathcal{A}$ outputs some expected group element $\xi_{\mathbb{G}_t}(f(x_1, \ldots, x_c))$, after a total number q of oracle queries.

Since some computational assumptions are defined with respect to problems whose expected answers are vectors of elements instead of single elements, in all generality one should really consider not a single polynomial $f \in \mathbb{F}_p[X_1, \ldots, X_c]$ but three vectors $V_{\mathbb{G}}$, $V_{\hat{\mathbb{G}}}$, $V_{\mathbb{G}_t}$ of polynomials in $\mathbb{F}_p[X_1, \ldots, X_c]$ (one vector for each of the three groups). The success condition for $\mathcal{A}$ will be that it outputs the expected vector of group elements,

$$\langle \xi_{\mathbb{G}}(V_{\mathbb{G}}(x_1, \ldots, x_c)), \xi_{\hat{\mathbb{G}}}(V_{\hat{\mathbb{G}}}(x_1, \ldots, x_c)), \xi_{\mathbb{G}_t}(V_{\mathbb{G}_t}(x_1, \ldots, x_c)) \rangle \ .$$

A (restricted) decisional vector version of the BBG framework is studied in [15].

6 Extending the Uber-Assumption

As universal as it attempts to be, the classic uber-assumption family still manages to exclude several potentially important family members.

The prime counterexample is the SDH assumption [7], which, despite being the first bilinear assumption justified in the generic-group model [7], is not covered by the classic uber-assumption because each SDH problem instance admits not one but exponentially many (and mutually irreducible) valid solutions.

Another counterexample is the HSDH assumption [14], which is a very simple variation of SDH, in some respects stronger and weaker in others, but with the additional twist that its problem instances contain powers of the generators not only with polynomial exponents but also rational ones.

It turns out that it is almost trivial to extend the regular uber-assumption framework to cover these cases, without compromising the tightness of the generic lower bounds for any of the assumptions that were already covered. The proposed extensions are:

1. To let the solver choose one out of many possible distinct challenges;
2. To let instances and answers have not polynomial but rational exponents;
3. To support composite-order groups of known or unknown factorization.

We emphasize that, unlike in cryptographic constructions where we seek to prove security from the *weakest possible assumptions*, here, we are are interested in a theoretical framework to analyze, not actual schemes, but other assumptions. Hence, our purpose is to devise the *strongest and broadest class of assumptions* that subsumes all the others. Indeed, the stronger and more encompassing the family, the more diverse the specific assumptions that we will be able to instantiate from it. The only delicate point is that our quest for the broadest family should not compromise the tightness of the concrete generic-group lower bounds deduced from it for specific instantiations.

6.1 Flexible Challenges

The first observation we make is that the proof of Theorem 1 (see Appendix A of [8]) only depends on the "target" polynomial $f \in \mathbb{F}_p[X_1, \dots, X_c]$ through its total degree d_f and not through its actual polynomial expansion. Thus, in order to capture a computational assumption such as SDH that admits exponentially many possible solutions, one could simply keep f undefined until the adversary $\mathcal{A}$ is ready to produce its final answer (or, as the case may be, request its final challenge), as long as the total degree d_f of f remains within certain boundaries.

Concretely, when applying our framework to study an assumption with multiple good answers, such as SDH, one would simply let $\mathcal{A}$ specify which expression of f it wishes to use from some pre-specified family $\mathcal{F}$, and define the maximal total degree of f as the maximum of d_f over the entire family: $d_{\mathcal{F}} = \max_{f \in \mathcal{F}}\{d_f\}$.

For SDH, we would take $\mathcal{F} = \{f_w | w \in \mathbb{F}_p\}$ where $f_w(X) = \frac{1}{w+X}$, and allow $\mathcal{A}$ to choose any $f_w \in \mathcal{F}$. (Of course, in the particular example of SDH we still have the problem that the f_w are not polynomial but rational; we shall address this next.)

As a last comment, observe that with SDH the adversary will announce the choice of w as part of the answer; with HSDH on the contrary the adversary keeps w computationally hidden, but still commits to it via g^w (and $g^{\beta w}$). It makes no difference for our purpose whether the choice of f is announced or merely committed, because the generic-group analysis is an information-theoretic one. The only requirement is that the choice of f be extractable from the adversary's answer (if valid); from there, the simulator can always recover it explicitly.

6.2 Rational Exponents

The second observation we make is that, to handle assumptions such as HSDH or Poly-SDH that involve rational exponents in the instance elements and/or the final answer, one can notionally replace all the rational exponents ρ_i by a ratio π_i/Δ, where the π_i are polynomials and Δ is the least common multiple of all the denominators.

The preceding generic-group analysis will carry through using the π_i, regardless of the numerical value of the denominator Δ, as long as Δ does not vanish. If Δ vanishes, then at least one of the group elements given to $\mathcal{A}$ is undefined (due to a division by zero), and it is thus a pathological case that the solver $\mathcal{A}$ can use to recover secret data. Hence, Theorem 1 can still be used to derive lower complexity bounds even in the case of rational exponents, but with some caveats:

1. The definition of the maximum total degree d must refer to the notional polynomials $\pi_i = \rho_i \Delta$ and not the actual exponents ρ_i; therefore the value of d will be affected by the total degree of the common denominator Δ.
2. When analyzing a particular complexity problem, such as the HSDH problem, one must show that the problem definition will not cause the denominator Δ to vanish identically (*i.e.*, force $\Delta \equiv 0$) for all values of the variables. If it does, then the uber-assumption framework will not apply (and the proposed problem may quite well have a generic easy solution).
3. Even if $\Delta \not\equiv 0$ for the proposed problem, there is still a chance that Δ will vanish for a random assignment of the variables $X_1, \ldots, X_c$ in $\mathbb{F}_p$. Since this event may entail an easy victory for $\mathcal{A}$, the upper bound on $\mathbf{Adv}_{\mathcal{A}}$ given by Theorem 1 must be increased accordingly. This is straightforward: if the total degree of Δ is d_Δ, then we have $\Pr[\Delta(X_1, \ldots, X_c) = 0] \leq d_\Delta / p$.

Again, to deal with computational problems such as HSDH and Poly-SDH whose answers consist of more than one group element, one should substitute for f three vectors $V_{\mathbb{G}}$, $V_{\hat{\mathbb{G}}}$, $V_{\mathbb{G}_t}$ as discussed in Section 5.4.

6.3 Composite Group Orders

The last generalization of the uber-assumption we consider concerns bilinear groups of composite order, *i.e.*, bilinear contexts $\mathbf{G} = \langle n, \mathbb{G}, \hat{\mathbb{G}}, \mathbb{G}_t, g, \hat{g}, \mathbf{e} \rangle$ where $|\mathbb{G}| = |\hat{\mathbb{G}}| = |\mathbb{G}_t| = n = p_1 p_2$ is a product of two or more (safe-)prime factors. Although those are typically constructed on supersingular curves which mandate $\mathbb{G} = \hat{\mathbb{G}}$, for maximum generality we ought to maintain the distinction $\mathbb{G} \neq \hat{\mathbb{G}}$.

The first and simplest assumption in composite-order bilinear groups is called Decision Subgroup [11], and states that it infeasible to decide, given $g \in \mathbb{G}$ of order n, and $g_1 \in \mathbb{G}$ of order p_1, whether some $v \in \mathbb{G}$ has order n or order p_1.

More complex assumptions have also been made, that posit the hardness of various types of Diffie-Hellman problems in the group $\mathbb{G}$ or some of its subgroups. Sometimes, the factorization of the composite-order groups may be revealed: such groups only serve as containers for prime-order subgroups of interest [16]. Sometimes, however, the factorization is supposed secret and is an inherent part, but not the only part, of the challenge faced by the adversary [24].

The latter kind of assumption is particularly worrisome, for it combines many intertwined and possibly compounding vulnerabilities within a single package: the hardness of pairing-friendly Diffie-Hellman problems, subgroup hiding, and factoring — not constructively put together from elementary assumptions — but bundled *non-separably* into a single take-it-or-leave-it hypothesis.

Nevertheless, the uber-assumption framework can let us "gain confidence" in such composite-order assumptions, much as it did for prime-order ones, at least in a relaxed notion of generic groups. To its credit, [24] offers such justification.

Coping with Unbounded Adversaries. The main difficulty is that the generic model assumes a computationally unbounded adversary, but such an adversary can of course factor the group order and thus break any assumption that relies on the secrecy of the factors. Since we cannot prevent $\mathcal{A}$ from factoring n, our only escape is to force $\mathcal{A}$ to expose a non-trivial factor of n in order to solve the given instance with non-negligible advantage in the generic-group model.

Of course, we cannot simply ask $\mathcal{A}$ for the factorization of n; we must extract it from the interactions of $\mathcal{A}$ with the generic-group simulator, or at least prove that $\mathcal{A}$ must have had access to it. *E.g.*, if, by interacting with the generic-group oracles, $\mathcal{A}$ manages to obtain the representation $\xi_{\mathbb{G}}(p_1)$ of g^{p_1}, supposedly an infeasible task, one must conclude that $\mathcal{A}$ must have necessarily leaked p_1 to the simulator, though perhaps unwittingly, in a way that is recoverable from the oracle transcripts.

Factorizationless Generic Simulation. The implementation of this intuition is quite simple. It is based on the fact that each one of the various group elements given to $\mathcal{A}$ is an element of some order $o \mid n$ in one of the groups, $\mathbb{G}$, $\hat{\mathbb{G}}$, $\mathbb{G}_t$; and these properties will be preserved or modified by the generic-group oracles according to simple rules that *normally* should not depend on the factors of n. Should an *anomaly* occur, this will be the signal that $\mathcal{A}$ has managed to exploit the factorization of n, and thus that a factor of n is recoverable from the oracle transcripts.

To apply the rules without needing the factorization, our simulator internally represents each element by its projection over all relevant subgroups, such as $h = g_1^{\eta_1} g_2^{\eta_2}$ for fictitious generators g_1 and g_2 of orders p_1 and p_2. Hence, if the problem instance called for $\mathcal{A}$ to receive an element h of order p_1, the simulator would construct it by fixing $\eta_2 = 0$ and setting $\eta_2 \in_\$ \mathbb{Z}_n$.

Two elements will thus be deemed equal by the simulator if and only if all their exponents are the same *modulo* n — not mod p_1 or mod p_2 as should be the case in reality. This leads to the possibility of equality-test false negatives (and no other errors), but only as a symptom of $\mathcal{A}$'s factorization of n. Simulating the generic oracles is done by performing arithmetic on matching exponents.

Error Handling and Reduction. As we just mentioned, since the simulator's exponent handling disregards the factorization of n, certain elements will be deemed distinct when in fact they are the same (which the simulator will find out at the end when it instantiates the variables): in this event, $\mathcal{A}$ will be considered to have won, because the simulation was flawed.

However, by definition, such event can happen only when $\mathcal{A}$ has managed to summon two distinct representations ξ and ξ' of the same group element, and whose respective simulator exponents are not all congruent mod n but are all congruent either mod p_1 or mod p_2. From this, the simulator can factor n.

7 Conclusion

To conclude, we shall reiterate that the framework described in this paper, just like the original from [8], is not intended to be used directly as a basis on which to prove the security of protocols, but merely as a convenient template, or shortcut, to analyze specific pairing assumptions that fall under its umbrella.

As we have seen, the original framework from [8] can be further broadened in many ways. These include the formal treatment of symmetric and asymmetric pairings alike, the extension to computational instances with flexible challenges, the support of polynomial and rational exponents, and the case of composite-order groups with public or secret factorization.

Acknowledgements

The author would like to thank Steven Galbraith and Kenneth Paterson for the kind invitation to speak at Pairing 2008, as well as for many useful suggestions.

References

1. Ateniese, G., Camenisch, J., Hohenberger, S., de Medeiros, B.: Practical group signatures without random oracles. Cryptology ePrint Archive, Report, 2005/385 (2005), `http://eprint.iacr.org/`
2. Barreto, P.S.L.M., Galbraith, S., O'hEigeartaigh, C., Scott, M.: Efficient pairing computation on supersingular abelian varieties. Designs, Codes and Cryptography 42(3), 239–271 (2007)
3. Barreto, P.S.L.M., Naehrig, M.: Pairing-friendly elliptic curves of prime order. In: Preneel, B., Tavares, S. (eds.) SAC 2005. LNCS, vol. 3897, pp. 319–331. Springer, Heidelberg (2006)
4. Blake, I., Seroussi, G., Smart, N.: Elliptic Curves in Cryptography. London Mathematical Society Lecture Note Series, vol. 265. Cambridge University Press, Cambridge (1999)
5. Blake, I.F., Seroussi, G., Smart, N.P. (eds.): Advances in Elliptic Curve Cryptography. London Mathematical Society Lecture Note Series, vol. 317. Cambridge University Press, Cambridge (2005)
6. Boneh, D., Boyen, X.: Efficient selective-ID secure identity based encryption without random oracles. In: Cachin, C., Camenisch, J.L. (eds.) EUROCRYPT 2004. LNCS, vol. 3027, pp. 223–238. Springer, Heidelberg (2004)
7. Boneh, D., Boyen, X.: Short signatures without random oracles. In: Cachin, C., Camenisch, J.L. (eds.) EUROCRYPT 2004. LNCS, vol. 3027, pp. 149–177. Springer, Heidelberg (2004); Journal of Cryptology, 21(2), 149–177 (2008)
8. Boneh, D., Boyen, X., Goh, E.-J.: Hierarchical identity based encryption with constant size ciphertext. In: Cramer, R. (ed.) EUROCRYPT 2005. LNCS, vol. 3494, pp. 440–456. Springer, Heidelberg (2005)
9. Boneh, D., Boyen, X., Shacham, H.: Short group signatures. In: Franklin, M. (ed.) CRYPTO 2004. LNCS, vol. 3152, pp. 41–55. Springer, Heidelberg (2004)
10. Boneh, D., Franklin, M.: Identity-based encryption from the Weil pairing. In: Kilian, J. (ed.) CRYPTO 2001. LNCS, vol. 2139. Springer, Heidelberg (2001); SIAM J. Computing, 32(4), 586–615 (2003)

11. Boneh, D., Goh, E.-J., Nissim, K.: Evaluating 2-DNF formulas on ciphertexts. In: Kilian, J. (ed.) TCC 2005. LNCS, vol. 3378. Springer, Heidelberg (2005)
12. Boyen, X.: Mesh signatures. In: Naor, M. (ed.) EUROCRYPT 2007. LNCS, vol. 4515, pp. 210–227. Springer, Heidelberg (2007)
13. Boyen, X., Waters, B.: Anonymous hierarchical identity-based encryption (without random oracles). In: Dwork, C. (ed.) CRYPTO 2006. LNCS, vol. 4117, pp. 290–307. Springer, Heidelberg (2006)
14. Boyen, X., Waters, B.: Full-domain subgroup hiding and constant-size group signatures. In: Okamoto, T., Wang, X. (eds.) PKC 2007. LNCS, vol. 4450, pp. 1–15. Springer, Heidelberg (2007)
15. Camenisch, J., Neven, G., Shelat, A.: Simulatable adaptive oblivious transfer. In: Naor, M. (ed.) EUROCRYPT 2007. LNCS, vol. 4515, pp. 573–590. Springer, Heidelberg (2007)
16. Chase, M., Lysyanskaya, A.: Simulatable VRFs with applications to multi-theorem NIZK. In: Menezes, A. (ed.) CRYPTO 2007. LNCS, vol. 4622, pp. 303–322. Springer, Heidelberg (2007)
17. Cohen, H., Frey, G., Avanzi, R. (eds.): Handbook of Elliptic and Hyperelliptic Curve Cryptography. CRC Press, Boca Raton (2006)
18. Dent, A.W.: Adapting the weaknesses of the random oracle model to the generic group model. In: Zheng, Y. (ed.) ASIACRYPT 2002. LNCS, vol. 2501, pp. 100–109. Springer, Heidelberg (2002)
19. Freeman, D., Scott, M., Teske, E.: A taxonomy of pairing-friendly elliptic curves. Cryptology ePrint Archive, Report 2006/372 (2006), `http://eprint.iacr.org/`
20. Galbraith, S.D.: Private communication (2008)
21. Galbraith, S.D., Paterson, K.G., Smart, N.P.: Pairings for cryptographers. Discrete Applied Mathematics (2007), Online version: doi:10.101/j.dam.2007.12.010
22. Joux, A.: A one round protocol for tripartite Diffie-Hellman. In: Bosma, W. (ed.) ANTS 2000. LNCS, vol. 1838, pp. 385–394. Springer, Heidelberg (2000); Full article in: Journal of Cryptology, 17(4), 263–276 (2004)
23. Joux, A., Nguyen, K.: Separating decision Diffie-Hellman from computational Diffie-Hellman in cryptographic groups. Journal of Cryptology 16(4) (2003)
24. Katz, J., Sahai, A., Waters, B.: Predicate encryption supporting disjunctions, polynomial equations, and inner products. In: Smart, N. (ed.) EUROCRYPT 2008. LNCS, vol. 4965, pp. 146–162. Springer, Heidelberg (2008)
25. Lynn, B.: On the Implementation of Pairing-Based Cryptosystems. PhD thesis, Stanford University (2007)
26. Maurer, U.: Abstract models of computation in cryptography. In: Smart, N. (ed.) Cryptography and Coding 2005. LNCS, vol. 3796, pp. 1–12. Springer, Heidelberg (2005)
27. Miller, V.: The Weil pairing, and its efficient calculation. Journal of Cryptology 17(4), 235–261 (2004)
28. Mitsunari, S., Sakai, R., Kasahara, M.: A new traitor tracing. IEICE Transactions on Fundamentals E85-A(2), 481–484 (2002)
29. Nechaev, V.I.: Complexity of a determinate algorithm for the discrete logarithm. Mathematical Notes 55(2), 165–172 (1994)
30. Okamoto, T., Pointcheval, D.: The gap-problems: A new class of problems for the security of cryptographic schemes. In: Kim, K.-c. (ed.) PKC 2001. LNCS, vol. 1992, pp. 104–118. Springer, Heidelberg (2001)
31. Shoup, V.: Lower bounds for discrete logarithms and related problems. In: Fumy, W. (ed.) EUROCRYPT 1997. LNCS, vol. 1233, pp. 256–266. Springer, Heidelberg (1997)

Homomorphic Encryption and Signatures from Vector Decomposition

Tatsuaki Okamoto[1] and Katsuyuki Takashima[2]

[1] NTT, 3-9-11 Midori-cho, Musashino-shi, Tokyo, 180-8585 Japan
okamoto.tatsuaki@lab.ntt.co.jp
[2] Mitsubishi Electric, 5-1-1, Ofuna, Kamakura, Kanagawa, 247-8501 Japan
Takashima.Katsuyuki@aj.MitsubishiElectric.co.jp

Abstract. This paper introduces a new concept, *distortion eigenvector space*; it is a (higher dimensional) vector space in which bilinear pairings and distortion maps are available. A distortion eigenvector space can be efficiently realized on a supersingular hyperelliptic curve or a direct product of supersingular elliptic curves. We also introduce an intractable problem (with trapdoor) on distortion eigenvector spaces, the higher dimensional generalization of the *vector decomposition problem (VDP)*. We define several computational and decisional problems regarding VDP, and clarify the relations among them. A *trapdoor bijective function* with algebraically rich properties can be obtained from the VDP on distortion eigenvector spaces. This paper presents two applications of this trapdoor bijective function; one is multivariate homomorphic encryption as well as a two-party protocol to securely evaluate 2DNF formulas in a higher dimensional manner, and the other is various types of signatures such as ordinary signatures, blind signatures, generically (selectively and universally) convertible undeniable signatures and their combination.

1 Introduction

Mathematically (or algebraically) rich structures should be useful to realize various types of cryptographic primitives and protocols. Up to now, fairly simple and elementary mathematical structures have been used in cryptography such as cyclic groups (genus 0) and pairings in genus 1 curves. Although higher genus curves have been investigated for application to cryptography, only a cyclic group and a doubly cyclic group with pairings have been applied. It was suggested in a few papers [8,9] to utilize richer algebraic structures in cryptography, but no concrete result has been reported except [6,7] where a higher dimensional ElGamal-type signature scheme was studied.

This paper develops a new methodology to employ richer mathematical structures in cryptography. First, this paper introduces a new concept, *distortion eigenvector space*, a generalization of the two-dimensional case (genus 1 curve) studied by Galbraith and Verheul [11]. A distortion eigenvector space is a (higher dimensional) vector space in which bilinear pairings and distortion maps are available; it can be efficiently constructed on a supersingular hyperelliptic curve by the theory recently developed by Takashima [18] (as well as [10]) or on a direct product

S.D. Galbraith and K.G. Paterson (Eds.): Pairing 2008, LNCS 5209, pp. 57–74, 2008.

of supersingular elliptic curves. Thanks to the algebraically rich structures, this concept should be useful for designing cryptographic primitives and protocols.

A promising candidate of computational problems on which distortion eigenvector spaces could be applied to cryptography is the higher dimensional generalization of the vector decomposition problem (VDP). VDP was originally introduced by Yoshida, Mitsunari and Fujiwara [20,21] and recently analyzed by Galbraith and Verheul [11]. The higher dimensional versions of VDP were studied by Duursma *et al.* [6,7].

This paper extends the study on the relationships of higher dimensional VDP with other problems such as the generalized (or higher dimensional) Diffie-Hellman (gDH) problem. Since the higher dimensional VDP is a newly introduced problem, it is important to characterize its intractability through its relations with (and equivalence to) well-studied problems such as the DH problems and the decisional linear problem (DLN) over bilinear groups. We show that computational VDP (CVDP) and computational gDH (gCDH) can be reduced to each other (under some conditions), decisional VDP (DVDP) and decisional subspace problem (DSP) can be reduced to each other, decisional gDH (gDDH) can be reduced to DSP/DVDP, and that DLN can be reduced to gDDH (i.e., DLN can be reduced to DSP/DVDP). Here, DSP is introduced to prove the semantic security of our homomorphic encryption.

We then present a trapdoor of the higher dimensional VDP over the distortion eigenvector space. That is, there is an efficient algorithm Deco that can solve the VDP problem by using a trapdoor, but under the VDP assumption it is intractable to solve the problem without the knowledge of the trapdoor.

This trapdoor leads to a *trapdoor bijective function*, which is, to the best of our knowledge, the first trapdoor bijective function, except the RSA function and its variants (the Rabin-Williams [19] and Paillier [16] functions) that are based on the integer factoring trick. Here, we say that f is a trapdoor bijective function if f is a trapdoor one-way function and bijection (one-to-one and onto, i.e., its domain and range are isomorphic). (See Section 4 for more details.)

Using this trapdoor bijective function, this paper proposes multivariate homomorphic encryption, which is semantically secure under the DSP (i.e., DVDP and gDDH) assumption. The encryption scheme is multivariate homomorphic in addition, i.e., multiple plaintexts, $\vec{m} \leftarrow (m_0, \ldots, m_{\ell_2-1}) \in (\mathbb{F}_r)^{\ell_2}$, can be encrypted to a single ciphertext $\boldsymbol{c} \leftarrow \mathsf{Enc}(\vec{m})$ and homomorphic transformation over ciphertexts on addition is available for each plaintext simultaneously. For example, given $\boldsymbol{c}_1 \leftarrow \mathsf{Enc}(\vec{m}_1)$ and $\boldsymbol{c}_2 \leftarrow \mathsf{Enc}(\vec{m}_2)$, $\boldsymbol{c}_1 + \boldsymbol{c}_2 = \mathsf{Enc}(m_{1,0} + m_{2,0} \bmod r, \ldots, m_{1,\ell_2-1} + m_{2,\ell_2-1} \bmod r)$. Note that our homomorphic encryption requires higher dimension than 2 (or higher genus than 1) to meet semantic security, i.e., it cannot be realized on elliptic curves (genus 1).[1]

[1] This is because: if a message, m (e.g., $m \in \mathbb{F}_r$), is embeded to a (one dimensional) subspace $\langle \boldsymbol{b}_0 \rangle$ (as $\boldsymbol{m}$, e.g., $\boldsymbol{m} \leftarrow m\boldsymbol{b}_0$) in a two dimensional space $\langle \boldsymbol{b}_0, \boldsymbol{b}_1 \rangle$, the correct message, $\boldsymbol{m}$, of ciphertext $\boldsymbol{c}$ can be publicly verifiable (i.e., not semantically secure) by verifying $\boldsymbol{m} \in \langle \boldsymbol{b}_0 \rangle$ and $\boldsymbol{c} - \boldsymbol{m} \in \langle \boldsymbol{b}_1 \rangle$ through pairing operations ($\boldsymbol{c}$ is a ciphertext such that $\boldsymbol{c} \leftarrow \boldsymbol{m} + \boldsymbol{r}$ and $\boldsymbol{m} \in \langle \boldsymbol{b}_0 \rangle$ and $\boldsymbol{r} \in \langle \boldsymbol{b}_1 \rangle$).

Based on our homomorphic encryption, we present a two-party protocol (between Alice and Bob) to securely evaluate a 2DNF formula ψ (over n variables) for higher dimensional n variables (assignment), where Bob knows secret input (assignment) $(\overrightarrow{m}_1, \ldots, \overrightarrow{m}_n)$ (to ψ), and Alice knows secret formula ψ. Here $\overrightarrow{m}_i \leftarrow (m_{i,0}, \ldots, m_{i,\ell_2-1}) \in \{0,1\}^{\ell_2}$ $(i = 1, \ldots, n)$. The protocol outputs $\psi(\overrightarrow{m}_1, \ldots, \overrightarrow{m}_n)$ $(\leftarrow (\psi(m_{1,0}, \ldots, m_{n,0}), \ldots, \psi(m_{1,\ell_2-1}, \ldots, m_{n,\ell_2-1})\ )$ while keeping the local secrets, i.e., Bob finally gets $\psi(\overrightarrow{m}_1, \ldots, \overrightarrow{m}_n)$ but learns nothing of ψ, and Alice learns nothing of $(\overrightarrow{m}_1, \ldots, \overrightarrow{m}_n)$.

The trapdoor bijective function also leads to various types of signatures. We present some examples of such signatures; ordinary signatures, blind signatures, generically (selectively and universally) convertible undeniable signatures. Note that our construction of undeniable signatures essentially requires a higher dimensional space greater than 2, and cannot be realized over an elliptic curve (two-dimensional space).[2]

The advantage of our approach is its flexibility in combining different types of signatures; for example, blind signatures and convertible undeniable signatures can be easily combined. To the best of our knowledge, the proposed scheme is the first efficient scheme that simultaneously realizes blind and generically (selectively and universally) convertible undeniable signatures where the confirmation protocol of undeniable signatures can be executed without signer's secret key.

Notations

When A is a random variable or distribution, $y \xleftarrow{\mathsf{R}} A$ denotes that y is randomly selected from A according to its distribution. When A is a set, $y \xleftarrow{\mathsf{U}} A$ denotes that y is uniformly selected from A. $y \leftarrow A$ denotes that y is set, defined or substituted by A. When a is a fixed value, $A(x) \rightarrow a$ (e.g., $A(x) \rightarrow 1$) denotes the event that machine (algorithm) A outputs a on input x.

A vector symbol denotes a vector representation over a finite field $\mathbb{F}_r$, e.g., $\overrightarrow{m}$ denotes $(m_0, \ldots, m_{\ell_2-1}) \in (\mathbb{F}_r)^{\ell_2}$. A bold face letter denotes an element of vector space $\mathbb{V}$, e.g., $\boldsymbol{m} \in \mathbb{V}$. When ψ is a (2DNF) formula over n variables and $\overrightarrow{m}_i \leftarrow (m_{i,0}, \ldots, m_{i,\ell_2-1}) \in \{0,1\}^{\ell_2}$ $(i = 1, \ldots, n)$, ψ is abused as $\psi(\overrightarrow{m}_1, \ldots, \overrightarrow{m}_n)$, that denotes $(\psi(m_{1,0}, \ldots, m_{n,0}), \ldots, \psi(m_{1,\ell_2-1}, \ldots, m_{n,\ell_2-1}))$.

When $\mathcal{A}$ is a machine, $t_{\mathcal{A}}$ denotes the running time of $\mathcal{A}$. When P and Q are computational problems, P $\leq_p$ Q informally denotes that P is reduced to Q by a probabilistic polynomial-time algorithm, and P $=_p$ Q denotes P $\leq_p$ Q and Q $\leq_p$ P.

---

[2] This is because: signature $\boldsymbol{s}$ in two dimensional space $\langle \boldsymbol{b}_0, \boldsymbol{b}_1 \rangle$, where $\boldsymbol{s}$ is a decomposed value to a one-dimensional subspace $\langle \boldsymbol{b}_0 \rangle$ from $\boldsymbol{h}$ determined by a hash value of message m, is publicly verifiable (i.e., not undeniable signatures) by verifying $\boldsymbol{s} \in \langle \boldsymbol{b}_0 \rangle$ and $\boldsymbol{h} - \boldsymbol{s} \in \langle \boldsymbol{b}_1 \rangle$ through pairing operations, while signature $\boldsymbol{s} \in \langle \boldsymbol{b}_0 \rangle$, in a three dimensional space $\langle \boldsymbol{b}_0, \boldsymbol{b}_1, \boldsymbol{b}_2 \rangle$, can not be publicly verifiable (i.e., may be undeniable signatures) since $\boldsymbol{s}_0 \in \langle \boldsymbol{b}_0 \rangle$ can be verified by pairing but $\boldsymbol{h} - \boldsymbol{s}_0 \in \langle \boldsymbol{b}_1, \boldsymbol{b}_2 \rangle$ may not be verified efficiently.

2 Distortion Eigenvector Spaces

We generalize the notion of distortion eigenvector basis introduced by [11] and introduce the notion of the *distortion eigenvector space.*

2.1 Definition

Let $\mathbb{F}_r$ be a finite field of odd order r. A *distortion eigenvector space* is a (higher dimensional) vector space over $\mathbb{F}_r$ that has efficiently computable distortion maps and bilinear pairing operations; it is formally defined as follows:

Definition 1. *"Distortion eigenvector space" $\mathbb{V}$ is a ℓ-dimensional vector space over $\mathbb{F}_r$ that satisfies the following conditions:*

1. *Let $\mathbb{A} \leftarrow (\boldsymbol{a}_0, \ldots, \boldsymbol{a}_{\ell-1})$ be a basis of $\mathbb{F}_r$-vector space $\mathbb{V}$ and F a polynomial-time computable automorphism of $\mathbb{V}$. The basis $\mathbb{A}$ is called a "distortion eigenvector basis" with respect to F, if each $\boldsymbol{a}_i$ is an eigenvector of F, their eigenvalues are different from each other, and there exist polynomial-time computable endomorphisms $\phi_{i,j}$ of $\mathbb{V}$ such that $\phi_{i,j}(\boldsymbol{a}_j) = \boldsymbol{a}_i$. We call $\phi_{i,j}$ a "distortion map". There exist such $\mathbb{A}$, F, and $\{\phi_{i,j}\}_{0 \le i,j \le \ell-1}$.*
2. *There exists a skew-symmetric nondegenerate bilinear pairing $e : \mathbb{V} \times \mathbb{V} \to \mu_r$ where μ_r is a multiplicative cyclic group of order r, i.e.,$e(\gamma \boldsymbol{u}, \delta \boldsymbol{v}) = e(\boldsymbol{u}, \boldsymbol{v})^{\gamma\delta}$ and $e(\boldsymbol{u}, \boldsymbol{u}) = 1$ for all $\boldsymbol{u}, \boldsymbol{v} \in \mathbb{V}$ and all $\gamma, \delta \in \mathbb{F}_r$, and if $e(\boldsymbol{u}, \boldsymbol{v}) = 1$ for all $\boldsymbol{v} \in \mathbb{V}$, then $\boldsymbol{u} = \boldsymbol{0}$.*
3. *There exists a polynomial-time computable automorphism ρ on $\mathbb{V}$ such that $e(\boldsymbol{v}, \rho(\boldsymbol{v})) \neq 1$ for any $\boldsymbol{v}$ except for $\boldsymbol{v}$ in a quadratic hypersurface of $\mathbb{V} \cong (\mathbb{F}_r)^\ell$.*

Lemma 1 (Projection Operators). *Let $\mathbb{A} \leftarrow (\boldsymbol{a}_0, \ldots, \boldsymbol{a}_{\ell-1})$ be a distortion eigenvector basis of $\mathbb{V}$, and $\boldsymbol{a}_i$ has its eigenvalue ν_i of F. A polynomial of F, $\mathsf{Pr}_j \leftarrow \left(\prod_{i \neq j}(\nu_j - \nu_i)\right)^{-1} \prod_{i \neq j}(F - \nu_i)$ gives the j-th projection operator w.r.t. $\mathbb{A}$, that is, $\mathsf{Pr}_j(\boldsymbol{a}_\kappa) = \boldsymbol{0}$ for $\kappa \neq j$ and $\mathsf{Pr}_j(\boldsymbol{a}_j) = \boldsymbol{a}_j$.*

2.2 Constructions

We will show two concrete examples of a distortion eigenvector space.

Jacobian Variety of a Supersingular Curve of Genus $g \geq 1$. We can realize a distortion eigenvector space by the Jacobian variety of a supersingular curve of genus $g \geq 1$ where $w \leftarrow 2g+1$ is a prime as follows: Let (p, r) be a pair of primes such that $\alpha \leftarrow p \bmod w$ is a generator of $\mathbb{F}_w^*$, and $r \mid p^g + 1$. We then use a supersingular curve $C/\mathbb{F}_p : Y^2 = X^w + 1$, and we see that $r \mid \sharp\mathrm{Jac}_C(\mathbb{F}_p) = p^g + 1$ for that curve C. We define $\mathbb{F}_r$-vector space $\mathbb{V}$ as $\mathrm{Jac}_C[r]$, which is isomorphic to $(\mathbb{F}_r)^{2g}$ and contained in $\mathrm{Jac}_C(\mathbb{F}_{p^{2g}})$.

A distortion eigenvector basis is efficiently constructed on the above Jacobian $\mathrm{Jac}_C[r]$. When $r > w$ as in typical cryptographic applications, such a basis on the Jacobian is given by Takashima [18]. The Jacobian has automorphism ρ that is induced from that of the curve, $(x, y) \mapsto (\zeta x, y)$ where ζ is a primitive w-th root of unity in $\mathbb{F}_{p^{2g}}$.

Let $\boldsymbol{a}^* \in \mathrm{Jac}_C(\mathbb{F}_p)[r]$ be a nonzero point on the Jacobian, i.e., $\boldsymbol{a}^* \neq \boldsymbol{0}$. We use operators $G_j \leftarrow \sum_{i=0}^{2g-1} (p^j)^i \rho^{\alpha^i}$ on $\mathrm{Jac}_C[r] \cong (\mathbb{F}_r)^{2g}$ where $\alpha \leftarrow p \bmod w$ and $j = 0, \ldots, 2g-1$. Let $\boldsymbol{a}_j \in \mathrm{Jac}_C[r]$ be $G_j(\boldsymbol{a}^*)$ for $j = 0, \ldots, 2g-1$. We then obtain a distortion eigenvector basis $\mathbb{A} \leftarrow (\boldsymbol{a}_0, \ldots, \boldsymbol{a}_{2g-1})$ w.r.t. the p-power Frobenius endomorphism F. $\boldsymbol{a}_j$'s eigenvalue of F is p^{-j}. A distortion map $\phi_{i,j}$ in Definition 1 is given by $(-1)^j w^{-1} G_i G_{-j} = (-1)^j w^{-1} J_{i,-j} G_{i-j}$ where the index of G is considered as in $\mathbb{Z}/2g\mathbb{Z}$ and $J_{i,-j} \in \mathbb{F}_r$ is a constant indexed by the pair $(i, -j)$.

We use the Weil pairing e. Since Corollary 2 in [18] showed that $e(\boldsymbol{a}^*, \rho(\boldsymbol{a}^*)) \neq 1$, we know that $e(\boldsymbol{v}, \rho(\boldsymbol{v})) \neq 1$ for any $\boldsymbol{v}$ except for $\boldsymbol{v}$ in a quadratic hypersurface of $\mathbb{V} \cong (\mathbb{F}_r)^\ell$. For the calculation of the Weil pairing on hyperelliptic curves, see [15], for example.

Given security parameter k and genus g, we can obtain such $\mathbb{V}$ by finding a pair of primes, (p, r), that satisfies the above conditions and the security level determined by k (e.g., $\lceil \log_2 r \rceil = k$).

Product of Supersingular Elliptic Curves. Among non-cyclic groups in Section 5 of [11], we can use a product of supersingular elliptic curves $E/\mathbb{F}_p$: $Y^2 = X^3 + 1$ where $p \equiv 2 \bmod 3$ for a distortion eigenvector space. Let (p, r) be a pair of primes and d a positive integer such that $p \equiv 2 \bmod 3$, $r \mid \sharp E(\mathbb{F}_p) = p + 1$, $2d < r$, and $r > 3$. Then, $\mathbb{V} \leftarrow \prod_{\kappa=0}^{d-1} E_\kappa[r] \cong (\mathbb{F}_r)^{2d}$ is a distortion eigenvector space where E_κ for each κ is (a copy of) E.

Automorphism F of $\mathbb{V}$ is defined as the diagonal action of $\prod_{\kappa=0}^{d-1} (\kappa+1) F_\kappa$ on $\prod_{\kappa=0}^{d-1} E_\kappa[r]$ where F_κ is the p-power Frobenius on E_κ. When $(\boldsymbol{a}_\kappa, \boldsymbol{a}_{\kappa+1})$ is a distortion eigenvector basis of $E_\kappa[r]$ for F_κ, $(\boldsymbol{a}_0, \ldots, \boldsymbol{a}_{2d-1})$ is a distortion eigenvector basis of $\mathbb{V} = \prod_{\kappa=0}^{d-1} E_\kappa[r]$ for F. The distortion maps for $\mathbb{V}$ are constructed from all the distortion maps for E_κ as follows: using projection operators on E_κ, we can decompose any vector $\boldsymbol{v} \in \mathbb{V}$ as $\boldsymbol{v} = \sum_{i=0}^{2d-1} \boldsymbol{v}'_i$ such that $\boldsymbol{v}'_i \in \langle \boldsymbol{a}_i \rangle$ (See Lemma 1), and then we can efficiently calculate distortion maps $\phi_{i,j}$ in Definition 1 (i.e. changes of components $\boldsymbol{v}'_i$'s). Automorphism ρ is the direct product of the automorphism ρ_κ of E_κ (defined as above), and a pairing e can be defined componentwise. In other words, for two vectors $\boldsymbol{u} \leftarrow (\boldsymbol{u}_\kappa) \in \mathbb{V}$ and $\boldsymbol{v} \leftarrow (\boldsymbol{v}_\kappa) \in \mathbb{V}$, $e(\boldsymbol{u}, \boldsymbol{v})$ is defined by $\prod_{\kappa=0}^{d-1} e_\kappa(\boldsymbol{u}_\kappa, \boldsymbol{v}_\kappa)$ where e_κ is the Weil pairing on E_κ.

3 Vector Decomposition Problems

This section introduces *vector decomposition problems* (VDPs) over a distortion eigenvector space. The computational VDP (CVDP) in Definition 2 is a generalization of 2-dimensional VDP, which was introduced in [20,21] and studied in [11]. We investigate relations between several problems regarding CVDP and decisional VDP (DVDP) .

3.1 Computational Vector Decomposition Problems

We now define CVDP and its assumption. The CVDP assumption is employed to ensure the one-wayness of the trapdoor bijective function introduced in Section 4 and to prove the security (in the random oracle model) of the signature schemes proposed in Section 6.

Definition 2 (CVDP$_{(\ell_1,\ell_2)}$: (ℓ_1,ℓ_2)-Computational Vector Decomposition Problem). *Let k be a security parameter and $\mathcal{G}_{\mathbb{V}}$ be an algorithm that outputs a description of a ℓ_1-dimensional $\mathbb{F}_r$-vector space $\mathbb{V}$ with security parameter k, and $\ell_1 > \ell_2$.*

Let $\mathcal{A}$ be a probabilistic polynomial-time machine. For all $k \in \mathbb{N}$, we define the CVDP$_{(\ell_1,\ell_2)}$ advantage of $\mathcal{A}$ as

$$\mathsf{Adv}_{\mathcal{A}}^{\mathsf{CVDP}_{(\ell_1,\ell_2)}}(k) \leftarrow$$
$$\Pr\Big[\, \omega = \textstyle\sum_{i=0}^{\ell_2-1} x_i \boldsymbol{b}_i \mid \mathbb{V} \xleftarrow{\mathsf{R}} \mathcal{G}_{\mathbb{V}}(1^k),\ (\boldsymbol{b}_0,\ldots,\boldsymbol{b}_{\ell_1-1}) \xleftarrow{\mathsf{U}} \mathbb{V}^{\ell_1},$$
$$(x_0,\ldots,x_{\ell_1-1}) \xleftarrow{\mathsf{U}} (\mathbb{F}_r)^{\ell_1}, \boldsymbol{v} \leftarrow \textstyle\sum_{i=0}^{\ell_1-1} x_i \boldsymbol{b}_i, \omega \leftarrow \mathcal{A}(1^k, \mathbb{V}, \boldsymbol{b}_0,\ldots,\boldsymbol{b}_{\ell_1-1}, \boldsymbol{v})\Big].$$

The CVDP$_{(\ell_1,\ell_2)}$ assumption is: For any polynomial-time adversary $\mathcal{A}$, the advantage $\mathsf{Adv}_{\mathcal{A}}^{\mathsf{CVDP}_{(\ell_1,\ell_2)}}(k)$ is negligible.

Remark: In the experiment of the definition of $\mathsf{Adv}_{\mathcal{A}}^{\mathsf{CVDP}_{(\ell_1,\ell_2)}}(k)$, a linearly-dependent tuple, $(\boldsymbol{b}_0,\ldots,\boldsymbol{b}_{\ell_1-1})$, may be selected, but the case occurs with a negligible probability in k. So, it does not affect whether the assumption holds or not.

A specific class of the CVDP instances that are specified over distortion eigenvector basis $\mathbb{A}$ are tractable as follows:

Lemma 2. *Let $\mathbb{A}$ be a distortion eigenvector basis of $\mathbb{V}$, and CVDP$^{\mathbb{A}}_{(\ell_1,\ell_2)}$ be a specific class of CVDP$_{(\ell_1,\ell_2)}$ in which $(\boldsymbol{b}_0,\ldots,\boldsymbol{b}_{\ell_1-1})$ is replaced by $\mathbb{A}$. Then the projection operators Pr_j $(j = 0,\ldots,\ell_1-1)$ in Lemma 1 can solve CVDP$^{\mathbb{A}}_{(\ell_1,\ell_2)}$ in polynomial time.*

3.2 Trapdoor

Although CVDP$_{(\ell_1,\ell_2)}$ is expected to be intractable in general, the efficient algorithm, Deco given in Fig. 1, can solve it by using a trapdoor X.

The input is $(\boldsymbol{v}, \langle \boldsymbol{b}_0,\ldots,\boldsymbol{b}_{\ell_2-1}\rangle, X, \mathbb{B})$ such that $\boldsymbol{v} \leftarrow \sum_{i=0}^{\ell_1-1} y_i \boldsymbol{b}_i$ is a target vector for decomposition, $\langle \boldsymbol{b}_0,\ldots,\boldsymbol{b}_{\ell_2-1}\rangle$ is a subspace to be decomposed into, X is a trapdoor (matrix), and $\mathbb{B} \leftarrow (\boldsymbol{b}_0,\ldots,\boldsymbol{b}_{\ell_1-1})$ is a basis generated by using X.

$\mathsf{Deco}(\boldsymbol{v}, \langle \boldsymbol{b}_0,\ldots,\boldsymbol{b}_{\ell_2-1}\rangle, X, \mathbb{B})$:
$(t_{i,j}) \leftarrow X^{-1}, \boldsymbol{u} \leftarrow \sum_{i=0}^{\ell_1-1}\sum_{j=0}^{\ell_2-1}\sum_{\kappa=0}^{\ell_1-1} t_{i,j} x_{j,\kappa} \phi_{\kappa,i}(\mathsf{Pr}_i(\boldsymbol{v}))$.
return $\boldsymbol{u}$.

Fig. 1. Decomposition Algorithm Deco

Lemma 3 proves that Deco works correctly. Deco is the key tool to construct a trapdoor function and cryptosystems in Sections 4, 5 and 6.

Lemma 3. *Algorithm* Deco *solves $CVDP_{(\ell_1,\ell_2)}$ by using matrix $X \leftarrow (x_{i,j})$ such that $\boldsymbol{b}_i = \sum_{j=0}^{\ell_1-1} x_{i,j}\boldsymbol{a}_j$.*

Proof. We show that output $\boldsymbol{u}$ is $\sum_{i=0}^{\ell_2-1} y_i\boldsymbol{b}_i$ for input vector $\boldsymbol{v} \leftarrow \sum_{i=0}^{\ell_1-1} y_i\boldsymbol{b}_i$ for Deco. Let a row vector $(z_0,\dots,z_{\ell_1-1})$ be $(y_0,\dots,y_{\ell_1-1})X$. Then, $y_j = \sum_{i=0}^{\ell_1-1} z_i t_{i,j}$ and $\boldsymbol{v} = \sum_{i=0}^{\ell_1-1} z_i\boldsymbol{a}_i$. We then obtain

$$y_j\boldsymbol{a}_j = \sum_{i=0}^{\ell_1-1} t_{i,j} z_i(\phi_{j,i}(\boldsymbol{a}_i)) = \sum_{i=0}^{\ell_1-1} t_{i,j}\phi_{j,i}(z_i\boldsymbol{a}_i) = \sum_{i=0}^{\ell_1-1} t_{i,j}\phi_{j,i}(\mathsf{Pr}_i(\boldsymbol{v})) \quad (1)$$

where Pr_i are projection operators with basis $\mathbb{A}$. Using $\phi_{\kappa,j}(y_j\boldsymbol{a}_j) = y_j\boldsymbol{a}_\kappa$, $\boldsymbol{b}_j = \sum_{\kappa=0}^{\ell_1-1} x_{j,\kappa}\boldsymbol{a}_\kappa$, and Eq.(1), we obtain

$$\begin{aligned} y_j\boldsymbol{b}_j &= \sum_{\kappa=0}^{\ell_1-1} x_{j,\kappa} y_j\boldsymbol{a}_\kappa = \sum_{\kappa=0}^{\ell_1-1} x_{j,\kappa}\phi_{\kappa,j}\Big(\sum_{i=0}^{\ell_1-1} t_{i,j}\phi_{j,i}(\mathsf{Pr}_i(\boldsymbol{v}))\Big) \\ &= \sum_{\kappa=0}^{\ell_1-1}\sum_{i=0}^{\ell_1-1} x_{j,\kappa}t_{i,j}\phi_{\kappa,j}(\phi_{j,i}(\mathsf{Pr}_i(\boldsymbol{v}))) = \sum_{\kappa=0}^{\ell_1-1}\sum_{i=0}^{\ell_1-1} t_{i,j}x_{j,\kappa}\phi_{\kappa,i}(\mathsf{Pr}_i(\boldsymbol{v})). \end{aligned}$$

Therefore, output $\boldsymbol{u}$ is $\sum_{j=0}^{\ell_2-1} y_j\boldsymbol{b}_j$. □

3.3 Relations between CVDP and the Generalized DH Problem

We show that $\mathrm{CVDP}_{(\ell_1,1)}$ is reduced to $\mathrm{CVDP}_{(\ell_1,\ell_2)}$ and vice versa if $\ell_2 = O(1)$. Moreover, we introduce a new computational problem $\mathrm{gCDH}_{(\ell_1,\ell_2)}$, which is a generalization of CDH (computational Diffie-Hellman problem) to a higher dimensional space (Definition 3), and show its relations with CVDP.

Definition 3 ($\mathbf{gCDH}_{(\ell_1,\ell_2)}$: (ℓ_1,ℓ_2)-generalized Computational Diffie-Hellman Problem). *Let k be a security parameter and $\mathcal{G}_{\mathbb{V}}$ be an algorithm that outputs description of ℓ_1-dimensional $\mathbb{F}_r$-vector space $\mathbb{V}$ with security parameter k, and $\ell_1 > \ell_2$. Let $\mathcal{A}$ be a probabilistic polynomial-time machine. For all $k \in \mathbb{N}$, we define the $gCDH_{(\ell_1,\ell_2)}$ advantage of $\mathcal{A}$ as*

$$\begin{aligned} &\mathsf{Adv}_{\mathcal{A}}^{\mathsf{gCDH}_{(\ell_1,\ell_2)}}(k) \leftarrow \\ &\Pr\Big[\omega = \sum_{i=\ell_2}^{\ell_1-1} x_i\boldsymbol{b}'_i \mid \mathbb{V} \xleftarrow{\mathsf{R}} \mathcal{G}_{\mathbb{V}}(1^k), (\boldsymbol{b}_{\ell_2},\dots,\boldsymbol{b}_{\ell_1-1},\boldsymbol{b}'_{\ell_2},\dots,\boldsymbol{b}'_{\ell_1-1}) \xleftarrow{\mathsf{U}} \mathbb{V}^{2(\ell_1-\ell_2)}, \\ &\qquad (x_{\ell_2},\dots,x_{\ell_1-1}) \xleftarrow{\mathsf{U}} (\mathbb{F}_r)^{\ell_1-\ell_2},\ \boldsymbol{v} \leftarrow \sum_{i=\ell_2}^{\ell_1-1} x_i\boldsymbol{b}_i, \\ &\qquad \omega \leftarrow \mathcal{A}(1^k,\mathbb{V},\boldsymbol{b}_{\ell_2},\dots,\boldsymbol{b}_{\ell_1-1},\boldsymbol{b}'_{\ell_2},\dots,\boldsymbol{b}'_{\ell_1-1},\boldsymbol{v})\Big]. \end{aligned}$$

The $gCDH_{(\ell_1,\ell_2)}$ assumption is: For any polynomial-time adversary $\mathcal{A}$, $\mathsf{Adv}_{\mathcal{A}}^{\mathsf{gCDH}_{(\ell_1,\ell_2)}}(k)$ is negligible.

Theorem 1. *($CVDP_{(\ell_1,\ell_2)} =_p CVDP_{(\ell_1,1)}$ for $\ell_2 = O(1)$, $gCDH_{(\ell_1,\ell_2)} \le_p CVDP_{(\ell_1,\ell_2)}$, for $\ell_1/\ell_2 = O(1)$, and $CVDP_{(\ell_1,\ell_2)} \le_p gCDH_{(\ell_1,0)}$)*

– *There is an adversary* $\mathcal{A}$ *with* $\mathsf{Adv}_{\mathcal{A}}^{\mathsf{CVDP}_{(\ell_1,1)}}(k)$, *if and only if there is an adversary* $\mathcal{B}$ *with* $\mathsf{Adv}_{\mathcal{B}}^{\mathsf{CVDP}_{(\ell_1,\ell_2)}}(k)$ *such that*
  - *(if part:)* $\mathsf{Adv}_{\mathcal{A}}^{\mathsf{CVDP}_{(\ell_1,1)}}(k) \geq \left(\mathsf{Adv}_{\mathcal{B}}^{\mathsf{CVDP}_{(\ell_1,\ell_2)}}(k)\big/2\right)^{\ell_2+1}$ *and* $t_{\mathcal{A}} = (\ell_2 + 1) \cdot t_{\mathcal{B}} + O(\ell_1^3 \cdot \ell_2 \cdot k^3)$,
  - *(only if part:)* $\mathsf{Adv}_{\mathcal{B}}^{\mathsf{CVDP}_{(\ell_1,\ell_2)}}(k) \geq \left(\mathsf{Adv}_{\mathcal{A}}^{\mathsf{CVDP}_{(\ell_1,1)}}(k)\big/2\right)^{\ell_2}$ *and* $t_{\mathcal{B}} = \ell_2 \cdot t_{\mathcal{A}} + O(\ell_1^3 \cdot \ell_2 \cdot k^3)$.

– *If there is an adversary* $\mathcal{A}$ *with* $\mathsf{Adv}_{\mathcal{A}}^{\mathsf{CVDP}_{(\ell_1,\ell_2)}}(k)$, *then there is an adversary* $\mathcal{B}$ *with* $\mathsf{Adv}_{\mathcal{B}}^{\mathsf{gCDH}_{(\ell_1,\ell_2)}}(k)$, *such that* $\mathsf{Adv}_{\mathcal{B}}^{\mathsf{gCDH}_{(\ell_1,\ell_2)}}(k) \geq \left(\mathsf{Adv}_{\mathcal{A}}^{\mathsf{CVDP}_{(\ell_1,\ell_2)}}(k)\big/2\right)^{c}$ *and* $t_{\mathcal{B}} = c \cdot t_{\mathcal{A}} + O(c \cdot \ell_1^3 \cdot k^3)$ *where* c *is* $\lceil \log_2(\ell_1/\ell_2) \rceil$.

– *If there is an adversary* $\mathcal{B}$ *with* $\mathsf{Adv}_{\mathcal{B}}^{\mathsf{gCDH}_{(\ell_1,0)}}(k)$, *then there is an adversary* $\mathcal{A}$ *with* $\mathsf{Adv}_{\mathcal{A}}^{\mathsf{CVDP}_{(\ell_1,\ell_2)}}(k)$, *such that* $\mathsf{Adv}_{\mathcal{A}}^{\mathsf{CVDP}_{(\ell_1,\ell_2)}}(k) \geq \left(\mathsf{Adv}_{\mathcal{B}}^{\mathsf{gCDH}_{(\ell_1,0)}}(k)\big/2\right)^{2}$ *and* $t_{\mathcal{A}} = 2 \cdot t_{\mathcal{B}} + O(\ell_1^3 \cdot k^3)$.

3.4 Decisional Problems

This section introduces several decisional problems regarding $\mathrm{VDP}_{(\ell_1,\ell_2)}$ and their relations. For example, our reduction result implies that the $\mathrm{DSP}_{(\ell_1,\ell_1-s)}$ assumption (Definition 5) is true if the DLN_s assumption (Definition 7) is true, where the DLN_2 (decisional linear) assumption has been widely employed and investigated recently. The $\mathrm{DSP}_{(\ell_1,\ell_2)}$ assumption is employed to prove the semantic security of the homomorphic encryption proposed in Section 5.

Definition 4 ($\mathbf{DVDP}_{(\ell_1,\ell_2)}$: (ℓ_1,ℓ_2)-Decisional Vector Decomposition Problem). *Let* k *be a security parameter and* $\mathcal{G}_{\mathbb{V}}$ *be an algorithm that outputs a description of* ℓ_1*-dimensional* $\mathbb{F}_r$*-vector space* $\mathbb{V}$ *with security parameter* k, *and* $\ell_1 > \ell_2 + 1$.

For all $k \in \mathbb{N}$ *we define two distributions,* $\mathbb{D}_1$ *and* $\mathbb{R}_1$, *as follows:*

$$\begin{aligned}\mathbb{D}_1(k) \leftarrow\ & \{(\mathbb{V}, \boldsymbol{b}_0, \ldots, \boldsymbol{b}_{\ell_1-1}, \boldsymbol{v}, \boldsymbol{u}) \mid \mathbb{V} \xleftarrow{\mathsf{R}} \mathcal{G}_{\mathbb{V}}(1^k),\ (\boldsymbol{b}_0, \ldots, \boldsymbol{b}_{\ell_1-1}) \xleftarrow{\mathsf{U}} \mathbb{V}^{\ell_1},\\ & (x_0, \ldots, x_{\ell_1-1}) \xleftarrow{\mathsf{U}} (\mathbb{F}_r)^{\ell_1},\ \boldsymbol{v} \leftarrow \textstyle\sum_{i=0}^{\ell_1-1} x_i \boldsymbol{b}_i,\ \boldsymbol{u} \leftarrow \textstyle\sum_{i=0}^{\ell_2-1} x_i \boldsymbol{b}_i\},\\ \mathbb{R}_1(k) \leftarrow\ & \{(\mathbb{V}, \boldsymbol{b}_0, \ldots, \boldsymbol{b}_{\ell_1-1}, \boldsymbol{v}, \boldsymbol{u}) \mid \mathbb{V} \xleftarrow{\mathsf{R}} \mathcal{G}_{\mathbb{V}}(1^k),\ (\boldsymbol{b}_0, \ldots, \boldsymbol{b}_{\ell_1-1}) \xleftarrow{\mathsf{U}} \mathbb{V}^{\ell_1},\\ & \boldsymbol{v} \xleftarrow{\mathsf{U}} \mathbb{V}, \boldsymbol{u} \xleftarrow{\mathsf{U}} \langle \boldsymbol{b}_0, \ldots, \boldsymbol{b}_{\ell_2-1} \rangle\}.\end{aligned}$$

Let $\mathcal{A}$ *be a probabilistic polynomial-time machine. For all* $k \in \mathbb{N}$, *we define the* $DVDP_{(\ell_1,\ell_2)}$ *advantage of* $\mathcal{A}$ *as*

$$\mathsf{Adv}_{\mathcal{A}}^{\mathsf{DVDP}_{(\ell_1,\ell_2)}}(k) \leftarrow \left| \mathsf{Pr}\left[\mathcal{A}(1^k,\eta) \rightarrow 1 \mid \eta \xleftarrow{\mathsf{R}} \mathbb{D}_1(k)\right] - \mathsf{Pr}\left[\mathcal{A}(1^k,\eta) \rightarrow 1 \mid \eta \xleftarrow{\mathsf{R}} \mathbb{R}_1(k)\right] \right|.$$

The $DVDP_{(\ell_1,\ell_2)}$ *assumption is: For any probabilistic polynomial-time adversary* $\mathcal{A}$, $\mathsf{Adv}_{\mathcal{A}}^{\mathsf{DVDP}_{(\ell_1,\ell_2)}}(k)$ *is negligible in* k.

Definition 5 (DSP$_{(\ell_1,\ell_2)}$: (ℓ_1,ℓ_2)-Decisional Subspace Problem). *Let k be a security parameter and $\mathcal{G}_{\mathbb{V}}$ be an algorithm that outputs a description of ℓ_1-dimensional $\mathbb{F}_r$-vector space $\mathbb{V}$ with security parameter k, and $\ell_1 > \ell_2 + 1$.*

For all $k \in \mathbb{N}$, we define two distributions, $\mathbb{D}_2$ and $\mathbb{R}_2$, as follows:

$$\mathbb{D}_2(k) \leftarrow \{(\mathbb{V}, \boldsymbol{b}_0, \ldots, \boldsymbol{b}_{\ell_1-1}, \boldsymbol{v}) \mid \mathbb{V} \xleftarrow{\mathsf{R}} \mathcal{G}_{\mathbb{V}}(1^k),\ (\boldsymbol{b}_0, \ldots, \boldsymbol{b}_{\ell_1-1}) \xleftarrow{\mathsf{U}} \mathbb{V}^{\ell_1},$$
$$\boldsymbol{v} \xleftarrow{\mathsf{U}} \langle \boldsymbol{b}_{\ell_2}, \ldots, \boldsymbol{b}_{\ell_1-1}\rangle\},$$
$$\mathbb{R}_2(k) \leftarrow \{(\mathbb{V}, \boldsymbol{b}_0, \ldots, \boldsymbol{b}_{\ell_1-1}, \boldsymbol{v}) \mid \mathbb{V} \xleftarrow{\mathsf{R}} \mathcal{G}_{\mathbb{V}}(1^k),\ (\boldsymbol{b}_0, \ldots, \boldsymbol{b}_{\ell_1-1}) \xleftarrow{\mathsf{U}} \mathbb{V}^{\ell_1}, \boldsymbol{v} \xleftarrow{\mathsf{U}} \mathbb{V}\}.$$

The DSP$_{(\ell_1,\ell_2)}$ advantage, $\mathsf{Adv}_{\mathcal{A}}^{\mathsf{DSP}_{(\ell_1,\ell_2)}}(k)$, of a probabilistic polynomial-time machine $\mathcal{A}$ and the DSP$_{(\ell_1,\ell_2)}$ assumption are defined similarly as in Definition 4.

Definition 6 (gDDH$_{(\ell_1,\ell_2)}$: (ℓ_1,ℓ_2)-generalized Decisional Diffie-Hellman Problem). *Let k be a security parameter and $\mathcal{G}_{\mathbb{V}}$ be an algorithm that outputs a description of ℓ_1-dimensional $\mathbb{F}_r$-vector space $\mathbb{V}$ with security parameter k, and $\ell_1 > \ell_2 + 1$.*

For all $k \in \mathbb{N}$ we define two distributions, $\mathbb{D}_3$ and $\mathbb{R}_3$, as follows:

$$\mathbb{D}_3(k) \leftarrow \{(\mathbb{V}, \boldsymbol{b}_{\ell_2}, \ldots, \boldsymbol{b}_{\ell_1-1}, \boldsymbol{b}'_{\ell_2}, \ldots, \boldsymbol{b}'_{\ell_1-1}, \boldsymbol{v}, \boldsymbol{u}) \mid \mathbb{V} \xleftarrow{\mathsf{R}} \mathcal{G}_{\mathbb{V}}(1^k),$$
$$(\boldsymbol{b}_{\ell_2}, \ldots, \boldsymbol{b}_{\ell_1-1}, \boldsymbol{b}'_{\ell_2}, \ldots, \boldsymbol{b}'_{\ell_1-1}) \xleftarrow{\mathsf{U}} \mathbb{V}^{2(\ell_1-\ell_2)},$$
$$(x_{\ell_2}, \ldots, x_{\ell_1-1}) \xleftarrow{\mathsf{U}} (\mathbb{F}_r)^{\ell_1-\ell_2}, \boldsymbol{v} \leftarrow \textstyle\sum_{i=\ell_2}^{\ell_1-1} x_i \boldsymbol{b}_i,\ \boldsymbol{u} \leftarrow \sum_{i=\ell_2}^{\ell_1-1} x_i \boldsymbol{b}'_i\},$$
$$\mathbb{R}_3(k) \leftarrow \{(\mathbb{V}, \boldsymbol{b}_{\ell_2}, \ldots, \boldsymbol{b}_{\ell_1-1}, \boldsymbol{b}'_{\ell_2}, \ldots, \boldsymbol{b}'_{\ell_1-1}, \boldsymbol{v}, \boldsymbol{u}) \mid \mathbb{V} \xleftarrow{\mathsf{R}} \mathcal{G}_{\mathbb{V}}(1^k),$$
$$(\boldsymbol{b}_{\ell_2}, \ldots, \boldsymbol{b}_{\ell_1-1}, \boldsymbol{b}'_{\ell_2}, \ldots, \boldsymbol{b}'_{\ell_1-1}) \xleftarrow{\mathsf{U}} \mathbb{V}^{2(\ell_1-\ell_2)}, \boldsymbol{v} \xleftarrow{\mathsf{U}} \langle \boldsymbol{b}_{\ell_2}, \ldots, \boldsymbol{b}_{\ell_1-1}\rangle, \boldsymbol{u} \xleftarrow{\mathsf{U}} \mathbb{V}\}.$$

The gDDH$_{(\ell_1,\ell_2)}$ advantage, $\mathsf{Adv}_{\mathcal{A}}^{\mathsf{gDDH}_{(\ell_1,\ell_2)}}(k)$, of a probabilistic polynomial-time machine $\mathcal{A}$ and the gDDH$_{(\ell_1,\ell_2)}$ assumption are defined similarly as in Definition 4.

We define the decisional linear problem as follows, where DLN$_1$ corresponds to DDH, DLN$_2$ to the decisional linear problem in [3] and DLN$_s$ in general to the problems in [13,17] (note that s is independent from dimension ℓ_1 of $\mathbb{V}$):

Definition 7 (DLN$_s$: s-Decisional Linear Problem). *Let k be a security parameter and $\mathcal{G}_{\mathbb{V}}$ be an algorithm that outputs a description of ℓ_1-dimensional $\mathbb{F}_r$-vector space $\mathbb{V}$ with security parameter k.*

For all $k \in \mathbb{N}$ we define two distributions, $\mathbb{D}_4$ and $\mathbb{R}_4$, as follows:

$$\mathbb{D}_4(k) \leftarrow \{(\mathbb{V}, \boldsymbol{u}_1, \ldots, \boldsymbol{u}_s, \boldsymbol{u}^*, \boldsymbol{v}_1, \ldots, \boldsymbol{v}_s, \boldsymbol{v}^*) | \mathbb{V} \xleftarrow{\mathsf{R}} \mathcal{G}_{\mathbb{V}}(1^k), \boldsymbol{u}^* \xleftarrow{\mathsf{U}} \mathbb{V}, (\boldsymbol{u}_1, \ldots, \boldsymbol{u}_s) \xleftarrow{\mathsf{U}} \langle \boldsymbol{u}^* \rangle^s,$$
$$(x_1, \ldots, x_s) \xleftarrow{\mathsf{U}} (\mathbb{F}_r)^s, \boldsymbol{v}_i \leftarrow x_i \boldsymbol{u}_i\ (i = 1, \ldots, s), \boldsymbol{v}^* \leftarrow (\textstyle\sum_{i=1}^{s} x_i)\boldsymbol{u}^*\},$$
$$\mathbb{R}_4(k) \leftarrow \{(\mathbb{V}, \boldsymbol{u}_1, \ldots, \boldsymbol{u}_s, \boldsymbol{u}^*, \boldsymbol{v}_1, \ldots, \boldsymbol{v}_s, \boldsymbol{v}^*) \mid \mathbb{V} \xleftarrow{\mathsf{R}} \mathcal{G}_{\mathbb{V}}(1^k), \boldsymbol{u}^* \xleftarrow{\mathsf{U}} \mathbb{V},$$
$$(\boldsymbol{u}_1, \ldots, \boldsymbol{u}_s, \boldsymbol{v}_1, \ldots, \boldsymbol{v}_s, \boldsymbol{v}^*) \xleftarrow{\mathsf{U}} \langle \boldsymbol{u}^* \rangle^{2s+1}\}.$$

The DLN$_s$ advantage, $\mathsf{Adv}_{\mathcal{A}}^{\mathsf{DLN}_s}(k)$, of a probabilistic polynomial-time machine $\mathcal{A}$ and the DLN$_s$ assumption are defined similarly as in Definition 4.

Theorem 2. *($DVDP_{(\ell_1,\ell_2)} =_p DSP_{(\ell_1,\ell_2)}$, $DSP_{(\ell_1,\ell_2)} \geq_p gDDH_{(\ell_1,\ell_2)}$, $gDDH_{(\ell_1,\ell_1-s)} \geq_p DLN_s$)*

- *There is an adversary $\mathcal{A}$ with $\mathsf{Adv}_{\mathcal{A}}^{\mathsf{DVDP}_{(\ell_1,\ell_2)}}(k)$, if and only if there is an adversary $\mathcal{B}$ with $\mathsf{Adv}_{\mathcal{B}}^{\mathsf{DSP}_{(\ell_1,\ell_2)}}(k)$, such that $\mathsf{Adv}_{\mathcal{A}}^{\mathsf{DVDP}_{(\ell_1,\ell_2)}}(k) = \mathsf{Adv}_{\mathcal{B}}^{\mathsf{DSP}_{(\ell_1,\ell_2)}}(k)$ and $|t_{\mathcal{A}} - t_{\mathcal{B}}| = O(\ell_1 \cdot \ell_2 \cdot k^3)$.*
- *If there is an adversary $\mathcal{A}$ with $\mathsf{Adv}_{\mathcal{A}}^{\mathsf{DSP}_{(\ell_1,\ell_2)}}(k)$, then there is an adversary $\mathcal{B}$ with $\mathsf{Adv}_{\mathcal{B}}^{\mathsf{gDDH}_{(\ell_1,\ell_2)}}(k)$, such that $\mathsf{Adv}_{\mathcal{B}}^{\mathsf{gDDH}_{(\ell_1,\ell_2)}}(k) = \mathsf{Adv}_{\mathcal{A}}^{\mathsf{DSP}_{(\ell_1,\ell_2)}}(k)$, and $t_{\mathcal{B}} = t_{\mathcal{A}} + O(\ell_1 \cdot \ell_2 \cdot k^3)$.*
- *Assume that $\ell_1 \geq s+1$. If there is an adversary $\mathcal{B}$ with $\mathsf{Adv}_{\mathcal{B}}^{\mathsf{gDDH}_{(\ell_1,\ell_1-s)}}(k)$, then there is an adversary $\mathcal{A}$ with $\mathsf{Adv}_{\mathcal{A}}^{\mathsf{DLN}_s}(k)$, such that $\mathsf{Adv}_{\mathcal{A}}^{\mathsf{DLN}_s}(k) \geq \mathsf{Adv}_{\mathcal{B}}^{\mathsf{gDDH}_{(\ell_1,\ell_1-s)}}(k) - \epsilon(k)$, and $t_{\mathcal{A}} = t_{\mathcal{B}} + O(s \cdot \ell_1^2 \cdot k^3)$, where $\epsilon(k)$ is negligible in k.*

4 Trapdoor Bijective Functions

The key technique to apply the VDP on distortion eigenvector spaces to cryptography is a *trapdoor bijective function* from the VDP, where we call f a *trapdoor bijective function* if f is a trapdoor one-way function and bijection (one-to-one and onto, i.e., its domain and range are isomorphic). In general, the representation of the domain is not always the same as that of the range, and trapdoor bijective function f is called a *trapdoor permutation* if the representation of the domain is equivalent to that of the range.

We will show two major applications of our trapdoor bijective functions; one is multivariate homomorphic encryption (in Section 5) and the other various types of signatures (in Section 6).

In this section, we introduce a trapdoor bijective function f from the VDP such that

$$f : \langle \boldsymbol{b}_0 \rangle \times \cdots \times \langle \boldsymbol{b}_{\ell-1} \rangle \to \mathbb{V}, \qquad f : (\boldsymbol{z}_0, \ldots, \boldsymbol{z}_{\ell-1}) \mapsto \textstyle\sum_{i=0}^{\ell-1} \boldsymbol{z}_i,$$

where $\boldsymbol{z}_i \in \langle \boldsymbol{b}_i \rangle$ for $i = 0, \ldots, \ell-1$. Here note that $\langle \boldsymbol{b}_0 \rangle \times \cdots \times \langle \boldsymbol{b}_{\ell-1} \rangle \cong \mathbb{V}(\cong (\mathbb{F}_r)^\ell)$ (i.e., the domain of f is isomorphic to the range), and that the representations of the domain and range are not equivalent in general.

A typical representation of $(\boldsymbol{z}_0, \ldots, \boldsymbol{z}_{\ell-1}) \in \langle \boldsymbol{b}_0 \rangle \times \cdots \times \langle \boldsymbol{b}_{\ell-1} \rangle$ is to represent each $\boldsymbol{z}_i \in \mathbb{V}$ $(i = 0, \ldots, \ell-1)$ by some standard expression of an element of $\mathbb{V}$. (If $\mathbb{V}$ is a Jacobian, an expression of an element of $\mathbb{V}$ can be a standard reduced form of an element of the Jacobian.) Another representation of $(\boldsymbol{z}_0, \ldots, \boldsymbol{z}_{\ell-1})$ is $(x_0, \ldots, x_{\ell-1}) \in (\mathbb{F}_r)^\ell$ such that $\boldsymbol{z}_i = x_i \boldsymbol{b}_i$ for $i = 0, \ldots, \ell-1$.

There is an efficient algorithm to evaluate f (to compute $\sum_{i=0}^{\ell-1} \boldsymbol{z}_i$ from $(\boldsymbol{z}_0, \ldots, \boldsymbol{z}_{\ell-1})$), but under the CVDP assumption, it is intractable to compute f^{-1} (to compute $(\boldsymbol{z}_0, \ldots, \boldsymbol{z}_{\ell-1})$ from $\boldsymbol{v} \in \mathbb{V}$ such that $\boldsymbol{v} = \sum_{i=0}^{\ell-1} \boldsymbol{z}_i$ and $\boldsymbol{z}_i \in \langle \boldsymbol{b}_i \rangle$ for

$i = 0, \ldots, \ell - 1$) without the knowledge of trapdoor X. There is an efficient algorithm Deco for computing f^{-1} if X (trapdoor) is available (see Section 3).

To the best of our knowledge, this function f is the first trapdoor bijective function except the RSA function and its variants (the Rabin-Williams [19] and Paillier [16] functions) that are based on the integer factoring trick (hereafter, we call these functions the RSA family functions). Note that the RSA family functions are not only trapdoor bijective functions but also permutations, while our trapdoor bijective function is not a permutation.

In contrast to the RSA family functions, it is not so easy, in general, to (decodably) embed a bit string in $\{0,1\}^*$ to the domain, $\langle \boldsymbol{b}_0 \rangle \times \cdots \times \langle \boldsymbol{b}_{\ell-1} \rangle$, of f. That is, a value (or a bit string through binary expression) $(x_0, \ldots, x_{\ell-1})$ in $(\mathbb{F}_r)^\ell$ can be easily embedded to an element $(\boldsymbol{z}_0, \ldots, \boldsymbol{z}_{\ell-1})$ of the domain of f, such that $\boldsymbol{z}_i = x_i \boldsymbol{b}_i$ for $i = 0, \ldots, \ell - 1$, but, in general, x_i cannot be efficiently decoded from $\boldsymbol{z}_i$ (unless the discrete logarithm problem is easy). If x_i is selected from a logarithmically small space, it is efficiently decodable.

On the other hand, in a manner similar to the RSA family functions, we can construct both public-key encryption and digital signatures directly from the trapdoor bijective function f. If plaintext m is (decodably) embedded to the domain of f in some manner, we can realize a basic (OW-CPA) public-key encryption scheme. If message m is embedded to the range, $\mathbb{V}$, of f, we can realize digital signatures (a way of embedding a bit string to an element of the range, $\mathbb{V}$, is shown in the footnote of Section 6.1).

Based on this strategy, *multivariate homomorphic* encryption is presented in Section 5, where a message from a logarithmic space is embedded to a subspace of the domain and the remaining subspace is used for randomization. Signatures as well as blind signatures, convertible undeniable signatures are presented in Section 6.

5 Multivariate Homomorphic Encryption

In this section, we propose a multivariate homomorphic encryption scheme that is constructed on the trapdoor bijective function introduced in Section 4. The scheme is a generalization of the Galbraith-Verheul scheme [11].

Our scheme is constructed on ℓ_1-dimensional distortion eigenvector space $\mathbb{V}$, and the message space is ℓ_2-dimensional, where $(\ell_1 - \ell_2)$-dimensional space is used for randomness.

5.1 Proposed Homomorphic Encryption Scheme

We assume that plaintext $(m_0, \ldots, m_{\ell_2-1})$ is bounded by some small integer τ such that $0 \leq m_i < \tau$ $(i = 0, \ldots, \ell_2 - 1)$ and τ is the logarithmic order of k. Such a small plaintext space is sufficient for many applications as shown in [4,12], See Fig. 2 for the proposed multivariate homomorphic encryption scheme.

An advantage of our scheme is its homomorphic property for multiple plaintexts (See Subsection 5.4).

$\mathsf{Gen}(1^k)$:
  $\mathbb{V} \xleftarrow{\mathsf{R}} \mathcal{G}_{\mathbb{V}}(1^k)$ with distortion eigenvector basis $\mathbb{A} \to (\boldsymbol{a}_0, \ldots, \boldsymbol{a}_{\ell_1-1})$
  $X \leftarrow (x_{i,j}) \xleftarrow{\mathsf{U}} (\mathbb{F}_r)^{\ell_1 \times \ell_1}$, $\boldsymbol{b}_i = \sum_{j=0}^{\ell_1-1} x_{i,j}\boldsymbol{a}_j$, $\mathbb{B} \leftarrow (\boldsymbol{b}_0, \ldots, \boldsymbol{b}_{\ell_1-1})$.
  $\mathsf{sk} \leftarrow X$, $\mathsf{pk} \leftarrow (\mathbb{V}, \mathbb{A}, \mathbb{B})$.
  return sk, pk.
$\mathsf{Enc}(\mathsf{pk}, (m_0, \ldots, m_{\ell_2-1}) \in \{0, \ldots, \tau-1\}^{\ell_2})$:
  $(r_{\ell_2}, \ldots, r_{\ell_1-1}) \xleftarrow{\mathsf{U}} (\mathbb{F}_r)^{\ell_1-\ell_2}$, $\quad \boldsymbol{c} \leftarrow \sum_{i=0}^{\ell_2-1} m_i \boldsymbol{b}_i + \sum_{i=\ell_2}^{\ell_1-1} r_i \boldsymbol{b}_i$.
  return ciphertext $\boldsymbol{c}$.
$\mathsf{Dec}(\mathsf{sk}, \boldsymbol{c})$:
  $\boldsymbol{b}'_i \leftarrow \mathsf{Deco}(\boldsymbol{c}, \langle \boldsymbol{b}_i \rangle, X, \mathbb{B})$. $m'_i \leftarrow \mathrm{Dlog}_{\boldsymbol{b}_i}(\boldsymbol{b}'_i)$ for $i = 0, \ldots, \ell_2 - 1$.
  return plaintext $(m'_0, \ldots, m'_{\ell_2-1})$.

Fig. 2. Proposed Multivariate Homomorphic Encryption Scheme

5.2 Security

Theorem 3. *The public key encryption scheme in Fig. 2 is semantically secure (IND-CPA secure) under the $DSP_{(\ell_1,\ell_2)}$ assumption.*

5.3 Two-Party Protocol to Securely Evaluate a 2DNF Formula

As an application of our multivariate homomorphic encryption, we present a two-party protocol to securely evaluate a 2DNF formula (over n variables) for higher dimensional n variables (assignments). (See [4] for some application of a protocol to securely evaluate a 2DNF formula.)

A 2DNF formula, ψ, over $y_1, \ldots, y_n$ is of the form $\bigvee_{i=1}^{h}(\lambda_{i,1} \wedge \lambda_{i,2})$ where $\lambda_{i,1}, \lambda_{i,2} \in \{y_1, \ldots, y_n, \bar{y}_1, \ldots, \bar{y}_n\}$.

We consider a two-party protocol between Alice and Bob, where Bob knows ℓ_2-dimensional secret input $(\overrightarrow{m}_1, \ldots, \overrightarrow{m}_n)$ (to formula ψ), and Alice knows secret 2DNF formula ψ. Here, $\overrightarrow{m}_i \leftarrow (m_{i,0}, \ldots, m_{i,\ell_2-1}) \in \{0,1\}^{\ell_2}$, for $i = 1, \ldots, n$. The protocol outputs $\psi(\overrightarrow{m}_1, \ldots, \overrightarrow{m}_n)(= (\psi(m_{1,0}, \ldots, m_{n,0}), \ldots, \psi(m_{1,\ell_2-1}, \ldots, m_{n,\ell_2-1})$) while keeping the local secrets, $(\overrightarrow{m}_1, \ldots, \overrightarrow{m}_n)$ and ψ (except for the number of disjunctive clauses of ψ). We will now describe a semi-honest protocol between Alice and Bob to securely evaluate 2DNF formula ψ over ℓ_2-dimensional n inputs.

1. (Input:) Alice holds a 2DNF formula, $\psi(y_1, \ldots, y_n) \leftarrow \bigvee_{i=1}^{h}(\lambda_{i,1} \wedge \lambda_{i,2})$ where $\lambda_{i,1}, \lambda_{i,2} \in \{y_1, \ldots, y_n, \bar{y}_1, \ldots, \bar{y}_n\}$, and Bob holds an ℓ_2-dimensional assignment to the formula, $(\overrightarrow{m}_1, \ldots, \overrightarrow{m}_n)$, where $\overrightarrow{m}_i \leftarrow (m_{i,0}, \ldots, m_{i,\ell_2-1}) \in \{0,1\}^{\ell_2}$, for $i = 1, \ldots, n$.
2. Bob executes $\mathsf{Gen}(1^k)$ to compute sk, pk, and sends pk to Alice. Bob also computes $\mathsf{Enc}(\mathsf{pk}, (m_{1,0}, \ldots, m_{1,\ell_2-1}))$, $\ldots, \mathsf{Enc}(\mathsf{pk}, (m_{n,0}, \ldots, m_{n,\ell_2-1}))$, and sends them to Alice.
3. Alice computes an arithmetization Ψ of ψ by replacing $\vee$ by $+$, $\wedge$ by $\cdot$ and $\bar{m}_i$ by $(1 - m_i)$. So, $\Psi(y_1, \ldots, y_n) = \sum_{i=1}^{h}(\lambda_{i,1} \cdot \lambda_{i,2})$, where $\lambda_{i,1}, \lambda_{i,2} \in \{y_1, \ldots, y_n, 1-y_1, \ldots, 1-y_n\}$. For ℓ_2-dimensional assignment, $\Psi(\overrightarrow{m}_1, \ldots, \overrightarrow{m}_n)$

$= (\sum_{i=1}^{h}(\lambda_{i,1,0} \cdot \lambda_{i,2,0}), \ldots, \sum_{i=1}^{h}(\lambda_{i,1,\ell_2-1} \cdot \lambda_{i,2,\ell_2-1}))$, where $\lambda_{i,1,j}, \lambda_{i,2,j} \in \{m_{1,j}, \ldots, m_{n,j}, 1 - m_{1,j}, \ldots, 1 - m_{n,j}\}$, $j = 0, \ldots, \ell_2 - 1$.

4. For each ℓ_2-dimensional disjunctive clause, $((\lambda_{i,1,0} \cdot \lambda_{i,2,0}), \ldots, (\lambda_{i,1,\ell_2-1} \cdot \lambda_{i,2,\ell_2-1}))$ $(i = 1, \ldots, h)$, Alice sets the corresponding ciphertexts, $\boldsymbol{c}_{i,1} \leftarrow \mathsf{Enc}(\mathsf{pk}, (\lambda_{i,1,0}, \ldots, \lambda_{i,1,\ell_2-1}))$ and $\boldsymbol{c}_{i,2} \leftarrow \mathsf{Enc}(\mathsf{pk}, (\lambda_{i,2,0}, \ldots, \lambda_{i,2,\ell_2-1}))$. If $(\lambda_{i,j,0}, \ldots, \lambda_{i,j,\ell_2-1})$ is $(1 - m_{s,0}, \ldots, 1 - m_{s,\ell_2-1})$, then $\boldsymbol{c}_{i,j}$ is set by $\mathsf{Enc}(\mathsf{pk}, (1, \ldots, 1)) - \mathsf{Enc}(\mathsf{pk}, (m_{s,0}, \ldots, m_{s,\ell_2-1}))$.
5. Alice computes $\boldsymbol{c}^*_{i,1}$, $\boldsymbol{c}^*_{i,2}$ and E_κ as follows:

$$t_{i,j,\kappa} \xleftarrow{\mathsf{U}} \mathbb{F}_r \quad (i = 1, \ldots, h; j = 1, 2; \kappa = 0, \ldots, \ell_1 - 1),$$
$$u_{\kappa,\mu} \xleftarrow{\mathsf{U}} \mathbb{F}_r \quad (\kappa = 0, \ldots, \ell_2 - 1, \mu = 0, \ldots, \ell_1 - 1),$$
$$\boldsymbol{c}^*_{i,1} \leftarrow \boldsymbol{c}_{i,1} + \sum_{\kappa=0}^{\ell_1-1} t_{i,1,\kappa}\boldsymbol{b}_\kappa, \quad \boldsymbol{c}^*_{i,2} \leftarrow \boldsymbol{c}_{i,2} + \sum_{\kappa=0}^{\ell_1-1} t_{i,2,\kappa}\boldsymbol{b}_\kappa,$$
$$E_\kappa \leftarrow \sum_{i=1}^{h}(t_{i,1,\kappa}\boldsymbol{c}_{i,2} + t_{i,2,\kappa}\boldsymbol{c}_{i,1} + t_{i,1,\kappa}t_{i,2,\kappa}\boldsymbol{b}_\kappa) + \sum_{\mu\neq\kappa,\mu=0}^{\ell_1-1} u_{\kappa,\mu}\boldsymbol{b}_\mu,$$
$$\text{for} \quad \kappa = 0, \ldots, \ell_2 - 1.$$

6. Alice sends $\boldsymbol{c}^*_{i,1}$, $\boldsymbol{c}^*_{i,2}$ for $i = 1, \ldots, h$ and $(E_0, \ldots, E_{\ell_2-1})$ to Bob.
7. Bob computes

$$Z_\kappa \leftarrow \prod_{i=1}^{h} e(\mathsf{Deco}(\boldsymbol{c}^*_{i,1}, \langle \boldsymbol{b}_\kappa \rangle), \rho(\mathsf{Deco}(\boldsymbol{c}^*_{i,2}, \langle \boldsymbol{b}_\kappa \rangle)))/e(\mathsf{Deco}(E_\kappa, \langle \boldsymbol{b}_\kappa \rangle), \rho(\boldsymbol{b}_\kappa))$$

for $\kappa = 0, \ldots, \ell_2 - 1$, where $\mathsf{Deco}(\cdot, \langle \boldsymbol{b}_\kappa \rangle)$ denotes $\mathsf{Deco}(\cdot, \langle \boldsymbol{b}_\kappa \rangle, X, \mathbb{B})$. Bob then computes w_κ such that $Z_\kappa = e(\boldsymbol{b}_\kappa, \rho(\boldsymbol{b}_\kappa))^{w_\kappa}$ for $\kappa = 0, \ldots, \ell_2 - 1$.
8. Bob outputs $(w_0, \ldots, w_{\ell_2-1})$.

Lemma 4. *The output of the protocol is correct, i.e., $w_j = \Psi(m_{1,j}, \ldots, m_{n,j})$ for $j = 0, \ldots, \ell_2 - 1$.*

Proof.

$$\prod_{i=1}^{h} e(\mathsf{Deco}(\boldsymbol{c}^*_{i,1}, \langle \boldsymbol{b}_\kappa \rangle), \rho(\mathsf{Deco}(\boldsymbol{c}^*_{i,2}, \langle \boldsymbol{b}_\kappa \rangle)))$$
$$= e(\boldsymbol{b}_\kappa, \rho(\boldsymbol{b}_\kappa))^{\sum_{i=1}^{h} \lambda_{i,1,\kappa}\lambda_{i,2,\kappa}} \cdot e(\boldsymbol{b}_\kappa, \rho(\boldsymbol{b}_\kappa))^{\sum_{i=1}^{h} \lambda_{i,1,\kappa}t_{i,2,\kappa} + \lambda_{i,2,\kappa}t_{i,1,\kappa} + t_{i,1,\kappa}t_{i,2,\kappa}}.$$
$$e(\mathsf{Deco}(E_\kappa, \langle \boldsymbol{b}_\kappa \rangle), \rho(\boldsymbol{b}_\kappa)) = e(\sum_{i=1}^{h}((\lambda_{i,1,\kappa}t_{i,2,\kappa} + \lambda_{i,2,\kappa}t_{i,1,\kappa} + t_{i,1,\kappa}t_{i,2,\kappa})\boldsymbol{b}_\kappa, \rho(\boldsymbol{b}_\kappa))$$
$$= e(\boldsymbol{b}_\kappa, \rho(\boldsymbol{b}_\kappa))^{\sum_{i=1}^{h} \lambda_{i,1,\kappa}t_{i,2,\kappa} + \lambda_{i,2,\kappa}t_{i,1,\kappa} + t_{i,1,\kappa}t_{i,2,\kappa}}.$$

Therefore,

$$\prod_{i=1}^{h} e(\mathsf{Deco}(\boldsymbol{c}^*_{i,1}, \langle \boldsymbol{b}_\kappa \rangle), \rho(\mathsf{Deco}(\boldsymbol{c}^*_{i,2}, \langle \boldsymbol{b}_\kappa \rangle)))/e(\mathsf{Deco}(E_\kappa, \langle \boldsymbol{b}_\kappa \rangle), \rho(\boldsymbol{b}_\kappa))$$
$$= e(\boldsymbol{b}_\kappa, \rho(\boldsymbol{b}_\kappa))^{\sum_{i=1}^{h} \lambda_{i,1,\kappa}\lambda_{i,2,\kappa}}, \quad \text{and} \quad w_\kappa = \sum_{i=1}^{h} \lambda_{i,1,\kappa}\lambda_{i,2,\kappa}.$$ □

Lemma 5. *The protocol is secure against semi-honest Alice and Bob under the $DSP_{(\ell_1,\ell_2)}$ assumption, where the security definition follows that in [4] except that the number of disjunctive clauses, h, of ψ can be revealed in our definition.*

Remark: To prevent the disclosure of the exact number of disjunctive clauses, h, of ψ, Alice can send Bob additional dummy $poly(k)$ pairs of $(\boldsymbol{c}^*_{i,1}, \boldsymbol{c}^*_{i,2})$ $(i = h + 1, \ldots, h + poly(k))$ for $\boldsymbol{c}_{i,j} \leftarrow \mathsf{Enc}(\mathsf{pk}, (0, \ldots, 0))$.

5.4 Comparison with the BGN Encryption Scheme

The proposed homomorphic encryption shares some properties with Boneh-Goh-Nissim (BGN) encryption [4]. An advantage of our scheme is: the ciphertext size of our encryption scheme can be shorter than that of the BGN scheme, since the BGN scheme requires a composite number order subgroup. A disadvantage is: the homomorphic operation on a multiplication cannot be executed over ciphertexts in our scheme, while the BGN can do this operation over ciphertexts. Therefore, when we realize a secure two-party protocol for evaluating a 2DNF formula on the BGN scheme or on our scheme, the communication complexity from Bob to Alice on our scheme is shorter than that on the BGN scheme, but the communication complexity from Alice to Bob on our scheme is much greater than that on the BGN scheme. So, our scheme is not suitable for an application where the communication complexity from Alice to Bob is more important (e.g., PIR), while our scheme is suitable for an application where that from Alice to Bob is more important.

A major advantage of our scheme is that it has a richer algebraic structure than the BGN scheme. For example, our encryption scheme is multivariate and homomorphic encryption with distortion maps as well as bilinear pairings. Such an algebraic structure may imply new applications to various cryptographic protocols using higher dimensional secrets along with the homomorphic property and 2DNF formula protocol.

6 Signatures

In this section, we present ordinary, blind and convertible undeniable signatures, and their combination, as another cryptographic application of the trapdoor bijective function shown in Section 4.

6.1 Basic Signature Scheme

Public key: $(\mathbb{V}, \mathbb{A}, \mathbb{B}, h)$ such that $\mathbb{V} \xleftarrow{\mathsf{R}} \mathcal{G}_{\mathbb{V}}(1^k)$ is a ℓ_1-dimensional $\mathbb{F}_r$-vector space, $X \leftarrow (x_{i,j}) \xleftarrow{\mathsf{U}} (\mathbb{F}_r)^{\ell_1 \times \ell_1}$, $\boldsymbol{b}_i = \sum_{j=0}^{\ell_1 - 1} x_{i,j} \boldsymbol{a}_j$, $\mathbb{B} \leftarrow (\boldsymbol{b}_0, \ldots, \boldsymbol{b}_{\ell_1 - 1})$, and h is a hash function with $h : \{0,1\}^* \mapsto (\mathbb{F}_r)^{\ell_1}$, where $\mathbb{A}$ is a distortion eigenvector basis of $\mathbb{V}$.

Secret key: X.

Signing: $m \in \{0,1\}^*$ is a message to be signed. Hashed value $h(m)$ is embedded to $\mathbb{V}$ as $\boldsymbol{h}$. [3]

$\boldsymbol{s}_i \leftarrow \mathsf{Deco}(\boldsymbol{h}, \langle \boldsymbol{b}_i \rangle, X, \mathbb{B})$, for $i = 0, \ldots, \ell_1 - 2$.

$(\boldsymbol{s}_0, \ldots, \boldsymbol{s}_{\ell_1 - 2})$ is the signature of m.

[3] When we use the Jacobian shown in Section 2.1, there is a subspace, $\langle \boldsymbol{a}_0 \rangle$, to which embedding a string is easy, and there are effectively computable distortion maps, $\phi_{i,0}$, from $\langle \boldsymbol{a}_0 \rangle$ to other subspaces $\langle \boldsymbol{a}_i \rangle$ $(i = 1, \ldots, \ell_1 - 1)$. In this case, $m \leftarrow (m_0, \ldots, m_{\ell_1 - 1}) \in (\mathbb{F}_r)^{\ell_1}$ can be effectively embedded to $\mathbb{V} \cong \langle \boldsymbol{a}_0 \rangle \times \cdots \times \langle \boldsymbol{a}_{\ell_1 - 1} \rangle$ by embedding m_i to $\langle \boldsymbol{a}_0 \rangle$ first and then applying $\phi_{i,0}$ to map to $\langle \boldsymbol{a}_i \rangle$ for $i = 1, \ldots, \ell_1 - 1$.

Verification: Given $(\boldsymbol{s}_0, \ldots, \boldsymbol{s}_{\ell_1-2})$ and m, verifier $\mathcal{V}$ computes $\boldsymbol{h} \in \mathbb{V}$ from $h(m)$ and checks whether $e(\boldsymbol{s}_i, \boldsymbol{b}_i) = 1$ for $i = 0, \ldots, \ell_1 - 2$, and $e(\boldsymbol{h} - \sum_{i=0}^{\ell_1-2} \boldsymbol{s}_i, \boldsymbol{b}_{\ell_1-1}) = 1$ hold.

Security in the Random Oracle Model. If a message m is embedded to $\mathbb{V}$ through hashed value, $h(m)$, and h is modeled as a random oracle, the security (existential unforgeability against chosen message attacks) can be proven under the CVDP assumption, in a manner similar to that of the full domain hash RSA signatures [2,5].

6.2 Blind Signatures

The public key $(\mathbb{V}, \mathbb{A}, \mathbb{B}, h)$ and secret key X of a signer $\mathcal{S}$ are the same as those of the basic signature scheme.

Blinding: $m \in \{0,1\}^*$ is a message to be signed. Hashed value $h(m)$ is embedded to $\mathbb{V}$ as $\boldsymbol{h}$. (See Section 6.1 for how to embed.)
User $\mathcal{U}$ selects $\gamma_i \xleftarrow{\mathsf{U}} \mathbb{F}_r$ $(i = 0, \ldots, \ell_1 - 1)$ and computes a blinded message $\boldsymbol{d} \leftarrow \boldsymbol{h} + \sum_{i=0}^{\ell_1-1} \gamma_i \boldsymbol{b}_i$.

Signing: $\mathcal{U}$ gives $\boldsymbol{d}$ to signer $\mathcal{S}$. $\mathcal{S}$ computes $\boldsymbol{t}_i \leftarrow \mathsf{Deco}(\boldsymbol{d}, \langle \boldsymbol{b}_i \rangle, X, \mathbb{B})$, for $i = 0, \ldots, \ell_1 - 2$, and returns $(\boldsymbol{t}_0, \ldots, \boldsymbol{t}_{\ell_1-2})$ to $\mathcal{U}$.

Unblinding: $\mathcal{U}$ computes $\boldsymbol{s}_i \leftarrow \boldsymbol{t}_i - \gamma_i \boldsymbol{b}_i$ for $i = 0, \ldots, \ell_1 - 2$.
$(\boldsymbol{s}_0, \ldots, \boldsymbol{s}_{\ell_1-2})$ is the signature of (m_1, m_2).

Verification: Given $(\boldsymbol{s}_0, \ldots, \boldsymbol{s}_{\ell_1-2})$ and m, verifier $\mathcal{V}$ computes $\boldsymbol{h} \in \mathbb{V}$ from $h(m)$, and checks whether $e(\boldsymbol{s}_i, \boldsymbol{b}_i) = 1$ for $i = 0, \ldots, \ell_1 - 2$, and $e(\boldsymbol{h} - \sum_{i=0}^{\ell_1-2} \boldsymbol{s}_i, \boldsymbol{b}_{\ell_1-1}) = 1$ hold.

Here, $\boldsymbol{s}_i = \mathsf{Deco}(\boldsymbol{h} + \sum_{i=0}^{\ell_1-1} \gamma_i \boldsymbol{b}_i, \langle \boldsymbol{b}_i \rangle, X, \mathbb{B}) - \gamma_i \boldsymbol{b}_i = \mathsf{Deco}(\boldsymbol{h}, \langle \boldsymbol{b}_i \rangle, X, \mathbb{B})$.

The blind signature scheme is perfectly blind and unforgeable under a one-more-CVDP assumption, which is defined in a manner similar to [1], in the random oracle model.

6.3 Convertible Undeniable Signatures

This section presents generically (selectively and universally) convertible undeniable signatures (see [14] for the notion and security requirements.)

The public key $(\mathbb{V}, \mathbb{A}, \mathbb{B}, h)$ and secret key X are the same as those of the basic signature scheme.

Signing: $m \in \{0,1\}^*$ is a message to be signed. Hashed value $h(m)$ is embedded to $\mathbb{V}$ as $\boldsymbol{h}$.
$\boldsymbol{s}_i \leftarrow \mathsf{Deco}(\boldsymbol{h}, \langle \boldsymbol{b}_i \rangle, X, \mathbb{B})$, for $i = 0, \ldots, \ell_1 - 1$.
$(\boldsymbol{s}_0, \ldots, \boldsymbol{s}_{\ell_2-1})$ is the signature of m, where $\ell_2 < \ell_1 - 1$. (The signer secretly keeps $(\boldsymbol{s}_{\ell_2}, \ldots, \boldsymbol{s}_{\ell_1-1})$ for the confirmation protocol.)

Confirmation: Given signature $(\boldsymbol{s}_0, \ldots, \boldsymbol{s}_{\ell_2-1})$ and message m, a verifier $\mathcal{V}$ computes $\boldsymbol{h} \in \mathbb{V}$ from $h(m)$ and checks whether $e(\boldsymbol{s}_i, \boldsymbol{b}_i) = 1$ for $i = 0, \ldots, \ell_2 - 1$ hold. If $(\boldsymbol{s}_0, \ldots, \boldsymbol{s}_{\ell_2-1})$ is a valid signature (i.e., $\boldsymbol{h} - (\sum_{i=0}^{\ell_2-1} \boldsymbol{s}_i) \in \langle \boldsymbol{b}_{\ell_2}, \ldots, \boldsymbol{b}_{\ell_1-1} \rangle$), signer $\mathcal{S}$ with $(\boldsymbol{s}_{\ell_2}, \ldots, \boldsymbol{s}_{\ell_1-1})$ can execute the confirmation protocol with $\mathcal{V}$ as follows:

1. $\mathcal{S}$ generates $\gamma_i \xleftarrow{\mathsf{U}} \mathbb{F}_r^*$ and computes $\boldsymbol{u}_i \leftarrow (1/\gamma_i)\boldsymbol{s}_i$, for $i = \ell_2, \ldots, \ell_1 - 1$. $\mathcal{S}$ gives $(\boldsymbol{u}_{\ell_2}, \ldots, \boldsymbol{u}_{\ell_1-1})$ to $\mathcal{V}$.
2. $\mathcal{V}$ checks whether $e(\boldsymbol{u}_i, \boldsymbol{b}_i) = 1$ holds for $i = \ell_2, \ldots, \ell_1 - 1$.
3. $\mathcal{S}$ executes a zero-knowledge protocol (based on the standard Σ-protocol) to prove to $\mathcal{V}$ that $\mathcal{S}$ knows $(\gamma_{\ell_2}, \ldots, \gamma_{\ell_1-1})$ such that $\boldsymbol{h} - (\sum_{i=0}^{\ell_2-1} \boldsymbol{s}_i) = \sum_{i=\ell_2}^{\ell_1-1} \gamma_i \boldsymbol{u}_i$.

Disavowal: Given signature $(\boldsymbol{s}_0, \ldots, \boldsymbol{s}_{\ell_2-1})$ and message m, a verifier $\mathcal{V}$ computes $\boldsymbol{h} \in \mathbb{V}$ from $h(m)$ and checks whether $e(\boldsymbol{s}_i, \boldsymbol{b}_i) = 1$ holds for $i = 0, \ldots, \ell_2 - 1$.
Signer $\mathcal{S}$ computes $\boldsymbol{t} \leftarrow \boldsymbol{h} - (\sum_{i=0}^{\ell_2-1} \boldsymbol{s}_i)$ and $\boldsymbol{v}_i \leftarrow \mathsf{Deco}(\boldsymbol{t}, \langle \boldsymbol{b}_i \rangle, X, \mathbb{B})$ for $i = 0, \ldots, \ell_1 - 1$. If there exists $i \in \{0, \ldots, \ell_2 - 1\}$ such that $\boldsymbol{v}_i \neq \boldsymbol{0}$, $(\boldsymbol{s}_0, \ldots, \boldsymbol{s}_{\ell_2-1})$ is an invalid signature (i.e., $\boldsymbol{t} \notin \langle \boldsymbol{b}_{\ell_2}, \ldots, \boldsymbol{b}_{\ell_1-1} \rangle$). Then, signer $\mathcal{S}$ can execute the disavowal protocol with $\mathcal{V}$ as follows:

1. $\mathcal{S}$ generates $\gamma_i \xleftarrow{\mathsf{U}} \mathbb{F}_r^*$ and computes $\boldsymbol{u}_i \leftarrow (1/\gamma_i)\boldsymbol{v}_i$, for $i = 0, \ldots, \ell_1 - 1$.
2. $\mathcal{S}$ selects $\delta \xleftarrow{\mathsf{U}} \mathbb{F}_r$ and computes $\boldsymbol{w} \leftarrow \sum_{i=0}^{\ell_2-1} \boldsymbol{v}_i + \delta \boldsymbol{b}_{\ell_2}$ $(= \sum_{i=0}^{\ell_2-1} \gamma_i \boldsymbol{u}_i + \delta \boldsymbol{b}_{\ell_2})$.
3. $\mathcal{S}$ gives $\boldsymbol{w}$ and $(\boldsymbol{u}_0, \ldots, \boldsymbol{u}_{\ell_1-1})$ to $\mathcal{V}$.
4. $\mathcal{V}$ checks whether $e(\boldsymbol{u}_i, \boldsymbol{b}_i) = 1$ for $i = 0, \ldots, \ell_1 - 1$ and $e(\boldsymbol{w}, \boldsymbol{b}_{\ell_2}) \neq 1$ hold.
5. $\mathcal{S}$ executes a zero-knowledge protocol (based on the standard Σ-protocol) to prove to $\mathcal{V}$ that $\mathcal{S}$ knows $(\gamma_0, \ldots, \gamma_{\ell_2-1}, \delta)$ such that $\boldsymbol{w} = \sum_{i=0}^{\ell_2-1} \gamma_i \boldsymbol{u}_i + \delta \boldsymbol{b}_{\ell_2}$.
6. $\mathcal{S}$ also executes a zero-knowledge protocol to prove to $\mathcal{V}$ that $\mathcal{S}$ knows $(\gamma_{\ell_2}, \ldots, \gamma_{\ell_1-1}, \delta)$ such that $\boldsymbol{t} - \boldsymbol{w} = \sum_{\ell_2}^{\ell_1-1} \gamma_i \boldsymbol{u}_i - \delta \boldsymbol{b}_{\ell_2}$.

Selective Conversion: To selectively convert an undeniable signature, $(\boldsymbol{s}_0, \ldots, \boldsymbol{s}_{\ell_2-1})$, to an ordinary signature, the signer additionally releases $(\boldsymbol{s}_{\ell_2}, \ldots, \boldsymbol{s}_{\ell_1-2})$. So, $(\boldsymbol{s}_0, \ldots, \boldsymbol{s}_{\ell_1-2})$ is the (ordinary) signature, which is equivalent to the basic signature.

Universal Conversion: To universally convert undeniable signatures, $(\boldsymbol{s}_0, \ldots, \boldsymbol{s}_{\ell_2-1})$, to ordinary signatures, the signer additionally releases $(x_{j,\kappa}, t_{i,j})$ for $\kappa, i = 0, \ldots, \ell_2 - 1$ and $j = \ell_2, \ldots, \ell_1 - 1$ as follows:

- Remind that $X \leftarrow (x_{i,j})$ $(i, j = 0, \ldots, \ell_1 - 1)$.
- For $i = \ell_2, \ldots, \ell_1 - 1$ and $j = 0, \ldots, \ell_1 - 1$, $y_{ij} \leftarrow 1$ if $i = j$, $y_{ij} \leftarrow 0$ if $i \neq j$.
  For $i = 0, \ldots, \ell_2 - 1$ and $j = 0, \ldots, \ell_1 - 1$, $y_{ij} \leftarrow 0$ if $j \geq \ell_2$, $y_{ij} \xleftarrow{\mathsf{U}} \mathbb{F}_r$ if $j < \ell_2$.
- $Z \leftarrow YX, \quad (t_{i,j}) \leftarrow Z^{-1}$,
- Output: $(x_{j,\kappa}, t_{i,j})$ for $\kappa, i = 0, \ldots, \ell_2 - 1$ and $j = \ell_2, \ldots, \ell_1 - 1$.

Let $\boldsymbol{b}_j^*$, $\boldsymbol{h}^*$ and $\boldsymbol{s}_j^*$ be the projection of $\boldsymbol{b}_j$, $\boldsymbol{h}$ and $\boldsymbol{s}_j$ to $\langle \boldsymbol{a}_0, \ldots, \boldsymbol{a}_{\ell_2-1} \rangle$ (e.g., $\boldsymbol{b}_j^* \leftarrow \sum_{i=0}^{\ell_2-1} \mathsf{Pr}_i(\boldsymbol{b}_j)$). Obtaining $(x_{j,\kappa}, t_{i,j})$, anyone can compute $(\boldsymbol{s}_{\ell_2}^*, \ldots, \boldsymbol{s}_{\ell_1-1}^*)$ by $\boldsymbol{s}_j^* = \sum_{i=0}^{\ell_2-1} \sum_{\kappa=0}^{\ell_2-1} t_{i,j} x_{j,\kappa} \phi_{\kappa,i} \mathsf{Pr}_i(\boldsymbol{h}^*)$ $(j = \ell_2, \ldots, \ell_1 - 1)$. So, anyone can check the validity of $(\boldsymbol{s}_0, \ldots, \boldsymbol{s}_{\ell_2-1})$ by computing $(\boldsymbol{s}_{\ell_2}^*, \ldots, \boldsymbol{s}_{\ell_1-1}^*)$ and checking $e(\boldsymbol{s}_j^*, \boldsymbol{b}_j^*) = 1$ $(j = 0, \ldots, \ell_1 - 1)$ and $\boldsymbol{h}^* = \sum_{j=0}^{\ell_1-1} \boldsymbol{s}_j^*$.

The convertible undeniable signature scheme is unforgeable under the CVDP assumption in the random oracle model. It is invisible under a variant of the DVDP assumption, in which it is hard to distinguish $(\mathbb{V}, \boldsymbol{b}_0, \ldots, \boldsymbol{b}_{\ell_1-1}, \sum_{i=0}^{\ell_1-1} x_i \boldsymbol{b}_i, (x_0 \boldsymbol{b}_0, \ldots, x_{\ell_2-1} \boldsymbol{b}_{\ell_2-1}))$ from $(\mathbb{V}, \boldsymbol{b}_0, \ldots, \boldsymbol{b}_{\ell_1-1}, \sum_{i=0}^{\ell_1-1} x_i \boldsymbol{b}_i, (y_0 \boldsymbol{b}_0, \ldots, y_{\ell_2-1} \boldsymbol{b}_{\ell_2-1}))$ where $x_i \stackrel{\mathsf{U}}{\leftarrow} \mathbb{F}_r (i = 0, \ldots, \ell_1 - 1)$, and $y_i \stackrel{\mathsf{U}}{\leftarrow} \mathbb{F}_r (i = 0, \ldots, \ell_2 - 1)$. The confirmation protocol is perfectly zero-knowledge and the disavowal protocol is computationally zero-knowledge under the DVDP assumption.

6.4 Combination

The above-mentioned signatures can be combined easily. To combine blind signatures and undeniable signatures, a user who requests a blind signing on message m to a signer obtains signature $(\boldsymbol{s}_0, \ldots, \boldsymbol{s}_{\ell_1-1})$ in a blinded manner. Then, the user can use $(\boldsymbol{s}_0, \ldots, \boldsymbol{s}_{\ell_2-1})$ as an undeniable signature and secretly keep $(\boldsymbol{s}_{\ell_2}, \ldots, \boldsymbol{s}_{\ell_1-1})$ for the confirmation protocol.

The confirmation protocol can be made in the same manner as the correct protocol. Selective conversion is also the same as the protocol. Here note that no secret key of the signer is needed for the user (prover) to execute the confirmation protocol and selective conversion. As for the disavowal and universal conversion, the user cannot make the procedures by himself, instead the signer (of the blind signature) can do the protocols for the user.

Acknowledgments

The authors would like to thank Steven Galbraith for his invaluable comments and suggestions on our preliminary manuscript. We also appreciate anonymous reviewers of Pairing 2008 for their valuable comments.

References

1. Bellare, M., Namprempre, C., Pointcheval, D., Semanko, M.: The one-more-RSA-inversion problems and the security of Chaum's blind signature scheme. Journal of Cryptology 16(3), 185–215 (2003)
2. Bellare, M., Rogaway, P.: The exact security of digital signatures—how to sign with RSA and Rabin. In: Maurer, U.M. (ed.) EUROCRYPT 1996. LNCS, vol. 1070, pp. 399–416. Springer, Heidelberg (1996)
3. Boneh, D., Boyen, X., Shacham, H.: Short group signatures. In: Franklin, M. (ed.) CRYPTO 2004. LNCS, vol. 3152, pp. 41–55. Springer, Heidelberg (2004)

4. Boneh, D., Goh, E.-J., Nissim, K.: Evaluating 2-DNF formulas on ciphertexts. In: Kilian, J. (ed.) TCC 2005. LNCS, vol. 3378, pp. 325–341. Springer, Heidelberg (2005)
5. Coron, J.S.: On the exact security of full domain hash. In: Bellare, M. (ed.) CRYPTO 2000. LNCS, vol. 1880, pp. 229–235. Springer, Heidelberg (2000)
6. Duursma, I., Kiyavash, N.: The vector decomposition problem for elliptic and Hyperelliptic Curves. J. Ramanujan Math. Soc. 20(1), 59–76 (2005)
7. Duursma, I., Park, S.: ElGamal type signature schemes for n-dimensional vector spaces, available at IACR ePrint Archive, 2006/312 (2006)
8. Freeman, D.: Constructing pairing-friendly genus 2 curves with ordinary Jacobians. In: Takagi, T., Okamoto, T., Okamoto, E., Okamoto, T. (eds.) Pairing 2007. LNCS, vol. 4575, pp. 152–176. Springer, Heidelberg (2007)
9. Galbraith, S.D., Hess, F., Vercauteren, F.: Hyperelliptic pairings. In: Takagi, T., Okamoto, T., Okamoto, E., Okamoto, T. (eds.) Pairing 2007. LNCS, vol. 4575, pp. 108–131. Springer, Heidelberg (2007)
10. Galbraith, S.D., Pujolàs, J., Ritzenthaler, C., Smith, B.: Distortion maps for genus two curves, available at arxiv math.NT/0611471 (2006)
11. Galbraith, S.D., Verheul, E.: An analysis of the vector decomposition problem. In: Cramer, R. (ed.) PKC 2008. LNCS, vol. 4939, pp. 308–327. Springer, Heidelberg (2008)
12. Groth, J., Ostrovsky, R., Sahai, A.: Perfect non-interactive zero-knowledge for NP. In: Vaudenay, S. (ed.) EUROCRYPT 2006. LNCS, vol. 4004, pp. 338–359. Springer, Heidelberg (2006)
13. Hofheinz, D., Kiltz, E.: Secure hybrid encryption from weakened key encapsulation. In: Menezes, A. (ed.) CRYPTO 2007. LNCS, vol. 4622, pp. 553–571. Springer, Heidelberg (2007)
14. Huang, X., Mu, Y., Susilo, W., Wu, W.: Provably secure pairing-based convertible undeniable signature with short signature length. In: Takagi, T., Okamoto, T., Okamoto, E., Okamoto, T. (eds.) Pairing 2007. LNCS, vol. 4575, pp. 367–391. Springer, Heidelberg (2007)
15. Okamoto, T., Sakurai, K.: Efficient algorithms for the construction of hyperelliptic cryptosystems. In: Feigenbaum, J. (ed.) CRYPTO 1991. LNCS, vol. 576, pp. 267–278. Springer, Heidelberg (1992)
16. Paillier, P.: A trapdoor permutation equivalent to factoring. In: Imai, H., Zheng, Y. (eds.) PKC 1999. LNCS, vol. 1560, pp. 219–222. Springer, Heidelberg (1999)
17. Shacham, H.: A Cramer-Shoup encryption scheme from the linear assumption and from progressively weaker linear variants, available at IACR ePrint Archive, 2007/074 (2007)
18. Takashima, K.: Efficiently computable distortion maps for supersingular curves. In: van der Poorten, A.J., Stein, A. (eds.) ANTS-VIII 2008. LNCS, vol. 5011, pp. 88–101. Springer, Heidelberg (2008)
19. Williams, H.C.: Some public-key crypto-functions as intractable as factorization. Cryptologia 9, 223–237 (1985)
20. Yoshida, M., Mitsunari, S., Fujiwara, T.: Vector decomposition problem and the trapdoor inseparable multiplex transmission scheme based the problem. In: Proceedings of the 2003 Symposium on Cryptography and Information Security (SCIS), 7B-1 (2003)
21. Yoshida, M.: Inseparable multiplex transmission using the pairing on elliptic curves and its application to watermarking. In: Fifth Conference on Algebraic Geometry, Number Theory, Coding Theory and Cryptography, Univ. of Tokyo (2003)

Hidden-Vector Encryption with Groups of Prime Order

Vincenzo Iovino[1] and Giuseppe Persiano[1]

Dipartimento di Informatica ed Applicazioni,
Università di Salerno, 84084 Fisciano (SA), Italy
{iovino,giuper}@dia.unisa.it

Abstract. Predicate encryption schemes are encryption schemes in which each ciphertext Ct is associated with a binary attribute vector $\boldsymbol{x} = (x_1, \ldots, x_n)$ and keys K are associated with predicates. A key K can decrypt a ciphertext Ct if and only if the attribute vector of the ciphertext satisfies the predicate of the key. Predicate encryption schemes can be used to implement fine-grained access control on encrypted data and to perform search on encrypted data.

Hidden vector encryption schemes [Boneh and Waters – TCC 2007] are encryption schemes in which each ciphertext Ct is associated with a binary vector $\boldsymbol{x} = (x_1, \ldots, x_n)$ and each key K is associated with binary vector $\boldsymbol{y} = (y_1, \cdots, y_n)$ with "don't care" entries (denoted with $\star$). Key K can decrypt ciphertext Ct if and only if $\boldsymbol{x}$ and $\boldsymbol{y}$ agree for all i for which $y_i \neq \star$.

Hidden vector encryption schemes are an important type of predicate encryption schemes as they can be used to construct more sophisticated predicate encryption schemes (supporting for example range and subset queries).

We give a construction for hidden-vector encryption from standard complexity assumptions on bilinear groups of *prime order*. Previous constructions were in bilinear groups of *composite order* and thus resulted in less efficient schemes. Our construction is both payload-hiding and attribute-hiding meaning that also the privacy of the attribute vector, besides privacy of the cleartext, is guaranteed.

1 Introduction

Traditional public key encryption schemes are well tailored for point-to-point security in which a sender wishes to send private messages to the owner of the public key. Recently, there has been a trend for private user data to be stored over the Internet by a third party server. It is then expected that user will encrypt the data so to preserve the privacy of the data itself. If a traditional encryption scheme is employed then user will not be able to search its data. Indeed, the user has to download and the decrypt its data and then perform the search; which can be very inconvenient.

This problem has been first studied by Boneh et al. [BDOP04] that introduced the concept of an encryption scheme supporting test equality. Roughly speaking,

S.D. Galbraith and K.G. Paterson (Eds.): Pairing 2008, LNCS 5209, pp. 75–88, 2008.

in such an encryption scheme, the owner of the public key can compute, for any message M, a trapdoor information K_M that allows the server that physically holds the data to check whether a given ciphertext encrypts message M without obtaining any additional information. Boneh et al. [BDOP04] suggested to use this system for storing encrypted e-mail messages on a server so that the user could decide to download only the e-mail messages with a given subject without having to compromise his privacy (and without having to download and decrypt all the messages).

Recently along this line of research, Goyal et al. [GPSW06] have introduced the concept of an attribute-based encryption scheme (ABE scheme). In an ABE scheme, a cyphertext is labeled with a set of attributes and private keys are associated with a predicate. A private key can decrypt a ciphertext iff the attributes of the ciphertext satisfy the predicate associated with the key. An ABE schem can thus been seen as a special encryption scheme for which, given the key associated with a predicate P, one can test whether a given ciphertext Ct carries a message M that satisfies predicates P without having to decrypt and without getting any additiocal information. The construction of [GPSW06] is very general as it supports any predicate that can be expressed as a circuit with threshold gates. On the other hand the construction only achieved what is called *payload security* which consists in guaranteeing the security of the cleartext. Indeed, in the construction of [GPSW06], the attribute vector associated with a ciphertext appears in clear in the ciphertext.

In several applications instead one would like to be able to encrypt a cleartext and label the ciphertext with attributes so that both the cleartext and the attributes are secure. This extra property is called *attribute hiding*. Indeed, it is an important research problem to design encryption schemes for large predicate classes that enjoy both the payload and the attribute hiding property. In [BW07], Boneh and Waters give construction for encryption schemes for several families of predicates including conjuctions, and subset and range predicates. This has been recently extended to disjunctions, polynomial equations and inner products [KSW08]. Both constructions are based on hardness assumptions regarding bilinear groups on *composite order*. More efficient schemes for range queries over encrypted data have been presented in [SBC+07].

Our results. In this paper we give a construction for *hidden vector encryption* schemes (HVE, in short). Roughly speaking, in a hidden vector encryption scheme ciphertexts are associated with binary vectors and private keys are associated with with binary vectors with "don't care" entries (denoted by $\star$). A private key can decipher a ciphertext if all entries of the key vector that are not $\star$ agree with the corresponding entries of the ciphertext vector (see Definition 1). The first construction for HVE has been given by [BW07] which also showed that HVE gives efficient encryption schemes supporting conjunctions of equality queries, range queries and subset queries. By applying the reductions of [BW07] to our construction we obtain encryption schemes supporting the same classe of predicates as [BW07].

Both the payload and the attribute security of our construction rely on standard computational assumptions on bilinear groups of *prime* order; namely, the Bilinear Decision Diffie-Hellman assumption and the Decision Linear assumption (used also in [BW06, GPSW06]). As already noted above, the security of the construction of [BW07] instead relies on the Composite Bilinear Decision Diffie-Hellman assumption and the Composite 3-Party Diffie-Hellman assumption. Both assumptions imply that the order of the group is difficult to factor and this results in larger group elements and thus more expensive operations.

2 The Symmetric Bilinear Setting

We have multiplicative groups $\mathbb{G}$ and $\mathbb{G}_T$ of prime order p and a non-degenerate bilinear pairing function $\mathbf{e} : \mathbb{G} \times \mathbb{G} \rightarrow \mathbb{G}_T$. That is, for all $g \in \mathbb{G}, g \neq 1$, we have $\mathbf{e}(g, g) \neq 1$ and $\mathbf{e}(g^a, g^b) = \mathbf{e}(g, g)^{ab}$. We denote by g and $\mathbf{e}(g, g)$ the generators of $\mathbb{G}$ and $\mathbb{G}_T$. We call a *symmetric bilinear* instance a tuple $\mathcal{I} = [p, \mathbb{G}, \mathbb{G}_T, g, \mathbf{e}]$ and assume that there exists an efficient generation procedure that, on input security parameter 1^k, outputs an instance with $|p| = \Theta(k)$.

In our constructions we make the following hardness assumptions.

Decision BDH. Given a tuple $[g, g^{z_1}, g^{z_2}, g^{z_3}, Z]$ for random exponents $z_1, z_2, z_3 \in \mathbb{Z}_p$ it is hard to distinguish between $Z = \mathbf{e}(g, g)^{z_1 z_2 z_3}$ and a random Z from $\mathbb{G}_T$. More specifically, for an algorithm $\mathcal{A}$ we define experiment $\mathsf{DBDHExp}_{\mathcal{A}}$ as follows.

$\mathsf{DBDHExp}^{\mathcal{A}}(1^k)$
Choose instance $\mathcal{I} = [p, \mathbb{G}, \mathbb{G}_T, g, \mathbf{e}]$ with security parameter 1^k;
Choose $a, b, c \in \mathbb{Z}_p$ at random;
Choose $\eta \in \{0, 1\}$ at random;
if $\eta = 1$ **then** choose $z \in \mathbb{Z}_p$ at random
  else set $z = abc$;
set $A = g^a, B = g^b, C = g^c$ and $Z = \mathbf{e}(g, g)^z$;
let $\eta' = \mathcal{A}(\mathcal{I}, A, B, C, Z)$;
if $\eta = \eta'$ **then** return 0 **else** return 1;

Assumption 1 (Decision Bilinear Diffie-Hellman). *For all probabilistic polynomial-time algorithms* $\mathcal{A}$,

$$\left|\text{Prob}[\mathsf{DBDHExp}^{\mathcal{A}}(1^k) = 1] - 1/2\right| = \nu(k)$$

for some negligible function ν.

Decision Linear. Given a tuple $[g, g^{z_1}, g^{z_2}, g^{z_1 z_3}, g^u, Z]$ for random exponents $z_1, z_2, z_3, u \in \mathbb{Z}_p$ it is hard to distinguish between $Z = g^{z_2(u-z_3)}$ and a random Z from $\mathbb{G}$. More specifically, for an algorithm $\mathcal{A}$ we define experiment $\mathsf{DLExp}_{\mathcal{A}}$ as follows.

$\mathsf{DLExp}^{\mathcal{A}}(1^k)$
Choose instance $\mathcal{I} = [p, \mathbb{G}, \mathbb{G}_T, g, \mathbf{e}]$ with security parameter 1^k;
Choose $z_1, z_2, z_3, u \in \mathbb{Z}_p$ at random;
Choose $\eta \in \{0, 1\}$ at random;
if $\eta = 1$ **then** choose $z \in \mathbb{Z}_p$ at random
    else set $z = z_2(u - z_3)$;
set $Z_1 = g^{z_1}, Z_2 = g^{z_2}, Z_{13} = g^{z_1 z_3}, U = g^u$, and $Z = g^z$;
let $\eta' = \mathcal{A}(\mathcal{I}, Z_1, Z_2, Z_{13}, U, Z)$;
if $\eta = \eta'$ **then** return 0 **else** return 1;

Assumption 2 (Decision Linear). *For all probabilistic polynomial-time algorithms $\mathcal{A}$,*

$$\left|\mathrm{Prob}[\mathsf{DLExp}^{\mathcal{A}}(1^k) = 1] - 1/2\right| = \nu(k)$$

for some negligible function ν.

Note that Decision Linear implies Decision Bilinear Diffie-Hellman and the Decision Linear assumption has been introduced in [BBS04] and used also in [BW06].

3 HVE Schemes

Let $\boldsymbol{x}$ be a string over the alphabet $\{0, 1\}$ and $\boldsymbol{y}$ be a string over the alphabet $\{0, 1, \star\}$. Assume $\boldsymbol{x}$ and $\boldsymbol{y}$ have the same length n and define predicate $P_{\boldsymbol{x}}(\boldsymbol{y})$ to be true if and only if for each $1 \le i \le n$ we have $x_i = y_i$ or $y_i = \star$. In other words, for $P_{\boldsymbol{x}}(\boldsymbol{y})$ to be true, the two strings must match in positions i where $y_i \neq \star$ and, intuitively, $\star$ is the "don't care" symbol.

Definition 1 (HVE). *A* Hidden Vector Encryption Scheme *(a* HVE scheme*) is a quadruple of probabilistic polynomial-time algorithms* (Setup, Enc, KeyGeneration, Dec) *such that:*

1. Setup *takes as input the security parameter 1^k and the* attribute length $n = \mathsf{poly}(k)$ *and outputs the* master public key Pk *and the* master secret key Msk.
2. KeyGeneration *takes as input the master secret key* Msk *and string $\boldsymbol{y} \in \{0, 1, \star\}^n$ and outputs the decryption key $K_{\boldsymbol{y}}$ associated with $\boldsymbol{y}$.*
3. Enc *takes as input the public key* Pk, attribute string $\boldsymbol{x} \in \{0, 1\}^n$ *and message M from the associated message space and returns ciphertext $\mathsf{Ct}_{\boldsymbol{x}}$.*
4. Dec *takes as input a secret key $K_{\boldsymbol{y}}$ and a ciphertext $\mathsf{Ct}_{\boldsymbol{x}}$ and outputs a message M.*

We require that for all k and $n = \mathsf{poly}(k)$, and for all strings $\boldsymbol{x} \in \{0, 1\}^n$ and $\boldsymbol{y} \in \{0, 1, \star\}^n$ such that $P_{\boldsymbol{x}}(\boldsymbol{y}) = 1$, it holds that:

$$\mathrm{Prob}[(\mathsf{Pk}, \mathsf{Msk}) \leftarrow \mathsf{Setup}(1^k, n); \quad K_{\boldsymbol{y}} \leftarrow \mathsf{KeyGeneration}(\mathsf{Msk}, \boldsymbol{y}); \quad \mathsf{Ct}_{\boldsymbol{x}} \leftarrow \mathsf{Enc}(\mathsf{Pk}, \boldsymbol{x}, M) : \mathsf{Dec}(K_{\boldsymbol{y}}, \mathsf{Ct}) = M] = 1.$$

We define two notions of security for our HVE scheme: semantic security that captures the payload-hiding property and the attribute hiding property that guarantees security of the attribute string. Both notions are in the selective models in which the adversary committs to the attribute vector at the beginning of the game. We note that this is same notion of security used in [BW07, KSW08].

Definition 2 (Semantic Security). *An HVE scheme* (Setup, Enc, KeyGeneration, Dec) *is* semantically secure *if for all PPT adversaries $\mathcal{A}$,*

$$\left|\text{Prob}[\mathsf{SemanticExp}_{\mathcal{A}}(1^k) = 1] - 1/2\right| = \nu(k)$$

for some negligible function ν, where $\mathsf{SemanticExp}_{\mathcal{A}}(1^k)$ *is the following experiment.*

Init. *The adversary $\mathcal{A}$ announces the vector $\boldsymbol{x}$ it wishes be challenged upon.*
Setup. *The public and the secret key* (Msk, Pk) *are generated using the* Setup *procedure and $\mathcal{A}$ receives* Pk.
Query Phase I. *$\mathcal{A}$ requests and gets private keys $K_{\boldsymbol{y}}$ relative to vectors $\boldsymbol{y}$ such that $P_{\boldsymbol{x}}(\boldsymbol{y}) = 0$. Key $K_{\boldsymbol{y}}$ is computed using the* KeyGeneration *procedure.*
Challenge. *$\mathcal{A}$ returns two different messages M_0, M_1 of the same length in the message space. η is chosen at random from $\{0,1\}$. $\mathcal{A}$ is given ciphertext* $\mathsf{Ct}_{\boldsymbol{x}} \leftarrow \mathsf{Enc}(\mathsf{Pk}, \boldsymbol{x}, M_\eta)$.
Query Phase II. *Identical to Query Phase I.*
Output. *$\mathcal{A}$ returns η'. If $\eta = \eta'$ then return 1 else return 0.*

We are now ready to define the notion of attribute hiding.

Definition 3. *An HVE scheme* (Setup, Enc, KeyGeneration, Dec) *is* attribute hiding *if for all PPT adversaries $\mathcal{A}$,*

$$\left|\text{Prob}[\mathsf{AttributeHidingExp}_{\mathcal{A}}(1^k) = 1] - 1/2\right| = \nu(k)$$

for some negligible function ν, where $\mathsf{AttributeHidingExp}_{\mathcal{A}}(1^k)$ *is the following experiment.*

Init. *The adversary $\mathcal{A}$ announces two attribute strings $\boldsymbol{x}_0 \neq \boldsymbol{x}_1$ it wishes be challenged upon.*
Setup. *The public and the secret key* (Msk, Pk) *are generated using the* Setup *procedure and $\mathcal{A}$ receives* Pk.
Query Phase I. *$\mathcal{A}$ requests and gets private keys $K_{\boldsymbol{y}}$ relative to vectors $\boldsymbol{y}$ such that $P_{\boldsymbol{x}_1}(\boldsymbol{y}) = P_{\boldsymbol{x}_2}(\boldsymbol{y}) = 0$. Key $K_{\boldsymbol{y}}$ is computed using the* KeyGeneration *procedure.*
Challenge. *$\mathcal{A}$ returns two different messages M_0, M_1 of the same length. η is chosen at random from $\{0,1\}$. $\mathcal{A}$ is given ciphertext* $\mathsf{Ct}_{\boldsymbol{x}} \leftarrow \mathsf{Enc}(\mathsf{Pk}, \boldsymbol{x}_\eta, M_\eta)$.
Query Phase II. *Identical to Query Phase I.*
Output. *$\mathcal{A}$ returns η'. If $\eta = \eta'$ then return 1 else return 0.*

If in the previous experiment we let $\boldsymbol{x}_0 = \boldsymbol{x}_1$ we have the definition of Semantic Security.

4 Our Construction

In this section we describe our construction for an HVE scheme.

Setup. Procedure Setup, on input security parameter 1^k and attribute length $n = \mathsf{poly}(k)$, computes the public key Pk and the master secret key Msk in the following way.

Choose a random instance $\mathcal{I} = [p, \mathbb{G}, \mathbb{G}_T, g, \mathbf{e}]$.

Choose y at random in $\mathbb{Z}_p$ and set $Y = \mathbf{e}(g, g)^y$.

For $1 \leq i \leq n$, choose t_i, v_i, r_i, m_i at random in $\mathbb{Z}_p$ and set $T_i = g^{t_i}, V_i = g^{v_i}$ and $R_i = g^{r_i}, M_i = g^{m_i}$.

Then, $\mathsf{Setup}(1^k, n)$ returns $[\mathsf{Pk}, \mathsf{Msk}]$ where

$$\mathsf{Pk} = [\mathcal{I}, Y, (T_i, V_i, R_i, M_i)_{i=1}^n] \text{ and } \mathsf{Msk} = [y, (t_i, v_i, r_i, m_i)_{i=1}^n].$$

Encryption. Procedure Enc takes as input cleartext $M \in \mathbb{G}_T$, attribute string $\boldsymbol{x}$ and public key Pk and computes ciphertext as follows.

Choose s at random in $\mathbb{Z}_p$, and, for $1 \leq i \leq n$, choose s_i at random in $\mathbb{Z}_p$ and compute ciphertext

$$\mathsf{Enc}(\mathsf{Pk}, \boldsymbol{x}, M) = [\Omega, C_0, (X_i, W_i)_{i=1}^n],$$

where $\Omega = M \cdot Y^{-s}$, $C_0 = g^s$ and

$$X_i = \begin{cases} T_i^{s-s_i}, & \text{if } x_i = 1; \\ R_i^{s-s_i}, & \text{if } x_i = 0. \end{cases} \quad \text{and } W_i = \begin{cases} V_i^{s_i}, & \text{if } x_i = 1; \\ M_i^{s_i}, & \text{if } x_i = 0. \end{cases}$$

Key Generation. Procedure $\mathsf{KeyGeneration}$ on input Msk and $\boldsymbol{y} \in \{0, 1, \star\}^n$ derives private key $K_{\boldsymbol{y}}$ relative to attribute string $\boldsymbol{y}$ in the following way.

If $\boldsymbol{y} = (\star, \star, \ldots, \star)$ then $K_{\boldsymbol{y}} = g^y$. Else, denote by $S_{\boldsymbol{y}}^1$ and $S_{\boldsymbol{y}}^0$ the set of indices i for which $y_i = 1$ and $y_i = 0$, respectively and let $S_{\boldsymbol{y}} = S_{\boldsymbol{y}}^1 \cup S_{\boldsymbol{y}}^0$ be the set of indices for $y_i \neq \star$. Then, for $i \in S_{\boldsymbol{y}}$, choose a_i at random in $\mathbb{Z}_p$ under the constraint that $\sum_{i \in S_{\boldsymbol{y}}} a_i = y$ and let $K_{\boldsymbol{y}} = (Y_i, L_i)_{i=1}^n$, where

$$Y_i = \begin{cases} g^{\frac{a_i}{t_i}}, & \text{if } y_i = 1; \\ g^{\frac{a_i}{r_i}}, & \text{if } y_i = 0; \\ \emptyset, & \text{if } y_i = \star. \end{cases} \quad \text{and } L_i = \begin{cases} g^{\frac{a_i}{v_i}}, & \text{if } y_i = 1; \\ g^{\frac{a_i}{m_i}}, & \text{if } y_i = 0; \\ \emptyset, & \text{if } y_i = \star. \end{cases}$$

Decryption. Procedure Dec decrypts cyphertext $\mathsf{Ct}_{\boldsymbol{x}}$ using secret key $K_{\boldsymbol{y}}$ such that $P_{\boldsymbol{x}}(\boldsymbol{y}) = 1$.

$$\mathsf{Dec}(\mathsf{Pk}, \mathsf{Ct}_{\boldsymbol{x}}, K_{\boldsymbol{y}}) = \Omega \cdot \prod_{i \in S_{\boldsymbol{y}}} \mathbf{e}(X_i, Y_i) \mathbf{e}(W_i, L_i)$$

where $S_{\boldsymbol{y}}$ is the set of indices i such that $y_i \neq \star$. If $S_{\boldsymbol{y}} = \emptyset$ then $K_{\boldsymbol{y}} = g^y$ and

$$\mathsf{Dec}(\mathsf{Pk}, \mathsf{Ct}_{\boldsymbol{x}}, K_{\boldsymbol{y}}) = \Omega \cdot \mathbf{e}(C_0, K_{\boldsymbol{y}}).$$

This ends the description of our construction. We remark that our construction can be extended to attribute vectors taken from a larger alphabet Σ (and not simply $\{0,1\}$) without increasing the length of the ciphertexts and of the secret keys but only the length of the public key Pk. We omit further details.

We next prove that the quadruple is indeed an HVE.

Theorem 1. *The quadruple of algorithms* (Setup, Enc, KeyGeneration, Dec) *specified above is an HVE.*

Proof. It is sufficient to verify that this procedure computes M correctly when $P_{\boldsymbol{x}}(\boldsymbol{y}) = 1$. The case in which $\boldsymbol{y} = (\star, \star, \cdots, \star)$ is obvious.

We remind the reader that $S^1_{\boldsymbol{y}}$ (respectively, $S^0_{\boldsymbol{y}}$) denotes the (possibly empty) set of indices i such that $y_i = 1$ (respectively, $y_i = 0$) and that $S_{\boldsymbol{y}} = S^1_{\boldsymbol{y}} \cup S^0_{\boldsymbol{y}}$.

Then we have

$$
\begin{aligned}
\mathsf{Dec}(\mathsf{Pk}, \mathsf{Ct}_{\boldsymbol{x}}, K_{\boldsymbol{y}}) &= \Omega \prod_{i \in S_{\boldsymbol{y}}} \mathbf{e}(X_i, Y_i)\mathbf{e}(W_i, L_i) \\
&= M\mathbf{e}(g,g)^{-ys} \cdot \prod_{i \in S^1_{\boldsymbol{y}}} \mathbf{e}(g^{t_i(s-s_i)}, g^{\frac{a_i}{t_i}})\mathbf{e}(g^{w_i s_i}, g^{\frac{a_i}{w_i}}) \\
&\quad \cdot \prod_{i \in S^0_{\boldsymbol{y}}} \mathbf{e}(g^{r_i(s-s_i)}, g^{\frac{a_i}{r_i}})\mathbf{e}(g^{m_i s_i}, g^{\frac{a_i}{m_i}}) \\
&= M\mathbf{e}(g,g)^{-ys} \prod_{i \in S^1_{\boldsymbol{y}}} \mathbf{e}(g,g)^{(s-s_i)a_i}\mathbf{e}(g,g)^{s_i a_i} \prod_{i \in S^0_{\boldsymbol{y}}} \mathbf{e}(g,g)^{(s-s_i)a_i}\mathbf{e}(g,g)^{s_i a_i} \\
&= M\mathbf{e}(g,g)^{-ys} \prod_{i \in S_{\boldsymbol{y}}} \mathbf{e}(g,g)^{(s-s_i)a_i}\mathbf{e}(g,g)^{s_i a_i} \\
&= M\mathbf{e}(g,g)^{-ys} \prod_{i \in S_{\boldsymbol{y}}} \mathbf{e}(g,g)^{s a_i} \\
&= M\mathbf{e}(g,g)^{-ys}\mathbf{e}(g,g)^{ys} = M.
\end{aligned}
$$

Efficiency. In our construction we have that, for an attribute string of length n, the ciphertext contains 1 element from $\mathbb{G}_T$ and $O(n)$ elements from $\mathbb{G}$. The secret key corresponding to vector $\boldsymbol{y}$ instead contains $O(\mathsf{weight}(\boldsymbol{y}))$ elements from $\mathbb{G}$, where $\mathsf{weight}(\boldsymbol{y})$ is the number of entries of $\boldsymbol{y}$ that are either 0 or 1. Thus our scheme has the same ciphertext and key-length as the constructions presented in [KSW08, BW07].

5 Proofs

In this section we prove that our construction is semantically secure and attribute hiding.

Theorem 2 (Semantic Security). *Assume BDDH holds. Then HVE scheme* (Setup, Enc, KeyGeneration, Dec) *described above is semantically secure.*

Proof. Suppose that there exists PPT adversary $\mathcal{A}$ which has success in experiment SemanticExp with probability non-negligibly larger than 1/2. We then construct an adversary $\mathcal{B}$ for the experiment DBDHExp that uses $\mathcal{A}$ as subroutine.

Input. $\mathcal{B}$ receives in input $[\mathcal{I}, A = g^a, B = g^b, C = g^c, Z]$, where Z is $\mathbf{e}(g,g)^{abc}$ or a random element of $\mathbb{G}_T$.

Init. $\mathcal{B}$ runs $\mathcal{A}$ and receives the attribute string $\boldsymbol{x}$ it wishes to be challenged upon.

Setup. Set $Y = \mathbf{e}(A, B)$. For every $1 \le i \le n$, $\mathcal{B}$ chooses $t'_i, v'_i, r'_i, m'_i \in \mathbb{Z}_p$ at random and set

$$T_i = \begin{cases} g^{t'_i}, & \text{if } x_i = 1; \\ B^{t'_i}, & \text{if } x_i = 0; \end{cases} \quad \text{and} \quad V_i = \begin{cases} g^{v'_i}, & \text{if } x_i = 1; \\ B^{v'_i}, & \text{if } x_i = 0; \end{cases}$$

$$R_i = \begin{cases} B^{r'_i}, & \text{if } x_i = 1; \\ g^{r'_i}, & \text{if } x_i = 0; \end{cases} \quad \text{and} \quad M_i = \begin{cases} B^{m'_i}, & \text{if } x_i = 1; \\ g^{m'_i}, & \text{if } x_i = 0; \end{cases}$$

$\mathcal{B}$ runs $\mathcal{A}$ on input $\mathsf{Pk} = [\mathcal{I}, Y, (T_i, V_i, R_i, M_i)_{i=1}^n]$.

Notice that Pk has the same distribution of a public key received by $\mathcal{A}$ in the Setup phase of SemanticExp with $y = a \cdot b$, and with $t_i = t'_i$, $v_i = v'_i$, $r_i = br'_i$, and $m_i = bm'_i$ for i with $x_i = 1$, and $t_i = bt'_i$, $v_i = bv'_i$, $r_i = r'_i$, and $m_i = m'_i$ for i with $x_i = 0$.

Query Phase I. $\mathcal{B}$ answers $\mathcal{A}$'s queries for $\boldsymbol{y}$ such that $P_{\boldsymbol{x}}(\boldsymbol{y}) = 0$ as follows. Let j be an index where $x_j \neq y_j$ and $y_j \neq \star$ (such an index always exists). For every $i \neq j$ such that $y_i \neq \star$, choose a'_i uniformly at random in $\mathbb{Z}_p$ and let $a' = \sum a'_i$.

Set Y_j and L_j as

$$Y_j = \begin{cases} A^{1/t'_j} g^{-a'/t'_j}, & \text{if } y_j = 1; \\ A^{1/r'_j} g^{-a'/r'_j}, & \text{if } y_j = 0. \end{cases} \quad \text{and} \quad L_j = \begin{cases} A^{1/v'_j} g^{-a'/v'_j}, & \text{if } y_j = 1; \\ A^{1/m'_j} g^{-a'/m'_j}, & \text{if } y_j = 0. \end{cases}$$

and, for $i \neq j$, set Y_i, L_i as follows

$$Y_i = \begin{cases} B^{a'_i/t'_i}; & \text{if } x_i = y_i = 1; \\ B^{a'_i/r'_i}; & \text{if } x_i = y_i = 0; \\ g^{a'_i/r'_i}; & \text{if } x_i = 1 \text{ and } y_i = 0; \\ g^{a'_i/t'_i}; & \text{if } x_i = 0 \text{ and } y_i = 1; \\ \emptyset; & \text{if } y_i = \star. \end{cases} \quad \text{and} \quad L_i = \begin{cases} B^{a'_i/v'_i}; & \text{if } x_i = y_i = 1; \\ B^{a'_i/m'_i}; & \text{if } x_i = y_i = 0; \\ g^{a'_i/m'_i}; & \text{if } x_i = 1 \text{ and } y_i = 0; \\ g^{a'_i/v'_i}; & \text{if } x_i = 0 \text{ and } y_i = 1; \\ \emptyset; & \text{if } y_i = \star. \end{cases}$$

Notice that $K_{\boldsymbol{y}}$ has the same distribution of the key returned by the KeyGeneration procedure. In fact, for $i \neq j$, set $a_i = ba'_i$ and set $a_j = b(a - a')$. Then we have that $\sum_{i \in S_{\boldsymbol{y}}} a_i = y$. Moreover, if $y_i = 1$ then $Y_i = g^{\frac{a_i}{t_i}}$ and $L_i = g^{\frac{a_i}{v_i}}$ and, if $y_i = 0$ then $Y_i = g^{\frac{a_i}{r_i}}$ and $L_i = g^{\frac{a_i}{m_i}}$.

Challenge. $\mathcal{A}$ returns two messages $M_0, M_1 \in \mathbb{G}_T$.

$\mathcal{B}$ chooses uniformly at random $\eta \in \{0,1\}$ and $s_i \in \mathbb{Z}_p$, for $i = 1, \cdots, n$. Then $\mathcal{B}$ constructs $\mathsf{Ct}_{\boldsymbol{x}} = (\Omega, C, (X_i, W_i)_{i=1}^n)$, where $\Omega = M_\eta Z^{-1}$, $C_0 = C$ and

$$X_i = \begin{cases} C^{t'_i} g^{-t'_i s_i}; & \text{if } x_i = 1; \\ C^{r'_i} g^{-r'_i s_i}; & \text{if } x_i = 0. \end{cases} \quad \text{and } W_i = \begin{cases} g^{-v'_i s_i}; & \text{if } x_i = 1; \\ g^{-m'_i s_i}; & \text{if } x_i = 0. \end{cases}$$

Observe that if $Z = \mathbf{e}(g,g)^{abc}$ then $\mathsf{Ct}_{\boldsymbol{x}}$ is an encryption of M_η with $s = c$. If instead Z is random in $\mathbb{G}_T$ then $\mathsf{Ct}_{\boldsymbol{x}}$ is independent from η.

Query Phase II. Identical to Query Phase I.

Output. $\mathcal{A}$ outputs η'. $\mathcal{B}$ returns 0 iff $\eta' = \eta$.

To conclude the proof observe that, if $Z = \mathbf{e}(g,g)^{abc}$ then, since $\mathcal{A}$ is a successful adversary for semantic security, the probability that $\mathcal{B}$ returns 0 is at least $1/2 + 1/\mathsf{poly}(k)$. On the other hand if Z is random in $\mathbb{G}_T$ the probability that $\mathcal{B}$ returns 0 is at most $1/2$. This contradicts the BDDH assumption.

We now turn our attention at the attribute hiding property. We stress that a crucial tool in achieving this property is the "linear splitting" technique first used to construct anonymous hierarchical identity-based encryption in [BW06]. As an effect of employing this technique our ciphertexts and keys roughly double in sizes. If one does not require attribute hiding then our scheme can be modified so that, for attribute vectors of length n, the ciphertext has $n+2$ elements and keys at most n elements.

To prove that the HVE scheme presented is attribute hiding we show that for any attribute string $\boldsymbol{x}$ and for any message M, an encryption of M with respect to attribute string $\boldsymbol{x}$ is computationally indistinguishable from the uniform distribution on $\mathbb{G}_T \times \mathbb{G}^{2n+1}$ to an adversary that has access to the key generation procedure for $\boldsymbol{y}$ such that $P_{\boldsymbol{x}}(\boldsymbol{y}) = 0$.

Specifically, for $j = 0, 1, \ldots, n$, we denote by $\mathsf{Dist}_j(\boldsymbol{x}, M)$ the following distribution.

$\mathsf{Dist}_j(\boldsymbol{x}, M)$

1. choose $\mathcal{I} = [p, \mathbb{G}, \mathbb{G}_T, g, \mathbf{e}]$ with security parameter 1^k;
2. compute $[\mathsf{Msk}, \mathsf{Pk}]$ by executing $\mathsf{Setup}(1^k, n)$;
3. choose R_0 uniformly at random from $\mathbb{G}_T$ and s uniformly at random from $\mathbb{Z}_p$; set $C_0 = g^s$;
4. for $i = 1, \cdots, j$ choose X_i, W_i uniformly at random from $\mathbb{G}$;
5. for $i = j+1, \cdots, n$
   choose s_i uniformly at random $\mathbb{Z}_p$ and set
   $$X_i = \begin{cases} T_i^{s-s_i}, & \text{if } x_i = 1; \\ R_i^{s-s_i}, & \text{if } x_i = 0. \end{cases} \quad \text{and } W_i = \begin{cases} V_i^{s_i}, & \text{if } x_i = 1; \\ M_i^{s_i}, & \text{if } x_i = 0. \end{cases}$$
6. **return:** $(R_0, C_0, (X_i, W_i)_{i=1}^n)$;

From the proof of semantic security it follows, that under the BDDH, distribution $\mathsf{Dist}_0(\boldsymbol{x}, M)$ is indistinguishable from the distribution of the legal ciphertexts $\mathsf{Enc}(\mathsf{Pk}, \boldsymbol{x}, M)$ of M with attribute string $\boldsymbol{x}$. Moreover, for all j, $\mathsf{Dist}_j(\boldsymbol{x}, M)$ is independent from M and $\mathsf{Dist}_n(\boldsymbol{x}, M)$ is the uniform distribution on $\mathbb{G}_T \times \mathbb{G}^{2n+1}$

and thus is independent from $\boldsymbol{x}$. Next lemma shows that distributions $\mathsf{Dist}_{\ell-1}$ and Dist_{ℓ} are computational indistinguishable even to an adversary that has access to the key generation oracle. This concludes the proof of the attribute hiding property.

Lemma 1. *Under the DL assumption, for $\ell = 1, 2, \ldots, n$ and for any $\boldsymbol{x} \in \{0,1\}^n$, we have that distributions $\mathsf{Dist}_{\ell-1}(\boldsymbol{x})$ and $\mathsf{Dist}_{\ell}(\boldsymbol{x})$ are computationally indistinguishable to an adversary that has access to the key generation oracle.*

Proof. Suppose that there exists PPT adversary $\mathcal{A}$ which distinguishes $\mathsf{Dist}_{\ell-1}$ from Dist_{ℓ}. We then construct an adversary $\mathcal{B}$ for the experiment DLExp.

Input. $\mathcal{B}$ takes in input $[\mathcal{I}, Z_1 = g^{z_1}, Z_2 = g^{z_2}, Z_{13} = g^{z_1 z_3}, U = g^u, Z]$, where either $Z = g^{z_2(u-z_3)}$ or Z is a random element of $\mathbb{G}$.

Init. $\mathcal{B}$ receives from $\mathcal{A}$ the attribute string $\boldsymbol{x}$ it wishes to be challenged upon.

Setup. $\mathcal{B}$ sets $Y = \mathbf{e}(Z_1, Z_2)$ and, for $i = 1 \cdots n$, $\mathcal{B}$ chooses t'_i, v'_i, r'_i, m'_i uniformly at random from $\mathbb{Z}_p$ and sets

$$T_\ell = \begin{cases} Z_2^{t'_\ell}, & \text{if } x_\ell = 1; \\ Z_1^{t'_\ell}, & \text{if } x_\ell = 0; \end{cases} \quad \text{and} \quad V_\ell = \begin{cases} Z_1^{v'_\ell}, & \text{if } x_\ell = 1; \\ Z_1^{v'_\ell}, & \text{if } x_\ell = 0; \end{cases}$$

$$R_\ell = \begin{cases} Z_1^{r'_\ell}, & \text{if } x_\ell = 1; \\ Z_2^{r'_\ell}, & \text{if } x_\ell = 0; \end{cases} \quad \text{and} \quad M_\ell = \begin{cases} Z_1^{m'_\ell}, & \text{if } x_\ell = 1; \\ Z_1^{m'_\ell}, & \text{if } x_\ell = 0; \end{cases}$$

Moreover, for $i \neq \ell$, $\mathcal{B}$ sets

$$T_i = \begin{cases} g^{t'_i}, & \text{if } x_i = 1; \\ Z_1^{t'_i}, & \text{if } x_i = 0; \end{cases} \quad \text{and} \quad V_i = \begin{cases} g^{v'_i}, & \text{if } x_i = 1; \\ Z_1^{v'_i}, & \text{if } x_i = 0; \end{cases}$$

$$R_i = \begin{cases} Z_1^{r'_i}, & \text{if } x_i = 1; \\ g^{r'_i}, & \text{if } x_i = 0; \end{cases} \quad \text{and} \quad M_i = \begin{cases} Z_1^{m'_i}, & \text{if } x_i = 1; \\ g^{m'_i}, & \text{if } x_i = 0; \end{cases}$$

$\mathcal{B}$ runs $\mathcal{A}$ on input $\mathsf{Pk} = [\mathcal{I}, Y, (T_i, V_i, R_i, M_i)_{i=1}^n]$.

Notice that Pk has the same distribution of a public key computed using $\mathsf{KeyGeneration}$, with $y = z_1 \cdot z_2$, and $t_i = t'_i, v_i = v'_i, r_i = z_1 r'_i, m_i = z_1 m'_i$ for $i \neq \ell$ with $x_i = 1$, and $t_i = z_1 t'_i, v_i = z_1 v'_i, r_i = r'_i, m_i = m'_i$ for $i \neq \ell$ with $x_i = 0$; moreover, if $x_\ell = 1$ then we have $t_\ell = z_2 t'_\ell, v_\ell = z_1 v'_\ell, r_\ell = z_1 r'_\ell, m_\ell = z_1 m'_\ell$ whereas, if $x_\ell = 0$, we have $t_\ell = z_1 t'_\ell, v_\ell = z_1 v'_\ell, r_\ell = z_2 r'_\ell, m_\ell = z_1 m'_\ell$.

Query Phase I. $\mathcal{B}$ answers $\mathcal{A}$'s queries for $\boldsymbol{y}$ such that $P_{\boldsymbol{x}}(\boldsymbol{y}) = 0$ in the following way. We distinguish two cases.

Case 1: $x_\ell = y_\ell$ or $y_\ell = \star$. In this case there exists index $j \neq \ell$ such that $x_j \neq y_j$ and $y_j \neq \star$.

Then, for $i \neq j$ B chooses a'_i uniformly at random in $\mathbb{Z}_p$ and let us denote by a' the sum $a' = \sum_{i \neq j, \ell} a'_i$.

For $i \neq j$ and $i \neq \ell$, $\mathcal{B}$ sets

$$Y_i = \begin{cases} Z_1^{a'_i/t'_i}, & \text{if } x_i = y_i = 1; \\ Z_1^{a'_i/r'_i}, & \text{if } x_i = y_i = 0; \\ g^{a'_i/r'_i}, & \text{if } x_i = 1, y_i = 0; \\ g^{a'_i/t'_i}, & \text{if } x_i = 0, y_i = 1; \\ \emptyset, & \text{if } y_i = \star. \end{cases} \quad \text{and} \quad L_i = \begin{cases} Z_1^{a'_i/v'_i}, & \text{if } x_i = y_i = 1; \\ Z_1^{a'_i/m'_i}, & \text{if } x_i = y_i = 0; \\ g^{a'_i/m'_i}, & \text{if } x_i = 1, y_i = 0; \\ g^{a'_i/v'_i}, & \text{if } x_i = 0, y_i = 1; \\ \emptyset, & \text{if } y_i = \star. \end{cases}$$

Moreover, $\mathcal{B}$ sets

$$Y_\ell = \begin{cases} Z_1^{a'_\ell/t'_\ell}, & \text{if } y_\ell = 1; \\ Z_1^{a'_\ell/r'_\ell}, & \text{if } y_\ell = 0; \\ \emptyset, & \text{if } y_\ell = \star. \end{cases} \quad \text{and} \quad L_\ell = \begin{cases} Z_2^{a'_\ell/v'_\ell}, & \text{if } y_\ell = 1; \\ Z_2^{a'_\ell/m'_\ell}, & \text{if } y_\ell = 0; \\ \emptyset, & \text{if } y_\ell = \star. \end{cases}$$

Finally, $\mathcal{B}$ sets

$$Y_j = \begin{cases} Z_2^{(1-a'_\ell)/t'_j} g^{-a'/t'_j}, & \text{if } y_j = 1; \\ Z_2^{(1-a'_\ell)/r'_j} g^{-a'/r'_j}, & \text{if } y_j = 0. \end{cases} \quad \text{and} \quad L_j = \begin{cases} Z_2^{(1-a'_\ell)/v'_j} g^{-a'/v'_j}, & \text{if } y_j = 1; \\ Z_2^{(1-a'_\ell)/m'_j} g^{-a'/m'_j}, & \text{if } y_j = 0. \end{cases}$$

By the settings above we have that, for $i \neq j$ and $i \neq \ell$, $a_i = z_1 a'_i$, $a_\ell = z_1 z_2 a'_\ell$ and $a_j = z_1 z_2 - z_1 z_2 a'_\ell - z_1 a'$. Therefore, the a_i's are independently and randomly chosen in $\mathbb{Z}_p$ under the costraint that their sum is $z_1 z_2 = y$ and thus the key computed by $\mathcal{B}$ has the exact same distribution as the key computed by the KeyGeneration algorithm.

Case 2: $x_\ell \neq y_\ell$ and $y_\ell \neq \star$. In this case, for $i \neq \ell$, $\mathcal{B}$ chooses a'_i uniformly at random in $\mathbb{Z}_p$ and let us denote by a' the sum $a' = \sum_{i \neq \ell} a'_i$. Then for $i \neq \ell$, $\mathcal{B}$ sets Y_i and L_i exactly as in the previous case, whereas, $\mathcal{B}$ sets Y_ℓ and L_ℓ as follows

$$Y_\ell = \begin{cases} Z_2^{1/r'_\ell} g^{-a'/r'_\ell}, & \text{if } x_\ell = 1; \\ Z_2^{1/t'_\ell} g^{-a'/t'_\ell}, & \text{if } x_\ell = 0; \end{cases} \quad \text{and} \quad L_\ell = \begin{cases} Z_2^{1/m'_\ell} g^{-a'/m'_\ell}, & \text{if } x_\ell = 1; \\ Z_2^{1/v'_\ell} g^{-a'/v'_\ell}, & \text{if } x_\ell = 0; \end{cases}$$

By the settings above we have that $a_i = z_1 a'_i$ and $a_\ell = z_1 z_2 - z_1 a'$. Therefore, the a_i's are independently and randomly chosen in $\mathbb{Z}_p$ under the costraint that their sum is $z_1 z_2 = y$. Hence, also in this case, the key computed by $\mathcal{B}$ has the exact same distribution as the key returned by the KeyGeneration algorithm.

Challenge. $\mathcal{B}$ chooses R_0 uniformly at random $\mathbb{G}_T$ and, for $\ell \leq i \leq n$, chooses s'_i uniformly at random in $\mathbb{Z}_p$. $\mathcal{B}$ then constructs the tuple

$$D^* = (R_0, C_0, (X_i, W_i)_{i=1}^n)$$

where $C_0 = U$, and, for $i < \ell$, X_i and W_i are chosen uniformly from $\mathbb{G}$ whereas, for $i \geq \ell$, $\mathcal{B}$ computes

$$X_i = \begin{cases} Z^{t'_i}, & \text{if } i = \ell, x_i = 1; \\ Z^{r'_i}, & \text{if } i = \ell, x_i = 0; \\ U^{t'_i} g^{-t'_i s'_i}, & \text{if } i > \ell, x_i = 1; \\ U^{r'_i} g^{-r'_i s'_i}, & \text{if } i > \ell, x_i = 0; \end{cases} \quad \text{and} \quad W_i = \begin{cases} Z_{13}^{v'_i}, & \text{if } i = \ell, x_i = 1; \\ Z_{13}^{m'_i}, & \text{if } i = \ell, x_i = 0; \\ g^{v'_i s'_i}, & \text{if } i > \ell, x_i = 1; \\ g^{m'_i s'_i}, & \text{if } i > \ell, x_i = 0; \end{cases}$$

Now observe that if $Z = g^{z_2(u-z_3)}$ then $D^\star$ is distributed according to $\mathsf{Dist}_{\ell-1}(\boldsymbol{x})$, $s = u, s_\ell = z_3$, and $s_i = s'_i$ for $i > \ell$. On the other hand, if Z is random in $\mathbb{G}$, then $D^\star$ is distributed according to $\mathsf{Dist}_\ell(\boldsymbol{x})$ with $s = u$ and $s_i = s'_i$ for $i > \ell$.

Query Phase II. Identical to Query Phase I.

Output. $\mathcal{A}$ outputs η which represents a guess for the tuple in input ($\eta = 0$ for $D_{\ell-1}$ and $v = 1$ for D_ℓ). $\mathcal{B}$ forwards the same bit as its guess for the tuple of the experiment DLExp.

By the observation above, we observe that if $Z = g^{z_2(u-z_3)}$ then $\mathcal{A}$'s view is exactly the same as $\mathcal{A}$'s view (including the answers for queries for private keys) when receiving an input from $\mathsf{Dist}_{\ell-1}(\boldsymbol{x}, M)$; if Z is randomly and uniformly distributed in $\mathbb{G}$ then $\mathcal{A}$'s view (again this includes the replies obtained to the queries for private keys) is the same as when receiving an input from $\mathsf{Dist}_\ell(\boldsymbol{x}, M)$. Therefore, if $\mathcal{A}$ distinguishes between Dist_ℓ and $\mathsf{Dist}_{\ell-1}$ then the DL assumption is broken.

The above lemma implies the following theorem.

Theorem 3 (Attribute Hiding). *Assume DL holds. Then HVE scheme* (Setup, Enc, KeyGeneration, Dec) *described above is attribute hiding.*

6 Applications

As we have discussed in the introduction HVE schemes are a special type of *predicate encryption schemes.*

Definition 4. *A* predicate encryption scheme *for a class $\mathcal{F}$ of predicates over n-bit strings is quadruple of probabilistic polynomial-time algorithms* (Setup, Enc, KeyGeneration, Dec) *such that:*

1. Setup *takes as input the security parameter 1^k and attribute length n =* poly(k) *and outputs the* master public key Pk *and the* master secret key Msk.
2. KeyGeneration *takes as input the master secret key* Msk *and a predicate $f \in \mathcal{F}$ and outputs the decryption key K_f associated with f.*
3. Enc *takes as input the public key* Pk *and an* attribute string $\boldsymbol{x} \in \{0,1\}^n$ *and a message M in some associated message space and returns ciphertext* $\mathsf{Ct}_{\boldsymbol{x}}$.

4. Dec *takes as input a secret key* K_f *and a ciphertext* $\mathsf{Ct}_{\boldsymbol{x}}$ *and outputs a message* M.

We require that for all k *and* $n = \mathsf{poly}(k)$, *and for all strings* $\boldsymbol{x} \in \{0,1\}^n$ *and predicates* $f \in \mathcal{F}$ *such that* $f(\boldsymbol{x}) = 1$, *it holds that:*

$$\text{Prob}[(\mathsf{Pk}, \mathsf{Msk}) \leftarrow \mathsf{Setup}(1^k, n); \quad K_f \leftarrow \mathsf{KeyGeneration}(\mathsf{Msk}, f); \\ \mathsf{Ct}_{\boldsymbol{x}} \leftarrow \mathsf{Enc}(\mathsf{Pk}, \boldsymbol{x}, M) : \mathsf{Dec}(K_f, \mathsf{Ct}) = M] = 1.$$

The construction of searchable encryption of [BDOP04] can be seen an predicate encryption for the class $\mathcal{F}$ of predicates P_a defined as $P_a(x) = 1$ iff and only if $a = x$.

In [BW07], it is shown that HVE scheme can be used to construct predicate encryption for the class of *conjunctive comparison predicates* defined as follows $P_{a_1,\cdots,a_n}(x_1, \cdots, x_n) = 1$ if and only if $a_i \leq x_i$ for all i. Futhermore, in [BW07] it was shown how to construct predicate encryption schemes also for conjunctive range query predicates and subset query predicates starting from HVE. All reductions can be applied to our HVE thus yielding the following theorem.

Theorem 4. *Assume DL holds. Then there exist predicate encryption schemes for conjunctive comparison predicates, conjunctive range query predicates and subset query predicates that are semantically secure and attribute hiding.*

We expect there to be several other applications of HVE.

Acknowledgements. The work of the authors has been supported in part by the European Commission through Contract IST-2002-507932 ECRYPT and Contract FP6-1596 AEOLUS.

References

[BBS04] Boneh, D., Boyen, X., Shacham, H.: Short group signatures. In: Franklin, M. (ed.) CRYPTO 2004. LNCS, vol. 3152, pp. 41–55. Springer, Heidelberg (2004)

[BDOP04] Boneh, D., Di Crescenzo, G., Ostrovsky, R., Persiano, G.: Public key encryption with keyword search. In: Cachin, C., Camenisch, J.L. (eds.) EUROCRYPT 2004. LNCS, vol. 3027, pp. 506–522. Springer, Heidelberg (2004)

[BW06] Boyen, X., Waters, B.: Anonymous Hierarchical Identity-Based Encryption (Without Random Oracles). In: Dwork, C. (ed.) CRYPTO 2006. LNCS, vol. 4117, pp. 290–307. Springer, Heidelberg (2006)

[BW07] Boneh, D., Waters, B.: Conjunctive, subset and range queries on encrypted data. In: Vadhan, S.P. (ed.) TCC 2007. LNCS, vol. 4392, pp. 535–554. Springer, Heidelberg (2007)

[GPSW06] Goyal, V., Pandey, O., Sahai, A., Waters, B.: Attribute-Based Encryption for Fine-Grained Access Control for Encrypted Data. In: ACM CCS 2006 13th Conference on Computer and Communications Security, Alexandria, VA, USA, October 30 - November 3, 2006, pp. 89–98. ACM Press, New York (2006)

[KSW08] Katz, J., Sahai, A., Waters, B.: Predicate Encryption Supporting Disjunction, Polynomial Equations, and Inner Products. In: Smart, N. (ed.) EUROCRYPT 2008. LNCS, vol. 4965. Springer, Heidelberg (2008)

[SBC+07] Shi, E., Bethencourt, J., Chan, H., Song, D., Perrig, A.: Multi-Dimensional Range Query over Encrypted Data. In: 2007 IEEE Symposium on Security and Privacy, Oakland, CA. IEEE Computer Society Press, Los Alamitos (2007)

The Hidden Root Problem

Frederik Vercauteren*

Department of Electrical Engineering, University of Leuven,
Kasteelpark Arenberg 10, B-3001 Leuven-Heverlee, Belgium
frederik.vercauteren@esat.kuleuven.be

Abstract. In this paper we study a novel computational problem called the *Hidden Root Problem*, which appears naturally when considering fault attacks on pairing based cryptosystems. Furthermore, a variant of this problem is one of the main obstacles for efficient pairing inversion. We present an algorithm to solve this problem over extension fields and investigate for which parameters the algorithm becomes practical.[1]

Keywords: finite fields, subgroups, hidden root problem, pairing inversion.

1 Introduction

All known public key cryptosystems are based on a presumed hard mathematical problem, such as problems related to discrete logarithms [6], factoring [13] or finding short vectors in a lattice [11]. The security of the cryptographic protocol should ideally be implied directly by the hardness of the mathematical problem and the precise relation should be captured in a proof of security.

Since the inception of pairing based cryptography, a plethora of new supposedly hard problems has been introduced, but the main hard problem undoubtedly is pairing inversion [7], which has far-reaching implications [15,16]. In this paper we formally define a new computational problem called the Hidden Root Problem (HRP), which arises naturally as a cryptanalytic problem when studying pairing inversion [7] and side-channel or fault attacks on pairing implementations [10,17,18]. The problem resembles the well-known Hidden Number Problem [2,3,14], but is of a more algebraic nature, since the information returned about the hidden root consist of a projection into a subgroup of the multiplicative group of a finite field.

The remainder of this paper is organized as follows: Section 2 formally defines several variants of the HRP and points out the similarities with the Hidden Number Problem. Section 3 provides the motivation for studying this problem

* Postdoctoral Fellow of the Research Foundation - Flanders (FWO).

[1] The work described in this paper has been supported in part by the European Commission through the IST Programme under Contract IST-2002-507932 ECRYPT. The information in this document reflects only the author's views, is provided as is and no guarantee or warranty is given that the information is fit for any particular purpose. The user thereof uses the information at its sole risk and liability.

S.D. Galbraith and K.G. Paterson (Eds.): Pairing 2008, LNCS 5209, pp. 89–99, 2008.

and in Section 4 we describe an algorithm to solve the HRP over extension fields. In Section 5 we study the practicality of this algorithm for the linear version of the HRP by considering projection onto algebraic tori. Finally, Section 6 concludes the paper.

2 The Hidden Root Problem

In this section we define various versions of the Hidden Root Problem (HRP) over finite fields. We assume that the oracles appearing in the definitions are perfect, i.e. they always return the correct result. The following problem was first formulated in [7].

Definition 1 (Linear Hidden Root Problem). *Let $\mathbb{F}_q$ be a finite field with $q = p^n$ elements, where p is prime and let e be a positive integer with $e|(q-1)$. Let $\mathcal{O}_x(\cdot,\cdot)$ denote an oracle that on input $(a,b) \in \mathbb{F}_q^2$ returns*

$$\xi_{a,b} = (ax+b)^e$$

for a fixed secret $x \in \mathbb{F}_q$. The Linear Hidden Root Problem (LHRP) is to recover x in expected polynomial time in $\log q$ by querying the oracle repeatedly with chosen pairs (a_i, b_i).

The restriction $e|(q-1)$ can be explained as follows: let e' be a positive integer with $\gcd(e', q-1) = 1$, then e'-powering defines a permutation on $\mathbb{F}_q$ with inverse $(e'^{-1} \bmod (q-1))$-powering. The LHRP formulated for e' therefore is trivial to solve using only one query by computing

$$x = (\xi_{a,b}^{e'^{-1} \bmod (q-1)} - b)/a \,.$$

Similarly, if $d = \gcd(e', q-1)$, then (e'/d)-powering is a permutation which can easily be inverted, so the problem reduces to the LHRP with $e = d$ and thus $e|(q-1)$. In the remainder of this paper we will use the notation $h = (q-1)/e$, i.e. the cofactor of e.

Note that in the definition of the LHRP, the querying party can choose the pairs (a_i, b_i) himself. It would be possible to define a randomized version, where we are simply given a list of random pairs (a_i, b_i) with the corresponding responses $\mathcal{O}_x(a_i, b_i) = \xi_{a_i,b_i}$. Clearly, this randomized version cannot be easier than the LHRP as defined above, but the algorithms described in Section 4 work equally well in the randomized case. In fact, most cryptanalytic applications correspond to the randomized version, since the adversary does not control a, b directly. For instance, a, b could be related to the coordinates of a point R on an elliptic curve with $R = [m]P$, where the adversary is given P and can only choose m.

Obvious generalizations would be to allow degree d polynomials in x instead of a linear polynomial, or in fact any function in x of specified form, such as fractions of polynomials. Further, one can also replace x by any unknown quantity such as a vector of unknowns. This brings us to the following more general definition.

Definition 2 (Hidden Root Problem). *Let $\mathbb{F}_q$ be a finite field with $q = p^n$ elements, where p is prime and let e be a positive integer with $e|(q-1)$. Let $f_x(\cdot) : \mathcal{D}(\mathbb{F}_q) \to \mathbb{F}_q$ be a specified map, i.e. the precise description of which is given, depending on x from a domain $\mathcal{D}$ to $\mathbb{F}_q$. Let $\mathcal{O}_x(\cdot)$ denote an oracle that on input $\alpha \in \mathcal{D}$ returns*

$$\xi_\alpha = f_x(\alpha)^e$$

for a fixed secret $x \in \mathbb{F}_q$. The Hidden Root Problem is to recover x in expected polynomial time in $\log q$ by querying the oracle repeatedly for chosen $\alpha_i \in \mathcal{D}(\mathbb{F}_q)$.

An even further generalization would be to replace the multiplicative group $\mathbb{F}_q^*$ by any group G and the e-th powering by a projection into a proper subgroup of G.

The name "Hidden Root Problem" was chosen to point out the resemblance with the Hidden Number Problem [2,3] and its generalizations as described in [14]. Recall that the $\mathbb{F}_p$-Hidden Number Problem (HNP) is defined as: for a secret $x \in \mathbb{F}_p$, we are given the k pairs

$$(t_i, MSB_{l,p}(xt_i)) \qquad i = 1, \ldots, k$$

for k elements $t_1, \ldots, t_k \in \mathbb{F}_p^*$, chosen independently and uniformly at random, and for some $l > 0$, where $MSB_{l,p}$ denotes (roughly speaking) the l most significant bits. The problem then is to recover x.

Note that in the LHRP, for each query we obtain roughly $\log_2 h$ bits of information about x, so in this sense the HRP is similar to the HNP. Heuristically, we therefore expect a unique solution to the LHRP after roughly $\log_h q$ queries. The main difference with the HNP are the following two facts: firstly, the HNP is randomized, i.e. the querying party cannot choose the t_i for $i = 1, \ldots, k$ and secondly, hiding the information about x in the HRP corresponds to an algebraic operation, namely e-th powering.

Finally, a last version of the HRP is the **subfield HRP**, i.e. the HRP with the restriction that the secret x lies in a strict subfield of $\mathbb{F}_{p^d} \subsetneq \mathbb{F}_q$ for $d|n$.

3 Applications in Pairing Based Cryptography

The main motivation for our study of the Hidden Root Problem is undoubtedly cryptanalysis of pairings, more specifically, side-channel and fault attacks on pairings [10,17,18] and pairing inversion [7].

Recall that the general setting of pairings is the following: let E be an elliptic curve over a finite field $\mathbb{F}_q$ and let r be a large prime with $\gcd(r, q) = 1$ and $r \mid \#E(\mathbb{F}_q)$. By definition, the embedding degree k is the smallest positive integer with $r \mid (q^k - 1)$. All variants of the Tate pairing can then be described as functions of the form

$$e : \mathbb{G}_1 \times \mathbb{G}_2 \to \mu_r \subset \mathbb{F}_{q^k} : (P, Q) \mapsto e(P, Q) = f_{S,P}(Q)^L , \qquad (1)$$

where $\mathbb{G}_1, \mathbb{G}_2$ are given cyclic subgroups of $E(\mathbb{F}_{q^k})[r]$, $f_{S,P}$ is a Miller function, i.e. has divisor $S(P) - ([S]P) - (S-1)(\mathcal{O})$ and $L|(q^k - 1)$. Often, $L = (q^k - 1)/r$

with $r \mid \Phi_k(q)$, where Φ_k is the k-th cyclotomic polynomial. However, to speed up the final exponentiation, L is sometimes taken to have very low Hamming weight in base q, by moving some of the complexity into the scalar S (see for instance [1]). In most protocols, one of the input values to the pairing is public, e.g. the point P and so is the pairing value. The security of the protocol then relies on the inability of the adversary to recover the other input point Q.

3.1 Fault Attacks on Pairings

The function $f_{S,P}$ is computed using Miller's algorithm [9] and thus consists of a product of roughly $\log_2 S$ powers of evaluations of lines appearing in the scalar multiplication of P by S. Algorithm 1 below gives the pseudo-code of Miller's algorithm: $l_{T,P}$ denotes the line through the points T and P and v_T denotes the vertical line through T.

Algorithm 1. Miller's algorithm for elliptic curves

Inputs: $S \in \mathbb{N}$, $P, Q \in E[r]$
Outputs: $f_{S,P}(Q)$

```
Write s = Σ_{j=0}^{t} s_j 2^j, with s_j ∈ {0,1} and s_t = 1.
T ← P, f ← 1.
for j = t − 1 down to 0 do
  f ← c^2 · l_{T,T}(Q)/v_{[2]T}(Q).
  T ← [2]T
  if s_j = 1 then
    f ← f · l_{T,P}(Q)/v_{T⊕P}(Q).
    T ← T ⊕ P
  end if
end for
Return f.
```

The parameter S typically is public knowledge, since it's either equal or related to the group order r. In a much simplified setting (see [10,17,18] for more realistic attacks) we could mount a fault attack on S resulting in a bit-flip of the least significant bit of S. As such the adversary will have access to two pairing values: the correct value $f_{S,P}(Q)^L$ and a faulted one $f_{S\oplus 1,P}(Q)^L$. Note that all steps in Miller's algorithm, except the last, will be exactly the same in the computation of both values. By dividing both values the adversary will know the value of (assuming S is odd)

$$\xi_P = \left(\frac{l_{[S-1]P,P}(Q)}{v_{[S]P}(Q)} \right)^L . \tag{2}$$

Furthermore, in most cases we can ignore the last vertical line $v_{[S]P}(Q)$, either due to denominator elimination or simply because $[S]P = \mathcal{O}$. By repeating the attack (for different known P), several equations of the form (2) can be gathered and the adversary is left to solve an instance of the HRP (and in many cases the LHRP) with exponent L over $\mathbb{F}_{q^k}$.

3.2 Pairing Inversion

In [7], several approaches to invert the pairing function (1) itself were described. One of these approaches consists of solving two problems: firstly, inverting the final exponentiation and secondly, Miller inversion, i.e. given the evaluation $f_{S,P}(Q)$ and the point P, recover Q. Furthermore, it was shown that for some instances of the ate pairing [8], Miller inversion can be achieved in polynomial time and the security relies entirely on the final exponentiation. In these cases, the function $f_{S,P}$ is of low degree and to invert the final exponentiation, the adversary has to solve an instance of the HRP. Note that due to bilinearity of the pairing, the adversary can easily generate many equations from one given equation $z = f_{S,P}(Q)^L$ by the following simple rule

$$z^i = f_{S,[i]P}(Q)^L .$$

Finally, we note that in case of the ate pairing, the point Q is defined over the field $\mathbb{F}_q$, and pairing inversion corresponds to the subfield HRP.

4 An Algorithm for the HRP over Extension Fields

In this section we devise an algorithm to solve the HRP over extension fields $\mathbb{F}_{q^k}$ by combining Weil restriction and linearity of q-th powering. For simplicity we will focus on the LHRP, but the same technique applies to any algebraic function f_x. However, the feasibility of the algorithm will very much depend on the degree in x of f_x (see the paragraph "Complexity for HRP").

Version 0. Let $e|(q^k - 1)$ and write

$$e = \sum_{i=0}^{k-1} c_i q^i - \sum_{i=0}^{k-1} d_i q^i , \tag{3}$$

where $c_i, d_i \in \mathbb{N}_{\geq 0}$. Clearly there are many tuples c_i, d_i that give a valid expression for e, but we are only interested in those c_i, d_i that minimize the quantity

$$D_e := \max \left\{ \sum_{i=0}^{k-1} c_i,\ \sum_{i=0}^{k-1} d_i \right\} . \tag{4}$$

Consider $\mathbb{F}_{q^k}$ as a degree k extension over $\mathbb{F}_q$, i.e. $\mathbb{F}_{q^k} = \mathbb{F}_q[\theta]/(f(\theta))$ where $f \in \mathbb{F}_q[x]$ is an irreducible polynomial of degree k. In the LHRP we are given an equation $\xi_{a,b} = (ax+b)^e$ for some pair (a, b) and unknown x. By Weil restriction we will consider this as k equations over $\mathbb{F}_q$ in k unknowns. More precisely, consider the map

$$\psi : \mathbb{F}_{q^k} \to (\mathbb{F}_q)^k : \alpha = \sum_{i=0}^{k-1} \alpha_i \theta^i \mapsto [\alpha_0, \ldots, \alpha_{k-1}]$$

and note that q-th powering is a linear operation, i.e. there exists an easily computable $k \times k$ matrix F such that $\psi(\alpha^q) = F \cdot \psi(\alpha)$. By substituting the expression (3) for e, we obtain

$$\xi_{a,b} = \frac{(ax+b)^{\sum_{i=0}^{k-1} c_i q^i}}{(ax+b)^{\sum_{i=0}^{k-1} d_i q^i}} = \frac{\prod_{i=0}^{k-1}(ax+b)^{c_i q^i}}{\prod_{i=0}^{k-1}(ax+b)^{d_i q^i}} .$$

Exploiting linearity of q^i-th powering, the above equation can be rewritten as

$$\prod_{i=0}^{k-1}\left(a^{q^i}x^{q^i} + b^{q^i}\right)^{c_i} = \xi_{a,b}\prod_{i=0}^{k-1}\left(a^{q^i}x^{q^i} + b^{q^i}\right)^{d_i} . \tag{5}$$

Finally, apply ψ to both sides of this equation to obtain k non-linear equations over $\mathbb{F}_q$ in the unknowns $x_0, \ldots, x_{k-1}$. Furthermore, since $\psi(x^{q^i}) = F^i\psi(x)$, each of the factors in the product is linear in the unknowns x_i, so the degree of the non-linear system of equations is precisely D_e defined in (4). Solving a system of non-linear equations over finite fields is in general a very hard problem. A notable exception however is when the system is highly overdetermined, which can be easily obtained by repeatedly querying the oracle. In the latter case, Groebner basis techniques [4] are rather efficient or one could resort to relinearization [5]. For both algorithms, the complexity is given by the time to solve a linear system of equations of dimension equal to the total number of monomials of degree less than or equal to D_e in k variables, which is given by

$$M_e := \binom{D_e + k}{D_e} .$$

The complexity of both algorithms then is $O(M_e^\omega)$ with $\omega \leq 3$ the matrix multiplication exponent. Since k is given, minimizing D_e is crucial, since only for reasonably sized M_e and thus very small D_e, will it be possible to even write down the system of non-linear equations. The experiments in Section 5 show that for the algorithm to succeed we need D_e to be very small (depending on k), e.g. at most 4 for $k = 30$. This implies that for fixed k and growing q, version 0 of the algorithm will only be efficient for a constant number of exponents. This situation will be much improved in versions 1 and 2.

Complexity for HRP. Version 0 (and also 1 and 2) also works for any algebraic function f_x. Assume that $f_x = h(x)/g(x)$ with $h, g \in \mathbb{F}_{q^k}[x]$, then the equivalent of equation (5) simply is:

$$\prod_{i=0}^{k-1}\left(h(x)^{q^i}\right)^{c_i} \cdot \prod_{i=0}^{k-1}\left(g(x)^{q^i}\right)^{d_i} = \xi_{a,b}\prod_{i=0}^{k-1}\left(h(x)^{q^i}\right)^{d_i} \cdot \prod_{i=0}^{k-1}\left(g(x)^{q^i}\right)^{c_i} .$$

By applying ψ to both sides of this equation, we again obtain k non-linear equations over $\mathbb{F}_q$ in the unknowns $x_0, \ldots, x_{k-1}$. The main difference however is the degree of these non-linear equations, namely

$$D_{f,e} := \max\left\{\deg h\sum_{i=0}^{k-1} c_i + \deg g\sum_{i=0}^{k-1} d_i, \deg h\sum_{i=0}^{k-1} d_i + \deg g\sum_{i=0}^{k-1} c_i\right\} .$$

If $\deg h = \deg g = d$, we conclude that $dD_e \leq D_{f,e} \leq 2dD_e$, so the algorithm will only be efficient for f_x of very low degree.

Version 1. The main idea of version 1 is to drastically increase the applicability of version 0 of the algorithm by considering multiples of e. Indeed, each equation $\xi_{a,b} = (ax+b)^e$ gives rise to many other equations $\xi_{a,b}^m = (ax+b)^{me}$ for $m \in \mathbb{Z}$ and as long as $me \neq 0 \bmod (q^k - 1)$, these equations will be non-trivial. Note that by raising to the power m we are in fact ignoring information contained in the original equation. This is not a problem since we can query the oracle to obtain sufficient equations. The main advantage however is that a random looking exponent can be transformed in one with much more algebraic structure. The goal therefore is to find a multiple me of e with D_{me} as small as possible.

A first method to find good multiples of e is to exploit the algebraic factorization $x^k - 1 = \prod_{d|k} \Phi_d(x)$ with $\Phi_d \in \mathbb{Z}[x]$ the d-th cyclotomic polynomial. Since $e|(q^k-1)$, we can determine an index set I and a polynomial $\Pi(x) := \prod_{d|k, d \in I} \Phi_d(x)$ such that $e|\Pi(q)$ and $\Pi(x)$ has low D (i.e. sum of positive coefficients minus the sum of the negative coefficients). If $\Pi(x) \neq (x^k-1)$, we have found a good multiple $\Pi(q)$ of e.

Version 2. Although version 1 gives reasonable results for a wide variety of e, it is limited to finding multiples me of e that also divide $(q^k - 1)$. Version 2 relies on LLL [12] to automatically find the best multiple possible. Consider the lattice $L \subset \mathbb{Z}^k$ spanned by the vectors

$$L := \begin{pmatrix} e & 0 & 0 & \cdots & 0 \\ -q & 1 & 0 & \cdots & 0 \\ -q^2 & 0 & 1 & \cdots & 0 \\ \vdots & \vdots & & \ddots & \\ -q^{k-1} & 0 & \ldots & 0 & 1 \end{pmatrix} .$$

Note the inner product of all vectors in the lattice with the vector $[q^0, \ldots, q^{k-1}]$ is a multiple of e. By reducing the lattice L, we find a short vector $[s_0, \cdots, s_{k-1}]$ with $e | \sum_{i=0}^{k-1} s_i q^i$. Note that short vectors automatically have small D.

Subfield HRP. In case of the subfield HRP, we have the extra information that $x \in \mathbb{F}_q$, so all unknowns $x_i = 0$ for $i > 0$. The system of non-linear equations thus simplifies to a system of k univariate polynomials in $x = x_0$ of degree D_e (or $D_{f,e}$ in case of an algebraic function f_x). To find the solution x, it suffices to take the GCD of several of these polynomials, each of which takes $O(D^2)$ operations in $\mathbb{F}_q$. The subfield HRP therefore is fundamentally easier than the full HRP, since its complexity is polynomial in the parameter D and the dependence on k only appears in the first phase, i.e. writing down the system of univariate polynomials.

5 LHRP and Projection onto Algebraic Tori

In this section we analyze the effectiveness of the various versions of the algorithm given in Section 4 in the special case of LHRP, where the e-th powering equals projection onto the algebraic torus $T_k(\mathbb{F}_q)$. The main reason to consider this special case is that $T_k(\mathbb{F}_q)$ is the biggest subgroup which is really contained in $\mathbb{F}_{q^k}$ itself and not in a strict subfield. Furthermore, all known pairings map into (a strict subgroup of) $T_k(\mathbb{F}_q)$, so the torus $T_k(\mathbb{F}_q)$ can be considered as the base case.

Recall that by definition we have

$$T_k(\mathbb{F}_q) = \{\alpha \in \mathbb{F}_{q^k} \mid N_{\mathbb{F}_{q^k}/\mathbb{F}_{q^d}}(\alpha) = 1 \text{ for all } d|k, d < k\},$$

and $|T_k(\mathbb{F}_q)| = \Phi_k(q)$ with Φ_k the k-th cyclotomic polynomial. Therefore, we choose the e in the LHRP equal to

$$e = (q^k - 1)/\Phi_k(q).$$

This implies that we can take $\Pi(x)$ in version 1 of the algorithm to be equal to $(x^k - 1)/\Phi_k(x)$. To investigate the degree D of the resulting non-linear system, we prove the following lemma.

Lemma 1. *Assume k has prime factorization $k = \prod_{i=1}^t p_i^{s_i}$ with $p_i \neq p_j$ for $i \neq j$ and $s_i > 0$. Define $\hat{k} = \prod_{i=1}^t p_i$, then*

$$\Phi_k(x) = \Phi_{\hat{k}}(x^{k/\hat{k}}).$$

Proof. This follows immediately from the explicit expression

$$\Phi_k(x) = \prod_{d|k}(x^d - 1)^{\mu(k/d)},$$

with μ the Möbius function: for $n \in \mathbb{N}_{>0}$, $\mu(n) = 0$ if n is not squarefree and $(-1)^k$ if n is the product of k distinct primes. Note $\mu(k/d) = 0$ except for d that are multiples of $k/\hat{k}$, i.e. d is of the form $d = v \cdot (k/\hat{k})$ for $v|\hat{k}$.

Corollary 1. *Using the notation of Lemma 1 we have $\Pi_k(x) = \Pi_{\hat{k}}(x^{k/\hat{k}})$, where $\Pi_k(x) = (x^k - 1)/\Phi_k(x)$.*

The above corollary implies that the degree D of the non-linear system only depends on $\hat{k}$ and not on k itself. A further analysis gives the following refinement: if $\hat{k} = p$, a prime, then $D = 1$ and thus the system of equations becomes linear. For $\hat{k} = pq$, we have $D = \min\{p, q\}$ which follows from the equality

$$\Pi_{pq}(x) = \frac{(x^p - 1)(x^q - 1)}{(x - 1)}.$$

Indeed, assume that $p < q$, then $(x^p - 1)/(x - 1)$ is the all-one polynomial of degree $(p-1)$ and since $p < q$, multiplication by $(x^q - 1)$ does not cancel out any non-zero terms. For three or more primes, the pattern becomes more intricate, but for $k = 2pq$ with $p < q$, we have $D = 2p$. These cases cover all k up to 100 which is sufficient for practical applications.

To compare version 1 and version 2 of the algorithm described in Section 4, we ran several tests using MAGMA for some interesting cases of k, namely $k = 3, 6, 12, 15, 30$. For each k, we determined the best multiple of $e = (q^k - 1)/\Phi_k(q)$ using version 1 and 2. We then derived the system of non-linear equations as described in Section 4 for the LHRP using a predetermined number of queries to the oracle. Finally, the system of non-linear equations was solved using the Groebner basis command in MAGMA. In these tests, the prime field $\mathbb{F}_p$ was kept constant for some fixed random 32-bit prime p, since the size of p only has a minor influence on the feasibility of the tests. Table 1 gives a summary of the test results. Especially the last two entries are interesting since version 2 outperforms version 1 considerably. For example, for $k = 30$, version 1 leads to a system of non-linear equations of degree $D = 6$ corresponding to

$$\Pi(x) = (x^{30}-1)/\Phi_{30}(x) = x^{22}-x^{21}+x^{20}+x^{17}-x^{16}+x^{15}-x^{7}+x^{6}-x^{5}-x^{2}+x-1$$

so it would be nearly impossible to even write down the system of equations, since each non-linear equation contains 1947792 terms. However, version 2 finds a much better multiple of e corresponding to

$$(x^5 + x^4 - x^2 - x - 1)\Pi(x) = x^{27} - x^{20} - x^{17} - x^{15} - x^{12} + x^5 + x^2 + 1\,,$$

leading to $D = 4$ and thus non-linear equations of only 46376 terms, which is just manageable, but already used up around 3 Gb of memory.

In each case, we made roughly $2\log_h q^k$ queries to the oracle. The column "Min # queries" contains the minimum number of queries to the oracle that gives a unique solution to the system of non-linear equations. This number approximates the expected $\log_h q^k$ from the information theoretical viewpoint.

For most values of $k < 100$, the algorithm performs fairly well since D will often be very small. However, for k having many prime factors, the number of terms in the non-linear equations simply becomes too large, e.g. for $k = 70$, version 2 gives $D = 4$ and thus $M = 1150626$.

In stark contrast, the subfield LHRP with projection onto $T_k(\mathbb{F}_q)$ is easy for all practical k (for instance all $k < 1000$) and the algorithm runs very fast in

Table 1. Comparison of version 1 and 2 for the LHRP and projection on $T_k(\mathbb{F}_q)$ for various extension degrees k

k	D (v.1)	D (v.2)	$M = \binom{D+k}{D}$	Min # queries	Time (s)
3	1	1	4	2	0.01
6	2	2	28	4	0.01
12	2	2	91	4	0.04
15	3	2	816 / 136	5	0.15
30	6	4	1947792 / 46376	16	2420

a matter of seconds, e.g. for $k = 665$ we have $D = 85$ and solving the subfield LHRP by using GCD's only takes 240 seconds, most of which is spent in computing the system of univariate equations. For the full LHRP, the number of terms in the non-linear equations would be $M \simeq 10^{114}$, which is clearly infeasible.

The results in this section show that the LHRP will be solvable for moderately sized k if e is a divisor of $(q^k - 1)/\Phi_k(q)$, since then it can be reduced to projection onto $T_k(\mathbb{F}_q)$. However, when e is of the form $(q^k - 1)/s$ with s a proper divisor of $\Phi_k(q)$ with large cofactor (e.g. 2^{80}), i.e. e-th powering corresponds to projection into a strict subgroup of $T_k(\mathbb{F}_q)$ with large cofactor, then the HRP (and also the subfield HRP) remains hard. In implementing pairing based cryptosystems in a side-channel resistant manner, it is therefore paramount to ensure that the final exponentiation does not succumb to the algorithms described in Section 4. In particular, this implies that final exponentiations of low Hamming weight in base q should be avoided. The implications for the more general problem of pairing inversion are currently unclear: it is possible to define pairings with "easy" final exponentiation from an HRP point of view, but the obvious candidates all lead to Miller functions $f_{S,P}$ of restrictively high degree.

6 Conclusion

In this paper we studied a new computational problem called the Hidden Root Problem, motivated by immediate applications to cryptanalysis of pairing based cryptosystems. We have given the first algorithm to solve this problem over extension fields and concluded that for exponents e which are divisors of $(q^k - 1)/\Phi_k(q)$ the problem can be solved efficiently for moderately sized k. However, for exponents e of the form $(q^k - 1)/s$ with s a proper divisor of $\Phi_k(q)$ with large cofactor, the problem remains hard. In implementing pairing based cryptosystems, it is therefore advisable to avoid using Miller functions $f_{S,P}$ where S is a multiple of $\Phi_k(q)$.

Acknowledgements

The author would like to thank the anonymous referees of Pairing 2008 for suggesting several improvements to the exposition of the paper.

References

1. Barreto, P.S.L.M., Galbraith, S.D., Ó'hÉigeartaigh, C., Scott, M.: Efficient pairing computation on supersingular abelian varieties. Designs, Codes and Cryptography 42(3), 239–271 (2007)
2. Boneh, D., Venkatesan, R.: Hardness of computing the most significant bits of secret keys in Diffie-Hellman and related schemes. In: Koblitz, N. (ed.) CRYPTO 1996. LNCS, vol. 1109, pp. 129–142. Springer, Heidelberg (1996)
3. Boneh, D., Venkatesan, R.: Rounding in lattices and its cryptographic applications. In: Proc. 8th Annual ACM-SIAM Symp. on Discr. Algorithms, pp. 675–681. ACM, New York (1997)

4. Buchberger, B.: A theoretical basis for the reduction of polynomials to canonical forms. SIGSAM Bull 10(3), 19–29 (1976)
5. Courtois, N., Klimov, A., Patarin, J., Shamir, A.: Efficient algorithms for solving overdefined systems of multivariate polynomial equations. In: Preneel, B. (ed.) EUROCRYPT 2000. LNCS, vol. 1807, pp. 392–407. Springer, Heidelberg (2000)
6. ElGamal, T.: A public key cryptosystem and a signature scheme based on discrete logarithms. IEEE Trans. Inform. Theory 31(4), 469–472 (1985)
7. Galbraith, S.D., Hess, F., Vercauteren, F.: Aspects of pairing inversion (preprint 2008), `http://eprint.iacr.org/2007/256`
8. Hess, F., Smart, N., Vercauteren, F.: The Eta-pairing revisited. IEEE Transactions on Information Theory 52(10), 4595–4602 (2006)
9. Miller, V.S.: The Weil pairing, and its efficient calculation. J. Cryptology 17(4), 235–261 (2004)
10. Page, D., Vercauteren, F.: A Fault Attack on Pairing Based Cryptography. IEEE Transactions on Computers 55(9), 1075–1080 (2006)
11. Hoffstein, J., Pipher, J., Silverman, J.H.: NTRU: a ring-based public key cryptosystem. In: Buhler, J.P. (ed.) ANTS 1998. LNCS, vol. 1423, pp. 267–288. Springer, Heidelberg (1998)
12. Lenstra, A.K., Lenstra, H.W., Lovász, L.: Factoring polynomials with rational coefficients. Mathematische Annalen 261(4), 515–534 (1982)
13. Rivest, R.L., Shamir, A., Adleman, L.: A method for obtaining digital signatures and public-key cryptosystems. Comm. ACM 21(2), 120–126 (1978)
14. Shparlinski, I.E.: Playing "Hide-and-Seek" in finite fields: Hidden number problem and its applications. In: Proc. 7th Spanish Meeting on Cryptology and Information Security, vol. 1, pp. 49–72. Univ. of Oviedo (2002)
15. Verheul, E.: Evidence that XTR Is More Secure than Supersingular Elliptic Curve Cryptosystems. In: Pfitzmann, B. (ed.) EUROCRYPT 2001. LNCS, vol. 2045, pp. 195–210. Springer, Heidelberg (2001)
16. Verheul, E.: Evidence that XTR is more secure than supersingular elliptic curve cryptosystems. J. Crypt. 17(4), 277–296 (2004)
17. Whelan, C., Scott, M.: The Importance of the Final Exponentiation in Pairings when Considering Fault Attacks. In: Takagi, T., Okamoto, T., Okamoto, E., Okamoto, T. (eds.) Pairing 2007. LNCS, vol. 4575, pp. 225–246. Springer, Heidelberg (2007)
18. Whelan, C., Scott, M.: Side Channel Analysis of Practical Pairings: Which Path is more Secure? In: Nguyên, P.Q. (ed.) VIETCRYPT 2006. LNCS, vol. 4341, pp. 99–114. Springer, Heidelberg (2006)

Evaluating Large Degree Isogenies and Applications to Pairing Based Cryptography

Reinier Bröker, Denis Charles, and Kristin Lauter

Microsoft Research, One Microsoft Way, Redmond, WA 98052, USA
reinierb@microsoft.com, cdx@microsoft.com, klauter@microsoft.com

Abstract. We present a new method to evaluate large degree isogenies between elliptic curves over finite fields. Previous approaches all have *exponential* running time in the logarithm of the degree. If the endomorphism ring of the elliptic curve is 'small' we can do much better, and we present an algorithm with a running time that is *polynomial* in the logarithm of the degree. We give several applications of our techniques to pairing based cryptography.

1 Introduction

Various algorithms using elliptic curves rely on the efficient computation of *isogenies* between them. A noteworthy example is the 'Schoof-Elkies-Atkin' algorithm [10] to compute the group order of an elliptic curve over a finite field. Here, it is crucial that we are able to efficiently compute small degree isogenies. The known algorithms to evaluate an isogeny are all exponential time algorithms (in the logarithm of the degree), and the practicality of these algorithms is therefore limited to relatively small degrees. A lot of effort has gone into speeding up the algorithms [2]. In this paper we propose an algorithm to evaluate an isogeny between ordinary elliptic curves over finite fields that, in special cases, has a run time that is *polynomial* in the logarithm of its degree.

In Section 2 we explain how to represent certain prime degree l isogenies that are defined over $\mathbf{F}_q$ with at most $3 \log l$ bits. This is a big contrast with the representation by rational functions or by its 'kernel polynomial' as these represenations take roughly l bits. We show that our representation applies to almost all isogenies: the only condition is that not all subgroups of order l are defined over the base field $\mathbf{F}_q$. As the l-torsion has l^2 elements for $l \neq \mathrm{char}(\mathbf{F}_q)$, this condition is harmless for large l.

We present our approach to evaluate an isogeny $\varphi : E \to E'$ in Sections 3 and 4. The run time is polynomial in the class number of the endomorphism ring of E, and is therefore only fast when this class group is *small*. This certainly limits the practicality of the method since randomly chosen elliptic curves over $\mathbf{F}_q$ will have an associated class group of size roughly $\sqrt{q}$. However, if the elliptic curve in question is contructed by *complex multiplication techniques* then the class group will always be small. In particular, our method applies to the curves with prescribed prime order constructed in [3], the curves with prime order of

S.D. Galbraith and K.G. Paterson (Eds.): Pairing 2008, LNCS 5209, pp. 100–112, 2008.

prescribed size constructed in e.g. [7] and the pairing friendly curves constructed in e.g. [9].

Section 5 gives several examples of the new evaluation algorithm. Our approach is so fast that isogenies of degree $l \approx 10^{100}$ are easily computed. We focus on applications to pairing-based cryptography in Section 6. We first describe a variant of the 'BLS signature scheme' [1] where two different isogenous elliptic curves are used instead of a single elliptic curve as in the basic BLS scheme. As another application, we show how our technique can be used in the isogeny variant of BLS which was proposed by Jao and Venkatesan [5]. Their scheme replaces a secret integer by a secret isogeny. For the security of the scheme, the secret isogeny must have degree of cryptographic size. Until the present paper, efficient evaluation of such large degree isogenies was only possible in special cases such as integer multiplication, or integer multiplication composed with a small degree isogeny.

2 Representation of Isogenies

Let E, E' be two elliptic curves defined over some field F. An *isogeny* φ between E and E' is a non-constant morphism $\varphi : E \to E'$. It is well known that isogenies are geometrically surjective, i.e., for every point $P \in E'(F)$ there exists a point $Q \in E(\overline{F})$ with $\varphi(Q) = P$. We say that φ is defined over F if the kernel of φ is as a group defined over F, meaning that the absolute Galois group of F maps the kernel of φ into itself. This does not mean that all the points of the kernel of φ are F-rational. Indeed, the multiplication by n-map is F-rational, yet most n-torsion points will not be defined over F.

An isogeny φ induces an inclusion $F(E') \subset F(E)$ of function fields, and the degree $[F(E) : F(E')]$ is called the *degree* $\deg(\varphi)$ of φ. If $\deg(\varphi)$ is coprime to the characteristic of F, the extension $F(E)/F(E')$ is separable and the degree of φ equals the number of points in its kernel. Most of the isogenies we consider in this article are separable.

The 'standard' way to represent an isogeny φ is to give 3 homogeneous polynomials $f_1, f_2, f_3 \in \overline{F}[X, Y, Z]$ satisfying $\varphi((x : y : z)) = (f_1(x, y, z) : f_2(x, y, z) : f_3(x, y, z)) \in \mathbf{P}^2(\overline{F})$. If l denotes the degree of φ, then usually one of these polynomials will have degree roughly l, and this representation takes *exponential* time in $\log l$ to write down. In this section we explain a representation of isogenies between elliptic curves over finite fields whose length is *polynomial* in $\log l$.

Assume that E/F has complex multiplication, meaning that the endomorphism ring $\mathrm{End}_F(E)$ is isomorphic to the imaginary quadratic order $\mathcal{O}_\Delta$ for some $\Delta < 0$. By writing $\mathcal{O}_\Delta = \mathbf{Z}[\alpha]$ and fixing a root in F of the minimal polynomial of α, we view F as an $\mathcal{O}_\Delta$-algebra. There are $|\mathcal{O}_\Delta^*| > 1$ isomorphisms $\mathrm{End}_F(E) \xrightarrow{\sim} \mathcal{O}_\Delta$ and throughout this article we assume that we have *fixed* the normalized isomorphism, i.e., the unique isomorphism ι with the property that $\iota^*(x)\omega = x\omega$ for all invariant differentials ω and all $x \in \mathcal{O}_\Delta$. In particular, we will identify the rings $\mathrm{End}_F(E)$ and $\mathcal{O}_\Delta$.

We let $\mathrm{Ell}_\Delta(F)$ be the set of $\overline{F}$-isomorphism classes of elliptic curves over F whose endomorphism ring equals $\mathcal{O}_\Delta$. It is well known that for $F = \mathbf{C}$, the set $\mathrm{Ell}_\Delta(\mathbf{C})$ is a finite set of cardinality h_Δ, the class number of the order $\mathcal{O}_\Delta$. The key to this result is that the class group acts in a natural way on $\mathrm{Ell}_\Delta(\mathbf{C})$. Indeed, if we let

$$E[\mathfrak{L}] = \{P \in E(\mathbf{C}) \mid \forall \alpha \in \mathfrak{L} : \alpha(P) = 0\}$$

denote the group of '$\mathfrak{L}$-torsion points' for an $\mathcal{O}_\Delta$-ideal $\mathfrak{L}$, then the map

$$j(E) \mapsto j(E/E[\mathfrak{L}]) = j(E)^{\mathfrak{L}}$$

factors through the class group. One then proves that this action is transitive and free [11, Prop. II.1.2].

As there are only finitely many isomorphism classes of complex elliptic curves with endomorphism ring equal to $\mathcal{O}_\Delta$, the j-invariant $j(E)$ is algebraic for $j(E) \in \mathrm{Ell}_\Delta(\mathbf{C})$. In fact, we have $\mathrm{Ell}_\Delta(\mathbf{C}) = \mathrm{Ell}_\Delta(H_\mathcal{O})$ where $H_\mathcal{O}$ is the ring class field associated to $\mathcal{O}_\Delta$, i.e., the unique abelian extension inside $\mathbf{C}$ of $\mathbf{Q}(\sqrt{\Delta})$ whose Galois group is isomorphic to the class group $\mathrm{Pic}(\mathcal{O}_\Delta)$ under the Artin map. If p is a prime that does not ramify in $H_\mathcal{O}/\mathbf{Q}$, then we get a natural injection

$$g : \mathrm{Ell}_\Delta(H_\mathcal{O}) \to \mathrm{Ell}_\Delta(\mathbf{F}_q).$$

Here, $\mathbf{F}_q$ is the finite field with $q = p^f$ elements and f equals the residue class degree of a prime lying over p. In particular, if p splits completely we get an injection $\mathrm{Ell}_\Delta(H_\mathcal{O}) \to \mathrm{Ell}_\Delta(\mathbf{F}_p)$. By the Deuring lifting theorem [8, Th. 13.12], the map g is surjective as well. Furthermore, the class group action in characteristic zero respects the reduction map, and we get a natural action of $\mathrm{Pic}(\mathcal{O}_\Delta)$ on $\mathrm{Ell}_\Delta(\mathbf{F}_q)$. Just like in characteristic zero, an $\mathcal{O}_\Delta$-ideal $\mathfrak{L}$ acts on $j(E) \in \mathrm{Ell}_\Delta(\mathbf{F}_q)$ by $j(E) \mapsto j(E/E[\mathfrak{L}]) = j(E)^{\mathfrak{L}}$. Since the Frobenius endomorphism of E commutes with all endomorphisms in $\mathfrak{L}$, the group $E[\mathfrak{L}]$ is $\mathbf{F}_q$-rational.

Lemma 1. *Let $E/\mathbf{F}_q$ be an ordinary elliptic curve and let $\varphi : E \to E'$ be an $\mathbf{F}_q$-isogeny of prime degree $l \neq \mathrm{char}(\mathbf{F}_q)$. Let π_q be the Frobenius morphism of E and let $\mathfrak{L} \subset \mathrm{End}(E)$ be an ideal of norm l. If l does not divide the index $[\mathrm{End}(E) : \mathbf{Z}[\pi_q]]$ then the kernel of φ equals either $E[\mathfrak{L}]$ or $E[\overline{\mathfrak{L}}]$.*

Proof. The kernel of φ is a subgroup of order l of the l-torsion of E. We have $E[l] \cong \mathbf{Z}/l\mathbf{Z} \times \mathbf{Z}/l\mathbf{Z}$ and there are $l+1$ subgroups of order l. A slight generalization of [6, Prop. 23] gives that only $\left(\frac{\Delta}{l}\right) + 1 \in \{0, 1, 2\}$ of those are $\mathbf{F}_q$-rational if l does not divide $[\mathrm{End}(E) : \mathbf{Z}[\pi_q]]$. As φ is defined over $\mathbf{F}_q$, the group $E[\mathfrak{L}]$ is $\mathbf{F}_q$-rational and the lemma follows. □

This lemma shows that 'most' of the $\mathbf{F}_q$-rational prime degree isogenies between ordinary elliptic curves over finite fields have a kernel of the form $E[\mathfrak{L}]$. Every $\mathcal{O}_\Delta$-ideal of prime norm l not dividing $[\mathrm{End}(E) : \mathbf{Z}[\pi_q]]$ can be written in the form

$$\mathfrak{L} = (l, c + d\pi_q),$$

and we can therefore represent the *kernel* of $E \to E/E[\mathfrak{L}]$ by specifying the $\mathrm{End}(E)$-ideal $\mathfrak{L} = (l, c + d\pi_q)$. This representation requires only $3 \log l$ bits.

The kernel C of a separable isogeny $\varphi : E \to E'$ does not uniquely determine φ. Indeed, if we compose φ with an isomorphism $E' \xrightarrow{\sim} E''$ then the kernel is unchanged. To keep track of isomorphisms, we choose Weierstraß equations for E and E' and note that the pull back $\varphi^*(\omega_{E'})$ of the invariant differential of E' equals a constant multiple of the invariant differential ω_E of E. If we have

$$\varphi^*(\omega_{E'}) = \omega_E$$

then the isogeny φ is said to be *normalized*. It is easy to see that a subgroup $C \subset E[l]$ of order l defines a unique elliptic curve E' such that there exists a normalized isogeny $E \to E'$ with kernel C. The isogeny $E \to E'$ is uniquely determined up to automorphisms of the curve E'. We conclude that a subgroup $C \subset E[l]$ determines a well-defined map $E \to E'/\mathrm{Aut}(E')$. The quotient $E'/\mathrm{Aut}(E')$ is isomorphic to the projective line $\mathbf{P}^1$ and in practice we will often map a point $P \in E'(\mathbf{F}_q)$ to its x-coordinate in $\mathbf{P}^1(\mathbf{F}_q)$. If E' has endomorphism ring $\mathbf{Z}[i]$ or $\mathbf{Z}[\zeta_3]$ we need to consider the square resp. cube of the x-coordinate. With this convention, the main result of the paper is the following.

Theorem 1. *Let $E/\mathbf{F}_q$ be an ordinary elliptic curve with Frobenius π_q, given by a Weierstraß equation, and let $P \in E(\mathbf{F}_{q^n})$ be a point on E. Let $\Delta = \mathrm{disc}(\mathrm{End}(E))$ be given. Assume that $[\mathrm{End}(E) : \mathbf{Z}[\pi_q]]$ and $\#E(\mathbf{F}_{q^n})$ are coprime, and let $\mathfrak{L} = (l, c + d\pi_q)$ be an $\mathrm{End}(E)$-ideal of prime norm $l \neq \mathrm{char}(\mathbf{F}_q)$ not dividing the index $[\mathrm{End}(E) : \mathbf{Z}[\pi_q]]$. Then Algorithm 4.1 computes the unique elliptic curve E' such that there exists a normalized isogeny $\varphi : E \to E'$ with kernel $E[\mathfrak{L}]$. Furthermore, it computes the x-coordinate of $\varphi(P)$ if $\mathrm{End}(E)$ does not equal $\mathbf{Z}[i]$ or $\mathbf{Z}[\zeta_3]$ and the square resp. cube of the x-coordinate of $\varphi(P)$ otherwise. The running time of the algorithm is polynomial in $\log l$, $\log q$, n and $|\Delta|$.*

Although the run time algorithm is polynomial in the discriminant Δ of the endomorphism ring $\mathrm{End}(E)$, the description of the algorithm in Section 4 shows that this 'bottleneck' disappears once $\mathfrak{L}$ is *principal*. Hence, it gives a polynomial time algorithm to evaluate all endomorphisms of the curve, regardless of the size of endomorphism ring of E.

3 Evaluating Small Degree Isogenies

Throughout this section, $E/\mathbf{F}_q$ is a fixed ordinary elliptic curve and $\mathfrak{L} = (l, c + d\pi_q)$ is an $\mathrm{End}(E)$-ideal of prime norm $l \neq \mathrm{char}(\mathbf{F}_q)$ not dividing the index $[\mathrm{End}(E) : \mathbf{Z}[\pi_q]]$. In this section we explain two methods to compute the image $\varphi(P) \in E'/\mathrm{Aut}(E') \cong \mathbf{P}^1$ of a point $P \in E(\mathbf{F}_{q^n})$ under 'the' normalized isogeny $\varphi : E \to E'$ defined by $\mathfrak{L}$. As the run time of these approaches is polynomial in l, the prime l should be small for these methods to be practical.

The first method is strongly based on the techniques that Atkin and Elkies used to improve Schoof's original point counting algorithm [10, Sec. 6–8]. It does not work in some special cases and we will make assumptions while describing the method. The second method works in general, but is typically slower.

3.1 Atkin-Elkies Techniques

We assume $p = \mathrm{char}(\mathbf{F}_q) > l \geq 3$ in this subsection, and we let E be given by a Weierstraß equation $Y^2 = X^3 + aX + b$. We assume that $\mathrm{End}(E)$ does not equal $\mathbf{Z}[i]$ or $\mathbf{Z}[\zeta_3]$. We will compute a polynomial $f_{\mathfrak{L}} \in \mathbf{F}_q[X]$ with the property that its roots are the x-coordinates of the points in $E[\mathfrak{L}]$. Once we know $f_{\mathfrak{L}}$ it is an easy matter to compute the image $\varphi(P)$. Indeed, Vélu's formulas [13] give us the normalized isogeny φ as rational function and we can simply evaluate at the point P.

To compute $f_{\mathfrak{L}} \in \mathbf{F}_q[X]$, we start by computing the j-invariant of E'. As E' is l-isogenous to E, we know that $j(E')$ is a root of the l-th modular polynomial $\Phi_l(j(E), X) \in \mathbf{F}_p[X]$ specialized in $j(E)$. The modular polynomial has degree $l+1$, but the assumption $l \nmid [\mathrm{End}(E) : \mathbf{Z}[\pi_q]]$ ensures that it has either 1 (if $\mathfrak{L}$ is ramified) or 2 (if $\mathfrak{L}$ splits) roots in $\mathbf{F}_q$. We fix a root $h \neq 0, 1728$. If $\mathfrak{L}$ splits, then h is either $j(E') = j(E)^{\mathfrak{L}}$ or $j(E)^{\overline{\mathfrak{L}}}$. We do not know which one yet.

The 'Atkin-Elkies techniques' only work if the partial derivative Φ_Y of $\Phi_l \in \mathbf{F}_q[X, Y]$ with respect to Y does not vanish when evaluated in $(X, Y) = (j(E), h)$. Using some algebraic geometry, one can prove [10, Sec. 7] that this only happens when l is larger than $4|\Delta|$, with Δ the discriminant of $\mathrm{End}(E)$. Hence, it only fails for 'large' l. In the examples we computed, this hardly caused any problems. If it does happen, we switch to the second method described below. For the remainder of this subsection we assume that $\Phi_Y(j(E), h)$ is not zero.

Next we compute an elliptic curve E_1 with j-invariant h such that the isogeny $E \to E_1$ with kernel $E[\mathfrak{L}]$ or $E[\overline{\mathfrak{L}}]$ is normalized. As in [10, Sec. 7], we put

$$s = -\frac{18}{l}\frac{b}{a}\frac{\Phi_X(j(E,h)}{\Phi_Y(j(E),h)}j(E) \in \mathbf{F}_q$$

and with

$$a' = -\frac{1}{48}\frac{s^2}{h(h-1728)} \in \mathbf{F}_q$$

$$b' = -\frac{1}{864}\frac{s^3}{h^2(h-1728)} \in \mathbf{F}_q,$$

the equation for E_1 is given by $Y^2 = X^3 + a'X + b'$.

Let C be the kernel of the normalized isogeny $E \to E_1$, i.e., C is either $E[\mathfrak{L}]$ or $E[\overline{\mathfrak{L}}]$. Theoretically, the hard part is computing the constant term p_1 of the kernel polynomial f_C describing C. The formulas are rather involved and can be found in [10, Sec. 8]. The other coefficients of f_C can now be found using a recursive relation involving the coefficients of the Laurent series of the Weierstraß-$\wp$ function. The key point is that computing f_C involves nothing more than simple arithmetic in $\mathbf{F}_q$. Once we have the equation for E_1, there are other methods as well to find f_C; we refer to [2] for an overview.

Knowing the polynomial f_C, it remains to check if our initial guess h was correct. We either have $f_C = f_{\mathfrak{L}}$ or $f_C = f_{\overline{\mathfrak{L}}}$ and to check in which case we are, we note that with $\mathfrak{L} = (l, c + d\pi_q)$, the Frobenius π_q acts as multiplication by

$-c/d \in \mathbf{F}_l$ on the points in $E[\mathfrak{L}]$. We test if $(X^q, Y^q) = (-c/d) \cdot (X, Y)$ holds for the points in C, i.e., we compute both (X^q, Y^q) and $(-c/d) \cdot (X, Y)$ in the ring

$$\mathbf{F}_q[X, Y]/(f_C(X), Y^2 - X^3 - aX - b).$$

Note that the $\cdot$ means repeated adding *on the curve* and $(-c/d) \cdot (X, Y)$ can be computed by employing division polynomials.

If we find that f_C does not equal $f_{\mathfrak{L}}$ we know that the unique other zero $h_2 \in \mathbf{F}_q$ of

$$\gcd(X^q - X, \Phi_l(j(E), X)) \in \mathbf{F}_q[X]$$

must be the j-invariant of E' and we repeat the computation with h replaced by h_2 to find the polynomial $f_C = f_{\mathfrak{L}} \in \mathbf{F}_q[X]$.

3.2 General Technique

The approach described in this subsection works for any prime power q and any prime $l \neq \mathrm{char}(\mathbf{F}_q)$. Let Ψ_l be the division polynomial for $E/\mathbf{F}_q$. For $l > 2$, the polynomial Ψ_l has degree $(l^2 - 1)/2$. By computing roots of Ψ_l, we compute two generators G_1, G_2 of the group $E[l] \cong \mathbf{Z}/l\mathbf{Z} \times \mathbf{Z}/l\mathbf{Z}$. The points will typically be defined over an extension of $\mathbf{F}_q$ of degree close to l. Indeed, if L denotes the field of definition of the l-torsion, then the degree $[L : \mathbf{F}_q]$ equals the order of π_q in the group $(\mathcal{O}_\Delta/l)^*$, and this order is usually close to l.

The goal is to find a point Q in the kernel $E[\mathfrak{L}]$. With $\mathfrak{L} = (l, c + d\pi_q)$ we need to find an l-torsion point Q with $\pi_q(Q) = (-c/d)Q$. We can simply list the generators $\alpha G_1 + \beta G_2$ of the $l+1$ subgroups of order l of $E[l]$ and check for each generator if Frobenius acts as multiplication by $-c/d$.

Once we find Q, we compute the subgroup generated by Q and use Vélu's formulas [13] to evaluate the isogeny.

4 Evaluating Large Degree Isogenies

The method described in Section 3 is intended for relatively small primes l. In this section we explain how to use the class group of the endomorphism ring $\mathrm{End}(E) = \mathcal{O}_\Delta$ to reduce the computation of a large degree isogeny to the computation of small degree isogenies. As before, $E/\mathbf{F}_q$ is an ordinary curve and $\mathfrak{L} = (l, c + d\pi_q)$ is an $\mathrm{End}(E)$-ideal of prime norm $l \nmid [\mathrm{End}(E) : \mathbf{Z}[\pi_q]]$. Let $P \in E(\mathbf{F}_{q^n})$ be a point. For reasons to become clear, we demand in this section that $[\mathrm{End}(E) : \mathbf{Z}[\pi_q]]$ and $\#E(\mathbf{F}_{q^n})$ are coprime. The goal is to compute $\varphi(P) \in E'/\mathrm{Aut}(E')$ with $\varphi : E \to E'$ an isogeny with kernel $E[\mathfrak{L}]$.

We have an equality

$$[\mathfrak{L}] = [\mathfrak{p}_1]^{e_1} \ldots [\mathfrak{p}_k]^{e_k} \tag{4.1}$$

inside the class group $\mathrm{Pic}(\mathcal{O}_\Delta)$ for some suitable choice of generators $\mathfrak{p}_i$. The key observation is that the norms of $\mathfrak{p}_i$ can be *much* smaller than the norm of $\mathfrak{L}$. Indeed, the size of $\mathfrak{p}_i$ depends only on the discriminant of $\mathrm{End}(E)$ and not of the norm of $\mathfrak{L}$. We can write $\mathfrak{L} = \mathfrak{p}_1^{e_1} \cdots \mathfrak{p}_k^{e_k}(\alpha)$ for some fractional principal

$\mathcal{O}_\Delta$-ideal (α). To find α, we compute the integral ideal $\mathfrak{L}\overline{\mathfrak{p}}_1^{e_1}\ldots\overline{\mathfrak{p}}_k^{e_k}$ and use Cornacchia's algorithm [4, Sec. 1.5.2] to find a generator $\beta \in \mathcal{O}_\Delta$. The choice $\alpha = \beta/m$, with m the product of the norms of the ideals occuring in (4.1), works.

To evaluate 'the' isogeny φ associated to $\mathfrak{L}$, it suffices to evaluate the isogenies associated to the $\mathfrak{p}_i$'s and to (α). If the $\mathfrak{p}_i$'s don't divide $[\mathrm{End}(E) : \mathbf{Z}[\pi_q]]$, we can use the method from Section 3 in the following way. We compute the isogeny

$$E \longrightarrow E_1 = E/E[\mathfrak{p}_1]$$

and note that we have a canonical isomorphism $\mathrm{End}(E) \xrightarrow{\sim} \mathcal{O}_\Delta \xrightarrow{\sim} \mathrm{End}(E_1)$ that allows us to interpret the 'next' ideal occuring in (4.1) as an $\mathrm{End}(E_1)$-ideal. Multiplication of $\mathcal{O}_\Delta$-ideals and composition of isogenies is compatible in the sense that we have

$$E/[\mathfrak{p}_1\mathfrak{p}_2] \cong E_1/[\mathfrak{p}_1].$$

By applying the method from Section 3 iteratively, we compute the normalized isogeny $\phi_c : E \to E_c = E/[\mathfrak{p}_1^{e^1}\cdots\mathfrak{p}_k^{e_k}]$.

We now explain how to deal with the ideal (α). The element β will typically *not* lie in the subring $\mathbf{Z}[\pi_q]$ of $\mathcal{O}_\Delta$. However, we can write $\alpha = (u + v\pi_q)/(mz)$ with $z \in \mathbf{Z}$ dividing the index $[\mathrm{End}(E) : \mathbf{Z}[\pi_q]]$. The curves E_c and $E' = E/E[\mathfrak{L}]$ are $\mathbf{F}_q$-isomorphic because (α) is a principal ideal. The space of invariant differentials for E' is a 1-dimensional $\mathbf{F}_q$-vector space, and because π_q is inseparable we have $\pi_q^*(\omega_{E'}) = 0$. Hence, the invariant differentials for the Weierstraß equations of E_c and E' satisfy

$$\omega_{E'} = (u/mz)\omega_{E_c}$$

if m is non-zero in $\mathbf{F}_q$. To find the equation for E', we need to apply an isomorphism $\eta : E_c \xrightarrow{\sim} E'$ with $\eta^*(\omega_{E'}) = (u/mz)\omega_{E_c}$. This is easy: if E_c is given by $Y^2 = X^3 + a'X + b'$ then for $\lambda \in \mathbf{F}_q^*$ the isomorphism $(X,Y) \mapsto (\lambda^2 X, \lambda^3 Y)$ multiplies ω_{E_1} by $1/\lambda$. Hence, the curve E' is given by $Y^2 = X^3 + (u/mz)^4 a'X + (u/mz)^6 b'$.

Having found the equation for E', we need to compute the action of (α) on the image $\eta(\phi_c(P)) \in E'(\mathbf{F}_{q^n})$. By assumption, the integer z in the denominator of α is coprime to $\#E(\mathbf{F}_{q^n})$. If m is also coprime to the group order of $E(\mathbf{F}_{q^n})$ then we can simply compute the inverse of zm modulo $\#E(\mathbf{F}_{q^n})$ and compute $R = ((zm)^{-1}(u + v\pi_q))(Q) \in E'(\mathbf{F}_{q^n})$. A suitable power of the x-coordinate of R is the value we are looking for. Summarizing everything, we have the following algorithm.

Algorithm 4.1

Input: a discriminant Δ, an elliptic curve $E/\mathbf{F}_q$ with $\mathrm{End}(E) = \mathcal{O}_\Delta$ and a point $P \in E(\mathbf{F}_{q^n})$ such that $[\mathrm{End}(E) : \mathbf{Z}[\pi_q]]$ and $\#E(\mathbf{F}_{q^n})$ are coprime, an $\mathrm{End}(E)$-ideal $\mathfrak{L} = (l, c + d\pi_q)$ of prime norm $l \neq \mathrm{char}(\mathbf{F}_q)$ not dividing $[\mathrm{End}(E) : \mathbf{Z}[\pi_q]]$.

Output: the elliptic curve E' such that an isogeny $\varphi : E \to E'$ with kernel $E[\mathfrak{L}]$ is normalized and the x-coordinate of $\varphi(P)$ for $\Delta \neq -3, 4$ and the cube resp. square of the x-coordinate otherwise.

1. Compute the direct sum decomposition $\text{Pic}(\mathcal{O}_\Delta) = \bigotimes \langle [\mathfrak{p}_i] \rangle$ of $\text{Pic}(\mathcal{O}_\Delta)$ into cyclic groups generated by the degree 1 prime ideals $\mathfrak{p}_i$ of smallest norm that are coprime to the product $p \cdot \#E(\mathbf{F}_{q^n}) \cdot [\text{End}(E) : \mathbf{Z}[\pi_q]]$.
2. Write $\mathfrak{L} = \mathfrak{p}_1^{e_1} \cdot \ldots \cdot \mathfrak{p}_k^{e_k} \cdot (\alpha)$ with the $\mathfrak{p}_i$'s as in Step 1.
3. Compute a sequence of isogenies $(\phi_1, \ldots, \phi_s)$ such that the composition $\phi_c : E \to E_c$ has kernel $E[\mathfrak{p}_1^{e_1} \ldots \mathfrak{p}_k^{e_k}]$ using the method from Section 3. Evaluate $\phi_c(P) \in E_c(\mathbf{F}_{q^n})$.
4. Write $\alpha = (u + v\pi_q)/(zm)$. Compute an isomorphism $\eta : E_c \xrightarrow{\sim} E'$ with $\eta^*(\omega_{E'}) = (u/zm)\omega_{E_c}$. Compute $Q = \eta(\phi_c(P))$.
5. Compute the inverse $(zm)^{-1}$ of zm modulo $\#E(\mathbf{F}_{q^n})$ and compute $R = ((zm)^{-1}(u + v\pi_q))(Q)$.
6. Put $r = x(R)^{|\mathcal{O}_\Delta|^*/2}$ and return (E', r).

An analysis of the algorithm yields Theorem 1:

Proof of Theorem 1. To prove the correctness of the algorithm, it suffices to show that we can take the generators in Step 1 coprime to $p \cdot \#E(\mathbf{F}_{q^n}) \cdot [\text{End}(E) : \mathbf{Z}[\pi_q]]$. This follows from the fact that every element in the class group is represented by infinitely many ideals.

The exact run time of Step 1 depends on the method we choose and what we are willing to assume, i.e., whether we want a probabilistic/deterministic algorithm and whether we are willing to assume GRH. We refer to [4, Sec. 5.4–5.5] for an overview. It can be done in deterministic polynomial time in $|\Delta|$, and the primes $\mathfrak{p}_i$ can be taken of polynomial size in $|\Delta|$. If we are willing to assume GRH, then we may even take $\mathfrak{p}_i$ to be of size $O((\log|\Delta|)^2)$. However, as we possibly have very large exponents in relation (4.1) this does not affect the total run time.

The computation of the exponents e_i in Step 1 can be done in various ways. The most naïve way of looping over all elements $I \in \text{Pic}(\mathcal{O}_\Delta)$ and checking whether $I^{-1}\mathfrak{L}$ is principal using Cornacchia's algorithm already has a run time that is polynomial in $\log l$ and $|\Delta|$ and this suffices for the proof of Theorem 1. This computation yields α as a by product.

Computing the cycle in Step 3 takes time polynomial in the norms of the $\mathfrak{p}_i$'s using the method in subsection 3.2. As the norms are of polynomial size in $|\Delta|$, this step takes polynomial time in $|\Delta|$. The computation of $\phi_c(P)$ takes polynomial time in $n \log q$ and $|\Delta|$. Steps 4–6 take time polynomial in $n \log q$ and the theorem follows. □

5 Examples

In this section we give two examples of Algorithm 4.1. The first example is rather small, and we check the result of the computation by employing the method of Section 3 directly. In the second example we use an isogeny of degree roughly 10^{21}, and checking the result using the method from Section 3 is impossible in this case.

5.1 Small Example

We fix $q = p = 101$ for this subsection. The elliptic curve $E : Y^2 = X^3 + 79X + 44$ has j-invariant $93 \in \mathbf{F}_p$ and we will show how to evaluate an isogeny of degree $l = 31$ using the class group algorithm from Section 4. An easy computation shows that E has trace of Frobenius $t = 15$, and as $\Delta = t^2 - 4p = -179$ is prime, we have $\mathrm{End}(E) \cong \mathcal{O}_\Delta$. By fixing a root π_p in $\mathcal{O}_\Delta$ of the polynomial $X^2 - tX + p$, we identify the rings $\mathrm{End}(E)$ and $\mathcal{O}_\Delta$.

We will compute the normalized isogeny φ corresponding to the $\mathcal{O}_\Delta$-ideal $\mathfrak{L} = (31, \pi_p + 3)$ lying over 31. The class group $\mathrm{Pic}(\mathcal{O}_\Delta)$ is cyclic of order 5. To find a suitable generator, we compute $\#E(\mathbf{F}_p) = 101 + 1 - 15 = 87 = 3 \cdot 29$. We see that we cannot use a prime lying over 3 to generate $\mathrm{Pic}(\mathcal{O}_\Delta)$, and we choose

$$\mathrm{Pic}(\mathcal{O}_\Delta) = \langle [\mathfrak{p}_5] \rangle$$

with $\mathfrak{p}_5 = (5, -2\pi_p + 1)$. We have $\mathfrak{L} = \mathfrak{p}_5(\alpha)$ with $\alpha = \frac{-3-\pi_p}{5}$.

Using the method from Section 3.1, we compute the kernel polynoimal $f_{\mathfrak{p}_5} = X^2 + 59X + 81 \in \mathbf{F}_p[X]$ associated to $\mathfrak{p}_5$. By applying Vélu's formulas, we find that the isogenous curve $E_c = E/E[\mathfrak{p}_5]$ has Weierstraß equation $Y^2 = X^3 + 30X + 63$. To find the Weierstraß equation for $E' = E/E[\mathfrak{L}]$, we compute $-3/5 = 60 \in \mathbf{F}_p$ and compute

$$Y^2 = X^3 + 30 \cdot 60^4 X + 63 \cdot 60^6$$

to find the equation $Y^2 = X^3 + 96X + 75$ for E'. We let $\eta : E_c \to E'$ be an isomorphism.

Take a random point $P = (68, 53) \in E(\mathbf{F}_p)$. We apply the isogeny ϕ_c associated to $\mathfrak{p}_5$ and find $\phi_c(P) = (30, 17) \in E_c(\mathbf{F}_p)$. The point $Q = \eta(\phi_c(P)) = (31, 44) \in E'(\mathbf{F}_p)$ lies on the right curve. As it lies in the base field, the Frobenius acts as the identity on this point and we multiply Q by $(-3-1)/5 = 34 \in \mathbf{Z}/87\mathbf{Z}$ to find the image $R = (46, 25) \in E'(\mathbf{F}_p)$. The output of the algorithm is $(Y^2 = X^3 + 96X + 75, 46)$.

The degree $l = 31$ is small enough that we can check this output by using the method from Section 3 directly. The kernel polynomial associated to $\mathfrak{L}$ is $f_{\mathfrak{L}} = X^{15} + 39X^{14} + 88X^{13} + \ldots + 17X^2 + 65X + 4 \in \mathbf{F}_p[X]$ and we compute the image $\varphi(P) = (46, 25)$ for the isogeny $\varphi : E \to E'$ directly from Vélu's formulas.

5.2 Medium-Sized Example

Our algorithm is capable of handling much larger inputs than the $l = 31$ from section 5.1. Evaluating isogenies of degree roughly 10^{100} is no problem. As displaying large numbers is not especially pleasing to the human eye, we give a 'medium sized' example in this section. Using the method from [3], we construct a curve with small endomorphism ring having exactly $10^{20} + 39 = \mathrm{nextprime}(10^{20})$ points.

With $p = 99999999980010207001$, the elliptic curve $E/\mathbf{F}_p$ defined by

$$Y^2 = X^3 + 93111780581619358815X + 13776438796781696372$$

has $10^{20}+39$ points. The endomorphism ring $\mathrm{End}(E)$ is isomorphic to $\mathcal{O}_\Delta$ for $\Delta=-3635$. The prime $l=10^{21}+117=\mathrm{nextprime}(10^{21})$ splits in $\mathcal{O}_\Delta$ and we take the $\mathcal{O}_\Delta$-ideal $\mathfrak{L}=(l,\pi_p+469155077064851443344)$. Here, π_p is the image of the Frobenius morphism under the normalized isomorphism $\mathrm{End}(E)\xrightarrow{\sim}\mathcal{O}_\Delta$.

The smallest prime not dividing $[\mathrm{End}(E):\mathbf{Z}[\pi_p]]=3^4\cdot 19^2\cdot 31^2\cdot 1999^2$ that splits in $\mathcal{O}_\Delta$ is 37 and we have $\mathrm{Pic}(\mathcal{O}_\Delta)\cong\mathbf{Z}/10\mathbf{Z}\cong\langle[\mathfrak{p}_{37}]\rangle$ with $\mathfrak{p}_{37}=(37,\pi_p+15)$. An easy computation yields the equality $\mathfrak{L}=\mathfrak{p}_{37}(\alpha)$ with

$$\alpha=\frac{-2947049\pi_p-708893381093724965}{3\cdot 19\cdot 31\cdot 1999\cdot 37}.$$

The primes in the denominator of α are 37 and the primes dividing the index $[\mathrm{End}(E):\mathbf{Z}[\pi_p]]$.

We compute the isogeny ϕ_c corresponding to $\mathfrak{p}_{37}$ using the method from Section 3. The kernel polynomials equals $X^{18}+67504589328326227502X^{17}+\ldots+35418368365443750601\in\mathbf{F}_p[X]$ and the isogenous curve E_c has Weierstraß equation

$$Y^2=X^3+80827651155168177778X+51575975418311029503.$$

We multiply the coefficients of this equation by the 4th resp. 6th power of $-708893381093724965/(3\cdot 19\cdot 31\cdot 1999\cdot 37)=98412218672392141083\in\mathbf{F}_p$ to find the Weierstraß equation

$$Y^2=X^3+83032917062416905069X+31170711888319926172$$

for E'. We let $\eta:E_c\xrightarrow{\sim}E'$ be an isomorphism.

Take a random point $P=(73931099962253475826,29177286940991158970)$ on E. We compute $Q=\eta(\phi_c(P))\in E'(\mathbf{F}_p)$ and multiply this by $(-2947049-708893381093724965)/(3\cdot 19\cdot 31\cdot 1999\cdot 37)=89908927599601102372\in\mathbf{Z}/(10^{20}+39)\mathbf{Z}$ to find

$$R=(95529214469768926304,49609901207400538475)\in E'(\mathbf{F}_p).$$

The output of the algorithm is the equation for E' and 95529214469768926304.

6 Applications to Pairing-Based Cryptography

In the last decade, bilinear pairings have been used to enable new cryptographic functionality and have been proposed as the basis for a wide variety of cryptographic protocols, from Identity Based Encryption (IBE) to tri-partite Diffie-Hellman to shorter digital signatures. The first digital signature scheme (BLS) based on bilinear pairings was introduced in 2001 by Boneh, Lynn, and Shacham [1].

6.1 BLS Digital Signatures

Here is an informal description of how the basic BLS signature scheme works on an elliptic curve E with the Weil pairing.

Public parameters. Let E be an elliptic curve over a field $\mathbf{F}_q$ of characteristic p. Let m be a positive integer and let $e_m(P, Q)$ denote the Weil pairing of two points P and Q in the group of m-torsion points $E[m]$. The Tate pairing or other modified pairings, such as the squared Tate pairing, can also be substituted in the scheme and in its security assumptions. The set-up for the scheme includes a public point $Q \in E[m]$. We assume that m is prime.

Public/Private Key. Each user has a secret key which is an integer, s, and a corresponding public key, sQ, which is published.

Signing. A message, M, to be transmitted and signed with signature σ is signed as follows. The message is first hashed to a point $P \in E[m]$, following for example the procedure outlined in [1, Section 3.2]. The signer has a secret integer s, and signs the message by computing $\sigma = sP$.

Verifying. To verify the signature $\sigma = sP$ on a message M, the verifier uses the same hashing procedure as above to hash M to the point P on the elliptic curve. Then the verifier computes two Weil pairings $e_m(P, sQ)$ and $e_m(\sigma, Q)$ and checks that they are equal.

Note: For ordinary elliptic curves with m co-prime to p, the group $E[m]$ has rank 2, and the points P and Q in the above scheme are chosen to be linearly independent when using the Weil pairing, since otherwise the pairing would be trivial. For efficiency reasons, E is usually chosen or constructed [9] to be such that all the m-torsion is defined over a small degree extension of $\mathbf{F}_q$, and messages are hashed into the smallest possible field, to minimize the bit-length of the signature.

Security. In order for the above scheme to be secure, it is assumed that the groups generated by P and Q are a *co-GDH pair* ([1, Definition 2.1]), meaning that the co-Gap Diffie Hellman problem is hard for the two pieces of the m-torsion. The security proof models the hash function which maps messages to points as a random oracle.

6.2 Isogeny Variants of BLS

The techniques described in Algorithm 4.1 can be used to enable several different variants of the BLS signature scheme. These variants require expanded security assumptions and depend on the ability to efficiently evaluate a large degree isogeny (the degree should be of cryptographic size, such as on the order of 2^{160}). Isogenies of such large degree were previously impossible to evaluate in a reasonable amount of time, other than multiplication by an integer, possibly composed with a small degree isogeny.

A. One extension of the basic BLS scheme described above is to use points P and Q on two different isogenous elliptic curves.

In other words, let $E_1/\mathbf{F}_p$ be an ordinary elliptic curve with endomorphism ring $\mathrm{End}(E) \neq \mathbf{Z}[i], \mathbf{Z}[\zeta_3]$ and let $\varphi : E_1 \to E_2$ be specified by an ideal $\mathfrak{L}$ as in Section 2. The triple $(E_1, \mathfrak{L}, E_2)$ is public. Assume that the conditions from Theorem 1 are satisfied for the elliptic curve E_1. This is a rather harmless condition, since for pairing friendly curves, the degree l of φ does not divide $[\mathrm{End}(E) : \mathbf{Z}[\pi_p]]$. For a user with secret key $s \in \mathbf{Z}$, the public key is $sQ \in E_2[m]$. The message M, is hashed to a point $P \in E_1[m]$ and signed as above, with $\sigma = sP$, but the verification is accomplished by computing two pairings in $E_2[m]$ namely $e_m(\varphi(P), sQ)$ and $e_m(\varphi(\sigma), Q)$. As only the x-coordinate of $\varphi(P)$ is well-defined, we now accept the signature if $e_m(\varphi(P), sQ) = \pm e_m(\varphi(\sigma), Q)$ holds.

This scheme requires two evaluations of the isogeny in the verification step. Here the isogeny is public, and need not have large degree. Whereas it was essential to choose P and Q to be linearly independent in the original BLS-scheme, we now require P and Q to be such that $\varphi(P)$ and Q are lineary independent in $E_2[m]$. The security depends again on the co-Gap-Diffie-Hellman Assumption, this time for the two groups $G_1 = \langle P \rangle \subset E_1[m]$ and $G_2 = \langle Q \rangle \subset E_2[m]$.

B. Our original motivation for developing a polynomial time algorithm for evaluating large degree isogenies was for application to a BLS-variant proposed in [5] where the isogeny is the secret key of the user.

The set-up is as follows. Two ordinary elliptic curves E_1 and E_2 over a field $\mathbf{F}_p$ with isomorphic endomorphism rings of discriminant $\Delta < -4$, and a point Q in $E_2[m]$ are public parameters. A user has a secret key, which is an isogeny $\varphi : E_1 \to E_2$ specified by an ideal $\mathfrak{L}$ as in Theorem 1. Let $\hat{\varphi} : E_2 \to E_1$ denote the dual isogeny, i.e., $\hat{\varphi}$ corresponds to the complex conjugate $\overline{\mathfrak{L}} \subset \mathrm{End}(E_2) = \mathcal{O}_\Delta$ of $\mathfrak{L}$. The corresponding public key is the image $\hat{\varphi}(Q)$.

Signing. A user signs a message M by computing the hash of the message onto a point $P \in E_1[m]$, and then applying the secret isogeny φ to get the signature $\sigma = \varphi(P)$.

Verification. The verification step depends on the adjoint property of φ and $\hat{\varphi}$ with respect to the Weil pairing [12, Ch. 3, Prop. 8.2]. The verifier checks that $e_m(Q, \sigma) = \pm e_m(\hat{\varphi}(Q), P)$ holds.

This system also requires two applications of an isogeny, one for setting up the user's public key and one for signing. Verification does not require computation of an isogeny. Since the two elliptic curves are public, it is clear that the secret isogeny must have large degree to avoid exhaustive search attacks. We note that there are *many* isogenies of large degree that fit our theorem, since half of the primes split in the ring $\mathrm{End}(E)$ and lead to an ideal $\mathfrak{L}$ that we can use.

Acknowledgement. We thank René Schoof for helpful discussions.

References

[1] Boneh, D., Lynn, B., Shacham, H.: Short signatures from the Weil pairing. In: Boyd, C. (ed.) ASIACRYPT 2001. LNCS, vol. 2248, pp. 514–532. Springer, Heidelberg (2001)

[2] Bostan, A., Morain, F., Salvy, B., Schost, E.: Fast algorithms for computing isogenies between elliptic curves. Math. Comp. 77, 1755–1778 (2008)

[3] Bröker, R., Stevenhagen, P.: Constructing elliptic curves of prime order. Contemp. Math. 463, 17–28 (2008)
[4] Cohen, H.: A course in computational algebraic number theory. Springer Graduate Texts in Mathematics, vol. 138 (1993)
[5] Jao, D., Venkatesan, R.: Use of isogenies for design of cryptosystems, http://www.freepatentsonline.com/EP1528705.html
[6] Kohel, D.: Endomorphism Rings of Elliptic Curves over Finite Fields, PhD thesis, University of California at Berkeley (1996)
[7] Konstantinou, E., Stamatiou, Y.C., Zaroliagis, C.D.: On the construction of prime order elliptic curves. In: Johansson, T., Maitra, S. (eds.) INDOCRYPT 2003. LNCS, vol. 2904, pp. 309–322. Springer, Heidelberg (2003)
[8] Lang, S.: Elliptic functions, 2nd edn. Springer Graduate Texts in Mathematics, vol. 112 (1987)
[9] Miyaji, A., Nakabayashi, M., Takano, S.: New explicit conditions of elliptic curve traces for FR-reduction. IEICE Trans. on Fund. E84-A(5), 1234–1243 (2001)
[10] Schoof, R.: Counting points on elliptic curves over finite fields. J. Théor. Nombres Bordeaux 7, 219–254 (1995)
[11] Silverman, J.: Advanced topics in the arithmetic of elliptic curves. Springer Graduate Texts in Mathematics, vol. 151 (1994)
[12] Silverman, J.: The arithmetics of elliptic curves, 2nd edn. Springer Graduate Texts in Mathematics, vol. 106 (1992)
[13] Vélu, J.: Isogénies entre courbes elliptiques. C. R. Acad. Sci. Paris Sér. A–B 273, A238–A241 (1971)

Computing the Cassels Pairing on Kolyvagin Classes in the Shafarevich-Tate Group

Kirsten Eisenträger[1,⋆], Dimitar Jetchev[2], and Kristin Lauter[3]

[1] Department of Mathematics, The Pennsylvania State University, University Park, PA 16802
eisentra@math.psu.edu

[2] Department of Mathematics, University of California, Berkeley, CA 94720
jetchev@math.berkeley.edu

[3] Microsoft Research, One Microsoft Way, Redmond, WA 98052
klauter@microsoft.com

Abstract. Kolyvagin has shown how to study the Shafarevich-Tate group of elliptic curves over imaginary quadratic fields via Kolyvagin classes constructed from Heegner points. One way to produce explicit non-trivial elements of the Shafarevich-Tate group is by proving that a locally trivial Kolyvagin class is globally non-trivial, which is difficult in practice. We provide a method for testing whether an explicit element of the Shafarevich-Tate group represented by a Kolyvagin class is globally non-trivial by determining whether the Cassels pairing between the class and another locally trivial Kolyvagin class is non-zero. Our algorithm explicitly computes Heegner points over ring class fields to produce the Kolyvagin classes and uses the efficiently computable cryptographic Tate pairing.

1 Introduction

The Kolyvagin Euler system of cohomology classes constructed in [15] (see also [11] and [19]) is one of the most powerful tools for studying the Shafarevich-Tate group of an elliptic curve over a quadratic imaginary field with a Heegner discriminant. Under certain assumptions, it can be used to prove that the Shafarevich-Tate group of the elliptic curve is finite, and to determine its exact group structure (see [11], [15] and [19]). The cohomology classes constructed by Kolyvagin are locally trivial at all but a finite set of places. One can write down local divisibility conditions on the corresponding Heegner points which are equivalent to local triviality at the remaining finite set of places. Yet, even when those conditions are satisfied, there is no guarantee that the Kolyvagin cohomology class is a non-trivial element of the Shafarevich-Tate group.

One strategy for proving that such an element corresponds to a non-trivial element of the Shafarevich-Tate group is to prove that it pairs with another element of the Shafarevich-Tate group via the Cassels pairing to a nonzero element of

⋆ Partially supported by a National Science Foundation postdoctoral fellowship.

S.D. Galbraith and K.G. Paterson (Eds.): Pairing 2008, LNCS 5209, pp. 113–125, 2008.

$\mathbb{Q}/\mathbb{Z}$. The Cassels pairing is a skew-symmetric pairing on the Shafarevich-Tate group $\text{Ш}(K, E)$ which is non-degenerate when $\text{Ш}(K, E)$ is finite. To use this approach we need a way of explicitly computing the Cassels pairing. Our goal in this paper is to describe a method for deciding whether the Cassels pairing on two suitably chosen distinct locally trivial Kolyvagin cohomology classes is nonzero. The algorithm uses the cryptographic Tate pairing and the computation of Heegner points over ring class fields.

Our algorithm may have interesting applications to the study of $\text{Ш}(K, E)$. For instance, Mazur and Rubin ([18]) have suggested that $\text{Ш}(K, E)$ should be generated by locally trivial Kolyvagin cohomology classes. An interesting computational question is to test this conjecture experimentally and to observe in practice and predict bounds on the size of the Kolyvagin primes necessary to find generators for $\text{Ш}(K, E)[p]$. Furthermore, our algorithm may be used for producing examples related to visibility of $\text{Ш}(K, E)$ at higher level in the sense of Stein and the second author [14].

We first recall the definition of the Shafarevich-Tate group of an elliptic curve. Section 2 provides the necessary background on Heegner points and describes the construction of the Kolyvagin cohomology classes out of these points. In Section 3 we explain how to decide whether a Kolyvagin cohomology class is locally trivial. The algorithm takes as input the coordinates of the Heegner point in the corresponding ring class field.

In Section 4 we discuss the Cassels pairing in general and establish a formula for the pairing of two locally trivial Kolyvagin classes in terms of the Tate local pairing. In Section 5 we explain how the Tate local pairing is related to a pairing over finite fields, known in the cryptographic literature as the (cryptographic) Tate pairing. In Section 6 we apply the pairing for suitably chosen Kolyvagin classes to decide whether they pair non-trivially via the Cassels pairing. Finally, in Section 7 we explain the computation of the coordinates of Heegner points over ring class fields and give an example.

Notation and Background. For a field K, $\overline{K}$ denotes an algebraic closure of K. For a number field K, M_K denotes the set of places of K (both archimedean and non-archimedean). For $v \in M_K$, K_v is the completion of K at v. For a field K and a smooth commutative K-group scheme G, $\mathrm{H}^i(K, G)$ denotes the Galois cohomology group $\mathrm{H}^i(G_{K_s/K}, G(K_s))$ where K_s is a fixed separable closure of K.

Let E be an elliptic curve over a number field K. The *Shafarevich-Tate group* of E over K is

$$\text{Ш}(K, E) := \mathrm{Ker}\left(\mathrm{H}^1(K, E) \to \prod_{v \in M_K} \mathrm{H}^1(K_v, E)\right).$$

For a positive integer m we define the m-Selmer group $\mathrm{Sel}^{(m)}(K, E)$ of the elliptic curve as

$$\mathrm{Sel}^{(m)}(K, E) := \mathrm{Ker}\left(\mathrm{H}^1(K, E_m) \to \prod_{v \in M_K} \mathrm{H}^1(K_v, E)_m\right),$$

where each map $\mathrm{H}^1(K, E_m) \to \mathrm{H}^1(K_v, E)_m$ is the composition of the maps $\mathrm{H}^1(K, E_m) \to \mathrm{H}^1(K_v, E_m)$ and $\mathrm{H}^1(K_v, E_m) \to \mathrm{H}^1(K_v, E)_m$ (for more details see [24, Ch.X]). In general, if m is a positive integer and G is an abelian group object, we denote by either G_m or $G[m]$ the kernel of the multiplication-by-m map on G.

2 Heegner Points and the Kolyvagin Construction

Heegner Points over Ring Class Fields. The standard references for this section are [11], [15] and [19]. Let E be an elliptic curve over $\mathbb{Q}$ of conductor N. Let $K = \mathbb{Q}(\sqrt{-D})$, where $-D$ is a fundamental discriminant, $D \neq 3, 4$, and all prime factors of N are split in K, i.e. $(N) = \mathcal{N}\bar{\mathcal{N}}$ for an ideal $\mathcal{N}$ of the ring of integers $\mathcal{O}_K$ of K with $\mathcal{O}_K/\mathcal{N} \simeq \mathbb{Z}/N\mathbb{Z}$. We call such a discriminant a *Heegner discriminant* for $E/\mathbb{Q}$. By the modularity theorem [2], there is a modular parameterization $\varphi : X_0(N) \to E$. We view $\mathcal{O}_K$ and $\mathcal{N}$ as $\mathbb{Z}$-lattices of rank two in $\mathbb{C}$ and observe that $\mathbb{C}/\mathcal{O}_K \to \mathbb{C}/\mathcal{N}^{-1}$ is a cyclic isogeny of degree N between the elliptic curves $\mathbb{C}/\mathcal{O}_K$ and $\mathbb{C}/\mathcal{N}^{-1}$. Here $\mathcal{N}^{-1}$ denotes the fractional ideal of $\mathcal{O}_K$ for which $\mathcal{N}\mathcal{N}^{-1} = \mathcal{O}_K$. This isogeny corresponds to a complex point $x_1 \in X_0(N)$. According to the theory of complex multiplication [25, Ch.II], the point x_1 is defined over the Hilbert class field K_1 of K.

More generally, let $\mathcal{O}_n = \mathbb{Z} + n\mathcal{O}_K$ be the order of index n in $\mathcal{O}_K$ and let $\mathcal{N}_n = \mathcal{N} \cap \mathcal{O}_n$. Then $\mathcal{O}_n/\mathcal{N}_n \simeq \mathbb{Z}/N\mathbb{Z}$ and the map $\mathbb{C}/\mathcal{O}_n \to \mathbb{C}/\mathcal{N}_n^{-1}$ is a cyclic isogeny of degree N and thus it defines a point $x_n \in X_0(N)(\mathbb{C})$. Again, by the theory of complex multiplication, this point is defined over the ring class field K_n of conductor n over K.

One can use the parameterization $\varphi : X_0(N) \to E$ to obtain points $y_n = \varphi(x_n)$ on E. Define y_K to be the point $y_K = \mathrm{Tr}_{K_1/K}(y_1)$. If $\mathcal{N}'$ is another ideal with $\mathcal{O}/\mathcal{N}' \simeq \mathbb{Z}/N\mathbb{Z}$ and corresponding point y'_K, we have $y_K = \pm y'_K +$ (torsion). We refer to y_K as the Heegner point for the discriminant D (where $-D$ is negative).

Surjectivity of the Galois Representation and Choice of p. Let E and K be as above. For a rational prime p, let $\mathbb{Q}(E_p)$ be the extension generated by the p-torsion points of E in $\overline{K}$. If the elliptic curve E does not have complex multiplication then according to a theorem of Serre [23, Thm. 2], the extension $\mathbb{Q}(E_p)/\mathbb{Q}$ has Galois group $\mathrm{GL}_2(\mathbb{F}_p)$ for all but finitely many primes p; i.e., the associated mod p Galois representation is surjective for all but finitely many primes p. In practice, one tests the surjectivity of the Galois representation for a given E and p by using the algorithm of [10, §2.1], which has been implemented by Stein in the computer algebra system SAGE [26].

Construction of the Cohomology Classes. In [15], Kolyvagin uses the points y_n for suitably chosen indices n to define cohomology classes $d_{n,M} \in \mathrm{H}^1(K, E)_{p^M}$ which are locally trivial at all places coprime to n. We give a brief account of the construction here following the notation in [11] (see [11] and [15] for full details).

Definition 1. *Let E, K, p be as above, and let M be a positive integer. A prime number ℓ is called a* Kolyvagin prime *for (E, K, p, M) if the following three conditions are satisfied*

1. *ℓ does not divide $N \cdot D \cdot p$.*
2. *ℓ is inert in K.*
3. *$a_\ell \equiv \ell + 1 \equiv 0 \mod p^M$.*

For each such ℓ, let λ denote the unique prime of K above ℓ and let $G_\ell = G_{K_\ell/K_1}$. Then $G_\ell \simeq (\mathcal{O}_K/\ell\mathcal{O}_K)^\times/(\mathbb{Z}/\ell\mathbb{Z})^\times \simeq \mathbb{F}_\lambda^\times/\mathbb{F}_\ell^\times$ is cyclic of order $\ell + 1$, so one can choose a generator $\sigma_\ell \in G_\ell$. Let $\mathrm{Tr}_\ell = \sum_{i=0}^{\ell} \sigma_\ell^i$ and let $D_\ell \in \mathbb{Z}[G_\ell]$ be chosen in such a way that

$$(\sigma_\ell - 1) \cdot D_\ell = 1 + \ell - \mathrm{Tr}_\ell .$$

For instance, one can choose $D_\ell = \sum_{i=1}^{\ell} i \cdot \sigma_\ell^i$.

Now suppose that n is a square-free product of Kolyvagin primes for (E, K, p, M). Let $D_n = \prod_{\ell | n} D_\ell$, $\mathcal{G}_n = G_{K_n/K}$ and $G_n = G_{K_n/K_1}$. Suppose that $S \subset \mathcal{G}_n$ is a system of coset representatives for $\mathcal{G}_n/G_n$. One can show [11, Prop. 3.6] that the image of $D_n y_n$ in $E(K_n)/p^M E(K_n)$ is fixed by G_n. Thus, if we set

$$P_n = \sum_{\sigma \in S} \sigma(D_n y_n),$$

then the image of P_n in $E(K_n)/p^M E(K_n)$ will be fixed by $\mathcal{G}_n$. To define the classes, we consider the following commutative diagram

$$\begin{array}{ccccccccc}
 & & & & & & 0 & & \\
 & & & & & & \downarrow & & \\
 & & & & & & \mathrm{H}^1(K_n/K, E(K_n))_{p^M} & & \\
 & & & & & & \downarrow \mathrm{Inf} & & \\
0 & \longrightarrow & E(K)/p^M E(K) & \xrightarrow{\delta} & \mathrm{H}^1(K, E_{p^M}) & \longrightarrow & \mathrm{H}^1(K, E)_{p^M} & \longrightarrow & 0 \\
 & & \downarrow & & \downarrow \phi & & \downarrow \mathrm{Res} & & \\
0 & \rightarrow & (E(K_n)/p^M E(K_n))^{\mathcal{G}_n} & \xrightarrow{\delta} & \mathrm{H}^1(K_n, E_{p^M})^{\mathcal{G}_n} & \longrightarrow & \mathrm{H}^1(K_n, E)_{p^M}^{\mathcal{G}_n} & &
\end{array}$$

According to [11], the restriction map $\phi : \mathrm{H}^1(K, E_{p^M}) \to \mathrm{H}^1(K_n, E_{p^M})^{\mathcal{G}_n}$ is an isomorphism, so we can define the cohomology class

$$c_{n,M} = \phi^{-1}(\delta(P_n)) \in \mathrm{H}^1(K, E_{p^M}).$$

By Lemma 4.1 in [19], $c_{n,M}$ is represented by the cocycle

$$\sigma \mapsto -\frac{(\sigma - 1)P_n}{p^M} + \sigma\frac{P_n}{p^M} - \frac{P_n}{p^M}, \tag{1}$$

where $\frac{(\sigma-1)P_n}{p^M}$ is the unique p^M-division point of $(\sigma - 1)P_n$ in $E(K_n)$.

Let $d_{n,M}$ be the image of $c_{n,M}$ in $\mathrm{H}^1(K, E)_{p^M}$.

3 An Algorithm for Deciding Local Triviality

One of the basic properties of the class $d_{n,M}$ is that it is locally trivial at all but the places lying over the prime divisors of n. We denote by res_v the restriction $\mathrm{res}_v : H^1(K,E) \to H^1(K_v,E)$. The following proposition is proved in [11, Prop. 6.2].

Proposition 1. *(i) If v is a place of K such that $v \nmid n$ or if $v = \infty$ is the archimedean place then $\mathrm{res}_v(d_{n,M}) = 0$.*
(ii) If λ is a place of K above a prime divisor ℓ of n then $\mathrm{res}_\lambda(d_{n,M}) = 0$ if and only if $P_{n/\ell} \in p^M E(K_{n/\ell,\lambda'})$ for one (and hence all) places λ' of $K_{n/\ell}$ dividing λ. Here, $K_{n/\ell,\lambda'}$ denotes the completion of $K_{n/\ell}$ at λ'.

The following standard lemma will be used to provide an algorithm for deciding whether a point is divisible.

Lemma 1. *Let F be any number field and let F_v be the completion of F at a non-archimedean place v. Let E be an elliptic curve over F_v with good reduction. Let m be an integer which is relatively prime to the characteristic of the residue field k_v. We have*

1. *$E(F_v)/mE(F_v) \cong \tilde{E}(k_v)/m\tilde{E}(k_v)$, and*
2. *$E_m(F_v) \cong \tilde{E}_m(k_v)$.*

In particular, a point $Q \in E(F_v)$ is m-divisible if and only if its reduction $\tilde{Q} \in \tilde{E}(k_v)$ is m-divisible.

Proof. Consider the following commutative diagram

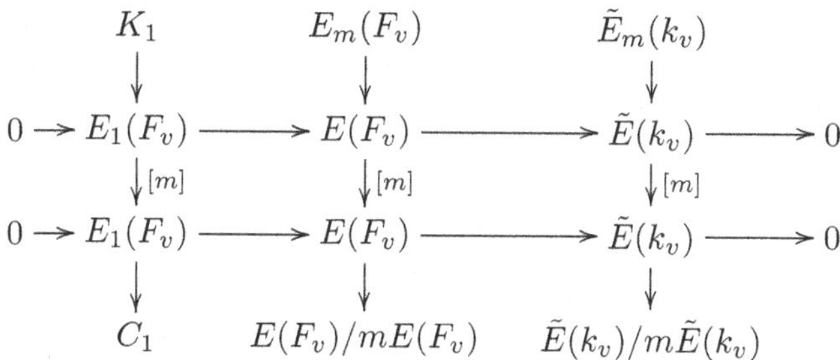

where K_1 and C_1 are the kernel and cokernel of the first map, and $E_1(F_v)$ is the kernel of reduction. We use the snake lemma to get an exact sequence

$$C_1 \to E(F_v)/mE(F_v) \to \tilde{E}(k_v)/m\tilde{E}(k_v) \to 0$$

Since m is prime to the characteristic of k_v the multiplication-by-m map on $E_1(F_v)$ is an isomorphism, i.e. $K_1 = C_1 = 0$ ([24, VII, Prop.2.2, IV,§3]). Thus, we get isomorphisms $E(F_v)/mE(F_v) \simeq \tilde{E}(k_v)/m\tilde{E}(k_v)$ and $E_m(F_v) \simeq \tilde{E}_m(k_v)$. This proves the lemma.

Remark 1. Applying this to the Kolyvagin setup, by Proposition 1, the class $d_{n,M}$ is locally trivial at λ if and only if $P_{n/\ell}$ is in $p^M E(K_{n/\ell,\lambda'})$. According to the above lemma, this is equivalent to $\tilde{P}_{n/\ell} \in p^M \tilde{E}(k_\lambda)$, since K_λ and $K_{n/\ell,\lambda'}$ have the same residue field. The last condition can be tested by computing the reduction of $P_{n/\ell}$ modulo λ'.

4 The Cassels Pairing on the Shafarevich-Tate Group

In this section we consider an elliptic curve E over an arbitrary number field K. We will describe a skew-symmetric pairing on the Shafarevich-Tate group $\text{Ш}(K, E)$, the Cassels pairing, which will be non-degenerate in the case when $\text{Ш}(K, E)$ is finite. The description follows [19, §2]. Let $d \in \text{Ш}(K, E)_m$ and $d' \in \text{Ш}(K, E)_{m'}$ be two elements of the Shafarevich-Tate group. Choose a lift $c' \in \text{Sel}^{(m')}(K, E)$ of d' and for each valuation $v \in M_K$ choose $y_v \in E(K_v)$, such that $c'_v = \delta(y_v)$ ($\delta : E(K_v) \to \mathrm{H}^1(K_v, E_{m'})$ is the connecting homomorphism). Also, assume that there exists a class $d_1 \in \mathrm{H}^1(K, E)_{mm'}$, such that $m'd_1 = d$. Since d is locally trivial, $\text{res}_v(d_1) \in \mathrm{H}^1(K_v, E)_{m'}$. We define the Cassels pairing $\langle\, , \rangle_C$ between d and d' as

$$\langle d, d' \rangle_C := \sum_{v \in M_K} \langle y_v, \text{res}_v(d_1) \rangle_{K_v},$$

where $\langle\, , \rangle_{K_v} : E(K_v) \times \mathrm{H}^1(K_v, E) \to \mathbb{Q}/\mathbb{Z}$ is the Tate local pairing. For more detail on Tate local duality and the pairing see [20].

Computing the Cassels Pairing on Kolyvagin Classes. The next proposition, which is proved in [19, Prop.4.7], specializes the above formula to locally trivial Kolyvagin cohomology classes.

Proposition 2. *Let E, K, p be as in Section 2. Let M and M' be positive integers and let n and n' be square-free products of Kolyvagin primes for $(E, K, p, M + M')$ and (E, K, p, M'), respectively. Suppose that the classes $d_{n,M} \in \mathrm{H}^1(K, E)_{p^M}$ and $d_{n',M'} \in \mathrm{H}^1(K, E)_{p^{M'}}$ are everywhere locally trivial (i.e. they lie in $\text{Ш}(K, E)$). Then the Cassels pairing is*

$$\langle d_{n,M}, d_{n',M'} \rangle_C = \sum_{\ell \mid n, (\ell, n')=1} \langle P_{n'}, \text{res}_\lambda(d_{n,M+M'}) \rangle_{K_\lambda}.$$

For simplicity of notation, we write the pairing as $\langle\ ,\ \rangle_{K_\lambda}$ rather than introducing $K_{n',\lambda'}$ as in Proposition 1. We use the definition of the pairing given in the previous section with $m = p^M$ and $m' = p^{M'}$. It follows from [19, Lemma 4.6] that $p^{M'}(d_{n,M+M'}) = d_{n,M}$, so the element $d_{n,M+M'}$ can be used as d_1 in the definition of the pairing given in the previous section.

5 Tate Local Pairing and Pairings over Finite Fields

The main reference for this section is [8]. Throughout this section, let K be any number field, and let K_v be the completion of K at a non-archimedean place v of K. Let k_v be the residue field of K_v, and let E be an elliptic curve over K_v with good reduction. Let m be an integer which is prime to the characteristic of k_v. Consider the Tate local pairing

$$\langle\, , \rangle_{K_v} : E(K_v)/mE(K_v) \times \mathrm{H}^1(K_v, E)_m \to \mathbb{Q}/\mathbb{Z}.$$

Since the Cassels pairing can be expressed as a sum of local Tate pairings by Proposition 2, we would like to compute the Tate local pairing to determine if two Kolyvagin classes pair non-trivially under the Cassels pairing. Unfortunately, the Tate local pairing in this form is quite hard to compute. We will now show how to relate the pairing $\langle\, ,\rangle_{K_v}$ to a pairing over finite fields, and we will then use this relationship to detect whether certain Kolyvagin classes pair to a non-trivial element.

Description of $\mathrm{H}^1(K_v, E)_m$. We will describe the group $\mathrm{H}^1(K_v, E)_m$ in a way which is more convenient for computations. Fix an algebraic closure $\overline{K}_v$ of K_v. Let π be a uniformizer of K_v and let ζ_m be a primitive m-th root of unity in $\overline{K}_v$. Consider the extensions $L_m = K_v(\zeta_m, \pi^{1/m})$ and $K_v(\zeta_m)$ of K_v, where $\pi^{1/m} \in \overline{K}_v$. The Galois group $G_{K_v(\zeta_m)/K_v}$ acts on $G_{L_m/K_v(\zeta_m)}$ by conjugation. Let $K_v(E_m)$ be the field obtained by adjoining the coordinates of all m-torsion points of E defined over $\overline{K}_v$. The extension $K_v(E_m)$ is unramified and therefore cyclic over K_v. Moreover, since the Weil pairing is Galois equivariant, it contains $K_v(\zeta_m)$. Therefore, $G_{K_v(E_m)/K_v}$ acts on $G_{L_m/K_v(\zeta_m)}$ through its quotient $G_{K_v(\zeta_m)/K_v}$. The following proposition (see also [8, Prop.3.15] and [4, Prop 6.5]) describes the group $\mathrm{H}^1(K_v, E)_m$ as the cohomology of a finite group acting on a finite module.

Proposition 3. $\mathrm{H}^1(K_v, E)_m \simeq \mathrm{Hom}_{G_{K_v(E_m)/K_v}}(G_{L_m/K_v(\zeta_m)}, E_m)$.
In particular, if $\mu_m \subset K_v$ then the isomorphism becomes

$$\mathrm{H}^1(K_v, E)_m \simeq \mathrm{Hom}_{G_{K_v(E_m)/K_v}}(G_{L_m/K_v}, E_m).$$

Proof. Let $G_{K_v} = G_{\overline{K}_v/K_v}$, let $I_{K_v} \subset G_{K_v}$ be the inertia group of G_{K_v}, and let $G_{k_v} = G_{\overline{k}_v/k_v}$. Consider the exact sequence

$$0 \to E_1(\overline{K}_v) \to E(\overline{K}_v) \to \tilde{E}(\overline{k}_v) \to 0,$$

where $E_1(\overline{K}_v)$ is the kernel of reduction. Consider the corresponding long exact sequence of Galois cohomology, where G_{K_v} acts on $\tilde{E}(\overline{k}_v)$ through its quotient $G_{K_v}/I_{K_v} \simeq G_{k_v}$.

$$\mathrm{H}^1(K_v, E_1) \to \mathrm{H}^1(K_v, E) \to \mathrm{H}^1(K_v, \tilde{E}) \to \mathrm{H}^2(K_v, E_1) \to \dots.$$

Since m is coprime to the characteristic of $\overline{k}_v$ it follows from [24, Prop. 3.1(a)] that the group $E_1(L)$ is m-primary for any finite extension L/K_v, $L \subset \overline{K}_v$, so E_1 is m-primary. It follows from the Kummer sequence for E_1 that $\mathrm{H}^1(K_v, E_1)[m] = \mathrm{H}^2(K_v, E_1)[m] = 0$. Therefore, the map

$$\mathrm{H}^1(K_v, E)[m] \to \mathrm{H}^1(K_v, \tilde{E})[m]$$

is an isomorphism.

Next, the inflation-restriction sequence for $I_{K_v} \subset G_{K_v}$ gives us

$$0 \to \mathrm{H}^1(k_v, \tilde{E}) \to \mathrm{H}^1(K_v, \tilde{E}) \to \mathrm{H}^1(I_{K_v}, \tilde{E})^{G_{K_v}/I_{K_v}} \to \mathrm{H}^2(k_v, \tilde{E})$$

By [16], $\mathrm{H}^1(k_v, \tilde{E}) = 0$, and by [22, p. 189] we also have $\mathrm{H}^2(k_v, \tilde{E}) = 0$. Since I_{K_v} acts trivially on $\tilde{E}$, we obtain an isomorphism

$$\mathrm{H}^1(K_v, \tilde{E}) \cong \mathrm{Hom}(I_{K_v}, \tilde{E})^{G_{K_v}/I_{K_v}}.$$

Finally, $\mathrm{Hom}(I_{K_v}, \tilde{E})^{G_{K_v}/I_{K_v}}[m] \cong \mathrm{Hom}(I_{K_v}, \tilde{E}_m)^{G_{K_v}/I_{K_v}}$, which is isomorphic to $\mathrm{Hom}_{G_{K_v(E_m)/K_v}}(G_{L_m/K_v(\zeta_m)}, E_m)$, by the fact that a homomorphism $I_{K_v} \to E_m$ factors through the tame inertia group since m is coprime to the residue characteristic. This finishes the proof.

Let $H := G_{K_v(E_m)/K_v}$. By Proposition 3, we obtain a modified pairing

$$\langle\, , \rangle_{K_v} : E(K_v)/mE(K_v) \times \mathrm{Hom}_H(G_{L_m/K_v(\zeta_m)}, E_m(K_v)) \to \mathrm{Br}(K_v)[m],$$

which is induced by the Tate pairing. Here we use the fact that $\mathrm{Br}(K_v) \cong \mathbb{Q}/\mathbb{Z}$, and that the image of the Tate local pairing (for m) lies in the m-torsion part of $\mathbb{Q}/\mathbb{Z}$.

Reducing to the Finite Field Case. In this section we assume that in addition $\mu_m \subset K_v$, or equivalently that $m \mid \#k_v^\times$. We have that $\mathrm{Br}(K_v)[m]$ is cyclic of order m. In this situation, we have a description of the Tate pairing (up to sign) due to Lichtenbaum (see [17], see also [9, 5.3.4]):

Theorem 1 (Lichtenbaum). *Let σ be a generator of G_{L_m/K_v} and let $P_1 \in E(K_v)$ and $P_2 \in E_m(K_v)$. Let $D_1 \sim (P_1) - (O)$ be such that D_1 is coprime to $D_2 = (P_2) - (O)$. Let $\varrho : G_{L_m/K_v} \to E_m(K_v)$ be the homomorphism sending σ to P_2 and let f_2 be a function on E whose divisor is equivalent to mD_2. Then*

$$\langle P_1 + mE(K_v), \varrho \rangle_{K_v} = f_2(D_1),$$

where $f_2(D_1)$ is considered as an element of $K_v^\times / N_{L_m/K_v}(L_m^\times)$. We have $K_v^\times / N_{L_m/K_v}(L_m^\times) \cong k_v^\times/(k_v^\times)^m$.

Since G_{L_m/K_v} is cyclic, an element $\psi \in \mathrm{Hom}(G_{L_m/K_v}, E_m(K_v))$ is uniquely determined by the image of its generator $\psi(\sigma)$ in $E_m(K_v)$. So $\mathrm{Hom}(G_{L_m/K_v}, E_m(K_v))$ is non-canonically (depending on the choice of π and the generator σ) isomorphic to $E_m(K_v)$.

Let $\tilde{E}/k_v$ be the reduction of E/K_v. We have $E_m(K_v) \simeq \tilde{E}_m(k_v)$ by Lemma 1. Together with the above argument this implies that $\mathrm{Hom}(G_{L_m/K_v}, E_m(K_v))$ is isomorphic to $\tilde{E}_m(k_v)$. Also, since $\mathrm{Br}(K_v)[m]$ is cyclic of order m, it is isomorphic to $k_v^\times/(k_v^\times)^m$.

By Lemma 1 we also have $E(K_v)/mE(K_v) \simeq \tilde{E}(k_v)/m\tilde{E}(k_v)$. Thus, as a corollary of Lichtenbaum's theorem, we obtain the well known cryptographic Tate pairing which can be efficiently computed.

Corollary 1. *There is a non-degenerate pairing*

$$\varphi_m : \tilde{E}(k_v)/m\tilde{E}(k_v) \times \tilde{E}_m(k_v) \to k_v^\times/(k_v^\times)^m,$$

which is given by the following rule: Let $P_1 \in \tilde{E}(k_v)$ and $P_2 \in \tilde{E}_m(k_v)$ be two points on the reduced elliptic curve. Let D_i be a divisor equivalent to $(P_i) - (O)$ $(i = 1, 2)$, such that D_1 and D_2 are coprime. Let f_2 be a function on $\tilde{E}$ with divisor mD_2. Then

$$\varphi_m(P_1 + m\tilde{E}(k_v), P_2) = f_2(D_1),$$

where $f_2(D_1)$ is considered as an element of $k_v^\times/(k_v^\times)^m$.

Remark 2. Let $P_1 \in E(K_v)$ and $P_2 \in E_m(K_v)$ be as in Theorem 1, and let ϱ be the homomorphism sending σ to P_2. Let $\tilde{P}_1$, $\tilde{P}_2$ be the reductions of P_1 and P_2, respectively. By Lemma 1, $E(K_v)/mE(K_v) \cong \tilde{E}(k_v)/m\tilde{E}(k_v)$, and $E_m(K_v) \cong \tilde{E}_m(k_v)$. This implies that $\varphi_m(\tilde{P}_1, \tilde{P}_2)$ is nonzero if and only if $\langle P_1 + mE(K_v), \varrho\rangle_{K_v}$ is nonzero. Hence we conclude that the Tate local pairing $\langle , \rangle_{K_v} : E(K_v)/mE(K_v) \times \mathrm{H}^1(K_v, E)_m \to \mathbb{Q}/\mathbb{Z}$ is nonzero if the cryptographic Tate pairing $\varphi_m(\tilde{P}_1, \tilde{P}_2)$ for the corresponding points $\tilde{P}_1, \tilde{P}_2$ is nonzero.

For efficient computation of the cryptographic Tate pairing, see [1] and [7].

6 Application to Kolyvagin Cohomology Classes

In this section we apply the results explained in the previous section to certain Kolyvagin cohomology classes. Let E, K, p and M be as in Section 2. Suppose that ℓ is a Kolyvagin prime for $(E, K, p, M + 1)$, such that the class $d_{\ell,M}$ is an element of the Shafarevich-Tate group $\text{Ш}(K, E)$, i.e. $d_{\ell,M}$ is locally trivial at the unique place $\lambda \mid \ell$.

We provide a method for testing whether the class $d_{\ell,M}$ is a non-trivial element of $\text{Ш}(K, E)$ by pairing the class $d_{\ell,M}$ with another everywhere locally trivial Kolyvagin cohomology class $d_{n',1}$ (which we call a *test class*) with $\ell \nmid n'$. If $\langle d_{\ell,M}, d_{n',1}\rangle_C \neq 0$, then both the class $d_{\ell,M}$ and the test class $d_{n',1}$ are nonzero elements of $\text{Ш}(K, E)$. However, if $\langle d_{\ell,M}, d_{n',1}\rangle_C = 0$, then we cannot conclude anything. So in practice, we will have to compute $\langle d_{\ell,M}, d_{n',1}\rangle_C$ for multiple test classes. Our algorithm takes as input the data (E, K, p, M, ℓ, n') for which both $d_{\ell,M}$ and $d_{n',1}$ are (possibly trivial) elements of $\text{Ш}(K, E)$. The output is TRUE or FALSE depending on whether the pairing is nonzero or zero, respectively.

We test whether the Cassels pairing is non-zero via the Tate pairing over finite fields by using Proposition 2 and the reduction in Section 5. First, we use the methods in Section 7 to compute the coordinates of the Heegner points y_ℓ, $y_{n'}$ and their Galois conjugates. Once we do this, we can compute the points P_ℓ and $P_{n'}$ in $E(K_\lambda)$ defined in Section 2. By Proposition 2,

$$\langle d_{\ell,M}, d_{n',1}\rangle_C = \langle P_{n'}, \mathrm{res}_\lambda(d_{\ell,M+1})\rangle_{K_\lambda}.$$

The field K_λ contains the p^{M+1}-th roots of unity since the Kolyvagin assumptions imply that $\#\mathbb{F}_{\ell^2}^\times = \ell^2 - 1 \equiv 0 \mod p^{M+1}$. Hence we can apply Proposition 3 with the local field K_λ and we let $m := p^{M+1}$. We obtain

$$\mathrm{H}^1(K_\lambda, E)_m \cong \mathrm{Hom}_{G_{K_\lambda(E_m)/K_\lambda}}(G_{L_m/K_\lambda}, E_m).$$

By the previous section, $\langle P_{n'}, \mathrm{res}_\lambda(d_{\ell,M+1})\rangle_{K_\lambda}$ is nonzero if $\varphi_m(\tilde{P}_{n'}, Q)$ is nonzero, where

$$\varphi_m : \tilde{E}(\mathbb{F}_{\ell^2})/m\tilde{E}(\mathbb{F}_{\ell^2}) \times \tilde{E}_m(\mathbb{F}_{\ell^2}) \to \mathbb{F}_{\ell^2}^\times/(\mathbb{F}_{\ell^2}^\times)^m$$

is the cryptographic Tate pairing, $\tilde{P}_{n'}$ is the image of $P_{n'}$ in $\tilde{E}(\mathbb{F}_{\ell^2})$, and Q is the image of $\mathrm{res}_\lambda(d_{\ell,M+1})$ in $\tilde{E}_m(\mathbb{F}_{\ell^2})$.

Since we have an explicit description for a 1-cocycle that represents the class $d_{\ell,M+1}$ associated to the Heegner point y_ℓ (see Equation (1) in Section 2), we can compute its image in $\mathrm{Hom}(G_{L_m/K_\lambda}, \tilde{E}_m(F_{\ell^2}))$ and therefore the corresponding point on the reduction $\tilde{E}_m(\mathbb{F}_{\ell^2})$. Thus, we can compute the pairing φ_m.

7 Computational Aspects and Implementation Issues

Choice of E, p, D, ℓ, n'. Choose a non-CM elliptic curve E over $\mathbb{Q}$ of conductor N and analytic rank 1 from Cremona's tables [5] with non-trivial conjectural p-part of $Ш(K, E)$ for some odd prime p such that $\rho_{E,p}$ is surjective, and some quadratic imaginary field K with a Heegner discriminant D coprime to p. In our case, *conjectural order of* $Ш(K, E)$ means the order predicted by a combination of the Birch and Swinnerton-Dyer conjecture and the Gross-Zagier formula for $E/\mathbb{Q}$ and the twist $E^D/\mathbb{Q}$. The precise conjecture (as stated in [19]) says that if the Heegner point $y_K \in E(K)$ has infinite order then

$$\#Ш(K,E) = \left(\frac{[E(K)/E(K)_{\mathrm{tors}} : \mathbb{Z}y_K]}{c \cdot \prod_{q|N} c_q}\right),$$

where c_q is the Tamagawa number for E at the prime q and c is a constant which depends on the modular parameterization (the Manin constant). The computation of the conjectural order of $Ш(K, E)$ has been implemented in SAGE and uses the Heegner point algorithm of Watkins (see [27]). We use the methods of Section 2 to test the surjectivity of the Galois representation $\rho_{E,p}$. For example, we check that the curve $E : y^2 + xy + y = x^3 - x^2$ (curve 53A1 in [5]) has non-trivial 3-torsion in $Ш(K, E)$ for $K = \mathbb{Q}(\sqrt{-43})$ and the representation $\rho_{E,3}$ is surjective.

Once E, p and D are chosen, pick Kolyvagin primes ℓ for $(E, p, D, 1)$ and n' for $(E, p, D, 2)$.

Computation of Heegner Points over Ring Class Fields. The computation of Heegner points for non-fundamental discriminants now exists in MAGMA, using the command `HeegnerPoints` and switching off the `RemovePoint` option. However during the time of writing of this article that functionality was not available or known to be available, so we implemented a package ourselves. This section explains some of the difficulties encountered during the implementation.

We compute the derived Heegner point $P_\ell = \sum_{\sigma \in S} \sigma(D_\ell y_\ell)$ over the ring class field K_ℓ by computing the minimal polynomial of its x-coordinate. The last step is achieved by running through all of the $G_{K_\ell/K} = \mathrm{Pic}(\mathcal{O}_\ell)$-conjugates, where

$\mathcal{O}_\ell$ is the order of conductor ℓ in K. Since ℓ is an inert prime in $K = \mathbb{Q}(\sqrt{D})$, the degree of the ring class field over K is $(\ell+1)h_K$, where h_K is the class number of K. We generate ideal class representatives for $\mathrm{Pic}(\mathcal{O}_\ell)$ by using Kluners and Pauli's package. This functionality is available in MAGMA as `RingClassGroup`.

To compute P_ℓ and its $G_{K_\ell/K}$-conjugates we use the reciprocity law. If $\mathfrak{c}$ represents a class in $\mathrm{Pic}(\mathcal{O}_\ell)$, then the ideal class $[\mathfrak{c}]$ acts on the Heegner point $[\mathbb{C}/I \to \mathbb{C}/J]$ by mapping it to the point $[\mathbb{C}/\mathfrak{c}^{-1}I \to \mathbb{C}/\mathfrak{c}^{-1}J]$. Once we compute τ in the upper half plane corresponding to each conjugate, we need to evaluate $\varphi(\tau)$ to sufficient accuracy, where $\varphi(\tau) = \sum_{n\geq 1} \frac{a_n}{n} q^n$, $q = e^{2\pi i \tau}$, and a_n is the n-th Fourier coefficient of the modular form corresponding to the elliptic curve E.

Evaluating φ to sufficient accuracy turns out to be computationally expensive. First of all, the number of digits of accuracy required is significant and can be estimated as in [3] and [6] in terms of the height of the Heegner point. Cohen and Elkies address the computation of Heegner points with coefficients in $\mathbb{Q}$, and use three tools which hold over $\mathbb{Q}$: (1) known bounds on the difference between the naïve and canonical heights of the Heegner point [3, Thm.8.1.18], (2) the Gross-Zagier formula for the canonical height of the Heegner point in terms of the derivative of the L-series [12], and (3) the conjectural BSD formula relating the derivative of the L-series to the order of Ш and the regulator (see [3, Alg.8.6.11, Step 1]). In our setting, the Heegner points are defined over the ring class field and we made rough estimates of the accuracy required using a conjectural generalization of the Gross-Zagier formula [21, Statement 2.6].

Secondly, the computation of the Heegner points requires not only large amounts of precision to evaluate and recognize, but also many terms from the Fourier expansion of the modular form. Cohen gives an estimate for the number of terms needed for computing Heegner points over $\mathbb{Q}$ in [3, Alg.8.6.11, Step 4] in terms of the binary quadratic forms representing the ideal classes.

The last few steps of the algorithm involve applying the Weierstrass $\wp$-function to the point $\varphi(\tau)$ on the complex uniformization $\mathbb{C}/\Lambda$ of E to obtain the x-coordinate of the Heegner point, forming the minimal polynomial of the x-coordinate over the field K, and using the continued fraction algorithm to recognize the coefficients of this polynomial.

Example: $D = -43$, $p = 3$, $\ell = 5$, $N = 53$. Let E be the elliptic curve 53A1 from Cremona's tables [5] with Weierstrass equation $E: y^2+xy+y = x^3-x^2$. We check that $D = -43$ is a Heegner discriminant for this elliptic curve and that the Galois representation $\rho_{E,3}$ is surjective. Conjecturally, $Ш(K,E)[3] \cong \mathbb{Z}/3 \times \mathbb{Z}/3$ for $K = \mathbb{Q}(\sqrt{-43})$.

By computing the coefficients of the Fourier expansion, we determine that $\ell = 5$ is a Kolyvagin prime for $(E, -43, 3, 1)$. Although the minimal polynomial for the x-coordinate of the Heegner point is small for this example, $z^6 - 12z^5 + 1980z^4 - 5855z^3 + 6930z^2 - 3852z + 864$, the minimal polynomial P_5 of the x-coordinate of the derived Heegner point required using 1,000 digits of accuracy and 20,000 Fourier coefficients to compute. Further computations of Kolyvagin classes from derived Heegner points are accomplished in [13], where a different method for proving non-triviality is given.

Acknowledgments. We would like to thank the anonymous referees for many helpful comments and corrections, and for pointing out the MAGMA Heegner points package for non-fundamental discriminants.

References

1. Barreto, P.S.L.M., Kim, H.Y., Lynn, B., Scott, M.: Efficient algorithms for pairing-based cryptosystems. In: Yung, M. (ed.) CRYPTO 2002. LNCS, vol. 2442, pp. 354–368. Springer, Heidelberg (2002)
2. Breuil, C., Conrad, B., Diamond, F., Taylor, R.: On the modularity of elliptic curves over **Q**: wild 3-adic exercises. J. Amer. Math. Soc. 14(4), 843–939 (2001)
3. Cohen, H.: Number Theory Volume I: Tools and Diophantine equations. Graduate Texts in Mathematics, vol. 239. Springer, New York (2007), `http://math.arizona.edu/~swc/notes/files/06CohenExtract.pdf`
4. Cohen, H., Frey, G.: Handbook of Elliptic and Hyperelliptic Curve Cryptography. Chapman and Hall, CRC (2006)
5. Cremona, J.E.: Elliptic curves of conductor ≤ 25000, `http://www.warwick.ac.uk/~masgaj/ftp/data/INDEX.html`
6. Elkies, N.: Heegner point computations. In: Huang, M.-D.A., Adleman, L.M. (eds.) ANTS 1994. LNCS, vol. 877, pp. 122–133. Springer, Heidelberg (1994)
7. Eisenträger, K., Lauter, K., Montgomery, P.L.: Improved Weil and Tate pairings for elliptic and hyperelliptic curves. In: Buell, D.A. (ed.) ANTS 2004. LNCS, vol. 3076, pp. 169–183. Springer, Heidelberg (2004)
8. Frey, G.: Applications of arithmetical geometry to cryptographic constructions. In: Finite fields and applications (Augsburg, 1999), pp. 128–161. Springer, Berlin (2001)
9. Frey, G., Lange, T.: Mathematical background of public key cryptography. In: Frey, G., Lange, T. (eds.) Arithmetic, geometry and coding theory (AGCT 2003), Soc. Math. France, Paris. Sémin. Congr., vol. 11, pp. 41–73 (2005)
10. Grigorov, G., Jorza, A., Patrikis, S., Stein, W., Tarniţă-Patraşcu, C.: Verification of the Birch and Swinnerton-Dyer conjecture for specific elliptic curves (preprint)
11. Gross, B.: Kolyvagin's work on modular elliptic curves. In: L-functions and arithmetic (Durham, 1989), pp. 235–256. Cambridge Univ. Press, Cambridge (1991)
12. Gross, B., Zagier, D.: Heegner points and derivatives of L-series. Invent. Math. 84(2), 225–320 (1986)
13. Jetchev, D., Lauter, K., Stein, W.: Explicit Heegner Points: Kolyvagin's Conjecture and Non-trivial Elements in the Shafarevich-Tate group (preprint, 2007)
14. Jetchev, D., Stein, W.: Visualizing elements of the Shafarevich-Tate group at higher level (preprint)
15. Kolyvagin, V.A.: Euler systems. In: The Grothendieck Festschrift, vol. II, pp. 435–483. Birkhäuser, Boston (1990)
16. Lang, S.: Algebraic groups over finite fields. Amer. J. Math. 78, 555–563 (1956)
17. Lichtenbaum, S.: Duality theorems for curves over p-adic fields. Inv. Math. 7, 120–136 (1969)
18. Mazur, B., Rubin, K.: private communication

19. McCallum, W.G.: Kolyvagin's work on Shafarevich-Tate groups. In: L-functions and arithmetic (Durham, 1989), pp. 295–316. Cambridge Univ. Press, Cambridge (1991)
20. Milne, J.S.: Arithmetic duality theorems. Academic Press Inc., Boston (1986)
21. Nekovar, J., Schappacher, N.: On the asymptotic behaviour of Heegner points. Turkish J. of Math. 23(4), 549–556 (1999)
22. Serre, J.-P.: Local fields, GTM, vol. 67. Springer, New York (1979)
23. Serre, J.-P.: Propriétés galoisiennes des points d'ordre fini des courbes elliptiques. Invent. Math. 15(4), 259–331 (1972)
24. Silverman, J.H.: The arithmetic of elliptic curves. Springer, New York (1992)
25. Silverman, J.H.: Advanced topics in the arithmetic of elliptic curves. Springer, New York (1994)
26. Stein, W.: SAGE, `http://modular.math.washington.edu/sage/`
27. Watkins, M.: Some remarks on Heegner point computations (preprint, 2004)

Constructing Brezing-Weng Pairing-Friendly Elliptic Curves Using Elements in the Cyclotomic Field

Ezekiel J. Kachisa[1], Edward F. Schaefer[2], and Michael Scott[1,⋆]

[1] School of Computing
Dublin City University
Ireland
ekachisa@computing.dcu.ie,
mike@computing.dcu.ie

[2] Department of Mathematics and Computer Science
of Santa Clara University
USA
eschaefer@scu.edu

Abstract. We describe a new method for constructing Brezing-Weng-like pairing-friendly elliptic curves. The new construction uses the minimal polynomials of elements in a cyclotomic field. Using this new construction we present new "record breaking" families of pairing-friendly curves with embedding degrees of $k \in \{16, 18, 36, 40\}$, and some interesting new constructions for the cases $k \in \{8, 32\}$.

1 Introduction

Standard cryptosystems such as the Elliptic Curve Digital Signature Algorithm, Elliptic Curve Diffie-Hellman and ElGamal Elliptic Curve Encryption require randomly generated elliptic curves for their implementation. On the other hand cryptosystems such as short digital signatures, identity-based encryption and one-round three-way key exchange, require so-called pairing-friendly elliptic curves. These curves have special properties which most randomly generated curves will not have. The interest in recent times is to explore various methods of constructing pairing-friendly elliptic curves with prescribed embedding degrees, ideally to make them readily available, more efficient and more secure. Many strategies have been proposed by different researchers to construct such curves ([1, 3, 4, 5, 7, 13]).

Of particular interest to our discussion is the strategy of constructing pairing friendly elliptic curves as proposed by Brezing and Weng [4]. This construction basically uses the Cocks and Pinch idea [5] over polynomials. The interesting point in the Brezing-Weng method is that it reduces the ratio between the bit lengths of the finite field p and the order r of the subgroup with embedding

⋆ These authors acknowledge support from the Science Foundation Ireland under Grant No. 06/MI/006.

S.D. Galbraith and K.G. Paterson (Eds.): Pairing 2008, LNCS 5209, pp. 126–135, 2008.

degree k. This is measured by using a parameter ρ, defined as $\frac{\log p}{\log r}$. For example the Cocks-Pinch method invariably produces curves with $\rho \sim 2$, which is rather inefficient. It is observed that small ρ-values are desirable in speeding up the arithmetic on the curves in the underlying field. Ideally we would prefer $\rho = 1$, which is already achieved by the MNT [13], BN [1] and Freeman [7] constructions, for the cases $k \in \{3, 4, 6, 10, 12\}$.

Let G_1 and G_2 be finite cyclic additive groups of prime order r and G_T be a finite cyclic multiplicative group of order r. A bilinear pairing is a map $e : G_1 \times G_2 \rightarrow G_T$ that satisfies the following properties:

1. (bilinear): $e(aP, bQ) = e(P, Q)^{ab}$, for all $P \in G_1$ and $Q \in G_2$ and for all $a, b, \in \mathbb{Z}_r$
2. (non-degenerate): there exists $P \in G_1$ and $Q \in G_2$ such that $e(P, Q) \neq 1$
3. (computable): e can be efficiently computed.

The traditional cryptographic pairings are the Weil and the Tate pairings. In terms of efficiency it is generally accepted that the Tate pairing is superior to the Weil pairing. The algorithm for the calculation of the Tate pairing requires a Miller loop, followed by a final exponentiation. Recently more efficient variants of the Tate pairing have been proposed like the Ate pairing [10], culminating in the recent discovery of the R-ate pairing [11], [9], [18]. These variants achieve their greater efficiency by requiring a much shorter (and thus faster) Miller loop.

Pairings change the elliptic curve discrete logarithm problem (ECDLP) on elliptic curves over a prime field $E(\mathbb{F}_p)$ into the discrete logarithm problem in some extension field $\mathbb{F}_{p^k}$. As such, for the pairing-based cryptosystems to be secure, the ECDLP in $E(\mathbb{F}_p)$ and the DLP in the multiplicative group $\mathbb{F}^*_{p^k}$ must both be computationally infeasible [8]. The parameter k is called the *embedding degree.*

On a non-supersingular elliptic curve while one parameter of the pairing may be a point over the base field $E(\mathbb{F}_p)$, the best that can be done for the second parameter is that it be a point on a twisted curve over an extension field $E(\mathbb{F}_{p^{k/d}})$, where $d \mid k$ and $d = 2$ for the quadratic twist is always possible for even k. The use of even k also enables the useful denominator elimination optimisation for the calculation of the pairing [2], and so this is generally regarded as a good idea. Note that for the optimal R-ate pairing, G_1 must be the group represented in the larger extension field.

The paper is organised as follows: In Section 2 we discuss pairing-friendly elliptic curves. The main contribution of this paper is presented in Sections 3 and 4 where we describe our method and where we give examples of the application of the new method to construct pairing-friendly elliptic curves with various embedding degrees k. We demonstrate the utility of the method by constructing new "record-breaking" families of pairing-friendly elliptic curves of embedding degrees 16, 18, 36 and 40.

2 Pairing-Friendly Elliptic Curves

The *embedding degree* in our context is defined as follows [7].

Definition 1. *Let E be an elliptic curve defined over a prime finite field $\mathbb{F}_p$. Let r be a prime dividing $\#E(\mathbb{F}_p)$. The embedding degree of E with respect to r is the smallest positive integer k such that $r \mid p^k - 1$.*

The definition explains that k is the smallest positive integer such that the extension field $\mathbb{F}_{p^k}$, contains a set of rth roots of unity. The problem is: given k, find a prime p and elliptic curve E, defined over the finite field $\mathbb{F}_p$, such that $\#E(\mathbb{F}_p)$ has a large prime factor r and the curve has embedding degree k with respect to r [7]. In pairing-based cryptography, when curves have small embedding degrees and a large prime-order subgroup they are known as *pairing-friendly elliptic curves.*

The number of points on an elliptic curve, E, is given by $\#E = p+1-t$, where t is the trace of the Frobenius; then by a simple substitution [2] the condition $r|p^k - 1$ is equivalent to

$$(t-1)^k \equiv 1 \bmod r,$$

so $t-1$ is a k-th root of unity modulo r. Note that it is not sufficient just to find values of r, p and t which satisfy these conditions but it is also necessary to be able to construct the associated elliptic curve. The only known method for doing this is the method of Complex Multiplication (CM). The CM method requires that $4p - t^2$ should be of the form Dy^2, where for practical reasons the discriminant D must be less than about 10^{10}. This is a very restrictive condition, and so pairing-friendly elliptic curves are not so easy to find.

Let us set down some definitions.

Definition 2. *Let $g(x)$ be a polynomial with rational coefficients. Then $g(x)$ represents integers if there exists $x_0 \in \mathbb{Z}$ such that $g(x_0)$ is an integer.*

Definition 3. *Let $g(x)$ be a polynomial of even degree with rational coefficients. Then $g(x)$ represents primes if:*

1. *it is a non-constant irreducible polynomial with a positive leading coefficient*
2. *it represents integers*
3. *there exists $x_1 \in \mathbb{Z}$ and $x_2 \in \mathbb{Z}$, for which $g(x)$ represents integers, such that $gcd\ (g(x_1), g(x_2)) = 1$.*

Note that if $n \cdot g(x) \in \mathbb{Z}[x]$ then we can verify the second condition by testing n consecutive integer values of x. In addition, if ℓ is a prime greater than the degree of $g(x)$, then we can test the third condition by testing ℓ consecutive integer values of x.

The following definition of pairing-friendly elliptic curves is an adaptation from [8]:

Definition 4. *Let $t(x)$, $r(x)$, and $p(x)$ be polynomials with rational coefficients. For a given positive integer k and square free integer D, the triple $(t(x), r(x), p(x))$ represents a family of elliptic curves with embedding degree k and CM discriminant D if the following conditions are satisfied:*

a. $p(x)$ represents primes.
b. $r(x)$ represents primes.
c. $t(x)$ represents integers.
d. $r(x)$ divides $p(x) + 1 - t(x)$.
e. $r(x)$ divides $\Phi_k(t(x) - 1)$, where Φ_k is the kth cyclotomic polynomial.
f. $Dy(x)^2 = 4p(x) - t(x)^2$ has infinitely many integer solutions in x.

Here the ρ-value for a family of curves is defined as follows:

Definition 5. *Let $t(x), r(x), p(x) \in \mathbb{Q}[x]$, and suppose (t, r, p) represents a family of elliptic curves with embedding degree k. The ρ-value of the family represented by (t, r, p) is given by $\rho = lim_{x\to\infty} \frac{\log(p(x))}{\log(r(x))} = \frac{\deg(p(x))}{\deg(r(x))}$.*

Note that the value of $p(x)$ is the size of the field while the value of $r(x)$ is the size of the group in which we wish to do our cryptography.

The algorithm for the Brezing-Weng construction is summarised in the following algorithm[8]:

Algorithm 2.4. For a fixed positive integer k and positive square-free integer D, execute the following steps:

1. Choose a number field K containing $\sqrt{-D}$ and a primitive kth root of unity ζ_k.
2. Find an irreducible (but not necessarily monic) polynomial $r(x) \in \mathbb{Z}[x]$ such that $\mathbb{Q}[x]/r(x) \cong K$.
3. Let $t(x) \in \mathbb{Q}[x]$ be a polynomial mapping to $\zeta_k + 1 \in K$.
4. Let $y(x) \in \mathbb{Q}[x]$ be a polynomial mapping to $\frac{\zeta_k - 1}{\sqrt{-D}} \in K$.
5. Let $p(x) \in \mathbb{Q}[x]$ be given by $(t(x)^2 + Dy(x)^2)/4$. If $p(x)$ and $r(x)$ represent primes, then the triple $(t(x), r(x), p(x))$ represents a family of curves with embedding degree k and discriminant D.

Pairing-friendly elliptic curves constructed using this method usually have their ρ-values less than 2 and closer to 1.

The challenging part in the Brezing-Weng construction is finding the polynomial $r(x)$ satisfying the following conditions, the *existence* conditions:

1. $\tilde{r}(x) = e \cdot r(x)$, where $r(x)$ represents primes and $e \in \mathbb{N}$ is a constant
2. $K \cong \mathbb{Q}[x]/\tilde{r}(x)$ contains ζ_k and $\sqrt{-D}$
3. $p(x)$ represents primes and
4. $t(x)$ represents integers.

In many cases the Brezing and Weng method results in curves with discriminant $D = 1$ or $D = 3$. Curves with these discriminants are not only easier to find using the CM method (as clearly D is very small), they also permit very efficient implementations, particularly of the R-ate pairing. For the case $D = 1$ the elliptic curve supports quartic twists ($d = 4$) if $4 \mid k$, and for the case $D = 3$ the curve supports cubic ($d = 3$) and sextic ($d = 6$) twists for $3 \mid k$ and $6 \mid k$ respectively. For example for the R-ate pairing, where $D = 3$ and $k = 12$ [1], it is possible to implement the group G_2 as points on $E(\mathbb{F}_p)$ over the base field and G_1 as points over the sextic twist, that is as points on $E'(\mathbb{F}_{p^{k/6}}) = E'(\mathbb{F}_{p^2})$.

3 The New Construction

We start by making a general, if rather obvious, point about working with polynomials with respect to an irreducible polynomial, rather than with integers with respect to a prime modulus. A power of a field element with respect to a prime modulus, will typically be a number the same size in bits as the modulus. However when working modulo an irreducible polynomial, the power of a field element will be a polynomial of degree *at least one less* than that of the irreducible polynomial. With some extra "luck" it may even be much less than this. Indeed it is exactly this kind of luck which results in Brezing and Weng curves often having a ρ value much less than 2, and closer to 1, (unlike the Cocks-Pinch method). This can also be exploited to reduce the workload of the pairing's final exponentiation [6].

In our construction we use a polynomial other than the cyclotomic polynomial $\Phi_l(x)$ to define the cyclotomic field $\mathbb{Q}(\zeta_l)$. In this construction we look for an element γ of the cyclotomic field $\mathbb{Q}(\zeta_l)$, where l is some multiple of the embedding degree k. Through experiments, we found that choosing γ to be a linear combination of a power basis $\{\zeta_l^i \mid 0 \leq i < \phi(l)\}$ with small integer coefficients, often led to success. So let L be a bound on the absolute size of the integer coefficients and allow a maximum of M non-zero coefficients. If γ is in $\mathbb{Q}(\zeta_l)$ but not in any proper subfield then we find the minimal polynomial of γ in $\mathbb{Q}(\zeta_l)$ which we set as $\tilde{r}(x)$. Otherwise γ gives a minimal polynomial whose degree is less than $\phi(l)$. If $D = 3$, we set $l = \text{lcm}(3, k)$; and if $D = 1$ we set $l = \text{lcm}(4, k)$. Then we proceed by using the Brezing-Weng construction to look for pairing-friendly elliptic curves, with predefined k and D, as follows:

3.1 Outline of Our Algorithm

Search through elements $\sum_{j=0}^{\phi(l)-1} m_i\zeta_l^j$ of $\mathbb{Q}(\zeta_l)$ where $m_i \in [-L, L]$. For each element, $\gamma \in \mathbb{Q}(\zeta_l)$ but not in any proper subfield of $\mathbb{Q}(\zeta_l)$ compute the minimal polynomial of γ and call it $\tilde{r}(x)$. Then for each primitive kth root of unity, ζ_k do:

1. Compute the polynomial $z(x)$ modulo $\tilde{r}(x)$ mapping to ζ_k.
2. Find $t(x) = z(x) + 1$, which maps to $\zeta_k + 1$ in $\mathbb{Q}[x]/\tilde{r}(x)$.
3. Using the algebraic relationship between ζ_k and $\sqrt{-D}$, find a polynomial $s(x)$ representing $\sqrt{-D}$ in $\mathbb{Q}[x]/\tilde{r}(x)$.
4. Compute the polynomial $y(x) = (t(x) - 2)s(x)/(-D)$.
5. Compute the polynomial $p(x) = (t(x)^2 + Dy(x)^2)/4$, and compute ρ. If $p(x)$ represents primes and the ρ-value is better than the best known, then
   (a) Find the smallest positive number $n \in \mathbb{Z}$, such that $n \cdot p(x) \in \mathbb{Z}[x]$.
   (b) Find the residue classes b modulo n such that $p(x) \in \mathbb{Z}$ for $x \equiv b \bmod n$.
   (c) Find the subset of those residue classes for which $t(x) \in \mathbb{Z}$ for $x \equiv b \bmod n$.
   If $\tilde{r}(nx + b) = e \cdot r(x)$ where e is a constant in $\mathbb{N}$ and $r(x)$ represents primes, then output $t(x), \tilde{r}(x), p(x), n, b, e$.

Thus for a given value of k, $(t(nx+b), \tilde{r}(nx+b)/e, p(nx+b))$ represents a family of pairing-friendly elliptic curves. The ρ-value for such a family of curves is then $\rho = \frac{\deg p(x)}{\deg r(x)}$. The elliptic curves found using this algorithm are pairing friendly by construction, and have an embedding degree of k.

3.2 Searching for New Families of Pairing-Friendly Curves

This algorithm is potentially very time consuming. Our approach is to restrict the search to integer coefficients with a limit L on their absolute size. We observe that smaller coefficients are more likely to lead to usable solutions. But even so the search space can quickly become huge for larger values of l. Therefore we have taken two approaches. The first performs an exhaustive search through all coefficients between $-L$ and $+L$. If this is not practical, the second approach is to limit the number of non-zero coefficients M to perhaps 2, 3 or 4. By trial and error we found that elements of $\mathbb{Q}(\zeta_l)$ of this form often produced good results. The search programs are written in a mixture of NTL [14] and PARI [15]. For comparision purposes a simple NTL program to generate Brezing and Weng families of pairing friendly curves can be found at Mike Scott's website [16].

4 Examples

The following examples demonstrate the construction of new families of pairing-friendly elliptic curves. Most of our examples also improve the existing ρ-values found in the literature. It is easy to verify that $(t(nx+b), \tilde{r}(nx+b)/e, p(nx+b))$ for a particular embedding degree, satisfy the conditions given in Definition 4.

Example 4.1. We start however with the case $k = 8$, where we set no records in terms of ρ, but nevertheless find some interesting new families of pairing friendly curves. For this embedding degree there is a known Brezing and Weng family of curves for $D = 3$ and $l = 24$ [4].

$$
\begin{aligned}
k &= 8,\ D = 3 \\
t(x) &= x^5 - x + 1 \\
p(x) &= (x^{10} + x^9 + x^8 - x^6 + 2x^5 - x^4 + x^2 - 32x + 1)/3 \\
r(x) &= x^8 - x^4 + 1 \\
\rho &= 5/4.
\end{aligned}
$$

Such a pairing suffers from the fact that we cannot use a higher order twist for G_1, which must therefore be represented by points on $E(\mathbb{F}_{p^4})$.

However for a family of curves with $k = 8$ and $D = 1$ the quartic twist for G_1 would be possible. Using our proposed method for $K \cong \mathbb{Q}(\zeta_8)$ we search through the range in which $m_i \in [-2, 2]$ and $M = 2$. We find that $\zeta_8 - 2\zeta_8^3 \in \mathbb{Q}(\zeta_8)$ has minimal polynomial $\tilde{r}(x) = x^4 - 8x^2 + 25$.

In this field we find that $(2x^3 - 11x)/15$ is a primitive 8th root of unity. So we let $t(x) = (2x^3 - 11x + 15)/15$. With this we get $y(x) = (x^3 + 5x^2 + 2x - 20)/15$

and $p(x) = (x^6+2x^5-3x^4+8x^3-15x^2-82x+125)/180$. When $x \equiv \pm 5 \bmod 30$, $t(x)$ represents integers, $p(x)$ and $\tilde{r}(x)/450$ represent primes. So both of $(t(30x\pm 5), \tilde{r}(30x\pm 5)/450, p(30x\pm 5))$ represent a family of curves with embedding degree 8. In both cases we have

$$\begin{aligned} k &= 8,\ D = 1 \\ t(x) &= (2x^3 - 11x + 15)/15 \\ p(x) &= (x^6 + 2x^5 - 3x^4 + 8x^3 - 15x^2 - 82x + 125)/180 \\ \tilde{r}(x) &= x^4 - 8x^2 + 25 \\ n &= 30,\ b = \pm 5,\ e = 450 \\ \rho &= 3/2. \end{aligned}$$

Here the ρ value is inferior to the previous case, but G_1 can now be represented by points over the smaller extension field $\mathbb{F}_{p^2}$. However this construction does not set any new records as similar families of curves are already reported in [8] Example 6.18, and in [17] and [12].

Interestingly our method finds the BN family of pairing friendly curves [1].

Example 4.2. Fix the embedding degree $k = 12$ and $D = 3$ and set $K \cong \mathbb{Q}(\zeta_{12})$. Searching through $m_i \in [-2, 2]$ and setting $M = 4$, we find $\zeta_{12}^3 - \zeta_{12}^2 + \zeta_{12} + 2 \in \mathbb{Q}(\zeta_{12})$ which has minimal polynomial $\tilde{r}(x) = x^4 - 6x^3 + 18x^2 - 36x + 36$. When $x \equiv 0 \bmod 6$, $t(x)$ represents integers, $p(x)$ and $\tilde{r}(x)/36$ represent primes. So $(t(6x), \tilde{r}(6x)/36, p(6x))$ represent a family of curves with embedding degree 12. In this case we have

$$\begin{aligned} k &= 12,\ D = 3 \\ t(x) &= (x^2 + 6)/6 \\ p(x) &= (x^4 - 6x^3 + 24x^2 - 36x + 36)/36 \\ \tilde{r}(x) &= x^4 - 6x^3 + 18x^2 - 36x + 36 \\ n &= 6,\ b = 0,\ e = 36 \\ \rho &= 1. \end{aligned}$$

Example 4.3. Fix the embedding degree $k = 16$ and $D = 1$ and set $K \cong \mathbb{Q}(\zeta_{16})$. Searching through $m_i \in [-2, 2]$ and $M = 2$, we find $-2\zeta_{16}^5 + \zeta_{16} \in \mathbb{Q}(\zeta_{16})$ which has minimal polynomial $\tilde{r}(x) = x^8 + 48x^4 + 625$. When $x \equiv \pm 25 \bmod 70$, $t(x)$ represents integers, $p(x)$ and $\tilde{r}(x)/61250$ represent primes. So both of $(t(70x\pm 25), \tilde{r}(70x\pm 25)/61250, p(70x\pm 25))$ represent a family of curves with embedding degree 16. In both cases we have

$$\begin{aligned} k &= 16,\ D = 1 \\ t(x) &= (2x^5 + 41x + 35)/35 \\ p(x) &= (x^{10} + 2x^9 + 5x^8 + 48x^6 + 152x^5 + 240x^4 + 625x^2 + 2398x + 3125)/980 \\ \tilde{r}(x) &= x^8 + 48x^4 + 625 \\ n &= 70,\ b = \pm 25,\ e = 61250 \\ \rho &= 5/4. \end{aligned}$$

This is an improvement over the old record value of $\rho = 11/8$.

Example 4.4. Fix the embedding degree $k = 18$ and $D = 3$ and set $K \cong \mathbb{Q}(\zeta_{18})$. With $m_i \in [-3, 3]$ and $M = 2$ we find $-3\zeta_{18}^5 + \zeta_{18}^2 \in \mathbb{Q}(\zeta_{18})$ has minimal polynomial $\tilde{r}(x) = x^6 + 37x^3 + 343$. When $x \equiv 14 \bmod 42$, $t(x)$ represents integers, $p(x)$ and $r(x)/343$ represent primes. So $(t(42x+14), \tilde{r}(42x+14)/343, p(42x+14))$ represents a family of curves with embedding degree 18. We have

$$
\begin{aligned}
k &= 18,\ D = 3\\
t(x) &= (x^4 + 16x + 7)/7\\
p(x) &= (x^8 + 5x^7 + 7x^6 + 37x^5 + 188x^4 + 259x^3 + 343x^2 + 1763x + 2401)/21\\
\tilde{r}(x) &= x^6 + 37x^3 + 343\\
n &= 42,\ b = 14,\ e = 343\\
\rho &= 4/3.
\end{aligned}
$$

This is a significant improvement in ρ over the old record value of 19/12.

Until now there has not been a good choice of pairing-friendly families of curves which are a good fit for the AES-256 level of security, for larger values of k.

Example 4.5. For the embedding degree $k = 32$, there is a Brezing and Weng family of curves with $\rho = 17/16$, but with $D = 3$, which is the "wrong" discriminant ($3 \nmid k$) for a simpler form of G_1 [8]. Here we suggest an alternative.

Fix embedding degree $k = 32$ and $D = 1$ and set $K \cong \mathbb{Q}(\zeta_{32})$. Searching through $m_i \in [-3, 3]$ and $M = 2$, we find $-3\zeta_{32} + 2\zeta_{32}^9 \in \mathbb{Q}(\zeta_{32})$ has minimal polynomial $\tilde{r}(x) = x^{16} + 57120x^8 + 815730721$. When $x \equiv \pm 325 \bmod 6214$, $t(x)$ represents integers, $p(x)$ and $\tilde{r}(x)/93190709028482$ represent primes. So both of $(t(6214x \pm 325), \tilde{r}(6214x \pm 325)/93190709028482, p(6214x \pm 325))$ represent a family of curves with embedding degree 32. In both cases we have

$$
\begin{aligned}
k &= 32,\ D = 1\\
t(x) &= (-2x^9 - 56403x + 3107)/3107\\
p(x) &= (x^{18} - 6x^{17} + 13x^{16} + 57120x^{10} - 344632x^9 + 742560x^8 + 815730721x^2\\
&\quad - 4948305594x + 10604499373)/2970292\\
\tilde{r}(x) &= x^{16} + 57120x^8 + 815730721\\
n &= 6214,\ b = \pm 325,\ e = 93190709028482\\
\rho &= 9/8.
\end{aligned}
$$

Example 4.6. Fix the embedding degree $k = 36$ and $D = 3$ and set $K \cong \mathbb{Q}(\zeta_{36})$. Searching through $m_i \in [-2, 2]$ with $M = 2$, we find $2\zeta_{36} + \zeta_{36}^7 \in \mathbb{Q}(\zeta_{36})$ has minimal polynomial $\tilde{r}(x) = x^{12} + 683x^6 + 117649$. When for example $x \equiv 287 \bmod 777$, $t(x)$ represents integers, $p(x)$ and $\tilde{r}(x)/161061481$ represent primes.(There are other classes mod 777 that work.) So $(t(777x+287), \tilde{r}(777x+287)/161061481, p(777x+287))$ represents a family of curves with embedding degree 36. In both cases we have

$$
\begin{aligned}
k &= 36,\ D = 3\\
t(x) &= (2x^7 + 757x + 259)/259
\end{aligned}
$$

$$\begin{aligned}
p(x) &= (x^{14} - 4x^{13} + 7x^{12} + 683x^8 - 2510x^7 \\
&\quad + 4781x^6 + 117649x^2 - 386569x + 823543)/28749 \\
\tilde{r}(x) &= x^{12} + 683x^6 + 117649 \\
n &= 777,\ b = 287,\ e = 161061481 \\
\rho &= 7/6.
\end{aligned}$$

Again this is an improvement in ρ over the old record value of 17/12.

Example 4.7 Fix the embedding degree $k = 40$ and $D = 1$ and set $K \cong \mathbb{Q}(\zeta_{40})$. Consider $-2\zeta_{40} + \zeta_{40}^{11} \in \mathbb{Q}(\zeta_{40})$. This element has minimal polynomial $\tilde{r}(x) = x^{16} + 8x^{14} + 39x^{12} + 112x^{10} - 79x^8 + 2800x^6 + 24375x^4 + 125000x^2 + 390625$. When for example $x \equiv \pm 1205 \bmod 2370$, $t(x)$ represents integers, $p(x)$ and $\tilde{r}(x)/2437890625$ represent primes. (There are other classes mod 2370 that work.) So both of $(t(2370x{\pm}1205), \tilde{r}(2370x{\pm}1205)/2437890625, p(2370x{\pm}1205))$ represent a family of curves with embedding degree 40. In both cases we have

$$\begin{aligned}
k &= 40,\ D = 1 \\
t(x) &= (2x^{11} + 6469x + 1185)/1185 \\
p(x) &= (x^{22} - 2x^{21} + 5x^{20} + 6232x^{12} - 10568x^{11} + 31160x^{10} \\
&\quad + 9765625x^2 - 13398638x + 48828125)/1123380 \\
\tilde{r}(x) &= x^{16} + 8x^{14} + 39x^{12} + 112x^{10} - 79x^8 \\
&\quad + 2800x^6 + 24375x^4 + 125000x^2 + 390625 \\
n &= 2370,\ b = \pm 1205,\ e = 2437890625 \\
\rho &= 11/8.
\end{aligned}$$

Again this is an improvement in ρ over the old record value of 23/16.

5 Conclusion

We have presented a new method of constructing pairing-friendly elliptic curves. Basically the construction extends ideas from the Brezing-Weng method. The main idea in the construction is to use minimal polynomials of the elements of the cyclotomic field other than the cyclotomic polynomial $\Phi_l(x)$ to define the cyclotomic field $\mathbb{Q}(\zeta_l)$. The potential of the method has been illustrated by constructing new families of pairing-friendly elliptic curves of degrees 8, 16, 18, 32, 36 and 40. In most of these cases the method improves the previously best known ρ-values. Interestingly our method also rediscovers the BN family of "ideal" $\rho = 1$ curves. This holds out the hope that by extending the search space, further families of ideal pairing friendly curves might be found.

Acknowledgement

Thanks are due to Michael Naehrig and David Freeman for useful comments on an early draft of this paper.

References

1. Barreto, P.S.L.M., Naehrig, M.: Pairing-friendly elliptic curves of prime order. In: Preneel, B., Tavares, S. (eds.) SAC 2005. LNCS, vol. 3897, pp. 319–331. Springer, Heidelberg (2006)
2. Barreto, P.S.L.M., Lynn, B., Kim, H., Scott, M.: Efficient Algorithms for Pairing-Based Cryptosystems. In: Yung, M. (ed.) CRYPTO 2002. LNCS, vol. 2442, pp. 354–368. Springer, Heidelberg (2002)
3. Barreto, P.S.L.M., Lynn, B., Scott, M.: Constructing elliptic curves with prescribed embedding degree. In: Cimato, S., Galdi, C., Persiano, G. (eds.) SCN 2002. LNCS, vol. 2576, pp. 263–273. Springer, Heidelberg (2002)
4. Brezing, F., Weng, A.: Elliptic curves suitable for pairing based cryptography. Designs Codes and Cryptography 37(1), 133–141 (2005)
5. Cocks, C., Pinch, R.G.E.: Identity-based cryptosystems based on the Weil pairing (unpublished manuscript, 2001)
6. Devegili, A.J., Scott, M., Dahab, R.: Implementing Cryptographic Pairings over Barreto-Naehrig Curves. Cryptography ePrint Archive, Report 2007/390 (2007), http://eprint.iacr.org/2007/390
7. Freeman, D.: Constructing pairing-friendly elliptic curves with embedding Degree 10. In: Hess, F., Pauli, S., Pohst, M. (eds.) ANTS-VII 2006. LNCS, vol. 4076, pp. 452–465. Springer, Heidelberg (2006)
8. Freeman, D., Scott, M., Teske, E.: A Taxonomy of pairing-friendly elliptic curves. Cryptography ePrint Archive, Report 2006/372 (2006), http://eprint.iacr.org/2006/372
9. Hess, F.: Pairing Lattices. Cryptography ePrint Archive, Report 2008/125 (2008), http://eprint.iacr.org/2008/125
10. Hess, F., Smart, N., Vercauteren, F.: The Eta Pairing revisited. IEEE Trans. Information Theory 52, 4595–4602 (2006)
11. Lee, E., Lee, H.S., Park, C.M.: Efficient and Generalized Pairing Computation on Abelian Varieties. Cryptography ePrint Archive, Report 2008/040 (2008), http://eprint.iacr.org/2008/040
12. Matsuda, S., Kanayama, N., Hess, F., Okamoto, E.: Optimised versions of the Ate and Twisted Ate Pairings. Cryptology ePrint Archive, Report 2007/013 (2007), http://eprint.iacr.org/2007/013
13. Miyaji, A., Nakabayashi, M., Takano, S.: New explicit conditions of elliptic curve traces for FR-reduction. IEICE Trans. Fundamentals E84, 1234–1243 (2001)
14. Shoup, V.: A Library for doing Number Theory (2006), http://www.shoup.net/ntl/
15. PARI-GP, version 2.3.2, Bordeaux (2006), http://pari.math.u-bordeaux.fr/
16. Scott, M.: An NTL program to find Brezing and Weng curves (2007), http://ftp.computing.dcu.ie/pub/crypto/bandw.cpp
17. Tanaka, S., Nakamula, K.: More constructing pairing-friendly elliptic curves for cryptography. Mathematics arXiv Archive, Report 0711.1942 (2007), http://arxiv.org/abs/0711.1942
18. Vercauteren, F.: Optimal Pairings. Cryptography ePrint Archive, Report 2008/096 (2008), http://eprint.iacr.org/2008/096

Constructing Pairing-Friendly Elliptic Curves Using Factorization of Cyclotomic Polynomials

Satoru Tanaka and Ken Nakamula

Department of Mathematics and Information Sciences
Tokyo Metropolitan University
1-1 Minami Osawa, Hachioji city, Tokyo, 192-0397 Japan
satoru@tnt.math.metro-u.ac.jp, nakamula@tnt.math.metro-u.ac.jp

Abstract. The problem of constructing pairing-friendly elliptic curves has received a lot of attention. To find a suitable field for the construction we propose a method to find a polynomial $u(x)$, by the method of indeterminate coefficients, such that $\Phi_k(u(x))$ factors. We also refine the algorithm by Brezing and Weng using a factor of $\Phi_k(u(x))$. As a result, we produce new families of parameters using our algorithm for pairing-friendly elliptic curves with embedding degree 8, and we compute some explicit curves as numerical examples.

1 Introduction

Research on pairing-based cryptographic schemes has received interest over the past few years. Recently many new and novel protocols have been proposed as in [2, 3, 8, 12]. These protocols need an elliptic curve with special properties, namely, the embedding degree is small enough and the curve has a large prime order subgroup. However, randomly chosen curves do not have these properties. The systematic construction of "pairing-friendly" elliptic curves, especially curves with higher embedding degrees is an attractive problem for cryptography in the future.

Let q be a large prime power and $E : y^2 = x^3 + Ax + B$ be an elliptic curve defined over a finite field $\mathbf{F}_q$. Let r be the largest prime dividing $\#E(\mathbf{F}_q) = q + 1 - t$, the order of the group of $\mathbf{F}_q$-rational points of E where t is the Frobenius trace. When q is a prime, we call the smallest positive integer k such that r divides $q^k - 1$ the *embedding degree*. The parameters required to construct pairing-friendly elliptic curves are t, r, q, k and the CM discriminant D for the CM method introduced in [1].

In this paper, we study the problem of computing suitable parameters t, r, q from given parameters k, D. We employ the method proposed in [4, 5, 11] which generates a family of pairing-friendly curves by considering t, r, q as polynomials $t(x), r(x), q(x)$ of a new parameter x. Our strategy is finding a polynomial $u(x)$ by the method of indeterminate coefficients so that $u(a) = \zeta_k$ for some $a \in \mathbf{Q}(\zeta_k)$ as in [6], and taking $r(x)$ to be an irreducible factor of $\Phi_k(u(x))$. We propose a refinement of the Brezing-Weng algorithm using the factorization of $\Phi_k(u(x))$,

S.D. Galbraith and K.G. Paterson (Eds.): Pairing 2008, LNCS 5209, pp. 136–145, 2008.

and we give explicit formulas to compute $u(x)$ for $\varphi(k) = 4$. As a result, we obtain new families of pairing-friendly curves which are given explicitly in Table 1 and Theorem 3 of Section 3. To the best of our knowledge, we give the first explicit numerical results as in Examples 1–3.

This paper is organized as follows. Section 2 gives a brief mathematical definition of curves suitable for pairing-based cryptography and the method of construction we use to generate our curves. The main contribution of this paper is presented in Sections 3 and 4 where we give our algorithm to construct curves and where we give numerical examples of curves that we generate using our parameters for $k = 8$ and whole formulas to solve the equation $u(x) = \zeta_k$ over $\mathbf{Q}(\zeta_k)$ for $\varphi(k) = 4$. Finally, we will discuss the conclusions that we will draw from our approach in Section 5.

2 Pairing-Friendly Curves

A survey on the construction of pairing-friendly elliptic curves is written by Freeman et al. [5]. We introduce several essential definitions from that paper to explain our algorithm. We will use the same notation that they used in [5]. Let lg be the base 2 logarithm.

2.1 Families of Curves for Pairing

Definition 1 ([5, Definition 2.3]). Suppose E is an elliptic curve defined over $\mathbf{F}_q$. We say that E is *pairing-friendly* if E satisfies the following conditions:

(1) there is a prime $r \geq \sqrt{q}$ such that $r \mid \#E(\mathbf{F}_q)$.
(2) the embedding degree of E with respect to r is less than $(\lg r)/8$.

To define a parametric family of curves, we introduce

Definition 2 ([5, Definition 2.5]). Let $f(x)$ be a polynomial with rational coefficients. We say f *represents primes* if the following conditions are satisfied.

(1) $f(x)$ is non-constant and irreducible.
(2) $f(x)$ has positive leading coefficient.
(3) $f(x) \in \mathbf{Z}$ for some $x \in \mathbf{Z}$.
(4) $\gcd(\{f(x) \mid x, f(x) \in \mathbf{Z}\}) = 1$.

Using this, we can now introduce

Definition 3 ([5, Definition 2.6]). For a given positive integer k and positive square-free integer D, the triple (t, r, q) *represents a family of elliptic curves with embedding degree k and CM discriminant D* if the following conditions are satisfied:

(1) $q(x) = p(x)^d$ for some $d \geq 1$ and $p(x)$ that represents primes.
(2) $r(x) = c \cdot \tilde{r}(x)$ with $c \in \mathbf{Z}_{\geq 1})$ and $\tilde{r}(x)$ that represents primes.
(3) $r(x) \mid q(x) + 1 - t(x)$.

(4) $r(x) \mid \Phi_k(t(x)-1)$, where Φ_k is the kth cyclotomic polynomial.
(5) The CM equation $4q(x) - t(x)^2 = Dy^2$ has infinitely many integer solutions (x, y).

For a family $(t(x), r(x), q(x))$, if the CM equation in (5) has a set of integer solutions (x_0, y_0) with both of $p(x_0)$ and $\tilde{r}(x_0)$ are primes, then we are able to construct curves E over $\mathbf{F}_{q(x_0)}$ where $E(\mathbf{F}_{q(x_0)})$ has a subgroup of order $\tilde{r}(x_0)$ and embedding degree k with respect to $\tilde{r}(x_0)$ by using the CM method in [1].

We introduce a parameter ρ that represents how close the curve is to the ideal curve by

Definition 4 ([5, Definition 2.7])

(1) Let $E/\mathbf{F}_q$ be an elliptic curve, and suppose E has a subgroup of order r. The *ρ-value of E (with respect to r)* is

$$\rho(E) = \frac{\log q}{\log r}.$$

(2) Let $t(x), r(x), q(x) \in \mathbf{Q}[x]$, and suppose (t, r, q) represents a family of elliptic curves with embedding degree k. The *ρ-value of the family represented by (t, r, q)* is

$$\rho(t, r, q) = \lim_{x \to \infty} \frac{\log q(x)}{\log r(x)} = \frac{\deg q(x)}{\deg r(x)}.$$

By Definition 1, a pairing-friendly curve E has $\rho(E) \leq 2$. On the other hand, the Hasse bound implies that $\rho(t, r, q)$ is always at least 1. Finding parameters efficiently with the same bit size of r and q, hence $\rho(E)$ is close to 1, is one of the important problems for cryptography (See [5, Section 1.1]).

2.2 Brezing-Weng Method

In this section, we briefly explain the construction of curves satisfying the condition of Definition 3 proposed by Brezing and Weng [4],[5, Section 6.1].

Theorem 1. *Fix a positive integer k and a positive square-free integer D. Execute the following steps:*

Step 1. *Choose a number field K containing a primitive kth root of unity ζ_k and $\sqrt{-D}$.*
Step 2. *Find an irreducible polynomial $r(x) \in \mathbf{Z}[x]$ such that $\mathbf{Q}[x]/(r(x)) \cong K$.*
Step 3. *Let $t(x) \in \mathbf{Q}[x]$ be a polynomial mapping to $\zeta_k + 1 \in K$.*
Step 4. *Let $y(x) \in \mathbf{Q}[x]$ be a polynomial mapping to $(\zeta_k - 1)/\sqrt{-D} \in K$. (So, if we discover a polynomial $s(x) \in \mathbf{Q}[x]$ mapping to $\sqrt{-D} \in K$, then $y(x) \equiv (2 - t(x))s(x)/D \pmod{r(x)}$.)*
Step 5. *Let $q(x) = (t(x)^2 + Dy(x)^2)/4$.*

If both of $r(x)$ and $q(x)$ represent primes, then the triple (t, r, q) represents a family of curves with embedding degree k and CM discriminant D.

The ρ-value for this family is

$$\rho(t, r, q) = \frac{2 \max\{\deg t(x), \deg y(x)\}}{\deg r(x)} < 2.$$

For more details, refer to [5, Section 6.1]. To find a family of pairing-friendly elliptic curves efficiently, we have to choose a good $r(x)$ satisfying $\zeta_k, \sqrt{-D} \in K$. The idea by Brezing and Weng (also see Freeman et al.) is as follows. Choose an integer multiple ℓ of k so that $\sqrt{-D} \in K = \mathbf{Q}(\zeta_\ell)$. Then let $r(x) = \Phi_\ell(x)$. Freeman et al. further give some sporadic families [5, Section 6.2]. Our idea given explicitly below is to construct such sporadic curves systematically.

3 Factorization of Cyclotomic Polynomial

3.1 The Background

When we use the Brezing-Weng method to construct families, the problem is how to determine the polynomial $r(x)$ in Theorem 1. One of the known strategies to find $r(x)$ and construct families is using the minimal polynomial of elements in $\mathbf{Q}(\zeta_k)$ by Kachisa, Schaefer and Scott [9]. On the other hand, if $\Phi_k(u(x))$ is reducible over $\mathbf{Q}$ with a factor of degree $\varphi(k)$ for some $u(x) \in \mathbf{Q}[x]$, we let $r(x)$ be one of the irreducible factors. Set $K = \mathbf{Q}[x]/(r(x))$ and we will get $u(x) \mapsto \zeta_k$. These factorizations are rare, however, it is possible to find them using the technique introduced by Galbraith, McKee and Valença [6, Lemma 1] for quadratic polynomials $u(x)$. In fact, it is effective for the general case as is easily seen from the proof there:

Lemma 1. *Let $u(x) \in \mathbf{Q}[x]$ and φ be the Euler function. Then, the degree of any irreducible factor of $\Phi_k(u(x))$ over $\mathbf{Q}$ is a multiple of $\varphi(k)$. Moreover, the polynomial $\Phi_k(u(x))$ has an irreducible factor of degree $\varphi(k)$ if and only if the equation*

$$u(x) = \zeta_k \tag{1}$$

has a solution in $\mathbf{Q}(\zeta_k)$.

3.2 Strategy of Factorization

Using Lemma 1, we can take $r(x)$ to be one of the irreducible factors of $\Phi_k(u(x))$. Assume $\sqrt{-D} \in \mathbf{Q}(\zeta_k)$. To obtain $u(x)$ such that $\Phi_k(u(x))$ has a factor of degree $\varphi(k)$, it is necessary and sufficient that

$$u(a(x)) \equiv x \pmod{\Phi_k(x)}$$

for some $a(x) \in \mathbf{Q}[x]$. We consider the case

$$u(x) = \sum_{i=0}^{\varphi(k)-1} u_i x^i, \quad a(x) = \sum_{i=0}^{\varphi(k)-1} a_i x^i.$$

Let $v(x)$ be the polynomial of degree $< \varphi(k)$ such that $v(x) \equiv u(a(x)) \pmod{\Phi_k(x)}$. Then $v(x)$ can be written in the form

$$v(x) = \sum_{i=0}^{\varphi(k)-1} \sum_{j=0}^{\varphi(k)-1} u_j v_{ij} x^i.$$

where v_{ij} are explicit polynomials of $a_0, \cdots, a_{\varphi(k)-1}$ of degree $< \varphi(k)$. Therefore, from given $a_0, \cdots, a_{\varphi(k)-1} \in \mathbf{Q}$, we should solve the linear equation

$$V \begin{pmatrix} u_0 \\ u_1 \\ u_2 \\ \vdots \\ u_{\varphi(k)-1} \end{pmatrix} = \begin{pmatrix} 0 \\ 1 \\ 0 \\ \vdots \\ 0 \end{pmatrix}, \qquad (2)$$

where $V = (v_{ij})$ is a $\varphi(k) \times \varphi(k)$ matrix with entries in $\mathbf{Q}$. It is well known that the general solution $u_0, \cdots, u_{\varphi(k)-1}$ can be written as explicit rational functions of $a_0, \cdots, a_{\varphi(k)-1}$ and exist if $\det(V) \neq 0$ by Cramer's rule. We now compute $d = \det(V)$ first, then take an irreducible factor $r(x)$ of $\Phi_k(u(x))$ when $d \neq 0$. The computation of $u(x)$ and $r(x)$ depends only on k. We can apply them for any D such that $\sqrt{-D} \in \mathbf{Q}(\zeta_k)$.

This leads to the following:

Algorithm 2

Input. Positive integers D, k such that $\sqrt{-D} \in \mathbf{Q}(\zeta_k)$ and a finite subset $S \subset \mathbf{Q}(\zeta_k)$.
Output. Families of elliptic curves with parameters $t(x), r(x), q(x)$.

Step 1. For each $a \in S$, compute $\det(V)$ of the equation (2). If $\det(V) = 0$ for all elements of S, then the algorithm fails. Otherwise, determine the coefficients of $u(x) \in \mathbf{Q}[x]$ by the equation (2) for each $a \in S$.
Step 2. For each $u(x)$ at Step 1, compute all irreducible factors $r(x)$ of the polynomial $\Phi_k(u(x))$.
Step 3. For each pair of $u(x), r(x)$ at Step 2, compute all polynomials $t(x) \in \mathbf{Q}[x]$ such that $\deg t(x) < \deg r(x)$ and $t(x) \equiv u(x)^m + 1 \pmod{r(x)}$ for all m with $1 \leq m < k, \gcd(m, k) = 1$.
Step 4. For each pair of $r(x), t(x)$ at Step 3, execute Step 4 and Step 5 of Theorem 1 to compute $q(x)$.
Step 5. For each triple $r(x), t(x), q(x)$ at Step 4, check whether $q(x), r(x)$ represent primes. If $q(x), r(x)$ represent primes, output a family $t(x), r(x), q(x)$.

4 Formulas for the Case $\varphi(k) = 4$

We solved the preceding linear equation for the case $\varphi(k) = 4$.

4.1 The Case $k = 8$

In this case, we have

$$V = \begin{pmatrix} 1 & a_0 & a_0{}^2 - a_2{}^2 - 2a_1a_3 & a_0{}^3 - 3a_2(a_0a_2 + a_1{}^2 - a_3{}^2) - 6a_0a_1a_3 \\ 0 & a_1 & 2a_0a_1 - 2a_2a_3 & a_3{}^3 - 3a_1(a_1a_3 + a_2{}^2 - a_0{}^2) - 6a_0a_2a_3 \\ 0 & a_2 & a_1{}^2 - a_3{}^2 + 2a_0a_2 & -a_2{}^3 + 3a_0(a_0a_2 + a_1{}^2 - a_3{}^2) - 6a_1a_2a_3 \\ 0 & a_3 & 2a_1a_2 + 2a_0a_3 & a_1{}^3 - 3a_3(a_1a_3 + a_2{}^2 - a_0{}^2) + 6a_0a_1a_2 \end{pmatrix}.$$

Let d and n_i be as follows:

$$\begin{aligned} d &:= (a_1{}^2 + a_3{}^2)((a_1 - a_3)^2 + 2a_2{}^2)((a_1 + a_3)^2 - 2a_2{}^2), \\ n_0 &:= a_2(5a_1{}^3a_2{}^2 - 5a_1a_2{}^2a_3{}^2 - 5a_1{}^4a_3 + 2a_2{}^4a_3 - 3a_3{}^5), \\ n_1 &:= a_1{}^5 - 4a_1{}^3a_3{}^2 + 9a_1{}^2a_2{}^2a_3 + a_1(2a_2{}^4 + 3a_3{}^4) + 3a_2{}^2a_3{}^3, \\ n_2 &:= 2a_2{}^3a_3 - a_1{}^3a_2 - 3a_1a_2a_3{}^2, \\ n_3 &:= a_3{}^3 - a_1{}^2a_3 + 2a_1a_2{}^2. \end{aligned}$$

Assume d is nonzero, the solution of the equation (1) is given by

$$u_i = \frac{1}{d} \sum_{j=i}^{\varphi(k)-1} (-a_0)^{j-i} n_j. \tag{3}$$

By Algorithm 2 with input $D = 1$ and $S = \{\omega \in \mathbf{Q}(\zeta_8) \mid \omega = \sum_{i=0}^{3} a_i x^i,\ a_i \in \mathbf{Z},\ 0 \le a_i \le 300\}$, the result of computations of families by MAGMA [10] for $\lg \mathrm{lc}\,(u) < 10$ is given in Table 1, where $\mathrm{lc}\,(u)$ denotes the leading coefficient of $u(x)$. In the actual computation to make polynomial coefficients small, we further transform $u(x)$ obtained by Step 3 of Algorithm 2 to $u(ax + b) \in \mathbf{Z}[x]$ with suitable $a, b \in \mathbf{Q}, a \neq 0$. We explain the symbols in Table 1. For the column $\deg q(x)$, the symbol (†) denotes that $q(x)$ does not represent primes for all pairs $t(x)$, $r(x)$. For the rows, the highlighted entries mean that there exists a family of curves such that both $q(x)$ and $r(x)$ are primes for many integers x, probably infinitely many. From this, we can expect to produce pairing-friendly curves of large bit size. We discovered many pairing-friendly families of curves with $\rho = 3/2$ and also rediscovered known families. We describe one of them in detail:

Theorem 3. *The polynomials $t(x)$, $r(x)$, $q(x) \in \mathbf{Z}[x]$ given as follows represent a family of elliptic curves with embedding degree $k = 8$ and CM discriminant $D = 1$. This family indeed generates pairing-friendly elliptic curves.*

$$\begin{aligned} t(x) &= -82x^3 - 108x^2 - 54x - 8, \\ r(x) &= 82x^4 + 108x^3 + 54x^2 + 12x + 1, \\ q(x) &= 379906x^6 + 799008x^5 + 705346x^4 \\ &\quad +333614x^3 + 88945x^2 + 12636x + 745. \end{aligned}$$

Proof. The first half is already proved by Algorithm 2, so we only need to prove the second half. We may verify both $q(x_0)$ and $r(x_0)$ are primes with some integer x_0. We take $x_0 = 104$, then we get $q(x_0) = 490506332802458249$ and $r(x_0) = 9714910817$. Both of these are primes, so we can generate pairing-friendly curves by them. □

Table 1. Sporadic families generated from cubic $u(x)$ with embedding degree 8

lc (u)	$u(x)$	$t(x)$	$\deg r(x)$	$\deg q(x)$	$\rho(t, r, q)$
2	$\mathbf{2x^3 + 4x^2 + 6x + 3}$	$\mathbf{u(x)^3 + 1}$	**4**	**6**	**3/2**
9	$9x^3 + 3x^2 + 2x + 1$	$u(x)^5 + 1$	4	6	3/2
17	$17x^3 + 32x^2 + 24x + 6$	$u(x)^3 + 1$	4	6	3/2
18	$18x^3 + 39x^2 + 31x + 7$	$u(x)^3 + 1$	4	6	3/2
64	$64x^3 + 112x^2 + 75x + 18$	$u(x)^5 + 1$	8	14	7/4
68	$68x^3 + 110x^2 + 65x + 15$	$u(x)^5 + 1$	4	6	3/2
82	$\mathbf{82x^3 + 108x^2 + 54x + 9}$	$\mathbf{u(x)^5 + 1}$	**4**	**6**	**3/2**
144	$144x^3 + 480x^2 + 539x + 202$	$u(x)^5 + 1$	8	14	7/4
144	$144x^3 + 96x^2 + 29x + 2$	$u(x)^5 + 1$	8	14	7/4
216	$216x^3 + 372x^2 + 263x + 69$	—	—	(†)	—
225	$225x^3 + 2x$	—	—	(†)	—
257	$257x^3 + 256x^2 + 96x + 12$	$u(x)^3 + 1$	4	6	3/2
388	$388x^3 + 798x^2 + 561x + 134$	$u(x)^5 + 1$	4	6	3/2
392	$392x^3 + 980x^2 + 821x + 231$	$u(x)^5 + 1$	8	14	7/4
450	$450x^3 + 11x$	—	—	(†)	—
626	$\mathbf{626x^3 + 500x^2 + 150x + 15}$	$\mathbf{u(x)^5 + 1}$	**4**	**6**	**3/2**
738	$\mathbf{738x^3 + 1488x^2 + 1006x + 229}$	$\mathbf{u(x)^5 + 1}$	**4**	**6**	**3/2**
800	$800x^3 + 9x$	$u(x)^5 + 1$	8	14	7/4
873	$873x^3 + 969x^2 + 379x + 53$	$u(x)^7 + 1$	4	6	3/2

From a family of curves, we can actually construct pairing-friendly curves. Find an integer x such that $q(x)$ is a prime and check whether $r(x)$ is a prime. To find such an integer x, we can restrict the candidates by the Chinese Remainder Theorem.

Lemma 2. *If an integer $q(x)$ in Theorem 3 is a prime, then $x \equiv 14, 24 \pmod{30}$.*

Proof. We can easily check that all $q(x)$ are even if x is odd. We see that

$$q(x) \equiv x^6 + 3x^5 + x^4 + 4x^3 + x \pmod{5}.$$

So $q(x) \equiv 0 \pmod{5}$ if $x \not\equiv 4 \pmod{5}$. In the same way we see that

$$q(x) \equiv x^6 + x^4 + 2x^3 + x^2 + 1 \pmod{3}.$$

So $q(x) \equiv 0 \pmod{5}$ if $x \not\equiv 1 \pmod{3}$. Then by the Chinese Remainder Theorem, $q(x)$ has no prime factor 2, 3 and 5 only if $x \equiv 14, 24 \pmod{30}$. □

From this family, we can easily compute explicit examples of elliptic curves. The elliptic curve $E/\mathbf{F}_q$ with CM discriminant $D = 1$ is represented as

$$E : Y^2 \equiv X^3 + AX \pmod{q} \quad (a \neq 0)$$

where A is a parameter. Recall Definition 1 and Definition 4.

Example 1. For $x = 24000000000010394$ ($\lg x \approx 54.4$), we get

$$
\begin{aligned}
q &= 7260116720044466049517034647917893289912173137766027688115053206975815675478784229870364764019632259006 9,\\
r &= 272056320000471307161600306182614014808404525177076771934 82845476817 \quad \text{(224-bit)},\\
t &= -113356800000147285043200063789391713609209096429146 0,\\
\#E(\mathbf{F}_q) &= 726011672004446604951703464791789328991217313776602780147 18532071231007186788480192620783732287286881530,\\
A &= 363005836002223302475851732395894664495608656888301384405 752660348790783773939211493518238200981612950 35.
\end{aligned}
$$

Then $\lg r \approx 224.0$, $\lg q \approx 345.0$ and $\rho(E) \approx 1.54$.

Example 2. For $x = 6130400000000029634$ ($\lg x \approx 62.4$), we get

$$
\begin{aligned}
q &= 201655015390974685980897990123384483374976 85 268073419312994695960148519299615127959281 95 249643154463102416170215935678 9,\\
r &= 115816144321490890478327891899665854763905 03 326918594658592037634937230763121 7 \quad \text{(256-bit)},\\
t &= -188921023622423219740582163008443944151642 9 1734047019380020,\\
\#E(\mathbf{F}_q) &= 201655015390974685980897990123384483374976 85 268073419312994714852250881541937102017498 25 334082595979531589574917873681 0,\\
A &= 100827507695487342990448995061692241687488 42 634036709656497347980074259649807563979640 97 624821577231551208085107967839 52.
\end{aligned}
$$

Then $\lg r \approx 256.0$, $\lg q \approx 393.0$ and $\rho(E) \approx 1.54$.

Example 3. For $x = -72057594037930756$ ($\lg x \approx 56.0$), we get

$$
\begin{aligned}
q &= 531807791263750413429276790125164740039557854 0 382773010005094121237143504602337266662859891 6 049952969199369,\\
r &= 221071562670669849137704118006392776209995893 1 7226038054748059074248 17 \quad \text{(230-bit)},\\
t &= 306798423708539154983403942081629850761644294 7 7994640,\\
\#E(\mathbf{F}_q) &= 531807791263750413429276790125164740039557854 0 382773006937109884151751954768297845846561383 9 885523491204730,\\
A &= 177269263754583471143092263375054913346519284 6 794257670001698040412381168200779088887619963 8 683317656399790.
\end{aligned}
$$

Then $\lg r \approx 230.4$, $\lg q \approx 354.5$ and $\rho(E) \approx 1.54$. For the Ate pairing [7], it is important that t has a low Hamming weight for computation. We tried to find a curve with r between 224 bit and 256 bit, we found that r has a Hamming weight 72 and t has a Hamming weight 45 in this example.

4.2 The Case $k = 5, 10, 12$

Similar to the case $k = 8$, we can write down V and determine d, n_i such that the solution of the equation (1) is given by the equation (3) for other $\phi(k)$. For convenience of explanation, we omit the representation of V.

In the case $k = 5$, we put as follows:

$$
\begin{aligned}
d :=& (a_1(a_1 - a_2 + a_3) - (a_2 - a_3)^2)((-a_1 + a_2 + a_3)^4 - (a_3 + 2a_2)a_1^3 \\
&-(-2a_2^2 - 2a_3^2 - 10a_2a_3)a_1^2 - (a_2^3 + 11a_2^2a_3 + 8a_2a_3^2 + 2a_3^3)a_1 \\
&+3a_2a_3^3 + 3a_2^3a_3 + 5a_2^2a_3^2), \\
n_0 :=& (-5a_2a_3 + 3a_3^2)a_1^4 + (-a_2^2a_3 + 5a_2^3 + 6a_2a_3^2 - 6a_3^3)a_1^3 \\
&+(-6a_2^4 - 3a_2^2a_3^2 + 3a_2^3a_3 + 5a_3^4)a_1^2 + (3a_2^5 + 5a_2^2a_3^3 - 3a_2^4a_3 \\
&-6a_2a_3^4)a_1 - 2a_2a_3^4(a_2 - a_3), \\
n_1 :=& a_1^5 - (3a_2 + 2a_3)a_1^4 + 5a_1^3a_2^2 + (5a_3^3 - 3a_2a_3^2)a_1^2 \\
&+(9a_2^2a_3^2 - 6a_2^3a_3 - 4a_2a_3^3)a_1 + 3a_2^3a_3(a_2 - a_3), \\
n_2 :=& -a_2a_1^3 + (3a_2^2 - 3a_2a_3)a_1^2 + a_3^3(3a_1 + a_2 - a_3), \\
n_3 :=& (a_2^2 - a_1^2)a_3 + 2a_1(a_3^2 - a_2a_3 + a_2^2) - a_2^3.
\end{aligned}
$$

Also in the case $k = 10$, we put as follows:

$$
\begin{aligned}
d :=& (a_1(a_1 + a_2 + a_3) - (a_2 + a_3)^2))((-a_1 + a_2 + a_3)^4 - (a_3 + 6a_2)a_1^3 \\
&+(2a_2^2 + 14a_2a_3 + 2a_3^2)a_1^2 + (-7a_2^3 - 11a_2^2a_3 - 16a_2a_3^2 - 2a_3^3)a_1 \\
&+5a_2a_3(a_2^2 + a_2a_3 + a_3^2)), \\
n_0 :=& (-5a_2a_3 - 3a_3^2)a_1^4 + (a_2^2a_3 + 5a_2^3 + 6a_2a_3^2 + 6a_3^3)a_1^3 \\
&+(3a_2^3a_3 - 5a_3^4 + 3a_2^2a_3^2 + 6a_2^4)a_1^2 + (3a_2^4a_3 - 5a_2^2a_3^3 \\
&-6a_2a_3^4 + 3a_2^5)a_1 + 2a_2a_3^4(a_2 + a_3), \\
n_1 :=& a_1^5 + (-2a_3 + 3a_2)a_1^4 + 5a_1^3a_2^2 + (3a_2a_3^2 + 5a_3^3)a_1^2 \\
&+(6a_2^3a_3 + 9a_2^2a_3^2 + 4a_2a_3^3)a_1 + 3a_2^3a_3(a_2 + a_3), \\
n_2 :=& -a_2a_1^3 + (-3a_2^2 - 3a_2a_3)a_1^2 + a_3^3(-3a_1 + a_2 + a_3), \\
n_3 :=& (a_2^2 - a_1^2)a_3 + 2a_1(a_2^2 + a_2a_3 + a_3^2) + a_2^3.
\end{aligned}
$$

Finally, in the case $k = 12$, we put as follows:

$$
\begin{aligned}
d :=& (a_1^2 + a_2^2)((a_1 + a_3)^2 - a_1a_3)((a_1 + 2a_3)^2 - 3a_2^2), \\
n_0 :=& -5a_1^4a_3a_2 + (5a_2^3 - 13a_2a_3^2)a_1^3 + (8a_2^3a_3 - 12a_3^3a_2)a_1^2 \\
&+(-4a_3^4a_2 - a_2^5 + a_2^3a_3^2)a_1 + a_2a_3(a_2^4 - 2a_2^2a_3^2 - 2a_3^4), \\
n_1 :=& a_1^5 + 5a_3a_1^4 + (6a_3^2 - 2a_2^2)a_1^3 + (3a_2^2a_3 - a_3^3)a_1^2 \\
&+(-2a_3^4 + 6a_3^2a_2^2 + 3a_2^4)a_1 + 5a_3^3a_2^2, \\
n_2 :=& -a_2(a_1^3 + 3a_1a_2^2 + 2a_3^3), \\
n_3 :=& 2(a_2^2 - a_3^2)a_1 + (a_2^2 - a_1^2)a_3.
\end{aligned}
$$

5 Conclusion

The method of the indeterminate coefficients and the factorization of cyclotomic polynomial gives us a chance to find more families of curves. It gives us an efficient method to find polynomials $u(x)$ such that $\Phi_k(u(x))$ has a factor of degree $\varphi(k)$. On the other hand, it will be hard to construct curves with ρ-values $\geq 3/2$, at least for $k = 8$. Since the size and complexity of elements of V are proportional to $\varphi(k)$, it is hard to find a formula for the coefficients for higher k. For constructing new families with higher embedding degrees and $\rho = 1$, we intend to find element of $\mathbf{Q}(\zeta_k)$ so that coefficients of higher degree of $u(x)$ vanish. Moreover, we also intend to apply our idea with other equations.

References

[1] Atkin, A.O.L., Morain, F.: Elliptic Curves and Primality Proving. Math. Comp. 61(203), 29–68 (1993)

[2] Boneh, D., Franklin, M.: Identity Based Encryption from the Weil Pairing. SIAM Journal of Computing 32(3), 586–615 (2003)

[3] Boneh, D., Lynn, B., Shacham, H.: Short Signatures from the Weil pairing. In: Boyd, C. (ed.) ASIACRYPT 2001. LNCS, vol. 2248, pp. 514–532. Springer, Heidelberg (2001)

[4] Brezing, F., Weng, A.: Elliptic Curves Suitable for Pairing Based Cryptography. Designs, Code and Cryptography 37(1), 133–141 (2005)

[5] Freeman, D., Scott, M., Teske, E.: A taxonomy of pairing-friendly elliptic curves. Cryptology ePrint Archive: 2006/372 (2006), `http://eprint.iacr.org/2006/372/`

[6] Galbraith, S., McKee, J., Valença, P.: Ordinary abelian varieties having small embedding degree. Finite Fields and Their Applications 13, 800–814 (2007)

[7] Hess, F., Smart, N., Vercauteren, F.: The Eta Pairing Revisited. IEEE Trans. Information Theory 52(10), 4595–4602 (2006)

[8] Joux, A.: A One Round Protocol for Tripartite Diffie-Hellman. Journal of Cryptology 17(4), 263–276 (2004)

[9] Kachisa, E.J., Schaefer, E.F., Scott, M.: Constructing Brezing-Weng pairing friendly elliptic curves using elements in the cyclotomic field. In: Galbraith, S.D., Paterson, K.G. (eds.) Pairing 2008. LNCS, vol. 5209. Springer, Heidelberg (2008)

[10] MAGMA Group: MAGMA Computational Algebra System, `http://magma.maths.usyd.edu.au`

[11] Miyaji, A., Nakabayashi, M., Takano, S.: New explicit conditions of elliptic curve traces for FR-reduction. IEICE Trans. Fundamentals E84, 1234–1243 (2001)

[12] Sakai, R., Ohgishi, K., Kasahara, M.: Cryptosystems based on pairing. In: 2000 Symposium on Cryptography and Information Security (SCIS 2000) (2000)

A Generalized Brezing-Weng Algorithm for Constructing Pairing-Friendly Ordinary Abelian Varieties

David Freeman

Computer Science Department
Stanford University
dfreeman@cs.stanford.edu

Abstract. We give an algorithm that produces families of Weil numbers for ordinary abelian varieties over finite fields with prescribed embedding degree. The algorithm uses the ideas of Freeman, Stevenhagen, and Streng to generalize the Brezing-Weng construction of pairing-friendly elliptic curves. We use our algorithm to give examples of pairing-friendly ordinary abelian varieties of dimension 2 and 3 that are absolutely simple and have smaller ρ-values than any previous such example.

Keywords: Abelian varieties, hyperelliptic curves, pairing-based cryptosystems, embedding degree, pairing-friendly varieties.

1 Introduction

In recent years, many new and useful cryptographic protocols have been proposed that make use of a bilinear map, or pairing [18]. For secure implementation, these protocols require an easily computable, nondegenerate pairing between finite groups in which the discrete logarithm problem is computationally infeasible. At present, the only known pairings with these properties are the Weil and Tate pairings on abelian varieties over finite fields. These pairings take as input points on an abelian variety defined over the field $\mathbb{F}_q$ and produce as output elements of an extension field $\mathbb{F}_{q^k}$. The degree of this extension is known as the *embedding degree.*

For a pairing-based cryptosystem on an abelian variety $A/\mathbb{F}_q$ to be secure and practical, the group of rational points $A(\mathbb{F}_q)$ should have a subgroup of large prime order r, and the embedding degree k should be large enough so that the discrete logarithm problems in $A[r]$ and $\mathbb{F}_{q^k}^{\times}$ are of roughly equal difficulty, yet small enough so that the pairing can be computed efficiently. As the embedding degree of a randomly chosen abelian variety over a field of cryptographic size is expected to be very large (see e.g., [1]), it is a difficult problem to construct "pairing-friendly" abelian varieties: those that have small embedding degree with respect to a large prime-order subgroup.

The problem of constructing pairing-friendly elliptic curves (i.e., one-dimensional abelian varieties) has been studied extensively; see [9] for a summary of

S.D. Galbraith and K.G. Paterson (Eds.): Pairing 2008, LNCS 5209, pp. 146–163, 2008.

the known constructions. In higher dimensions much less is known. Galbraith [11] and Rubin and Silverberg [19] have classified supersingular abelian varieties of dimension $g \geq 2$, and the latter have shown that for $g \leq 6$ the embedding degree always satisfies $k \leq 7.5g$. As the ratio k/g roughly measures the security level of pairings on the abelian variety, for high security levels we require larger ratios and must therefore turn to non-supersingular abelian varieties.

The only explicit constructions of pairing-friendly non-supersingular abelian varieties of dimension $g \geq 2$ are those of Freeman [8]; Kawazoe and Takahashi [14]; and Freeman, Stevenhagen, and Streng [10]. The first two constructions produce abelian surfaces ($g = 2$), while the last generalizes to arbitrary dimension the method of Cocks and Pinch [6] for constructing pairing-friendly elliptic curves.

For a pairing-friendly abelian variety $A/\mathbb{F}_q$, the ratio of the size (in bits) of the full group order $\#A(\mathbb{F}_q)$ to the size (in bits) of the subgroup order r is approximated by the parameter

$$\rho = \frac{g \log q}{\log r}. \tag{1.1}$$

The parameter ρ can be interpreted as measuring the ratio of an abelian variety's required bandwidth to its security level. In dimension $g = 2$ the constructions of Freeman and Freeman, Stevenhagen, and Streng both lead to ordinary, absolutely simple abelian varieties with $\rho \approx 8$. The construction of Kawazoe and Takahashi produces ordinary abelian surfaces with ρ-values between 3 and 4; however, these varieties are not absolutely simple, and thus the construction can be interpreted as producing pairing-friendly elliptic curves over some extension field of $\mathbb{F}_q$. In dimension $g = 3$ the best ρ-values produced by the Freeman-Stevenhagen-Streng method are $\rho \approx 18$.

In this paper, we demonstrate the first known examples of pairing-friendly ordinary abelian varieties of dimension $g = 2$ or 3 that are absolutely simple and have ρ-values significantly less than 8 or 18, respectively.

In Section 2 we give the conditions necessary for an abelian variety to be pairing-friendly and describe the approach of Brezing and Weng [4] to satisfying these conditions for elliptic curves. We then show how the ideas of Freeman, Stevenhagen, and Streng can be used to view the Brezing-Weng construction from a new perspective that admits a generalization to higher dimensions.

We give the details of this generalization in Section 3. The key idea is to parametrize the subgroup order r and the Frobenius element π of the pairing-friendly variety as polynomials of a single variable $r(x)$ and $\pi(x)$. The polynomial $r(x)$ has rational coefficients, while $\pi(x)$ has coefficients in a *CM field* K. Adapting the method of Freeman, Stevenhagen, and Streng, we construct $\pi(x)$ as the *extended type norm* of an element $\xi \in \widehat{K}[x]$, where $\widehat{K}$ is the *reflex field* of K, that is chosen to have specified residues modulo factors of $r(x)$ in $\widehat{K}[x]$.

Section 4 discusses how we use the polynomial $\pi(x)$ to construct explicit pairing-friendly abelian varieties. We first find an x_0 for which $q = \pi(x_0)\overline{\pi}(x_0)$ is prime and $r(x_0)$ has a large prime factor, and then use a *CM method* to construct an abelian variety over $\mathbb{F}_q$ whose Frobenius element is given by $\pi(x_0)$.

In Section 5 we discuss how to select the parameters in our algorithm to produce the optimal output, and provide a number of examples of families of ordinary abelian varieties produced by our method. These include several families of abelian surfaces ($g = 2$) with $\rho \leq 7$ and one with embedding degree 5 and $\rho \approx 4$, which could be a practical choice for certain security levels and which also answers (in one case) an open problem of Freeman, Stevenhagen, and Streng [10, Open Problem 3.5]. We also demonstrate a family of three-dimensional abelian varieties with $\rho \approx 12$. We conclude by discussing avenues for further research in this area.

2 Pairing-Friendly Abelian Varieties and the Brezing-Weng Method

Let A be a g-dimensional abelian variety defined over the finite field $\mathbb{F}_q$ of q elements. If the group of $\mathbb{F}_q$-rational points of A, denoted $A(\mathbb{F}_q)$, has a cyclic subgroup of order r with $\gcd(r, q) = 1$, then the *embedding degree of A with respect to r* is the smallest integer k such that the field $\mathbb{F}_{q^k}$ contains all rth roots of unity. Equivalently, the embedding degree is the order of q in $(\mathbb{Z}/r\mathbb{Z})^\times$. The embedding degree derives its name from the fact that $\mathbb{F}_{q^k}$ is the smallest field over which the Weil and Tate pairings take nontrivial values, and thus these pairings can be used to embed a cyclic, order-r subgroup of $A(\mathbb{F}_q)$ into $\mathbb{F}_{q^k}^\times$. (Note that if q is not prime then the image of these embeddings may be contained in a proper subfield of $\mathbb{F}_{q^k}$ [12].)

The embedding degree of an abelian variety $A/\mathbb{F}_q$ is determined by its *Frobenius endomorphism*, which acts by raising the coordinates of points on A to the qth power. The Frobenius endomorphism, denoted by π, satisfies a monic, integer polynomial h_A known as the *characteristic polynomial of Frobenius.* If A is simple then h_A is a power of an irreducible polynomial, and we can view π as an element of a number field $K = \mathbb{Q}(\pi)$. The field K is either a *CM field*, which is an imaginary quadratic extension of a totally real field, or the field $\mathbb{Q}(\sqrt{q})$ [21]; we will consider only the first case as the second corresponds to supersingular abelian varieties. By a theorem of Weil, all embeddings $K \hookrightarrow \mathbb{C}$ have $\pi\overline{\pi} = q$, where $\overline{\cdot}$ denotes complex conjugation. An algebraic integer π with this property is called a *q-Weil number.*

We will henceforth assume that A is simple, as the case of non-simple A can be reduced to the case of simple abelian varieties of lower dimension. We will further assume that $K = \mathbb{Q}(\pi)$ is the full endomorphism algebra $\mathrm{End}(A) \otimes \mathbb{Q}$; in particular, this is the case when A is ordinary. Under these assumptions, we have $[K : \mathbb{Q}] = 2 \cdot \dim A$ [22, Theorem 2], and the number of $\mathbb{F}_q$-rational points of A is given by

$$\#A(\mathbb{F}_q) = N_{K/\mathbb{Q}}(\pi - 1).$$

We can thus express the conditions for A being pairing-friendly as follows.

Proposition 2.1 ([10]). *Let $A/\mathbb{F}_q$ be a simple abelian variety with Frobenius endomorphism π, and assume $K = \mathbb{Q}(\pi)$ equals $\mathrm{End}(A) \otimes \mathbb{Q}$. Let k be a* *positive integer, Φ_k the kth cyclotomic polynomial, and r a square-free integer not dividing kq. If*

$$N_{K/\mathbb{Q}}(\pi - 1) \equiv 0 \pmod{r},$$
$$\Phi_k(\pi\overline{\pi}) \equiv 0 \pmod{r},$$

then A has embedding degree k with respect to r.

Proof. Since r is square free, the first condition tells us that $A(\mathbb{F}_q)$ has a cyclic subgroup of order r, the second that $\pi\overline{\pi} = q$ has order k in $(\mathbb{Z}/r\mathbb{Z})^\times$. □

The Brezing-Weng Method

If A is an ordinary elliptic curve over $\mathbb{F}_q$ with Frobenius endomorphism π, then $K = \mathbb{Q}(\pi) = \mathrm{End}(A) \otimes \mathbb{Q}$ is a quadratic imaginary field. In this case π can be described by its norm $q = \pi\overline{\pi}$ and its trace $t = \pi + \overline{\pi}$. The two conditions of Proposition 2.1 then become

$$r \mid q + 1 - t, \tag{2.1}$$
$$r \mid \Phi_k(q). \tag{2.2}$$

Furthermore, the condition $\pi \in K$ means that there is some integer y such that

$$t^2 - 4q = -Dy^2, \tag{2.3}$$

where D is the unique square-free positive integer such that $K = \mathbb{Q}(\sqrt{-D})$.

All of the known methods for constructing pairing-friendly ordinary elliptic curves involve fixing k and D and determining primes r and q and an integer t that satisfy (2.1)–(2.3) for some y. Many of these methods parametrize t, r, and q as polynomials $t(x)$, $r(x)$, $q(x)$ that produce valid curve parameters for many different inputs x. The advantage of such "families" is that the ρ-values (1.1) produced are often smaller than those produced by more general methods such as that of Cocks and Pinch [6]. One of the most successful such approaches is the method of Brezing and Weng [4]:

Algorithm 2.2 ([4])
Input: a positive integer k and a positive square-free integer D.

Output: polynomials $r(x) \in \mathbb{Z}[x]$ and $q(x) \in \mathbb{Q}[x]$ such that for any x_0 for which $q(x_0)$ is a prime integer, there is an ordinary elliptic curve E over $\mathbb{F}_{q(x_0)}$ such that $\mathrm{End}(E) \otimes \mathbb{Q} \cong \mathbb{Q}(\sqrt{-D})$ and E has embedding degree k with respect to $r(x_0)$.

1. Find an irreducible polynomial $r(x) \in \mathbb{Z}[x]$ such that $L = \mathbb{Q}[x]/(r(x))$ is a number field containing $\sqrt{-D}$ and the cyclotomic field $\mathbb{Q}(\zeta_k)$.
2. Choose a primitive kth root of unity $\zeta \in L$.
3. Let $t(x) \in \mathbb{Q}[x]$ be a polynomial mapping to $\zeta + 1$ in L.
4. Let $y(x) \in \mathbb{Q}[x]$ be a polynomial mapping to $(\zeta - 1)/\sqrt{-D}$ in L.
5. Set $q(x) \leftarrow (t(x)^2 + Dy(x)^2)/4$. Return $r(x)$ and $q(x)$. □

Note that while L is often chosen in practice to be isomorphic to $\mathbb{Q}(\zeta_k, \sqrt{-D})$, it can in fact be any number field — Galois or not — containing this field.

The key idea of the Brezing-Weng algorithm is that since elements of L are represented by polynomials modulo $r(x)$, we can always choose $t(x)$ and $y(x)$ to have degree strictly less than $\deg r(x)$. Thus we can always obtain $\deg q(x) \leq 2\deg r(x) - 2$, and in some cases we can do much better (see [9, §6]). As x grows the ρ-value (1.1) of the curve E approaches $\deg q/\deg r$; we call the latter quantity the ρ-value of the *family* $(t(x), r(x), q(x))$. We thus see that families generated by the Brezing-Weng method have ρ-values less than 2.

The Brezing-Weng algorithm is itself a generalization of an algorithm of Cocks and Pinch [6], which has the same form but works modulo a prime r instead of a polynomial $r(x)$. Freeman, Stevenhagen, and Streng [10] generalized the Cocks-Pinch algorithm to arbitrary CM fields K by demonstrating how the algorithm constructs a Frobenius element π with specified residues modulo certain primes over r in $\mathcal{O}_K$. We can use the same perspective to view the Brezing-Weng method in a new light.

Our new perspective starts with the fact that since $L = \mathbb{Q}[x]/(r(x))$ contains $K = \mathbb{Q}(\sqrt{-D})$, the polynomial $r(x)$ splits into two irreducible factors when viewed as an element of $K[x]$. We thus have $r(x) = r_1(x)\overline{r_1}(x)$ in $K[x]$, and $L \cong K[x]/(r_1(x)) \cong K[x]/(\overline{r_1}(x))$. Without loss of generality, we may assume that the map implied in Steps (3) and (4) of Algorithm 2.2 sends x to a root of $r_1(x)$.

If we compute $t(x)$ and $y(x)$ as in Algorithm 2.2 and let $\pi(x) = \frac{1}{2}(t(x) + y(x)\sqrt{-D})$, then $\pi(x) \equiv \zeta \bmod r_1(x)$. In addition, we see that $\overline{\pi}(x) = \frac{1}{2}(t(x) - y(x)\sqrt{-D}) \equiv 1 \bmod r_1(x)$. We thus see that $\pi(x)$ satisfies conditions analogous to those of Proposition 2.1:

$$(\pi(x) - 1)(\overline{\pi}(x) - 1) \equiv 0 \bmod r(x),$$
$$\Phi_k(\pi(x)\overline{\pi}(x)) \equiv 0 \bmod r(x).$$

The expression $\pi(x)\overline{\pi}(x)$ gives the $q(x)$ of the algorithm, so we conclude that for any $x_0 \in \mathbb{Q}$ for which $q(x_0)$ is a prime integer, $\pi(x_0) \in K$ is the Frobenius endomorphism of the elliptic curve E specified in the algorithm's description.

3 Generalizing the Brezing-Weng Method

If K is a CM field of degree $2g$, a *CM type* Φ of K is a set of g embeddings $\Phi = \{\phi_1, \ldots, \phi_g\}$ of K into its normal closure, one from each complex conjugate pair. A CM type is *primitive* if it is not induced from a CM type on a proper CM subfield of K. The *reflex type* of (K, Φ) consists of the *reflex field* $\widehat{K}$, which is a certain CM subfield of the normal closure of K, and a CM type Ψ of $\widehat{K}$. (For precise definitions of the reflex field and the reflex type, see [20, Section 8] or [10].) If Φ is primitive then the reflex of the reflex $(\widehat{K}, \Psi)$ is the original CM type (K, Φ). If K is Galois then $\widehat{K} = K$ and $\Psi = \{\phi^{-1} : \phi \in \Phi\}$; however for generic K the degree of $\widehat{K}$ will be much larger than the degree of K [10, Lemma 2.8].

The main algorithm of Freeman, Stevenhagen, and Streng [10, Algorithm 2.12] fixes a prime subgroup size r and uses the *type norm* from $\widehat{K}$ to construct a Frobenius element $\pi \in K$ that has specified residues modulo certain primes over r in $\mathcal{O}_K$. The type norm for a CM type (K, Φ) is the map

$$N_\Phi : \xi \mapsto \prod_{\phi \in \Phi} \phi(\xi).$$

The image of the type norm N_Φ is contained in the reflex field $\widehat{K}$ [10, Lemma 2.7], so the image of the reflex type norm N_Ψ is contained in K. If the CM type Φ is primitive, then for generic $\xi \in K$ we have $\widehat{K} = \mathbb{Q}(N_\Phi(\xi))$ (cf. [10, Theorem 3.1]).

To apply the ideas of Freeman, Stevenhagen, and Streng to the Brezing-Weng construction, we extend the type norm to a multiplicative map on polynomials in $K[x]$.

Definition 3.1. Let K be a CM field and Φ be a CM type of K, and let L be the normal closure of K. Define the *extended type norm* $\mathcal{N}_\phi : K[x] \to L[x]$ by

$$\mathcal{N}_\Phi(\xi) = \prod_{\phi \in \Phi} \phi(\xi),$$

where $\phi(\xi)$ is obtained by applying ϕ to the coefficients of ξ.

Lemma 3.2. *Let $\xi \in K[x]$, and let Φ be a CM type of K. Then $\mathcal{N}_\Phi(\xi) \in \widehat{K}[x]$, where $\widehat{K}$ is the reflex field of (K, Φ).*

Proof. Let L be the normal closure of K, and let $\sigma \in \mathrm{Gal}(L/\widehat{K})$. Then by definition of the reflex type, σ permutes the elements of Φ, so $\sigma(\prod_{\phi \in \Phi} \phi(\xi)) = \prod_{\phi \in \Phi} \phi(\xi)$. (Cf. [10, Lemma 2.7].) □

Remark 3.3. Similarly, for any extension of number fields L/K we can extend the field norm $N_{L/K}$ to a map of polynomials $\mathcal{N}_{L/K} : L[x] \to K[x]$ by setting $\mathcal{N}_{L/K}(f) = \prod_\phi \phi(f)$, where ϕ ranges over the set of embeddings of L in its normal closure that fix K.

Let K be a CM field of degree $2g$ with primitive CM type Φ. Let $(\widehat{K}, \Psi)$ be the reflex CM type, and let $\deg \widehat{K} = 2\widehat{g}$. Let $L = \mathbb{Q}[x]/(r(x))$ be a number field containing $\widehat{K}$ and $\mathbb{Q}(\zeta_k)$. In the case where $K = \widehat{K}$ is a quadratic imaginary field, the Brezing-Weng method constructs directly a polynomial $\pi(x)$ parametrizing Frobenius elements by prescribing the residues of $\pi(x)$ modulo each factor of $r(x)$ in $K[x]$. To generalize this construction along the lines of Freeman, Stevenhagen and Streng, we construct $\pi(x)$ as the extended type norm $\mathcal{N}_\Psi$ of an element $\xi \in \widehat{K}[x]$ with prescribed residues modulo factors of $r(x)$ in $\widehat{K}[x]$. The following proposition allows us to index the factors of $r(x)$ in $\widehat{K}[x]$ in a way that will be useful for our construction.

Proposition 3.4. *Let $\widehat{K}$ be a CM field and Ψ be a CM type on $\widehat{K}$. Let $r(x) \in \mathbb{Q}[x]$ be irreducible, and assume that $L = \mathbb{Q}[x]/(r(x))$ is Galois and contains*

$\widehat{K}$. Let $G = \text{Gal}(L/\mathbb{Q})$ and $H = \text{Gal}(L/\widehat{K})$. For each $\psi \in \Psi$ let $\psi' \in G$ be a representative of the left coset of H that induces the embedding ψ on $\widehat{K}$.

Fix a root $\gamma \in L$ of $r(x)$. For each $\psi \in \Psi$, define

$$r_\psi(x) = \mathcal{N}_{L/\widehat{K}}(x - \psi'^{-1}(\gamma)), \qquad \overline{r_\psi}(x) = \mathcal{N}_{L/\widehat{K}}(x - \overline{\psi'}^{-1}(\gamma)).$$

Then for each $\psi \in \Psi$, r_ψ and $\overline{r_\psi}$ are irreducible elements of $\widehat{K}[x]$, and the complete factorization of $r(x)$ in $\widehat{K}[x]$ is given by

$$r(x) = \prod\nolimits_{\psi \in \Psi} r_\psi(x)\overline{r_\psi}(x). \tag{3.1}$$

Proof. The fact that r_ψ and $\overline{r_\psi}$ are in $\widehat{K}[x]$ follows from Remark 3.3. Since L is Galois, any root $\delta \in L$ of $r_\psi(x)$ is also a root of $r(x)$, and thus $L = \mathbb{Q}(\delta) = \widehat{K}(\delta)$. It follows that the minimal polynomial of δ over $\widehat{K}$ has degree $[L : \widehat{K}]$, which by construction is the degree of $r_\psi(x)$. Therefore $r_\psi(x)$ is the minimal polynomial of δ over $\widehat{K}$ and is thus irreducible. The proof for $\overline{\psi}$ is analogous.

Since the elements of H induce the complete set of embeddings of $\widehat{K}$ in L, we have

$$r_\psi(x) = \prod\nolimits_{\sigma \in H}(x - \sigma\psi'^{-1}(\gamma)), \qquad \overline{r_\psi}(x) = \prod\nolimits_{\sigma \in H}(x - \sigma\overline{\psi'}^{-1}(\gamma)).$$

If we let $\Psi' = \{\psi' : \psi \in \Psi\}$, then the set of roots of the right hand side of (3.1) is exactly $\{\tau(\gamma) : \tau \in H(\Psi' \cup \overline{\Psi'})^{-1}\}$. Since $\Psi' \cup \overline{\Psi'}$ is a complete set of left coset representatives of H in G, its inverse is a complete set of *right* coset representatives of H in G, and thus $H(\Psi' \cup \overline{\Psi'})^{-1} = G$. We conclude that $\{\tau(\gamma) : \tau \in H(\Psi' \cup \overline{\Psi'})^{-1}\}$ consists of precisely the roots of $r(x)$ in L. □

We now obtain an analogue of the main theorem of Freeman, Stevenhagen, and Streng [10, Theorem 2.10]:

Theorem 3.5. *Let (K, Φ) be a CM type and $(\widehat{K}, \Psi)$ its reflex. Let $r(x) \in \mathbb{Q}[x]$ be an irreducible polynomial such that $L = \mathbb{Q}[x]/(r(x))$ is a Galois extension of $\mathbb{Q}$ containing $\widehat{K}$ and the cyclotomic field $\mathbb{Q}(\zeta_k)$.*

Let $\gamma \in L$ be a root of $r(x)$, and write the factorization of $r(x)$ in $\widehat{K}[x]$ as in Proposition 3.4. Given $\xi \in \widehat{K}[x]$, for each $\psi \in \Psi$ suppose $\alpha_\psi, \beta_\psi \in \mathbb{Q}[x]$ satisfy

$$\xi \equiv \alpha_\psi \bmod r_\psi(x) \qquad \text{and} \qquad \xi \equiv \beta_\psi \bmod \overline{r_\psi}(x). \tag{3.2}$$

Suppose that

$$\prod\nolimits_{\psi \in \Psi} \alpha_\psi(\gamma) = 1 \qquad \text{and} \qquad \prod\nolimits_{\psi \in \Psi} \beta_\psi(\gamma) = \zeta, \tag{3.3}$$

where $\zeta \in L$ is a primitive kth root of unity. Then $\pi(x) = \mathcal{N}_\Psi(\xi) \in K[x]$ satisfies

1. *$\pi(x)\overline{\pi}(x) \in \mathbb{Q}[x]$,*
2. *$\mathcal{N}_{K/\mathbb{Q}}(\pi(x) - 1) \equiv 0 \bmod r(x)$, and*
3. *$\Phi_k(\pi(x)\overline{\pi}(x)) \equiv 0 \bmod r(x)$.*

Proof. Statement (1) follows from Remark 3.3 and the fact that $\pi(x)\overline{\pi}(x) = \mathcal{N}_{\widehat{K}/\mathbb{Q}}(\xi)$. Next, (3.2) implies that $\xi - \alpha_\psi = f_\psi r_\psi$ for some $f_\psi \in \widehat{K}[x]$, so $\psi'^{-1}(\gamma) \in L$ is a root of $\xi - \alpha_\psi \in \widehat{K}[x]$. Applying ψ' to this expression and using the fact that $\alpha_\psi \in \mathbb{Q}[x]$, we see that γ is a root of $\psi(\xi) - \alpha_\psi \in L[x]$. It follows that $(\psi(\xi))(\gamma) = \alpha_\psi(\gamma)$, and by the same reasoning, $(\overline{\psi}(\xi))(\gamma) = \beta_\psi(\gamma)$. Now since $\pi(\gamma) = \prod_{\psi\in\Psi}(\psi(\xi))(\gamma)$ by definition of the extended type norm, we conclude from (3.3) that $\pi(\gamma) = 1$ and $\overline{\pi}(\gamma) = \zeta$, from which statements (2) and (3) follow. □

If $\pi(x)$ and $r(x)$ are as in Theorem 3.5, then by Proposition 2.1 for any $x_0 \in \mathbb{Q}$ for which $q = \pi(x_0)\overline{\pi}(x_0)$ is a prime, $\pi(x_0) \in \mathcal{O}_K$ is the Frobenius element of an abelian variety over $\mathbb{F}_q$ that has embedding degree k with respect to $r(x_0)$. We can thus view $\pi(x)$ as defining a one-parameter "family" of pairing-friendly Frobenius elements. The following definitions formalize this concept, generalizing the "families" of Freeman, Scott, and Teske [9, Definition 2.6].

Definition 3.6. Let $f(x) \in \mathbb{Q}[x]$ be a non-constant, irreducible polynomial with positive leading coefficient. We say f *represents primes* if (1) $f(x) \in \mathbb{Z}$ for some $x \in \mathbb{Z}$, and (2) $\gcd(\{f(x) : x, f(x) \in \mathbb{Z}\}) = 1$.

Definition 3.6 is motivated by the conjecture of Bateman and Horn [3], which gives a heuristic asymptotic formula for the number of prime values taken by a set of polynomials with integer coefficients.

Definition 3.7. Let K be a CM field of degree $2g$, let $\pi(x) \in K[x]$, and let $r(x) \in \mathbb{Q}[x]$. We say that (π, r) *represents a family of g-dimensional abelian varieties with embedding degree k* if:

1. $q(x) = \pi(x)\overline{\pi}(x)$ is in $\mathbb{Q}[x]$.
2. $q(x)$ represents primes (in the sense of Definition 3.6).
3. $r(x)$ is non-constant, irreducible, integer-valued, and has positive leading coefficient.
4. $\mathcal{N}_{K/\mathbb{Q}}(\pi(x) - 1) \equiv 0 \bmod r(x)$.
5. $\Phi_k(q(x)) \equiv 0 \bmod r(x)$, where Φ_k is the kth cyclotomic polynomial.

With our setup, we can now adapt [10, Algorithm 2.12] to our new context.

Algorithm 3.8
Input: a primitive CM type (K, Φ); its reflex type $(\widehat{K}, \Psi)$; a positive integer k; a polynomial $r(x) \in \mathbb{Q}[x]$, satisfying condition (3) of Definition 3.7, such that $\mathbb{Q}[x]/(r(x))$ is a Galois number field containing K and the cyclotomic field $\mathbb{Q}(\zeta_k)$; and a non-empty set $\Sigma \subset \mathbb{Q}[x]$.

Output: a polynomial $\pi(x) \in K[x]$ such that if $q(x) = \pi(x)\overline{\pi}(x)$ represents primes (in the sense of Definition 3.6), then (π, r) represents a family of abelian varieties with embedding degree k (in the sense of Definition 3.7).

1. Set $\widehat{g} \leftarrow \frac{1}{2}\deg \widehat{K}$ and write $\Psi = \{\psi_1, \psi_2, \ldots, \psi_{\widehat{g}}\}$. Set $L \leftarrow \mathbb{Q}[x]/(r(x))$.
2. Let $\gamma \in L$ be a root of $r(x)$. Compute the factorization of $r(x)$ in $\widehat{K}[x]$ as in Proposition 3.4.

3. Choose a primitive kth root of unity $\zeta \in L$.
4. Choose polynomials $\alpha_1, \ldots, \alpha_{\widehat{g}-1}, \beta_1, \ldots, \beta_{\widehat{g}-1} \in \mathbb{Q}[x]$ from Σ.
5. Compute $\alpha_{\widehat{g}} \in \mathbb{Q}[x]$ such that $\prod_{i=1}^{\widehat{g}} \alpha_i(\gamma) = 1$, and compute $\beta_{\widehat{g}} \in \mathbb{Q}[x]$ such that $\prod_{i=1}^{\widehat{g}} \beta_i(\gamma) = \zeta$.
6. Use the Chinese remainder theorem to compute $\xi \in \widehat{K}[x]$ such that $\xi \equiv \alpha_i \bmod r_{\psi_i}(x)$ and $\xi \equiv \beta_i \bmod \overline{r_{\psi_i}}(x)$ for $i = 1, 2, \ldots, \widehat{g}$.
7. Set $\pi(x) \leftarrow \mathcal{N}_{\Psi}(\xi)$, and return $\pi(x)$. □

We defer our discussion of how to choose the inputs $r(x)$ and Σ to Section 5. We note that if K is a quadratic imaginary field, then Step (4) is empty and setting $q(x) = \pi(x)\overline{\pi}(x)$ and $t(x) = \pi(x) + \overline{\pi}(x)$ recovers the Brezing-Weng algorithm.

4 From Families to Explicit Abelian Varieties

We now consider the problem of constructing the varieties represented by a family (π, r). Our strategy is to use a CM field K in Algorithm 3.8 such that abelian varieties A in characteristic zero with $\mathrm{End}(A) \otimes \mathbb{Q} \cong K$ are known or can be easily computed. We can then use the polynomial $\pi(x)$ to find primes q for which the reductions of A modulo primes over q are pairing-friendly.

The desired varieties A in characteristic zero are constructed via *CM methods*, which we will discuss shortly. We obtain the primes q by searching for values of x_0 for which $q = \pi(x_0)\overline{\pi}(x_0)$ is prime. If $\pi(x_0)$ generates K over $\mathbb{Q}$ and q is unramified in K — both of which occur with very high probability — then $\pi(x_0)$ is the Frobenius element of an ordinary, simple abelian variety of dimension $g = \frac{1}{2}[K : \mathbb{Q}]$ [10, Lemma 2.2]. For cryptographic applications, we also need $r(x_0)$ to be prime or very nearly prime. We use the following algorithm to search for values of x_0 with the desired properties.

Algorithm 4.1
Input: a CM field K, a pair of polynomials (π, r) that represents a family of abelian varieties with embedding degree k (in the sense of Definition 3.7), and a positive integer y_0.

Output: integers x_0 and h such that $q(x_0) = \pi(x_0)\overline{\pi}(x_0)$ is prime and $r(x_0)$ is h times a prime.

1. Set $q(x) \leftarrow \pi(x)\overline{\pi}(x)$.
2. Compute integers a, b such that $q(ax + b)$ is integer-valued and represents primes.
3. Set $h \leftarrow \gcd(\{q(ax + b)r(ax + b) : x \in \mathbb{Z}\})$.
4. Set $\widetilde{r}(x) \leftarrow r(x)/h$.
5. Set $x_1 \leftarrow y_0$.
6. Repeat $x_1 \leftarrow x_1 + 1$ until $q(ax_1 + b)$ and $\widetilde{r}(ax_1 + b)$ are prime.
7. Set $x_0 \leftarrow ax_1 + b$. Return h and x_0.

The input y_0 is the starting point for the search, and should be chosen so that $r(y_0)/h$ is at least the minimum size desired for security. Since h does not depend on the input y_0, if the h output by the algorithm is too large we can try again with a larger y_0.

Proposition 4.2. *Suppose the $\pi(x)$ and $r(x)$ input into Algorithm 4.1 have degrees d_1 and d_2 respectively. If the Bateman-Horn conjecture [3] is true, then the expected running time of the algorithm is $O(d_1 d_2 (\log \alpha\delta y_0)^2)$, where δ is the smallest integer such that $\delta q(x) \in \mathbb{Z}[x]$ and $\alpha = \max\{q(x)r(x) : |x| \leq \delta/2\}$.*

Proof. We first show that integers a, b as in Step (2) always exist. Write $q(x) = \widetilde{q}(x)/\delta$; then $\widetilde{q}(x) \in \mathbb{Z}[x]$. Write the prime factorization of δ as $\prod p^{e_p}$. Since $q(x)$ represents primes, for every prime p there exists an integer b_p such that $q(b_p)$ is an integer not divisible by p, and thus p^{e_p} divides $\widetilde{q}(b_p)$ exactly. Let a and b be integers such that $a = \prod_{p|\delta} p^{e_p+1}$ and $b \equiv b_p \pmod{p^{e_p+1}}$ for all $p \mid \delta$. Then $q(ax+b)$ is integer-valued and is nonzero mod p for every p dividing δ. For every p not dividing δ, $ax+b$ ranges through all residue classes mod p, so there is some residue class of x mod p for which p does not divide $\widetilde{q}(ax+b)$. Thus there is no prime p dividing $q(ax+b)$ for all x, which is equivalent to $q(ax+b)$ representing primes.

Let h be as in Step (3). Since $q(ax+b)$ and $r(ax+b)$ are integer-valued and $q(ax+b)$ represents primes, there is some c such that

$$\gcd\big(\{q(ax+b) : x \equiv c \bmod h\}\big) = 1,$$
$$\gcd\big(\{r(ax+b) : x \equiv c \bmod h\}\big) = h.$$

It follows that the values of the polynomials $q(ahx+ac+b)$ and $\widetilde{r}(ahx+ac+b)$ are integers with no common divisor. The Bateman-Horn conjecture implies that we should expect to test roughly $2d_1d_2(\log ahy_0)^2$ values of x_1 before we find one for which $q(ax_1+b)$ is prime and $r(ax_1+b)$ is h times a prime. Since $\log a = O(\log \delta)$ and $h \leq \alpha$, the result follows. □

We note (and find in practice) that the a computed in Step (2) of Algorithm 4.1 can be smaller than the a produced in the proof of Proposition 4.2, and that there may be multiple valid choices of b for a given a. In addition, the α given in Proposition 4.2 is usually a gross overestimate for h.

Once we have found an x_0 such that $q(x_0)$ is prime and $r(x_0)$ is nearly prime, the problem remains to construct an abelian variety over $\mathbb{F} = \mathbb{F}_{q(x_0)}$ whose Frobenius element is $\pi(x_0)$. This is achieved using *CM methods.* We use CM methods to construct all varieties over $\overline{\mathbb{Q}}$ whose endomorphism rings are isomorphic to the ring of integers $\mathcal{O}_K$ of the CM field K. Since any ordinary abelian variety over a finite field $\mathbb{F}$ arises as the reduction modulo a prime of a variety over $\overline{\mathbb{Q}}$ with the same endomorphism ring, we can produce a set of abelian varieties over $\mathbb{F}$ that includes representatives of all of the $\overline{\mathbb{F}}$-isomorphism classes of varieties A with $\mathrm{End}(A) \cong \mathcal{O}_K$. We test these candidates A, as well as all of their twists (varieties over $\mathbb{F}$ that are $\overline{\mathbb{F}}$-isomorphic to A), to see which is in the correct $\mathbb{F}$-isogeny class; this can be determined by seeing if the number of $\mathbb{F}$-rational points is equal to $N_{K/\mathbb{Q}}(\pi(x_0) - 1)$.

Even though counting the number of rational points on a g-dimensional abelian variety A over a field $\mathbb{F}$ of cryptographic size is in general infeasible for $g \geq 2$, we can quickly determine whether the number of points is equal to n by choosing a few random points $P_i \in A(\mathbb{F})$ and seeing if $[n]P_i$ is the identity on A for all i. In the case where A is the Jacobian of a hyperelliptic curve of the form $y^2 = x^n + a$, one can also use the algorithm of Buhler and Koblitz [5] to compute $\#A(\mathbb{F})$ directly.

Finally, a word is in order about the CM methods. Over an algebraically closed field, all principally polarized abelian varieties of dimension $g \leq 3$ are Jacobians of genus g curves. It thus suffices to produce all curves whose Jacobians have endomorphism ring isomorphic to $\mathcal{O}_K$; we say that these Jacobians have *CM by* $\mathcal{O}_K$. In dimension $g = 1$ we compute the *Hilbert class polynomial*, a polynomial in $\mathbb{Z}[x]$ whose roots are equal to the j-invariants of elliptic curves over $\overline{\mathbb{Q}}$ with CM by $\mathcal{O}_K$. In dimension $g = 2$ we compute the *Igusa class polynomials*, which are three polynomials in $\mathbb{Q}[x]$ whose roots are the *Igusa invariants* of genus 2 curves over $\overline{\mathbb{Q}}$ whose Jacobians have CM by $\mathcal{O}_K$. Methods for $g = 3$ are analogous but have only been developed for fields K containing i or ζ_3 [24,16]. Methods for $g \geq 4$ are completely undeveloped.

The class polynomials produced by the CM methods are very large: both the degree and the size of the coefficients grow very quickly with the class number of K, and in general the computation is only feasible for very small CM fields K. For $g = 1$ the upper limit is roughly class number 10^5 [7], while for $g = 2$ we can only achieve class numbers around 100 [15], and for $g = 3$ the methods are even more limited. Thus we must be careful to choose a field K as input to Algorithm 3.8 for which we know that the CM method is feasible.

5 Parameter Selection and Examples

The primary advantage of Algorithm 3.8 is that it leads to pairing-friendly ordinary, absolutely simple abelian varieties with smaller ρ-values than any previous such construction. Recall that the ρ-value (1.1) of a g-dimensional abelian variety over $\mathbb{F}_q$ with respect to a subgroup of order r is $\rho = g \log q / \log r$. If $q = q(x)$ and $r = r(x)$ are parametrized as polynomials, then for large x the ρ-value approaches $g \deg q / \deg r$. This motivates the definition of a ρ-value for a family of pairing-friendly abelian varieties.

Definition 5.1. Suppose (π, r) represents a family of g-dimensional abelian varieties with embedding degree k, and let $q(x) = \pi(x)\overline{\pi}(x)$. The *$\rho$-value of the family represented by* (π, r), denoted $\rho(\pi, r)$, is

$$\rho(\pi, r) = \lim_{x \to \infty} \frac{g \log q(x)}{\log r(x)} = \frac{g \deg q(x)}{\deg r(x)}.$$

The key feature of Algorithm 3.8 is that the polynomial ξ constructed via the Chinese remainder theorem in Step (6) can always be chosen to have degree strictly less than $\deg r$, and thus $\deg \pi \leq \widehat{g}(\deg r - 1)$. We thus obtain

$$\rho(\pi, r) = 2g\widehat{g} \frac{\deg \xi}{\deg r} \leq 2g\widehat{g} \frac{\deg r - 1}{\deg r}.$$

This asymptotic ρ-value is an improvement over the ρ-values produced by the algorithm of Freeman, Stevenhagen, and Streng, which gives $\rho \approx 2g\widehat{g}$ [10, Theorem 3.4].

To improve the ρ-values further we wish to choose the inputs to Algorithm 3.8 in some clever manner so that the π produced has degree significantly less than $\widehat{g} \deg r$. These choices include the ζ of Step (3), the α_i and β_i of Step (4) (which are chosen from the input Σ), and the input polynomial $r(x)$.

In dimension $g \geq 2$ we search for a $\pi(x)$ of low degree by following the model of Brezing and Weng [4]. We let $r(x)$ be a cyclotomic polynomial Φ_ℓ such that $k \mid \ell$ and $L \cong \mathbb{Q}(\zeta_\ell)$ contains the specified CM field K. Since L is abelian, in this case the CM field K must also be abelian, and thus equal to the reflex field $\widehat{K}$. Since $r(x)$ is the ℓth cyclotomic polynomial, x is a primitive ℓth root of unity in $\mathbb{Q}[x]/(r_\psi(x))$ for all $\psi \in \Psi$. We choose Σ to be the set of monomials x^i, all of which map to roots of unity (of some order) in L. Thus if we choose $\alpha_i, \beta_i \in \Sigma$ so that

$$(\alpha_1, \ldots, \alpha_g) \in \{(x^{a_1}, \ldots, x^{a_g}) : 0 \leq a_i < \ell,\ \textstyle\sum_{i=1}^g a_i = 0 \bmod \ell\}, \tag{5.1}$$
$$(\beta_1, \ldots, \beta_g) \in \left\{(x^{b_1}, \ldots, x^{b_g}) : 0 \leq b_i < \ell,\ \gcd(\ell, \textstyle\sum_{i=1}^g b_i) = \ell/k\right\}, \tag{5.2}$$

then $\prod \alpha_i = x^{\sum a_i} \equiv 1 \bmod r(x)$, and $\prod \beta_i = x^{\sum b_i}$ is a primitive kth root of unity mod $r(x)$.

For given CM type (K, Φ), embedding degree k, and cyclotomic polynomial $r(x) = \Phi_\ell(x)$, our implementation of Algorithm 3.8 searches through all α_i, β_i satisfying (5.1) and (5.2) and returns the ξ of smallest degree. We illustrate with a detailed example for $g = 2$ that produces ρ-values around 4, thus answering (in one case) an open problem of Freeman, Stevenhagen, and Streng [10, Open Problem 3.5].

Example 5.2 ($g = 2, k = 5, \rho = 4$)**.** We give a step-by-step account of the execution of Algorithm 3.8. We input the CM field $K = \mathbb{Q}(\zeta_5)$, the embedding degree $k = 5$, the polynomial $r(x) = \Phi_5(x)$, and the CM type $\Phi = \{\phi_1, \phi_2\}$ on K, where ϕ_1 is the identity and $\phi_2 : \zeta_5 \mapsto \zeta_5^2$. The reflex type of Φ is $\Psi = \{\psi_1, \psi_2\}$, where ψ_1 is the identity and $\psi_2 : \zeta_5 \mapsto \zeta_5^3$.

In Step (1) of the algorithm we set $L = \mathbb{Q}[x]/(r(x)) \cong \mathbb{Q}(\zeta_5)$. In Step (2) we use the root $\gamma = \zeta_5 \in L$ to factor $r(x)$ in $K[x]$ as in Proposition 3.4, obtaining

$$r(x) = r_1(x)r_2(x)\overline{r_1}(x)\overline{r_2}(x) = (x - \zeta_5)(x - \zeta_5^2)(x - \zeta_5^4)(x - \zeta_5^3).$$

In Steps (4) and (5) we set

$$\alpha_1 = x, \quad \alpha_2 = x^4, \quad \beta_1 = x, \quad \beta_2 = x^3,$$

which will be the residues modulo $r_1, r_2, \overline{r_1}, \overline{r_2}$, respectively, of an element $\xi \in K[x]$. Note that this choice of α_i, β_i satisfies (5.1) and (5.2). In Step (6) we use the Chinese remainder theorem to compute

$$\begin{aligned}\xi(x) = {} & \tfrac{1}{5}(-2\zeta_5^3 - 4\zeta_5^2 - \zeta_5 - 3)x^2 + \tfrac{1}{5}(-\zeta_5^3 - 2\zeta_5^2 + 2\zeta_5 + 1)x \\ & + \tfrac{1}{5}(-2\zeta_5^3 - 4\zeta_5^2 - \zeta_5 - 3).\end{aligned}$$

Finally, in Step (7) we take the extended type norm $\mathcal{N}_\Phi(\xi)$ (Definition 3.1) and obtain

$$\pi(x) = \tfrac{1}{5}(-\zeta_5^3 + \zeta_5^2 + \zeta_5 - 1)x^4 + \tfrac{1}{5}(\zeta_5^3 + 2\zeta_5 - 3)x^3 + \tfrac{1}{5}(3\zeta_5^2 + 4\zeta_5 - 2)x^2 \\ + \tfrac{1}{5}(\zeta_5^3 + 2\zeta_5 - 3)x + \tfrac{1}{5}(-\zeta_5^3 + \zeta_5^2 + \zeta_5 - 1). \quad (5.3)$$

The algorithm outputs $\pi(x)$, from which we compute

$$q(x) = \pi(x)\overline{\pi}(x) = \tfrac{1}{5}\left(x^8 + 2x^7 + 8x^6 + 9x^5 + 15x^4 + 9x^3 + 8x^2 + 2x + 1\right).$$

Since $q(x)$ is irreducible and $q(1) = 11$ and $q(-4) = 11941$ are distinct primes, $q(x)$ represents primes as in Definition 3.6, and thus (π, r) represents a family of abelian surfaces with embedding degree 5 (Definition 3.7).

Let us now construct an example abelian surface in this family. We use Algorithm 4.1 to find a value x_0 for which $q(x_0)$ is an integer prime and $r(x_0)$ has a large prime factor. We input $y_0 = 2^{54}$ to Algorithm 4.1. Using $a = 5$ and $b = 1$ in Step (2), the algorithm outputs $h = 5$ and $x_0 = 9007199254741082 6$. We then compute

$$r(x_0) = 5 \cdot 1316403645857017813158328592076236005067383734218583870028087952665 1$$
$$q(x_0) = 866459279412824385938792486752217637176780280467936882241506648125593 29726381 \backslash \\ 24680359567670957526027076700398139345585675165846688 47561 \quad \text{(449 bits)}.$$

Then $r(x_0)$ is 5 times a 224-bit prime r_0. The Frobenius element $\pi(x_0) \in \mathbb{Q}(\zeta_5)$ can be computed from (5.3), and the number of points n is

$$N_{K/\mathbb{Q}}(\pi(x_0) - 1) = 750751682880590880758711726029628178972094394801060386139877155309 9 \backslash \\ 700537243257397953153042637280430066227151584188526163207099845108 6 \backslash \\ 388168581855479229176414878143693605405281374968944086792908817931 7 \backslash \\ 241437357723407745744526071772162827435691962393142167332744537571 8 \backslash \\ 05.$$

Over any field $\mathbb{F}$ there is a single $\overline{\mathbb{F}}$-isomorphism class of abelian surfaces whose ring of $\overline{\mathbb{F}}$-endomorphisms is isomorphic to $\mathbb{Z}[\zeta_5]$. If $\operatorname{char}\mathbb{F}$ is prime to 10, then this abelian surface is isomorphic (over $\overline{\mathbb{F}}$) to the Jacobian of $C : y^2 = x^5 + 1$. Over $\mathbb{F}$ we must find the twist of C that is in the correct $\mathbb{F}$-isogeny class; i.e., has a Jacobian with the correct number of $\mathbb{F}$-rational points. By choosing a random point P on each twist and checking whether $[n]P = O$, we find that the correct curve over $\mathbb{F} = \mathbb{F}_{q(x_0)}$ is

$$C : y^2 = x^5 + 5.$$

The ρ-value (1.1) of $\mathrm{Jac}(C)$ with respect to the subgroup of order r_0 is 4.02. □

Remark 5.3. The abelian surface $A = \mathrm{Jac}(C)$ computed in Example 5.2 has the property that the bit size of the field $\mathbb{F}_{q^k}$ in which pairings on A take their values is $\rho k/g \approx 10$ times the bit size of the prime-order subgroup $A[r]$. It follows that A is suitable for applications with security level equivalent to a 112-bit symmetric-key system [9, §1.1]. In addition, since the curve C has a degree-10 twist, we expect that twisting methods such as those developed for elliptic curves [17] can be used to increase the speed of pairing computation on the Jacobian and reduce the size of the input.

We ran Algorithm 3.8 for all degree-4 CM fields K that are primitive (i.e., do not contain a quadratic imaginary subfield) and are contained in a cyclotomic field $\mathbb{Q}(\zeta_\ell)$ with $\varphi(\ell) \leq 16$. We let the inputs to the algorithm range over all such K and ℓ and embedding degrees k dividing ℓ. Given an η such that $K = \mathbb{Q}(\eta)$, we let Φ be the CM type that consists of embeddings ϕ_i such that $\phi_i(\eta)$ all have positive imaginary part. We tested all choices of α_i, β_i satisfying (5.1) and (5.2), and computed the ξ of smallest degree that produces a $q(x)$ that represents primes in the sense of Definition 3.6. Some examples appear below.

Example 5.4 $(g = 2, k = 10, \rho = 6)$**.** Let $K = \mathbb{Q}(\zeta_5)$, $k = 10$, $r(x) = \Phi_{10}(x) = x^4 - x^3 + x^2 - x + 1$. Algorithm 3.8 outputs

$$\pi(x) = \tfrac{1}{25}(\zeta_5^3 - \zeta_5^2 - \zeta_5 + 1)x^6 + \tfrac{1}{25}(-6\zeta_5^3 + 5\zeta_5^2 + 3\zeta_5 - 2)x^5 + \tfrac{1}{5}(2\zeta_5^3 - \zeta_5^2 - 2)x^4 + \tfrac{1}{5}(-2\zeta_5^3 - \zeta_5 + 4)x^3 + \tfrac{1}{5}(3\zeta_5^3 - 2\zeta_5^2 - 2)x^2 + \tfrac{1}{25}(-4\zeta_5^3 - \zeta_5^2 - \zeta_5 + 11)x + \tfrac{1}{25}(4\zeta_5^3 - 5\zeta_5^2 - 2\zeta_5 - 2).$$

The ρ-value of this family is 6. On input $y_0 = 2^{40}$, Algorithm 4.1 outputs $h = 5$ and $x_0 = 5497558154509$. We find that A is the Jacobian of the genus 2 curve

$$C : y^2 = x^5 + 15.$$

Then $r(x_0)$ is 5 times a 168-bit prime r_0. The ρ-value of A with respect to r_0 is within 10^{-10} of 6. □

Example 5.5 $(g = 2, k = 16, \rho = 7)$**.** Let $K = \mathbb{Q}(\eta)$, where $\eta = \sqrt{-2 + \sqrt{2}}$. Let $k = 16$ and $r(x) = \Phi_{16}(x) = x^8 + 1$. Algorithm 3.8 outputs

$$\pi(x) = \tfrac{1}{64}(-\eta^2 - 2)x^{14} + \tfrac{1}{32}(-\eta^2 - 3\eta - 2)x^{13} + \tfrac{1}{64}(\eta^2 - 4\eta - 16)x^{12} + \tfrac{1}{16}(-2\eta^3 + \eta^2 - 6\eta + 5)x^{11} + \tfrac{1}{64}(-8\eta^3 + \eta^2 - 28\eta)x^{10} + \tfrac{1}{32}(4\eta^3 - \eta^2 + 7\eta - 2)x^9 + \tfrac{1}{64}(8\eta^3 - \eta^2 + 16\eta - 34)x^8 + \tfrac{1}{8}(-\eta^3 - 2\eta + 4)x^7 + \tfrac{1}{64}(-8\eta^3 - \eta^2 - 16\eta - 2)x^6 + \tfrac{1}{32}(4\eta^3 - \eta^2 + 13\eta - 2)x^5 + \tfrac{1}{64}(8\eta^3 + \eta^2 + 28\eta - 16)x^4 + \tfrac{1}{16}(\eta^2 + 2\eta + 5)x^3 + \tfrac{1}{64}(\eta^2 + 4\eta)x^2 + \tfrac{1}{32}(-\eta^2 - \eta - 2)x + \tfrac{1}{64}(-\eta^2 - 2).$$

The ρ-value of this family is 7. The single $\overline{\mathbb{Q}}$-isomorphism class of genus 2 curves whose Jacobians have CM by $\mathcal{O}_K$ is given by van Wamelen [23]. On input $y_0 = 2^{18}$, Algorithm 4.1 outputs $h = 2$ and $x_0 = 1083939$. We find A to be the Jacobian of the genus 2 curve

$$C : y^2 = -x^5 + 3x^4 + 2x^3 - 6x^2 - 3x + 1.$$

Then $r(x_0)$ is 2 times a 160-bit prime r_0. The ρ-value of A with respect to r_0 is 6.91. □

Example 5.6 $(g = 2, k = 13, \rho = 20/3)$**.** Let $K = \mathbb{Q}(\eta)$, where $\eta = \sqrt{-13 + 2\sqrt{13}}$. Let $k = 13$ and let $r(x) = \Phi_{13}(x)$. Algorithm 3.8 outputs

$$\pi(x) = \tfrac{1}{4056}(-19\eta^3 + 183\eta^2 - 377\eta + 2301)x^{20} + \tfrac{1}{338}(-2\eta^3 + 7\eta^2 - 39\eta + 78)x^{19} + \tfrac{1}{4056}(23\eta^3 + 177\eta^2 + 481\eta + 2535)x^{18} + \tfrac{1}{1352}(7\eta^3 + 49\eta^2 + 65\eta + 767)x^{17} + \tfrac{1}{2028}(19\eta^3 + 141\eta^2 + 221\eta + 1755)x^{16} + \tfrac{1}{1352}(\eta^3 + 97\eta^2 - 65\eta + 1183)x^{15} + \tfrac{1}{2028}(31\eta^3 + 192\eta^2 + 377\eta + 2496)x^{14} + \tfrac{1}{1352}(13\eta^3 + 173\eta^2 + 195\eta + 2587)x^{13} + \tfrac{1}{26}(3\eta^2 - 2\eta + 39)x^{12} + \tfrac{1}{52}(\eta^3 + 8\eta^2 + 11\eta + 104)x^{11} + \tfrac{1}{312}(5\eta^3 + 33\eta^2 + 55\eta + 507)x^{10} + \tfrac{1}{78}(2\eta^3 + 9\eta^2 + 28\eta + 117)x^9 + \tfrac{1}{312}(5\eta^3 + 33\eta^2 + 55\eta + 507)x^8 + \tfrac{1}{4056}(97\eta^3 + 441\eta^2 + 1235\eta + 5811)x^7 + \tfrac{1}{338}(2\eta^3 + 32\eta^2 + 13\eta + 429)x^6 + \tfrac{1}{2028}(8\eta^3 + 165\eta^2 + 52\eta + 2535)x^5 + \tfrac{1}{1352}(19\eta^3 + 81\eta^2 + 273\eta + 923)x^4 + \tfrac{1}{338}(-\eta^3 + 9\eta^2 - 26\eta + 130)x^3 + \tfrac{1}{4056}(23\eta^3 + 99\eta^2 + 325\eta + 1521)x^2 + \tfrac{1}{2028}(8\eta^3 + 3\eta^2 + 130\eta + 39)x + \tfrac{1}{338}(-\eta^2 - 13).$$

The ρ-value of this family is 20/3. The single $\overline{\mathbb{Q}}$-isomorphism class of genus 2 curves whose Jacobians have CM by $\mathcal{O}_K$ is given by van Wamelen [23]. On input $y_0 = 7 \cdot 2^{15}$, Algorithm 4.1 outputs $h = 13$ and $x_0 = 3127658$. We find A to be the Jacobian of the genus 2 curve

$$C : y^2 = x^5 + 104x^4 + 5408x^3 + 140608x^2 + 1687296x + 7311616.$$

Then $r(x_0)$ is 13 times a 256-bit prime r_0. The ρ-value of A with respect to r_0 is 6.74. □

Some additional families we obtained for $g = 2$ are summarized in Table 1. The $\pi(x)$ produced by Algorithm 3.8 and example varieties of cryptographic size can be found online at `http://theory.stanford.edu/~dfreeman/papers/gen-bw-examples.pdf`

Table 1. Best ρ-values for families of abelian surfaces

k	CM field K	$r(x)$	ρ-value	k	CM field K	$r(x)$	ρ-value
6	$\mathbb{Q}(\sqrt{-6+3\sqrt{2}})$	$\Phi_{48}(x)$	7.5	30	$\mathbb{Q}(\zeta_5)$	$\Phi_{60}(x)$	7
8	$\mathbb{Q}(\sqrt{-5+\sqrt{5}})$	$\Phi_{40}(x)$	7.5	32	$\mathbb{Q}(\sqrt{-2+\sqrt{2}})$	$\Phi_{32}(x)$	7.5
15	$\mathbb{Q}(\zeta_5)$	$\Phi_{15}(x)$	7	40	$\mathbb{Q}(\zeta_5)$	$\Phi_{40}(x)$	6.5
20	$\mathbb{Q}(\zeta_5)$	$\Phi_{20}(x)$	6	60	$\mathbb{Q}(\zeta_5)$	$\Phi_{60}(x)$	7

We restrict to $r(x)$ of degree at most 16 because as the degree of $r(x)$ grows it becomes increasingly unlikely that we will find families with ρ-values significantly less than 8. For the same reason, we expect that non-Galois quartic CM fields K will not provide greatly improved ρ-values, as we must work in a field L that contains the compositum of the Galois closure of K and the cyclotomic field $\mathbb{Q}(\zeta_k)$.

In dimension $g = 3$, we used the same procedure for the degree-6 Galois CM field $\mathbb{Q}(\zeta_7)$. The family (π, r) we discovered leads to three-dimensional ordinary abelian varieties with ρ-values better than the best previously known examples, which have $\rho \approx 18$ [10, Example 5.3].

Example 5.7 ($g = 3, k = 7, \rho = 12$). Let $K = \mathbb{Q}(\zeta_7)$, $k = 7$, and $r(x) = \Phi_7(x)$. Algorithm 3.8 outputs

$$\begin{aligned}\pi(x) = & \tfrac{1}{49}(-2\zeta_7^5 - 2\zeta_7^3 - 2\zeta_7^2 + 6\zeta_7)x^{12} + \tfrac{1}{49}(-7\zeta_7^5 + 4\zeta_7^4 - 4\zeta_7^3 + 2\zeta_7^2 + 13\zeta_7 - 1)x^{11} + \tfrac{1}{49}(-9\zeta_7^5 \\ & + 10\zeta_7^4 - 2\zeta_7^3 + \zeta_7^2 + 23\zeta_7 + 5)x^{10} + \tfrac{1}{49}(-16\zeta_7^5 + 9\zeta_7^4 - 13\zeta_7^3 - 2\zeta_7^2 + 45\zeta_7 - 2)x^9 \\ & + \tfrac{1}{49}(-22\zeta_7^5 + 6\zeta_7^4 - 19\zeta_7^3 + 3\zeta_7^2 + 39\zeta_7 - 7)x^8 + \tfrac{1}{49}(-7\zeta_7^5 + 13\zeta_7^4 - 2\zeta_7^3 - 2\zeta_7^2 + 28\zeta_7 \\ & + 12)x^7 + \tfrac{1}{7}(-2\zeta_7^5 + \zeta_7^4 - 2\zeta_7^3 + \zeta_7^2 + 3\zeta_7 - 1)x^6 + \tfrac{1}{49}(-12\zeta_7^5 - 7\zeta_7^4 - 26\zeta_7^3 - 12\zeta_7^2 + 8\zeta_7)x^5 \\ & + \tfrac{1}{49}(-7\zeta_7^5 + 3\zeta_7^4 - 10\zeta_7^3 + 5\zeta_7^2 + 8\zeta_7 - 6)x^4 + \tfrac{1}{49}(2\zeta_7^5 + 4\zeta_7^4 + 2\zeta_7^3 - \zeta_7^2 + 5\zeta_7 + 9)x^3 \\ & + \tfrac{1}{49}(-5\zeta_7^5 - 2\zeta_7^4 - 8\zeta_7^3 + 2\zeta_7^2 - 3\zeta_7 - 5)x^2 + \tfrac{1}{49}(\zeta_7^5 + \zeta_7^4 - 2\zeta_7^3 - 3\zeta_7^2 + 3\zeta_7)x + \tfrac{1}{49}(\zeta_7^4 \\ & + 2\zeta_7^3 + 2\zeta_7^2 + 2)\end{aligned}$$

The ρ-value of this family is 12. The single $\overline{\mathbb{Q}}$-isomorphism class of genus 3 curves whose Jacobians have CM by $\mathcal{O}_K$ is given by $y^2 = x^7 + 1$. On input $y_0 = 2^{28}$, Algorithm 4.1 outputs $h = 7$ and $x_0 = 1879056152$. We find A to be the Jacobian of the genus 3 curve

$$C : y^2 = x^7 + 16.$$

Then $r(x_0)$ is 7 times a 183-bit prime r_0. The ρ-value of A with respect to r_0 is 12.10. □

We also ran our algorithm for the degree-6 CM field $\mathbb{Q}(\zeta_9)$ and found families with ρ-values of 15 for $k = 9$ and $k = 18$. Abelian varieties with CM by $\mathbb{Q}(\zeta_9)$ are Jacobians of Picard curves of the form $y^3 = x^4 + ax$ [16]. The $\pi(x)$ output by Algorithm 3.8 and example varieties of cryptographic size can be found online at `http://theory.stanford.edu/~dfreeman/papers/gen-bw-examples.pdf`

Future Directions

Our construction improves on the best known ρ-values of pairing-friendly ordinary abelian varieties of dimension $g \geq 2$ for many different choices of CM field K and embedding degree k. However, to make ordinary abelian varieties of dimension $g \geq 2$ competitive with elliptic curves in terms of performance, we must construct varieties with $\rho \leq 2$, with the ultimate goal of producing ρ-values close to 1. Achieving this goal is the most important problem for further work.

Our construction leaves a great deal of room for searching for better parameters. One direction would be to choose various Galois CM fields K and let $L = K(\zeta_k)$. Another approach would be to fix L and use the approach of Kachisa, Schaefer, and Scott [13] to search systematically through polynomials $r(x)$ such that $L = \mathbb{Q}[x]/(r(x))$. In the case where $g \geq 2$, one could also increase the size of the input Σ, which is the set from which we choose the residues α_i, β_i of ξ modulo factors of $r(x)$ in $\widehat{K}[x]$. In practice we find that when we use elements of Σ with large coefficients, the $q(x)$ computed have coefficients with large denominators and are thus unlikely to take integer values. However, even restricting Σ to contain only polynomials with small coefficients leaves many possible choices for α_i and β_i, and a program that searches systematically through these choices would have a good chance of finding improved ρ-values.

Acknowledgments

The author thanks Tanja Lange, Michael Naehrig, Edward Schaefer, and Marco Streng for helpful feedback on earlier drafts of this paper. This research was supported by a National Defense Science and Engineering Graduate Fellowship.

References

1. Balasubramanian, R., Koblitz, N.: The improbability that an elliptic curve has subexponential discrete log problem under the Menezes-Okamoto-Vanstone algorithm. Journal of Cryptology 11, 141–145 (1998)
2. Barreto, P.S.L.M., Lynn, B., Scott, M.: Constructing elliptic curves with prescribed embedding degrees. In: Cimato, S., Galdi, C., Persiano, G. (eds.) SCN 2002. LNCS, vol. 2576, pp. 263–273. Springer, Heidelberg (2002)
3. Bateman, P., Horn, R.: A heuristic asymptotic formula concerning the distribution of prime numbers. Math. Comp. 16, 363–367 (1962)
4. Brezing, F., Weng, A.: Elliptic curves suitable for pairing based cryptography. Designs, Codes and Cryptography 37, 133–141 (2005)
5. Buhler, J., Koblitz, N.: Lattice basis reduction, Jacobi sums and hyperelliptic cryptosystems. Bull. Austral. Math. Soc. 58, 147–154 (1998)
6. Cocks, C., Pinch, R.G.E.: Identity-based cryptosystems based on the Weil pairing (unpublished manuscript, 2001) (While this manuscript is unavailable, the main result appears as Theorem 4.1 of [9])
7. Enge, A.: The complexity of class polynomial computation via floating point approximations. Math. Comp. (to appear), `http://fr.arxiv.org/abs/cs.CC/0601104`
8. Freeman, D.: Constructing pairing-friendly genus 2 curves with ordinary Jacobians. In: Takagi, T., Okamoto, T., Okamoto, E., Okamoto, T. (eds.) Pairing 2007. LNCS, vol. 4575, pp. 152–176. Springer, Heidelberg (2007)
9. Freeman, D., Scott, M., Teske, E.: A taxonomy of pairing-friendly elliptic curves. Cryptology eprint 2006/371, `http://eprint.iacr.org`
10. Freeman, D., Stevenhagen, P., Streng, M.: Abelian varieties with prescribed embedding degree. In: van der Poorten, A.J., Stein, A. (eds.) ANTS-VIII 2008. LNCS, vol. 5011, pp. 60–73. Springer, Heidelberg (2008)
11. Galbraith, S.: Supersingular curves in cryptography. In: Boyd, C. (ed.) ASIACRYPT 2001. LNCS, vol. 2248, pp. 495–513. Springer, Heidelberg (2001)
12. Hitt, L.: On the minimal embedding field. In: Takagi, T., Okamoto, T., Okamoto, E., Okamoto, T. (eds.) Pairing 2007. LNCS, vol. 4575, pp. 294–301. Springer, Heidelberg (2007)
13. Kachisa, E., Schaefer, E., Scott, M.: Constructing Brezing-Weng pairing friendly elliptic curves using elements in the cyclotomic field. In: Galbraith, S.D., Paterson, K.G. (eds.) Pairing 2008. LNCS, vol. 5209. Springer, Heidelberg (2008), `http://eprint.iacr.org`
14. Kawazoe, M., Takahashi, T.: Pairing-friendly hyperelliptic curves of type $y^2 = x^5 + ax$. In: Galbraith, S.D., Paterson, K.G. (eds.) Pairing 2008. LNCS, vol. 5209. Springer, Heidelberg (2008), `http://eprint.iacr.org`
15. Kohel, D.: Quartic CM fields database, `http://echidna.maths.usyd.edu.au/kohel/dbs/complex_multiplication2.html`

16. Koike, K., Weng, A.: Construction of CM Picard curves. Math. Comp. 74, 499–518 (2004)
17. Naehrig, M., Barreto, P., Schwabe, P.: On compressible pairings and their computation. Cryptology eprint 2007/429, http://eprint.iacr.org
18. Paterson, K.: Cryptography from pairings. In: Blake, I.F., Seroussi, G., Smart, N.P. (eds.) Advances in Elliptic Curve Cryptography, pp. 215–251. Cambridge University Press, Cambridge (2005)
19. Rubin, K., Silverberg, A.: Supersingular abelian varieties in cryptology. In: Yung, M. (ed.) CRYPTO 2002. LNCS, vol. 2442, pp. 336–353. Springer, Heidelberg (2002)
20. Shimura, G.: Abelian Varieties with Complex Multiplication and Modular Functions. Princeton University Press, Princeton (1998)
21. Tate, J.: Classes d'isogénie des variétés abéliennes sur un corps fini (d'après T. Honda). In: Séminaire Bourbaki 1968/69. Springer Lect. Notes in Math, vol. 179, exposé 352, pp. 95–110. Springer, Heidelberg (1971)
22. Tate, J.: Endomorphisms of abelian varieties over finite fields. Invent. Math. 2, 134–144 (1966)
23. van Wamelen, P.: Examples of genus two CM curves defined over the rationals. Math. Comp. 68, 307–320 (1999)
24. Weng, A.: Hyperelliptic CM-curves of genus 3. Journal of the Ramanujan Mathematical Society 16(4), 339–372 (2001)

Pairing-Friendly Hyperelliptic Curves with Ordinary Jacobians of Type $y^2 = x^5 + ax$

Mitsuru Kawazoe and Tetsuya Takahashi

Faculty of Liberal Arts and Sciences
Osaka Prefecture University
1-1 Gakuen-cho Naka-ku Sakai Osaka 599-8531 Japan
{kawazoe,takahashi}@las.osakafu-u.ac.jp

Abstract. An explicit construction of pairing-friendly hyperelliptic curves with ordinary Jacobians was firstly given by D. Freeman. In this paper, we give other explicit constructions of pairing-friendly hyperelliptic curves with ordinary Jacobians based on the closed formulae for the order of the Jacobian of a hyperelliptic curve of type $y^2 = x^5 + ax$. We present two methods in this paper. One is an analogue of the Cocks-Pinch method and the other is a cyclotomic method. By using these methods, we construct a pairing-friendly hyperelliptic curve $y^2 = x^5 + ax$ over a finite prime field $\mathbb{F}_p$ whose Jacobian is ordinary and simple over $\mathbb{F}_p$ with a prescribed embedding degree. Moreover, the analogue of the Cocks-Pinch produces curves with $\rho \approx 4$ and the cyclotomic method produces curves with $3 \leq \rho \leq 4$.

Keywords: pairing-based cryptography, hyperelliptic curves.

1 Introduction

Pairing-based cryptography was proposed around 2000 by three important works due to Joux [15], Sakai, Ohgishi and Kasahara [20] and Boneh and Franklin [4]. In these last two papers, the authors constructed an identity-based encryption scheme by using the Weil pairing of elliptic curves. Pairing-based cryptosystem can be constructed by using the Weil or Tate pairing on abelian varieties over finite fields. The key idea is that for an abelian variety of dimension g defined over a finite field $\mathbb{F}_q$, its subgroup of prime order ℓ is embedded into the multiplicative group of some extension field $\mathbb{F}_{q^k}$ as the multiplicative group of ℓth roots of unity via the Weil pairing or some other pairing map. The ratio $g \log q / \log \ell$ and the extension degree k are important for the construction of pairing-based cryptosystem. This ratio $g \log q / \log \ell$ is denoted by ρ, and the extension degree k is called the embedding degree with respect to ℓ.

In cryptography, abelian varieties obtained as Jacobians of hyperelliptic curves are often used. The Jacobian variety of a hyperelliptic curve of genus g is an abelian variety of dimension g. Note that an elliptic curve is a hyperelliptic curve of genus one and also an abelian variety of dimension one. Suitable abelian varieties for pairing-based cryptography are called "pairing-friendly". Moreover,

S.D. Galbraith and K.G. Paterson (Eds.): Pairing 2008, LNCS 5209, pp. 164–177, 2008.

hyperelliptic curves whose Jacobians are suitable for pairing-based cryptography are also called "pairing-friendly". One of important conditions for being pairing-friendly is that the embedding degree should be in a appropriate size. It is known that supersingular abelian varieties have small embedding degree (cf. [19]). For example, for the case of dimension one (i.e. elliptic curves) it is at most 6, and for the case of dimension two it is at most 12. Hence, if we need a larger embedding degree, we need ordinary abelian varieties. Another important condition is that the value of ρ should be small. By the definition of ρ, its theoretical minimum is $\rho \approx 1$ for abelian varieties of any dimension.

For the case of elliptic curves, there are many results for constructing pairing-friendly ordinary elliptic curves: Miyaji, Nakabayashi and Takano [18], Cocks and Pinch [7], Brezing and Weng [5], Barreto and Naehrig [2], Scott and Barreto [21], Freeman, Scott and Teske [10] and so on. Using the above methods, we can construct pairing-friendly elliptic curves with $\rho \approx 1$ for the embedding degree less than or equal to 6 (cf. [18]), $\rho \approx 2$ (cf. [7]) or $1 < \rho < 2$ for many embedding degrees (cf. [10]). On the other hand, there are very few results for explicit constructions of pairing-friendly ordinary abelian varieties of higher dimension. The only known results are Freeman [8], Freeman, Stevenhagen and Streng [11] and Freeman [9]. The ρ-values in these results are $4 \leq \rho \leq 8$ for dimension two (one family with $\rho \approx 4$ is given in [9]) and $\rho \approx 12$ for dimension three.

In this paper, we give other explicit constructions of pairing-friendly hyperelliptic curves with ordinary Jacobians. One is an analogue of the Cocks-Pinch method and the other is a cyclotomic method. Both methods are based on the closed formulae for the order of the Jacobian of a hyperelliptic curve of type $y^2 = x^5 + ax$ over a finite prime field $\mathbb{F}_p$ which are given by E. Furukawa, M. Kawazoe and T. Takahashi [12] and M. Haneda, M. Kawazoe and T. Takahashi [14]. By using these methods, for a given embedding degree k, we construct a pairing-friendly hyperelliptic curve $y^2 = x^5 + ax$ over $\mathbb{F}_p$. Though Jacobians of curves constructed by our methods are not absolutely simple, our methods produce curves whose Jacobians are simple over defining fields with smaller ρ-values than previously obtained. In fact, the analogue of the Cocks-Pinch method produces curves with $\rho \approx 4$ for arbitrary embedding degree and the cyclotomic method produces curves with $3 \leq \rho \leq 4$. In particular, when the embedding degree equals 24, we obtain a cyclotomic family with $\rho \approx 3$.

2 Definition and Basic Facts on Hyperelliptic Curves and Pairing-Based Cryptography

In this section, we recall some basic facts on hyperelliptic curves and pairing-based cryptography.

2.1 Hyperelliptic Curves and Their Jacobians

First, we recall the relation between the order of the Jacobian and the Frobenius map. Let p be an odd prime and $\mathbb{F}_q$ a finite field with q elements where $q = p^r$ for a positive integer r.

Let C be a hyperelliptic curve of genus g defined over $\mathbb{F}_q$. Then the defining equation of C is given as $y^2 = f(x)$ where $f(x)$ is a polynomial in $\mathbb{F}_q[x]$ of degree $2g+1$ or $2g+2$. Let J_C be the Jacobian variety of a hyperelliptic curve C. The Jacobian variety J_C is an abelian variety of dimension g. Note that if $g=1$ (i.e. C is an elliptic curve), then C is isomorphic to J_C. The finite abelian group of $\mathbb{F}_q$-rational points on J_C is denoted by $J_C(\mathbb{F}_q)$ and called the Jacobian group of C. Let $\chi(t)$ be the characteristic polynomial of the qth power Frobenius endomorphism of C. We call $\chi(t)$ for C the characteristic polynomial of C. Then, it is well-known that the order $\#J_C(\mathbb{F}_q)$ is given by

$$\#J_C(\mathbb{F}_q) = \chi(1).$$

2.2 Pairing-Based Cryptography

Here we recall pairing-based cryptography using Jacobian varieties of hyperelliptic curves over finite fields. Let C be a hyperelliptic curve of genus g defined over $\mathbb{F}_q$. Assume that $J_C(\mathbb{F}_q)$ has a subgroup G of a large prime order. Let ℓ be the order of G. The group of ℓ-torsion points of $J_C(\overline{\mathbb{F}_q})$ is denote by $J_C[\ell]$ where $\overline{\mathbb{F}_q}$ is an algebraic closure of $\mathbb{F}_q$ and $J_C(\overline{\mathbb{F}_q})$ is a group of $\overline{\mathbb{F}_q}$-rational points on J_C.

For a positive integer ℓ coprime to the characteristic of $\mathbb{F}_q$, the Weil pairing is a non-degenerate bilinear map

$$e_\ell : J_C[\ell] \times J_C[\ell] \to \mu_\ell \subset \mathbb{F}_{q^k}^\times$$

where μ_ℓ is the multiplicative group of ℓth roots of unity in $\overline{\mathbb{F}_q}^\times$ and $\mathbb{F}_{q^k}$ is the smallest field extension of $\mathbb{F}_q$ containing μ_ℓ.

The key idea of pairing-based cryptography is based on the fact that the subgroup G of prime order ℓ is embedded to the group μ_ℓ via the Weil pairing or some other pairing map. The extension degree k of the field extension $\mathbb{F}_{q^k}/\mathbb{F}_q$ is called the *embedding degree* of J_C with respect to ℓ. The embedding degree with respect to ℓ equals the smallest positive integer k such that ℓ divides $q^k - 1$. In other words, q is a primitive kth root of unity modulo ℓ.

When C is an elliptic curve and k is the embedding degree of C with respect to ℓ, $\mathbb{F}_{q^k}$ is a field generated by coordinates of all ℓ-torsion points [1]. For the higher genus case, we refer to the following result for an abelian varieties due to Freeman [8].

Proposition 1 ([8]). *Let A be an abelian variety over a finite field $\mathbb{F}_q$, $\chi(t)$ the characteristic polynomial of the qth power Frobenius map of A. For a prime number $\ell \not| q$ and a positive integer k, suppose the following hold:*

$$\chi(1) \equiv 0 \pmod{\ell}$$
$$\Phi_k(q) \equiv 0 \pmod{\ell}$$

where Φ_k is the kth cyclotomic polynomial. Then A has the embedding degree k with respect to ℓ. Furthermore, if $k > 1$ then $A(\mathbb{F}_{q^k})$ contains two linearly independent ℓ-torsion points.

In pairing-based cryptography, for the Jacobian variety J_C defined over $\mathbb{F}_q$, the following conditions must be satisfied to make a system secure:

- the order ℓ of a prime order subgroup of $J_C(\mathbb{F}_q)$ should be large enough so that solving a discrete logarithm problem on the group is computationally infeasible and
- the order q^k of the field $\mathbb{F}_{q^k}$ should be large enough so that solving a discrete logarithm problem on the multiplicative group $\mathbb{F}_{q^k}^{\times}$ is computationally infeasible.

Moreover for an efficient implementation of a pairing-based cryptosystem, the following are important:

- the embedding degree k should be appropriately small and
- the ratio $\rho = g \log_2 q / \log_2 \ell$ should be appropriately small.

Jacobian varieties satisfying the above four conditions are called "pairing-friendly". Hyperelliptic curves whose Jacobian varieties are pairing-friendly are also called "pairing-friendly". In practice, it is currently recommended that ℓ should be larger than 2^{160} and q^k should be larger than 2^{1024}.

3 Formulae for the Order of the Jacobian of Hyperelliptic Curves of Type $y^2 = x^5 + ax$

Our methods are based on the closed formulae for the order of the Jacobian of a hyperelliptic curve of type $y^2 = x^5 + ax$ over a finite prime field $\mathbb{F}_p$ which were given by E. Furukawa, M. Kawazoe and T. Takahashi [12] and M. Haneda, M. Kawazoe and T. Takahashi [14]. Due to the results of [12] and [14], the characteristic polynomial of a hyperelliptic curve of type $y^2 = x^5 + ax$ over $\mathbb{F}_p$ are determined completely as follows. For the proof of the following theorem, see [14] for the proof of (1) and see [12] for others.

Theorem 1 ([12], [14]). *Let p be an odd prime, C a hyperelliptic curve defined by an equation $y^2 = x^5 + ax$ over $\mathbb{F}_p$, J_C the Jacobian variety of C and $\chi(t)$ the characteristic polynomial of the pth power Frobenius map of C. Then the following holds: (In the following, c and d denote integers such that $p = c^2 + 2d^2$ and $c \equiv 1 \pmod 4$. Note that such c and d exist if and only if $p \equiv 1, 3 \pmod 8$.)*

(1) If $p \equiv 1 \pmod 8$ and $a^{(p-1)/2} \equiv -1 \pmod p$, then $\chi(t) = t^4 - 4dt^3 + 8d^2t^2 - 4dpt + p^2$ where $f = (p-1)/8$ and $2(-1)^f d \equiv (a^f + a^{3f})c \pmod p$.
(2) If $p \equiv 1 \pmod 8$ and $a^{(p-1)/4} \equiv -1 \pmod p$, or if $p \equiv 3 \pmod 8$ and $a^{(p-1)/2} \equiv -1 \pmod p$, then $\chi(t) = t^4 + (4c^2 - 2p)t^2 + p^2$.
(3) If $p \equiv 1 \pmod{16}$ and $a^{(p-1)/8} \equiv 1 \pmod p$, or if $p \equiv 9 \pmod{16}$ and $a^{(p-1)/8} \equiv -1 \pmod p$, then $\chi(t) = (t^2 - 2ct + p)^2$.
(4) If $p \equiv 1 \pmod{16}$ and $a^{(p-1)/8} \equiv -1 \pmod p$, or if $p \equiv 9 \pmod{16}$ and $a^{(p-1)/8} \equiv 1 \pmod p$, then $\chi(t) = (t^2 + 2ct + p)^2$.
(5) If $p \equiv 3 \pmod 8$ and $a^{(p-1)/2} \equiv 1 \pmod p$, then $\chi(t) = (t^2 + 2ct + p)(t^2 - 2ct + p)$.

(6) If $p \equiv 5 \pmod 8$ *and* $a^{(p-1)/4} \equiv 1 \pmod p$, *or if* $p \equiv 7 \pmod 8$, *then* $\chi(t) = (t^2 + p)^2$.
(7) If $p \equiv 5 \pmod 8$ *and* $a^{(p-1)/4} \equiv -1 \pmod p$, *then* $\chi(t) = (t^2 - p)^2$.
(8) If $p \equiv 5 \pmod 8$ *and* $a^{(p-1)/2} \equiv -1 \pmod p$, *then* $\chi(t) = t^4 + p^2$.

Remark 1. For the convenience in the following argument, we replaced d in [14] by $(-1)^{f+1}d$ in Theorem 1 (1).

We remark that $\chi(t)$ for the case (3)-(7) are reducible over the ring $\mathbb{Z}$. Moreover, the case (6), (7) and (8) are the supersingular case. In the following we restrict our interest to the case (1) and (2), because these are the only cases that J_C is a simple ordinary Jacobian over $\mathbb{F}_p$. The above theorem leads to the closed formulae for the order of the Jacobian group $J_C(\mathbb{F}_p)$ by using $\#J_C(\mathbb{F}_p) = \chi(1)$.

4 Analogue of the Cocks-Pinch Method

By using the formulae given in Theorem 1 (1) and (2), we obtain an analogue of the Cocks-Pinch method for hyperelliptic curves $y^2 = x^5 + ax$. Let χ be $1 - 4d + 8d^2 - 4dp + p^2$ or $1 + 4c^2 - 2p + p^2$. Then we can construct pairing-friendly hyperelliptic curves of type $y^2 = x^5 + ax$ over $\mathbb{F}_p$ if we find integers c, d and odd primes p, ℓ satisfying the following conditions: (Note that $p \equiv 1, 3 \pmod 8$.)

$$\begin{aligned} \chi &\equiv 0 \pmod \ell \\ \Phi_k(p) &\equiv 0 \pmod \ell \\ p &= c^2 + 2d^2 \quad \text{with } c \equiv 1 \pmod 4. \end{aligned}$$

The first condition means that the order of the Jacobian of a constructed curve has a subgroup of prime order ℓ. The second condition means that the embedding degree with respect to ℓ of the Jacobian of a constructed curve is k. Note that the second condition implies that p is a primitive kth root of unity modulo ℓ and therefore it implies that $\ell - 1$ must be divisible by k. Moreover, in both cases of Theorem 1 (1) and (2), square roots of -1 and 2 are required to be contained in the ring $\mathbb{Z}/\ell\mathbb{Z}$ so that integers c and d satisfying the above conditions exist. Hence $\ell - 1$ is required to be divisible by 8.

According to Theorem 1 (1) and (2), we have the following theorems:

Theorem 2. *For a given positive integer k, execute the following procedure:*

(1) Let ℓ be a prime such that $\mathrm{LCM}(8, k) | (\ell - 1)$.
(2) Let α be a primitive kth root of unity in $(\mathbb{Z}/\ell\mathbb{Z})^\times$, *$\beta$ a positive integer such that* $\beta^2 \equiv -1 \pmod \ell$ *and γ a positive integer such that* $\gamma^2 \equiv 2 \pmod \ell$.
(3) Let c and d be integers such that

$$\begin{aligned} c &\equiv (\alpha + \beta)(\gamma(\beta + 1))^{-1} \pmod \ell \quad \textit{and} \quad c \equiv 1 \pmod 4, \\ d &\equiv (\alpha\beta + 1)(2(\beta + 1))^{-1} \pmod \ell. \end{aligned}$$

If $p = c^2 + 2d^2$ *is a prime satisfying* $p \equiv 1 \pmod 8$*, then for an integer* a *satisfying*

$$a^{(p-1)/2} \equiv -1 \pmod p$$
$$2(-1)^{(p-1)/8} d \equiv (a^{(p-1)/8} + a^{3(p-1)/8})c \pmod p,$$

the Jacobian group $J_C(\mathbb{F}_p)$ *of a hyperelliptic curve* C *defined by* $y^2 = x^5 + ax$ *over* $\mathbb{F}_p$ *has a subgroup of order* ℓ *and the embedding degree of* J_C *with respect to* ℓ *is* k.

Proof. First note that the condition $k|(\ell - 1)$ implies that a primitive kth root of unity is contained in the ring $\mathbb{Z}/\ell\mathbb{Z}$ and the condition $8|(\ell - 1)$ implies that square roots of -1 and 2 are contained in $\mathbb{Z}/\ell\mathbb{Z}$.

Let ℓ be a prime as in (1) and let α, β and γ be as in (2). Substituting $c \equiv (\alpha + \beta)(\gamma(\beta + 1))^{-1} \pmod \ell$ and $d \equiv (\alpha\beta + 1)(2(\beta + 1))^{-1} \pmod \ell$ into $p = c^2 + 2d^2$, we have

$$p \equiv \left((\alpha + \beta)^2 + (\alpha\beta + 1)^2\right) \left(2(\beta + 1)^2\right)^{-1} \equiv (4\alpha\beta)(4\beta)^{-1} \equiv \alpha \pmod \ell.$$

Since α is a primitive kth root of unity in $(\mathbb{Z}/\ell\mathbb{Z})^\times$, we have $\Phi_k(p) \equiv 0 \pmod \ell$.

Next we check the condition on the order of the Jacobian. From the condition $d \equiv (\alpha\beta + 1)(2(\beta + 1))^{-1} \pmod \ell$, we have

$$1 - 2d \equiv (2d - \alpha)\beta \pmod \ell.$$

Substituting this into the formula $\#J_C(\mathbb{F}_p) = 1 - 4d + 8d^2 - 4dp + p^2$ and using $p \equiv \alpha \pmod \ell$, we have

$$\#J_C(\mathbb{F}_p) = (1 - 2d)^2 + (2d - p)^2 \equiv -(2d - \alpha)^2 + (2d - p)^2 \equiv 0 \pmod \ell$$

Thus the Jacobian variety of a constructed curve $y^2 = x^5 + ax$ over $\mathbb{F}_p$ has a subgroup of order ℓ and its embedding degree with respect to ℓ is k. □

Theorem 3. *For a given positive integer* k*, execute the following procedure:*

(1) , (2) are as in Theorem 2.
(3) Let c *and* d *be integers such that*

$$c \equiv 2^{-1}(\alpha - 1)\beta \pmod \ell \quad \text{and} \quad c \equiv 1 \pmod 4,$$
$$d \equiv (\alpha + 1)(2\gamma)^{-1} \pmod \ell.$$

If $p = c^2 + 2d^2$ *is a prime satisfying* $p \equiv 1, 3 \pmod 8$*, take an integer* δ *satisfying* $\delta^{(p-1)/2} \equiv -1 \pmod p$ *and set an integer* a *as*

$$a = \delta^2 \quad \text{when } p \equiv 1 \pmod 8,$$
$$a = \delta \quad \text{when } p \equiv 3 \pmod 8.$$

Then the Jacobian group $J_C(\mathbb{F}_p)$ *of a hyperelliptic curve* C *defined by* $y^2 = x^5 + ax$ *over* $\mathbb{F}_p$ *has a subgroup of order* ℓ *and the embedding degree of* J_C *with respect to* ℓ *is* k.

Proof. As in the proof of Theorem 2, substituting $c \equiv 2^{-1}(\alpha-1)\beta \pmod{\ell}$ and $d \equiv (\alpha+1)(2\gamma)^{-1} \pmod{\ell}$ into $p = c^2 + 2d^2$, we have

$$p \equiv 4^{-1}\left((\beta(\alpha-1))^2 + (\alpha+1)^2\right) \equiv \alpha \pmod{\ell}.$$

In particular, we have $\Phi_k(p) \equiv 0 \pmod{\ell}$.

Next we check the condition on the order of the Jacobian. Substituting $c \equiv 2^{-1}(\alpha-1)\beta \pmod{\ell}$ into the formula $\#J_C(\mathbb{F}_p) = 1 + 4c^2 - 2p + p^2$ and using $p \equiv \alpha \pmod{\ell}$, we have

$$\#J_C(\mathbb{F}_p) = 4c^2 + (p-1)^2 \equiv -(\alpha-1)^2 + (p-1)^2 \equiv 0 \pmod{\ell}.$$

Thus the Jacobian variety of constructed curve $y^2 = x^5 + ax$ over $\mathbb{F}_p$ has a subgroup of order ℓ and its embedding degree with respect to ℓ is k. □

Theorem 2 and 3 give an analogue of the Cocks-Pinch method for a hyperelliptic curve of type $y^2 = x^5 + ax$. We call curves obtained by Theorem 2 "Type I", and curves obtained by Theorem 3 "Type II".

Since our method based on the closed formulae of the order of the Jacobian, we can construct a pairing-friendly hyperelliptic curve in a very short time. For the running time of our algorithm, see Section 5. Moreover, we remark that $\rho \approx 4$ in our construction. This ρ-value is smaller than previously obtained. (Recently, Freeman [9] proposed another method to construct pairing-friendly hyperelliptic curves and obtained one family with $\rho \approx 4$ for the embedding degree 5.)

We remark one more thing. As is shown in [12], Jacobians for curves of type I and II are isogenous to the product of two elliptic curves over the extension field which contains $a^{1/4}$.

Lemma 1 ([12]). *Let p be an odd prime and C a hyperelliptic curve defined by $y^2 = x^5 + ax$, $a \in \mathbb{F}_p^\times$ and $\mathbb{F}_q = \mathbb{F}_{p^r}$, $r \geq 1$. If $a^{1/4} \in \mathbb{F}_q$, then J_C is isogenous to the product of the following two elliptic curves E_1 and E_2 over $\mathbb{F}_q$:*

$$E_1 : Y^2 = X(X^2 + 4a^{1/4}X - 2a^{1/2}),$$
$$E_2 : Y^2 = X(X^2 - 4a^{1/4}X - 2a^{1/2}).$$

By the above lemma, we have the following: (1) Jacobian for type I splits over $\mathbb{F}_{p^4}$, (2) Jacobian for type II with $p \equiv 3 \pmod{8}$ splits over $\mathbb{F}_{p^4}$, and (3) Jacobian for type II with $p \equiv 1 \pmod{8}$ splits over $\mathbb{F}_{p^2}$.

Let C be a pairing-friendly hyperelliptic curve of type I or II with embedding degree k with respect to ℓ. We write the value $2\log_2 p/\log_2 \ell$ for C as $\rho(C)$. If C is of type I, or of type II with $p \equiv 3 \pmod{8}$, then E_1 or E_2 is a pairing-friendly elliptic curve over $\mathbb{F}_{p^4}$ with embedding degree $k/4$ with $\rho = \log_2 p^4/\log_2 \ell = 2\rho(C)$. If C is of type II with $p \equiv 1 \pmod{8}$, then E_1 or E_2 is a pairing-friendly elliptic curve over $\mathbb{F}_{p^2}$ with embedding degree $k/2$ with $\rho = \log_2 p^2/\log_2 \ell = \rho(C)$.

5 Result of Search for Pairing-Friendly Hyperelliptic Curves: The Analogue of the Cocks-Pinch Method

In Table 1 and Table 2, we show the number of pairing-friendly hyperelliptic curves of Type I, II for $13 \leq k \leq 32$ obtained by using our method. For the results of $k \leq 12$ and $k \geq 33$, see the extended version of this paper [17].

These tables show that we can find many pairing-friendly hyperelliptic curves with ordinary Jacobians by using our method. All computations have been done by Mathematica 6 on Mac OS X (1.66GHz Intel Core Duo with 1GB memory). For each k, the running time of the search is on average 90 seconds in Table 1 and 170 seconds in Table 2, respectively.

Here we show only one example of pairing-friendly hyperelliptic curves of type I with $k = 16$ obtained by the analogue of the Cocks-Pinch method. For examples of other type and other k, see the extended version of this paper [17].

Table 1. The number of pairing-friendly hyperelliptic curves obtained by the analogue of the Cocks-Pinch method for $\ell \in [2^{160}, 2^{160} + 2^{20}]$ with $|c| < \ell$ and $|d| < 2\ell$

k	Type I	Type II	
		$p \equiv 1 \pmod 8$	$p \equiv 3 \pmod 8$
13	44	42	39
14	34	38	40
15	42	43	38
16	149	163	169
17	33	42	46
18	29	39	48
19	32	42	44
20	78	75	81
21	34	29	30
22	35	50	34
23	64	46	45
24	141	152	124
25	33	47	32
26	43	35	36
27	41	45	31
28	82	90	69
29	31	40	36
30	32	31	30
31	29	26	37
32	143	161	164

Table 2. The number of pairing-friendly hyperelliptic curves obtained by the analogue of the Cocks-Pinch method for $\ell \in [2^{256}, 2^{256} + 2^{20}]$ with $|c| < \ell$ and $|d| < 2\ell$

k	Type I	Type II	
		$p \equiv 1 \pmod 8$	$p \equiv 3 \pmod 8$
13	16	19	12
14	6	13	18
15	16	13	18
16	55	59	81
17	9	16	19
18	14	14	10
19	18	28	26
20	30	27	29
21	15	7	18
22	15	17	26
23	21	13	17
24	70	67	61
25	21	12	24
26	26	17	12
27	16	13	17
28	34	25	26
29	17	14	10
30	15	13	14
31	6	10	17
32	64	59	47

k =16 (Type I)

ℓ =1461501637330902918203684832716283019655932840529 (161 bits)

α =81844167457893182397317622245688612690934307989

β =195562276567303320541291199692793181706146839127

γ =759224753535341599938962978629340510421546983720

c =44377152517514522371933429191352073808466251009

d =10989841417965341398489085346020251473054265996

p =2210884894346798442145165481525960184900817737075987357833399335\
226916051626079472576037262113 (311 bits)

a =3

$\#J_C(\mathbb{F}_p)$ = 48880120160508541101232277959462765729571682125818741808\
2910733116855655035560868542777327696362024706637568420695212814\
3139938957120301819393955637481342467018816294397128800020723098\
722 (621 bits)

ρ =3.88

6 Another Construction: Cyclotomic Families

Here we give another construction of pairing-friendly hyperelliptic curves of type $y^2 = x^5 + ax$. It is also based on the formulae given in Theorem 1 (1) and (2), but it is a hyperelliptic version of cyclotomic families.

Cyclotomic families for the case of elliptic curves have been investigated by Brezing and Weng [5], Freeman, Scott and Teske [10] and some other researchers. In a cyclotomic family, a cyclotomic polynomial is used to set a prime ℓ as $\ell = \Phi_k(t)$ or $\ell = \Phi_{ck}(t)$ for some $c > 1$ where k is the embedding degree and t is a positive integer. Though a prime ℓ is not taken arbitrarily, cyclotomic families have an advantage that the ρ-value of obtained curves can be smaller than the one obtained by the Cocks-Pinch method.

For a hyperelliptic curves of type $y^2 = x^5 + ax$, we require the condition that the embedding degree k is divisible by 8. Assume that the embedding degree k is divisible by 8 and $\ell - 1$ is divisible by k. Let α be a primitive kth root of unity modulo ℓ, β an integer such that $\beta^2 \equiv -1 \pmod{\ell}$ and γ an integer such that $\gamma^2 \equiv 2 \pmod{\ell}$. Then we have that $\beta = \pm\alpha^{k/4}$ and $\gamma = \pm\left(\alpha^{k/8} - \alpha^{3k/8}\right)$. Note that if $\gcd(k, h) = 1$, then α^h is also a primitive kth root of unity modulo ℓ.

6.1 A Cyclotomic Family of Type I

From Theorem 2, we have

$$c = \frac{\alpha+\beta}{\beta\gamma+\gamma} = \frac{(\alpha+\beta)(\beta\gamma-\gamma)}{(\beta\gamma+\gamma)(\beta\gamma-\gamma)} = \frac{\alpha(\gamma-\beta\gamma)+(\gamma+\beta\gamma)}{4}$$

$$d = \frac{\alpha\beta+1}{2(\beta+1)} = \frac{(\alpha\beta+1)(-\beta)\beta(1-\beta)}{2(1+\beta)(1-\beta)} = \frac{(\alpha-\beta)(\beta+1)}{4}.$$

Hence we obtain the following for curves of type I:

$$c = \begin{cases} \pm\frac{1}{2}\left(\alpha^{h+3k/8} - \alpha^{k/8}\right) & \text{when } \beta = \alpha^{k/4} \\ \pm\frac{1}{2}\left(\alpha^{h+k/8} - \alpha^{3k/8}\right) & \text{when } \beta = -\alpha^{k/4} \end{cases}$$

$$d = \begin{cases} \pm\frac{1}{4}\left(\alpha^{h} - \alpha^{k/4}\right)\left(\alpha^{k/4}+1\right) & \text{when } \beta = \alpha^{k/4} \\ \pm\frac{1}{4}\left(\alpha^{h} + \alpha^{k/4}\right)\left(-\alpha^{k/4}+1\right) & \text{when } \beta = -\alpha^{k/4} \end{cases}$$

where h is a positive integer such that $\gcd(k, h) = 1$. Here we consider all choices of primitive kth roots of unity modulo ℓ.

Let $\tilde{c}_i(t)$ and $\tilde{d}_i(t)$ for $i = 1, 2$ be polynomials of minimal degree satisfying the following conditions:

$$\begin{aligned} \tilde{c}_1(t) &\equiv t^{h+3k/8} - t^{k/8} \mod \Phi_k(t) \\ \tilde{d}_1(t) &\equiv \left(t^h - t^{k/4}\right)\left(t^{k/4}+1\right) \mod \Phi_k(t) \\ \tilde{c}_2(t) &\equiv t^{h+k/8} - t^{3k/8} \mod \Phi_k(t) \\ \tilde{d}_2(t) &\equiv \left(t^h + t^{k/4}\right)\left(-t^{k/4}+1\right) \mod \Phi_k(t). \end{aligned}$$

Set polynomials $\tilde{p}_i(t)$ for $i = 1, 2$ as

$$\tilde{p}_i(t) = 2\tilde{c}_i(t)^2 + \tilde{d}_i(t)^2.$$

Since $c = \pm\tilde{c}_i(\alpha)/2$ and $d = \pm\tilde{d}_i(\alpha)/4$, we have

$$\tilde{p}_i(\alpha) = 2\tilde{c}_i(\alpha)^2 + \tilde{d}_i(\alpha)^2 = 8(c^2 + 2d^2) = 8p.$$

It is necessary for $p = c^2 + 2d^2$ being prime with $p \equiv 1 \pmod 8$ and $c \equiv 1 \pmod 4$ that $\tilde{p}_i(x)$ is irreducible, $\tilde{c}_i(j) \equiv 2 \pmod 4$ and $\tilde{d}_i(j) \equiv 0 \pmod 4$ for some $i = 1, 2$ and $0 \le j \le 3$.

Searching suitable h which gives polynomials $\tilde{c}_i(t)$, $\tilde{d}_i(t)$ and $\tilde{p}_i(t)$ satisfying the above condition and $\rho < 4$, we find some pairs of (k, h) for $k \le 96$. Here we show only one example for $k = 56$. For other obtained pairs of (k, h) and examples of pairing-friendly curves with respect to them, see the extended version of this paper [17].

For $k = 56$, we have the following example with $h = 15$ ($t^h = t^{15}$):

$$\begin{aligned} &\tilde{c}_2(t) = -t^{21} + t^{22}, \quad \tilde{d}_2(t) = 1 + t + t^{14} + t^{15}, \\ &\tilde{p}_2(t) = 1 + 2t + t^2 + 2t^{14} + 4t^{15} + 2t^{16} + t^{28} + 2t^{29} + t^{30} + 2t^{42} - 4t^{43} + 2t^{44}. \end{aligned}$$

Since $\Phi_{56}(t) = 1 - t^4 + t^8 - t^{12} + t^{16} - t^{20} + t^{24}$, it is expected that $p \approx \ell^{11/6}$. Actually, using the above polynomials we obtain pairing-friendly hyperelliptic curves of type I with $p \approx \ell^{11/6}$ ($\rho \approx 11/3 = 3.667$). For example, we obtain the following curve $y^2 = x^5 + ax$ over $\mathbb{F}_p$:

$$
\begin{aligned}
t =& 17783 \\
\ell =& \Phi_{56}(t) \\
=& 1000277923068656865827189119874013991669139100253326573068816\backslash \\
& 6998268715367851559921840039393059855536 1\text{(339 bits)} \\
p =& 250099265879557406524307111682994614744774870053308144482663 09\backslash \\
& 218599942923741328818400016275808477589914035863072128327938 84\backslash \\
& 593036831026874212168508718320085925724310352568705063914008009 \\
& \text{(620 bits)} \\
a =& 16807 \\
\#J_C(\mathbb{F}_p) =& 625496427934935475676920579041449747666447378567370170\backslash \\
& 606687484528443885410658718388263509598094012137050323856584 54\backslash \\
& 398654158443811934859944230614043468506215122991870410804491 63\backslash \\
& 236403132204312221791227575314754184748247432490796790967125 85\backslash \\
& 209401227337987993157339071994114580909102063892721906341848 73\backslash \\
& 408553026752574955522629312042634216774627453552311605133808 25\backslash \\
& 632964082 \text{ (1238 bits)} \\
\rho =& 3.655.
\end{aligned}
$$

For some k, there is no h for which the necessary condition on the polynomials $\tilde{p}(t)$, $\tilde{c}_i(t)$ and $\tilde{d}_i(t)$ is satisfied. In such case, changing a choice of polynomials $\tilde{c}_i(t)$ and $\tilde{d}_i(t)$, we might obtain h for which the necessary condition is satisfied. For example, when $k = 8$, taking a polynomial $\tilde{d}_i(t)$ without modulo $\Phi_k(t)$, we obtain the following with $h = 1$ ($t^h = t$) which gives $\rho \approx 4$:

$$
\begin{aligned}
&\tilde{c}_1(t) = 1 + t, \quad \tilde{d}_1(t) = (t - t^2)(1 + t^2), \\
&\tilde{p}_1(t) = 2 + 4t + 3t^2 - 2t^3 + 3t^4 - 4t^5 + 3t^6 - 2t^7 + t^8.
\end{aligned}
$$

Since $\Phi_8(t) = 1 + t^4$, it is expected that $p \approx \ell^2$. Using the above polynomials we obtain pairing-friendly hyperelliptic curves of type I with $p \approx \ell^2$ ($\rho \approx 4$) when t is odd and $\ell = \Phi_8(t)/2$. For examples of pairing-friendly hyperelliptic curves, see the extended version of this paper [17].

6.2 A Cyclotomic Family of Type II

From Theorem 3, we have

$$
c = \frac{\beta(\alpha - 1)}{2}, \quad d = \frac{\alpha + 1}{2\gamma} = \frac{\gamma(\alpha + 1)}{4}.
$$

Hence we obtain the following for curves of type II:

$$
c = \pm\frac{\alpha^{k/4}\left(\alpha^h - 1\right)}{2}, \quad d = \pm\frac{\left(\alpha^{k/8} - \alpha^{3k/8}\right)\left(\alpha^h + 1\right)}{4}.
$$

Let $\tilde{c}(t)$ and $\tilde{d}(t)$ be polynomials of minimal degree satisfying

$$\tilde{c}(t) \equiv t^{k/4}\left(t^h - 1\right) \mod \Phi_k(t)$$
$$\tilde{d}(t) \equiv \left(t^{k/8} - t^{3k/8}\right)\left(t^h + 1\right) \mod \Phi_k(t).$$

As in Section 6.1, set a polynomial $\tilde{p}(t)$ as $\tilde{p}(t) = 2\tilde{c}(t)^2 + \tilde{d}(t)^2$. Since $c = \pm\tilde{c}(\alpha)/2$ and $d = \pm\tilde{d}(\alpha)/4$, we have

$$\tilde{p}(\alpha) = 2\tilde{c}(\alpha)^2 + \tilde{d}(\alpha)^2 = 8(c^2 + 2d^2) = 8p.$$

It is necessary for $p = c^2 + 2d^2$ being prime with $p \equiv 1, 3 \pmod 8$ and $c \equiv 1 \pmod 4$ that $\tilde{p}(x)$ is irreducible, $\tilde{c}(j) \equiv 2 \pmod 4$ and $\tilde{d}(j) \equiv 0 \pmod 4$ for $0 \le j \le 3$.

Searching suitable h which gives polynomials $\tilde{c}(t)$, $\tilde{d}(t)$ and $\tilde{p}(t)$ satisfying the above condition and $\rho < 4$, we find $(k, h) = (24, 11)$, $(24, 23)$. Here we show the detail only for $(k, h) = (24, 11)$:

$$h = 11, \quad t^h \equiv -t^3 + t^7 \pmod{\Phi_{24}(t)},$$
$$\tilde{c}(t) = -t^5 - t^6, \quad \tilde{d}(t) = -1 + t - t^2 + t^3 + t^4 - t^5,$$
$$\tilde{p}(t) = 1 - 2t + 3t^2 - 4t^3 + t^4 + 2t^5 - 3t^6 + 4t^7 - t^8 - 2t^9 + 3t^{10} + 4t^{11} + 2t^{12}.$$

Since $\Phi_{24}(t) = 1 - t^4 + t^8$, it is expected that $p \approx \ell^{3/2}$. Actually, using the above polynomials we obtain pairing-friendly hyperelliptic curves of type I with $p \approx \ell^{3/2}$ ($\rho \approx 3$). For example, we obtain the following curves. For other examples, see the extended version of this paper [17].

$$\begin{aligned}
t =& 1049085 \\
\ell =& \Phi_{24}(t) = 146718682892712893651454019963417202720810469000\,1 \text{(161 bits)} \\
p =& 44429248363784108259841001566549397808327738548422271126757160 08\backslash \\
& 30352907 \quad \text{(239 bits, } p \equiv 3 \mod 8) \\
a =& 2 \\
\#J_C(\mathbb{F}_p) =& 1973958110170812861009526020744941419898595856880875470 12\backslash \\
& 2279618528223042149310835451476496597264838239740989139215118496 8\backslash \\
& 78309880008204167073 36 \text{ (477 bits)} \\
\rho =& 2.975.
\end{aligned}$$

7 Conclusion

In this paper, we present the analogue of the Cocks-Pinch method and the cyclotomic method by which we can construct pairing-friendly hyperelliptic curves of type $y^2 = x^5 + ax$ with ordinary Jacobians for a prescribed embedding degree. These methods produce pairing-friendly hyperelliptic curves with small ρ-values.

More precisely, we obtain pairing-friendly hyperelliptic curves with $\rho \approx 4$ for arbitrary embedding degree by using the analogue of the Cocks-Pinch method and with $3 \leq \rho \leq 4$ by using the cyclotomic method.

Constructing pairing-friendly ordinary abelian varieties of higher dimension with smaller ρ-values are still in progress. The current best ρ-values are still large compared with elliptic curves. Thus the problem is still open.

References

1. Balasubramanian, R., Koblitz, N.: The improbability that an elliptic curve has subexponential discrete log problem under the Menezes-Okamoto-Vanstone algorithm. J. Cryptology 11(2), 141–145 (1998)
2. Barreto, P.S.L.M., Naehrig, M.: Pairing-friendly elliptic curves of prime order. In: Preneel, B., Tavares, S. (eds.) SAC 2005. LNCS, vol. 3897, pp. 319–331. Springer, Heidelberg (2006)
3. Blake, I.-F., Seroussi, G., Smart, N.-P.: Advances in Elliptic Curve Cryptography. Cambridge University Press, Cambridge (2005)
4. Boneh, D., Franklin, M.: Identity-based encryption from the Weil pairing. SIAM Journal of Computing 32(3), 586–615 (2003)
5. Brezing, F., Weng, A.: Elliptic curves suitable for pairing based cryptography. Design, Codes and Cryptography 37, 133–141 (2005)
6. Cardona, G., Nart, E.: Zeta Function and Cryptographic Exponent of Supersingular Curves of Genus 2. In: Takagi, T., Okamoto, T., Okamoto, E., Okamoto, T. (eds.) Pairing 2007. LNCS, vol. 4575, pp. 132–151. Springer, Heidelberg (2007)
7. Cocks, C., Pinch, R.G.E.: Identity-based cryptosystems based on the Weil pairing (unpublished manuscript, 2001)
8. Freeman, D.: Constructing pairing-friendly genus 2 curves with ordinary Jacobians. In: Takagi, T., Okamoto, T., Okamoto, E., Okamoto, T. (eds.) Pairing 2007. LNCS, vol. 4575, pp. 152–176. Springer, Heidelberg (2007)
9. Freeman, D.: A generalized Brezing-Weng method for constructing pairing-friendly ordinary abelian varieties. In: Galbraith, S.D., Paterson, K.G. (eds.) Pairing 2008. LNCS, vol. 5209, pp. 146–163. Springer, Heidelberg (2008)
10. Freeman, D., Scott, M., Teske, E.: A taxonomy of pairing-friendly elliptic curves, Cryptology ePrint Archive, Report 2006/372 (2006), `http://eprint.iacr.org/`
11. Freeman, D., Stevenhagen, P., Streng, M.: Abelian varieties with prescribed embedding degree. In: van der Poorten, A.J., Stein, A. (eds.) ANTS-VIII 2008. LNCS, vol. 5011, pp. 60–73. Springer, Heidelberg (2008)
12. Furukawa, E., Kawazoe, M., Takahashi, T.: Counting Points for Hyperelliptic Curves of Type $y^2 = x^5 + ax$ over Finite Prime Fields. In: Matsui, M., Zuccherato, R.J. (eds.) SAC 2003. LNCS, vol. 3006, pp. 26–41. Springer, Heidelberg (2004)
13. Galbraith, S., McKee, J., Valença, P.: Ordinary abelian varieties having small embedding degree. Finite Fields and Their Applications 13, 800–814 (2007)
14. Haneda, M., Kawazoe, M., Takahashi, T.: Suitable Curves for Genus-4 HCC over Prime Fields: Point Counting Formulae for Hyperelliptic Curves of Type $y^2 = x^{2k+1} + ax$. In: Gaires, L., Italiano, G.F., Monteiro, L., Palamidessi, C., Yung, M. (eds.) ICALP 2005. LNCS, vol. 3580, pp. 539–550. Springer, Heidelberg (2005)
15. Joux, A.: A one round protocol for tripartite Diffie-Hellman. In: Bosma, W. (ed.) ANTS-IV 2000. LNCS, vol. 1838, pp. 385–393. Springer, Heidelberg (2000); Full version: Journal of Cryptology 17, 263–276 (2004)

16. Kawazoe, M., Sakaeyama, R., Takahashi, T.: Pairing-friendly Hyperelliptic Curves of type $y^2 = x^5 + ax$. In: 2008 Symposium on Cryptography and Information Security (SCIS 2008), Miyazaki, Japan (2008)
17. Kawazoe, M., Takahashi, T.: Pairing-friendly Hyperelliptic Curves with Ordinary Jacobians of Type $y^2 = x^5 + ax$, Extended version of the present paper, Cryptology ePrint Archive, Report 2008/026 (2008), `http://eprint.iacr.org/`.
18. Miyaji, A., Nakabayashi, M., Takano, S.: New explicit conditions of elliptic curve traces for FR-reduction. IEICE Transactions on Fundamentals E84-A(5), 1234–1243 (2001)
19. Rubin, K., Silverberg, A.: Supersingular abelian varieties in cryptology. In: Yung, M. (ed.) CRYPTO 2002. LNCS, vol. 2442, pp. 336–353. Springer, Heidelberg (2002)
20. Sakai, R., Ohgishi, K., Kasahara, M.: Cryptosystem based on pairing. In: 2000 Symposium on Cryptography and Information Security (SCIS 2000), Okinawa, Japan (2000)
21. Scott, M., Barreto, P.S.L.M.: Generating more MNT elliptic curves. Designs, Codes and Cryptography 38, 209–217 (2006)
22. Wolfram Research, Inc., Mathematica, Version 6.0, Champaign, IL (2007)

Integer Variable χ–Based Ate Pairing

Yasuyuki Nogami[1], Masataka Akane[2], Yumi Sakemi[1], Hidehiro Kato[1], and Yoshitaka Morikawa[1]

[1] Graduate School of Natural Science and Technology, Okayama University
3-1-1, Tsushima-naka, Okayama, Okayama 700-8530, Japan
[2] Mitsubishi Electric Corporation, Inazawa Works,
1, Hishimachi, Inazawashi, Aichi, 492-8161, Japan
{nogami,sakemi,kato,morikawa}@trans.cne.okayama-u.ac.jp

Abstract. In implementing an efficient pairing calculation, it is said that the lower bound of the number of iterations of Miller's algorithm is $\log_2 r/\varphi(k)$, where $\varphi(\cdot)$ is the Euler's function. Ate pairing reduced the number of the loops of Miller's algorithm of Tate pairing from $\lfloor \log_2 r \rfloor$ to $\lfloor \log_2 (t-1) \rfloor$. Recently, it is known to systematically prepare a pairing–friendly elliptic curve whose parameters are given by a polynomial of integer variable "χ". For the curve, this paper gives *integer variable χ–based* Ate pairing that achieves the lower bound by reducing it to $\lfloor \log_2 \chi \rfloor$.

Keywords: Ate pairing, Miller's algorithm.

1 Introduction

Recently, pairing–based cryptographic applications such as ID–based cryptography [4] and group signature authentication [16] have received much attentions. In order to make these applications practical, pairing calculation needs to be efficiently carried out. For this purpose, several efficient pairings such as Tate, Ate [5], twisted Ate [15], and *subfield–twisted* Ate [6],[1] have been proposed. In this paper, Barreto–Naehrig (BN) curve, that is a typical class of non–supersingular (ordinary) pairing–friendly elliptic curves of embedding degree 12, is mainly dealt with. As a typical feature of BN curve, its characteristic p and Frobenius trace t are given by using *integer variable* χ as

$$p(\chi) = 36\chi^4 - 36\chi^3 + 24\chi^2 - 6\chi + 1, \tag{1a}$$
$$t(\chi) = 6\chi^2 + 1. \tag{1b}$$

Pairings can be roughly classified by the inputs for Miller's algorithm [10]. In general, as the inputs, Miller's algorithm needs two rational points and the number of iterations. Let us suppose a prime order BN curve of embedding degree 12 as $E : y^2 = x^3 + b,\ b \in \mathbb{F}_p$, where p is the characteristic and let

S.D. Galbraith and K.G. Paterson (Eds.): Pairing 2008, LNCS 5209, pp. 178–191, 2008.

the order be a prime number r. Since the embedding degree is 12, r divides $p^{12}-1$ and then r^2 divides $\#E(\mathbb{F}_{p^{12}})$. Tate pairing $\tau(\cdot,\cdot)$ uses rational points $P \in E(\mathbb{F}_p)$ and $Q \in E(\mathbb{F}_{p^{12}})/rE(\mathbb{F}_{p^{12}})$, the number of iterations of Miller's algorithm is $\lfloor \log_2 r \rfloor$. Tate pairing mainly uses P for calculation. The output of Miller's algorithm is denoted by $f_{r,P}(Q)$. Ate pairing $\alpha(\cdot,\cdot)$ uses rational points $P \in E(\mathbb{F}_p)$ and $Q \in E[r] \cap \mathrm{Ker}(\phi - [p])$, but the number of iterations is $\lfloor \log_2(t-1) \rfloor$, where ϕ is Frobenius map for rational point, $E[r]$ is the subgroup of rational points of order r in $E(\mathbb{F}_{p^{12}})$, and t is the Frobenius trace of $E(\mathbb{F}_p)$, that is $\#E(\mathbb{F}_p) = r = p+1-t$. The number of iterations is about half of that of Tate pairing; however, Ate pairing mainly uses Q for calculation. The output of Miller's algorithm is denoted by $f_{t-1,Q}(P)$ and thus plain Ate pairing is slower than Tate pairing.

Devegili et al.'s work [6] accelerated Ate pairing by using *subfield–twisted* BN curve $E'(\mathbb{F}_{p^2})$, where the twisted BN curve is given by $E' : y^2 = x^3 + bv^{-1}$ and v is a quadratic non residue and cubic non residue in $\mathbb{F}_{p^2}$. In detail, in addition to $P \in E(\mathbb{F}_p)$, it mainly uses $Q' \in E'(\mathbb{F}_{p^2})$ for calculation. The authors have also improved Ate pairing so as to substantially use subfield arithmetic operations [1]. In what follows, it is called *improved subfield–twisted* Ate (improved St–Ate) pairing. Both of these works [6],[1] have $\lfloor \log_2(t-1) \rfloor$ iterations in Miller's algorithm. According to [2], integer variable χ of small Hamming weight is efficient for Ate pairing with BN curve.

Let k be the embedding degree, it is said that the lower bound of the number of iterations of Miller's algorithm is $\log_2 r/\varphi(k)$, where $\varphi(\cdot)$ is the Euler's function. Ate pairing reduced the number of the iterations of Miller's algorithm from $\lfloor \log_2 r \rfloor$ to $\lfloor \log_2(t-1) \rfloor$. By reducing it to $\lfloor \log_2 \chi \rfloor$, this paper gives a bilinear map that achieves the lower bound. In detail, using Frobenius map and BN curve whose embedding degree is 12, this paper proposes *integer variable χ–based* Ate (Xate) pairing. First, based on Eqs.(1), the following relation is shown.

$$6\chi \equiv 1 + p + p^3 + p^{10} \pmod{r}. \tag{2}$$

Though plain Ate pairing calculates $f_{t-1,Q}(P)$ by using Miller's algorithm, where $P \in E(\mathbb{F}_p)$ and $Q \in E[r] \cap \mathrm{Ker}(\phi - [p])$, based on Eq.(2), the proposed Xate pairing calculates $f_{\chi,Q}(P)$. Noting that $\lfloor \log_2 \chi \rfloor$ is about half of $\lfloor \log_2(t-1) \rfloor$, Miller's part of Xate pairing is about twice more efficient than that of plain Ate pairing. The idea of [6] or improved St–Ate pairing [1] can be efficiently applied for Xate pairing. The authors simulated Xate pairing and also improved St–Xate pairing on Pentium4 (3.0GHz) with C language and GMP library [9]. Then, it is shown that, when r is a 254–bit prime number, improved St–Xate pairing that includes so–called *final exponentiation* is calculated within 11.0 milli–seconds. After that, it is shown that improved St–Xate is applied not only for BN curve but also Freeman's curve of embedding degree 10. Then, some recent works [19],[14] are introduced and compared to Xate pairing. Note that Eq.(2) is also efficient for scalar multiplications in $E(\mathbb{F}_{p^{12}})$ and *subfield-twisted* BN curve $E'(\mathbb{F}_{p^2})$, moreover exponentiation in $\mathbb{F}_{p^{12}}$.

Throughout this paper, p and k denote characteristic and extension degree, respectively. $\mathbb{F}_{p^k}$ denotes k-th extension field over $\mathbb{F}_p$ and $\mathbb{F}^*_{p^k}$ denotes the multiplicative group in $\mathbb{F}_{p^k}$. $X \mid Y$ and $X \nmid Y$ mean that X divides and does not divide Y, respectively.

2 Fundamentals

We briefly go over elliptic curve, Tate, Ate, improved St–Ate pairings, and divisor theorem. For instance, we mainly consider Barreto–Naehrig (BN) curve of embedding degree 12, that is a class of *ordinary pairing–friendly curves* [7].

2.1 Elliptic Curve and Barreto–Naehrig Curve

Let $\mathbb{F}_p$ be a prime field and E be an elliptic curve over $\mathbb{F}_p$. $E(\mathbb{F}_p)$ that is a set of rational points on the curve, including the *infinity point* $\mathcal{O}$, forms an additive Abelian group. Let $\#E(\mathbb{F}_p)$ be its order, consider a large prime r that divides $\#E(\mathbb{F}_p)$. The smallest positive integer k such that r divides $p^k - 1$ is especially called *embedding degree*. One can consider pairings such as Tate and Ate pairings by using $E(\mathbb{F}_{p^k})$. In general, $\#E(\mathbb{F}_p)$ is given as

$$\#E(\mathbb{F}_p) = p + 1 - t, \tag{3}$$

where t is the Frobenius trace of $E(\mathbb{F}_p)$. The characteristic p and Frobenius trace t of Barreto–Naehrig (BN) curve [3] are given by using an integer variable χ as Eqs.(1). In addition, the BN curve E is given by

$$E : y^2 = x^3 + b,\ b \in \mathbb{F}_p \tag{4}$$

whose embedding degree is 12. In this paper, let $\#E(\mathbb{F}_p)$ be a prime r.

2.2 Tate Pairing

Let $P \in E(\mathbb{F}_p)$ and $Q \in E(\mathbb{F}_{p^k})/rE(\mathbb{F}_{p^k})$, Tate pairing $\tau(\cdot,\cdot)$ is defined as

$$\tau(\cdot,\cdot) : \begin{cases} E(\mathbb{F}_p) \times E(\mathbb{F}_{p^k})/rE(\mathbb{F}_{p^k}) & \to \mathbb{F}^*_{p^k}/(\mathbb{F}^*_{p^k})^r \\ (P, Q) & \mapsto f_{r,P}(Q)^{(p^k-1)/r}. \end{cases} \tag{5}$$

In general, $A = f_{r,P}(Q)$ is calculated by Miller's algorithm [5], then so–called *final exponentiation* $A^{(p^k-1)/r}$ follows. The number of iterations of Miller's algorithm for Tate pairing is determined by r, in detail $\lfloor \log_2 r \rfloor$. Twisted Ate pairing [15] reduced the number of iterations of Miller's algorithm. Let d be the twist degree such as $d = 2, 3, 4, 6$, it is reduced to $(t-1)^{k/d} \pmod r$.

2.3 Ate Pairing

Let ϕ be Frobenius endomorphism, i.e.,

$$\phi : E \to E : (x, y) \mapsto (x^p, y^p), \tag{6}$$

where x and y are x–coordinate and y–coordinate of rational point, respectively. Then, let $\mathbb{G}_1$ and $\mathbb{G}_2$ be

$$\mathbb{G}_1 = E[r] \cap \mathrm{Ker}(\phi - [1]), \tag{7a}$$

$$\mathbb{G}_2 = E[r] \cap \mathrm{Ker}(\phi - [p]), \tag{7b}$$

and let $P \in \mathbb{G}_1$ and $Q \in \mathbb{G}_2$, Ate pairing $\alpha(\cdot,\cdot)$ is defined as

$$\alpha(\cdot,\cdot) : \begin{cases} \mathbb{G}_2 \times \mathbb{G}_1 & \to \mathbb{F}^*_{p^k}/(\mathbb{F}^*_{p^k})^r \\ (Q, P) & \mapsto f_{T,Q}(P)^{(p^k-1)/r}, \end{cases} \tag{8}$$

where $T = t - 1$, $E[r]$ denotes a subgroup of rational points of order r in $E(\mathbb{F}_{p^k})$, and $[i]$ denotes $[i] : P \mapsto iP$. The number of iterations of Miller's algorithm for Ate pairing is determined by $t - 1$, in detail $\lfloor \log_2(t-1) \rfloor$.

In the case of using BN curve for Ate pairing, $\mathbb{G}_1$ and $\mathbb{G}_2$ become as

$$\mathbb{G}_1 = E(\mathbb{F}_p), \tag{9a}$$

$$\mathbb{G}_2 = (\phi - [1]) \left\{E(\mathbb{F}_{p^{12}})/rE(\mathbb{F}_{p^{12}})\right\}, \tag{9b}$$

therefore, compared to Tate paring, the number of iterations becomes about half; however, Miller's algorithm needs a lot of calculations in the above defined $\mathbb{G}_2$. Thus, plain Ate pairing is not superior to Tate pairing.

Devegili et al.'s work [6] and the authors [1],[2] have improved Ate pairing with BN curve by using subfield–twisted elliptic curve $E'(\mathbb{F}_{p^2})$ over $\mathbb{F}_{p^2}$. It is given as

$$E' : y^2 = x^3 + bv^{-1}, \ b \in \mathbb{F}_p, \ v \in \mathbb{F}_{p^2}, \tag{10}$$

where v is a quadratic non residue and cubic non residue in $\mathbb{F}_{p^2}$. It is also called *sextic twisted* curve. In this case, we have the following isomorphism.

$$\psi : E'(\mathbb{F}_{p^2}) \to E(\mathbb{F}_{p^{12}}) \tag{11a}$$

$$Q'(x_{Q'}, y_{Q'}) \mapsto Q(x_{Q'}v^{\frac{1}{3}}, y_{Q'}v^{\frac{1}{2}}), \tag{11b}$$

Thus, noting that $E(\mathbb{F}_{p^{12}})$ and $E'(\mathbb{F}_{p^{12}})$ are isomorphic to each other, [6] efficiently used *subfield–twisted elliptic curve* $E'(\mathbb{F}_{p^2})$. In detail, it calculates $f_{T,\psi(Q')}$ (P) by using Q'. The authors have also improved Ate pairing so as to substantially use subfield arithmetic operations [1]. In what follows, it is called improved St–Ate pairing. In our previous work [2], it is shown that integer variable χ of small Hamming weight is quite efficient for Ate pairing. It is also efficient for twisted Ate pairing [18].

2.4 Relation between Tate and Ate Pairings

Table 1 shows the parameter settings for Miller's algorithm using BN curve of embedding degree 12.

Table 1. Parameter settings for calculating $f_{A,B}(C)$ with Miller's algorithm

pairing	A	group for B	group for C
plain Tate	r	$E(\mathbb{F}_p)$	$E(\mathbb{F}_{p^{12}})$
Twisted Ate [15]	$(t-1)^2 \pmod r$	$E(\mathbb{F}_p)$	$E(\mathbb{F}_{p^{12}})$
plain Ate	$t-1$	$E(\mathbb{F}_{p^{12}})$	$E(\mathbb{F}_p)$
Devegili et al.'s Ate [6]	$t-1$	$E'(\mathbb{F}_{p^2})$	$E(\mathbb{F}_p)$
improved St–Ate [2]	$t-1$	$E'(\mathbb{F}_{p^2})$	$E(\mathbb{F}_p)$

Between Tate and Ate pairings, we have the following relation [10].

$$\tau(Q,P)^L = f_{T,Q}(P)^{c(p^k-1)/N}, \tag{12}$$

where $c \equiv kp^{k-1} \pmod r$ and

$$N = \gcd(T^k-1, p^k-1),\ T^k-1 = LN,\ T = t-1. \tag{13}$$

Thus, let $N = ru$, according to Eq.(8), we have

$$\tau(Q,P)^{uL} = \alpha(Q,P)^c. \tag{14}$$

$r \nmid L$ is needed for Ate pairing to be nondegenerate.

2.5 Divisor

Let D be the principal divisor of $Q \in E$ given as

$$D = (Q) - (\mathcal{O}) = (Q) - (\mathcal{O}) + div\,(1)\,. \tag{15}$$

For scalars $a, b \in Z$, let aD and bD be written as

$$aD = (aQ) - (\mathcal{O}) + div\,(f_{a,Q})\,, \tag{16a}$$

$$bD = (bQ) - (\mathcal{O}) + div\,(f_{b,Q})\,, \tag{16b}$$

where $f_{a,Q}$ and $f_{b,Q}$ are the rational functions for aD and bD, respectively. Then, addition for divisors is carried out as

$$aD + bD = (aQ) + (bQ) - 2(\mathcal{O}) + div\,(f_{a,Q} \cdot f_{b,Q} \cdot g_{aQ,bQ})\,, \tag{17a}$$

where $g_{aQ,bQ} = l_{aQ,bQ}/v_{aQ+bQ}$, $l_{aQ,bQ}$ denotes the line passing through two points aQ, bQ, and v_{aQ+bQ} denotes the vertical line passing through $aQ + bQ$. Moreover, the following relation holds.

$$a(bD) = \sum_{i=0}^{a-1}(bQ) - a(\mathcal{O}) + div\left(f_{b,Q}^{a} \cdot f_{a,bQ}\right). \tag{17b}$$

Thus, let $(a + b)D$ and $(ab)D$ be written as

$$(a + b)D = ((a + b)Q) - (\mathcal{O}) + div\left(f_{a+b,Q}\right), \tag{18a}$$

$$(ab)D = (abQ) - (\mathcal{O}) + div\left(f_{ab,Q}\right), \tag{18b}$$

we have the following relation.

$$f_{a+b,Q} = f_{a,Q} \cdot f_{b,Q} \cdot g_{aQ,bQ}, \tag{19a}$$

$$f_{ab,Q} = f_{b,Q}^{a} \cdot f_{a,bQ} = f_{a,Q}^{b} \cdot f_{b,aQ}. \tag{19b}$$

Consider Frobenius map $\phi(Q)$ for rational point $Q \in E(\mathbb{F}_{p^k})$ as

$$\phi(\cdot) : \begin{cases} E(\mathbb{F}_{p^k}) \to E(\mathbb{F}_{p^k}) \\ (x_Q, y_Q) \mapsto (x_Q^p, y_Q^p). \end{cases} \tag{20}$$

In the case of Ate pairing, according to Eq.(7b) we have

$$\phi(Q) = pQ, \text{ where } Q \in \mathbb{G}_2. \tag{21}$$

Thus, for $Q \in \mathbb{G}_2$, let $f_{p,Q}$ be given as

$$pD = (pQ) - \mathcal{O} + div\left(f_{p,Q}\right), \tag{22}$$

Eq.(19b) with $b = p$ leads to

$$f_{ap,Q} = f_{p,Q}^{a} f_{a,pQ} = f_{p,Q}^{a} f_{a,\phi(Q)} = f_{p,Q}^{a} f_{a,Q}^{p}. \tag{23}$$

Iteratively applying the above relation from $a = p^{i-1}$, we have

$$f_{p^i,Q} = f_{p,Q}^{ip^{i-1}}. \tag{24}$$

3 Main Proposal

In this section, using BN curve of embedding degree 12, *integer variable χ–based* Ate pairing (Xate pairing) is proposed. First, derive $\sum_j d_j\chi^i = \sum_i c_i p^i$ with small coefficients c_j and d_j. Based on the p–adic expansion, then consider efficient bilinear map with Frobenius map, namely Xate pairing.

3.1 Frobenius Expansion with χ

Note that $\mathbb{G}_2$ is defined as $E[r] \cap \mathrm{Ker}(\phi - [p])$, the parameter settings of BN curve are Eq.(1), and $\#E(\mathbb{F}_p)$ is a prime number r. First, this section considers p–adic (Frobenius) expansion with respect to χ. In the case of BN curve, the expansion of 6χ is systematically obtained. According to Eq.(1b) and Eq.(3),

$$6\chi^2 \equiv t - 1 \equiv p \pmod{r}. \tag{25}$$

Then, substituting it to Eq.(1a), we have

$$p \equiv p^2 - 6\chi(p+1) + 4p + 1 \pmod{r}, \tag{26}$$

$$6\chi(1+p) \equiv (p^2 + 3p + 1) \pmod{r}. \tag{27}$$

Then, based on cyclotomic polynomial $p^4 - p^2 + 1 \equiv 0 \pmod{r}$ [1] and using extended Euclidean algorithm, $(1+p)^{-1}$ is calculated as

$$p^2(1-p)(1+p) \equiv 1 \pmod{r}, \tag{28}$$

$$(1+p)^{-1} \equiv p^2(1-p) \pmod{r}. \tag{29}$$

Then, substituting Eq.(29) and $p^6 \equiv -1 \pmod{r}$ to Eq.(27), we have

$$\begin{aligned} 6\chi &\equiv (1+p)^{-1}\left\{(1+p)^2 + p\right\} \\ &\equiv 1 + p + p^3 + p^{10} \pmod{r}. \end{aligned} \tag{30}$$

As introduced in **Sec.**3.4, for other pairing–friendly curves, such a p–adic (Frobenius) expansion can be obtained in the same way. In the next section, based on the above relation Eq.(30), we consider an efficient bilinear map that achieves the number of calculations of Miller's algorithm $\log_2 r/\varphi(k)$ with BN curve.

3.2 Integer Variable χ–Based Ate (Xate) Pairing

First, for $Q \in \mathbb{G}_2$, we consider the following relation.

$$f_{6\chi^2,Q} = f_{T,Q}. \tag{31}$$

Of course, for $\forall P \in \mathbb{G}_1$, we have

$$f_{6\chi^2,Q}(P)^{(p^{12}-1)/r} = f_{T,Q}(P)^{(p^{12}-1)/r} = \alpha(Q,P). \tag{32}$$

In order to apply Eq.(30), according to **Sec.**2.5, we rewrite $f_{6\chi^2,Q}$ as

$$f_{6\chi^2,Q}^{(p^{12}-1)/r} = f_{6\chi\cdot\chi,Q}^{(p^{12}-1)/r} = f_{(1+p+p^3+p^{10})\chi,Q}^{(p^{12}-1)/r}. \tag{33}$$

from which we can obtain

$$f_{(1+p+p^3+p^{10})\chi,Q}^{(p^{12}-1)/r} = \{f_{\chi,Q} \cdot f_{\chi,Q}^{p} \cdot g_{\chi Q,p\chi Q} \cdot f_{\chi,Q}^{p^3} \cdot f_{\chi,Q}^{p^{10}} \cdot g_{p^3\chi Q,p^{10}\chi Q} \\ \cdot g_{\chi Q+p\chi Q,p^3\chi Q+p^{10}\chi Q} \cdot f_{p,\chi Q}^{1+3p^2+10p^9}\}^{(p^{12}-1)/r}. \tag{34a}$$

Then, we have $f_{6\chi\cdot\chi,Q}^{(p^{12}-1)/r} = AB^{(p^{12}-1)/r}$ with

$$A = f_{p,\chi Q}^{1+3p^2+10p^9}, \tag{35a}$$

$$B = \hat{f}_{\chi,Q}, \tag{35b}$$

$$\text{where } \hat{f}_{\chi,Q} = f_{\chi,Q}^{1+p+p^3+p^{10}} \cdot g_{\chi Q,p\chi Q} \cdot g_{p^3\chi Q,p^{10}\chi Q} \\ \cdot g_{\chi Q+p\chi Q,p^3\chi Q+p^{10}\chi Q}. \tag{35c}$$

As shown in **App.**A, $f_{p,\chi Q}^{(p^{12}-1)/r} = f_{p,Q}^{\chi(p^{12}-1)/r}$. Thus, Eq.(35a) becomes

$$A^{(p^{12}-1)/r} = \{f_{p,Q}^{(1+3p^2+10p^9)\chi}\}^{(p^{12}-1)/r}. \tag{36}$$

and then $A^{(p^{12}-1)/r}$ gives a bilinear map. According to Eq.(30), we can consider the right–hand side of Eq.(30) as a polynomial of variable p such as

$$h(p) = 1 + p + p^3 + p^{10}. \tag{37}$$

Then, A is given with its *formal derivative* $h'(p)$ with respect to p as

$$A^{(p^{12}-1)/r} = \{f_{p,Q}^{\chi h'(p)}\}^{(p^{12}-1)/r}. \tag{38}$$

Finally, using A, B, Eqs.(31), (33), and (53), we have

$$\hat{f}_{\chi,Q}^{(p^{12}-1)/r} = \left\{f_{T,Q} \cdot A^{-1}\right\}^{(p^{12}-1)/r}, \tag{39}$$

we find that the right–hand side of the above equation gives a bilinear map. In what follows, we consider the following bilinear map referring as *integer variable χ–based* Ate (Xate) pairing $\zeta(\cdot,\cdot)$.

$$\zeta(\cdot,\cdot) : \begin{cases} \mathbb{G}_2 \times \mathbb{G}_1 & \rightarrow \mathbb{F}_{p^{12}}^*/(\mathbb{F}_{p^{12}}^*)^r \\ (Q,P) & \mapsto \hat{f}_{\chi,Q}(P)^{(p^{12}-1)/r}. \end{cases} \tag{40}$$

According to Eq.(35c), we find that the major computation of Xate pairing is $f_{\chi,Q}$ that achieves the lower bound $\log_2 r/\varphi(k)$. The others are efficiently calculated with Frobenius map. When one uses Miller's algorithm, the Hamming weight of χ directly affects the efficiency of calculating $\hat{f}_{\chi,Q}(P)$. Moreover, one can apply improved St–Ate pairing technique [1] to Xate paring, namely improved *subfield–twisted* Xate (improved St–Xate) pairing.

3.3 Nondegeneracy of Xate Pairing

Based on the nondegeneracies of Tate and Ate pairings, the condition that Xate pairing needs to satisfy is given as follows.

Let $N = \gcd(T^k - 1, q^k - 1)$, $T^k - 1 = LN$, and $T = t - 1$, $r \nmid L$ is needed for the nondegeneracy of Ate pairing. In the same, according to Eqs.(14), (35a), (39), (52), and (53), the following condition is needed for that of Xate pairing.

$$r \nmid uL - \chi(uL + c)h'(p), \tag{41}$$

where $c \equiv 12p^{11} \pmod r$ and $N = ru$. (*see* **App**.B)

3.4 For Other Pairing–Friendly Curves

In the case of Freeman's curve [7] whose embedding degree k is 10, the parameter settings become as

$$p(\chi) = 25\chi^4 + 25\chi^3 + 25\chi^2 + 10\chi + 3, \tag{42a}$$
$$t(\chi) = 10\chi^2 + 5\chi + 3, \tag{42b}$$

in the same way of Eq.(30), we have

$$5\chi \equiv -2p^2 + p - 2 \pmod r. \tag{43}$$

In the case of embedding degree 8 and parameter settings as follows [7],

$$p(\chi) = \chi^8 + \chi^5 - \chi^4 - \chi + 1, \tag{44a}$$
$$t(\chi) = \chi^5 - \chi + 1, \tag{44b}$$

then we have

$$\chi^2 \equiv -\chi p^5 - p^2 \pmod r. \tag{45}$$

As shown in Eq.(30), Eq.(43), and Eq.(45), the highest degree term of χ can be replaced to the other lower terms with powers of p from which the efficiency of Xate pairing comes. Thus, Eq.(30), Eq.(43), and Eq.(45) are efficient not only for constructing bilinear maps such as Eq.(40) but also scalar multiplication in $\mathbb{G}_2$ and exponentiation in $\mathbb{F}^*_{p^k}/(\mathbb{F}^*_{p^k})^r$. Furthermore, the authors have found that Eq.(30) leads to more improvement of Twisted Ate pairing.

4 Simulation

This section discusses the implementation of Xate pairing and then shows simulation result. In the discussion, suppose the following conditions.

- BN curve of prime order r and Frobenius trace t is given over $\mathbb{F}_p$ as Eq.(4).
- Using a certain integer χ, p and t are given by Eqs.(1).

- Its quadratic and cubic twisted curve is given by Eq.(10).
- $\mathbb{G}_1 = E(\mathbb{F}_p)$, $\mathbb{G}_2 = E[r] \cap \mathrm{Ker}(\phi - [p])$, and $\mathbb{G}'_2 = \psi^{-1}(\mathbb{G}_2)$, where ψ is defined as Eqs.(11) and ψ^{-1} is its inverse map.
- $P \in \mathbb{G}_1$, $Q \in \mathbb{G}_2$, $Q' \in \mathbb{G}'_2$, and Eq.(41) is satisfied.
- In this case, $v_{aQ+bQ}(P)$ becomes 1 at *final exponentiation* (*see* **App.**A).

4.1 Implementation

As shown in Eq.(35c), the major calculation of Xate pairing is $f_{\chi,Q}(P)$. It is efficiently calculated by using Miller's algorithm. In the Miller's algorithm for calculating $f_{\chi,Q}(P)$, we obtain $\chi Q \in E'(\mathbb{F}_{p^2})$. Then, efficiently using χQ and Frobenius map, the other parts of Xate pairing are calculated. By the way, *final exponentiaion* for f given by Eq.(46) is calculated by Algorithm 1 [6].

$$f = \hat{f}_{\chi,Q}(P). \tag{46}$$

Algorithm 1. Final exponentiation $f^{(p^6-1)(p^2+1)(p^4-p^2+1)/r}$

Input : f given by Eq.(46), χ, p
Output : $f^{(p^6-1)(p^2+1)(p^4-p^2+1)/r}$

Procedure :

1. $f \leftarrow f^{p^6} \cdot f^{-1}$
2. $f \leftarrow f^{p^2} \cdot f$
3. $a \leftarrow (f^6)^{\chi} \cdot (f^5)^{p^6}$
4. $b \leftarrow a^p$
5. $b \leftarrow a \cdot b$
6. compute f^p, f^{p^2}, and f^{p^3}
7. $c \leftarrow b \cdot (f^p)^2 \cdot f^{p^2}$
8. $f \leftarrow f^{p^3} \cdot (c^6)^{\chi^2} \cdot c \cdot b \cdot (f^p \cdot f)^9 \cdot a \cdot f^4$
9. Return f

4.2 Simulation Result

Using the following positive integer χ of small Hamming weight,

$$\chi = 2^{62} + 2^{55} + 1, \tag{47}$$

by which the order r becomes 254–bit prime number and the size of $\mathbb{F}_{p^{12}}$ becomes 3048–bit, the authors simulated improved St–Xate pairing. For constructing $\mathbb{F}_{p^{12}}$, the authors used the previous work [12] and tower field technique as $\mathbb{F}_{(p^4)^3}$ [17]. The detail of the implementation is introduced in **App.**C.

According to Eq.(1b), $\log_2(t-1) \approx 2\log_2(\chi)$. Therefore, it is understood that Miller's part of improved St–Xate pairing is about twice faster than that of improved St–Xate pairing. The simulation result also shows it.

Table 2. Comparison of timings of pairings with BN curve of 254–bit prime order

[unit:ms]

pairing	Miller's part	final exponentiation	total
plain Tate	22.1	5.1	27.2
Twisted Ate [18]	13.9		19.0
plain Ate	26.5		31.6
improved St–Ate [2]	10.5		15.6
Xate	**13.6**		**18.7**
improved St–Xate	**5.4**		**10.5**
Devegili et al.'s Ate [6]	NA	NA	23.2

Remark : Pentium4 (3.0GHz), C language, and GMP [9] are used.
The authors did not use 64–bit mode of Pentium4.

4.3 Some Recent Works and Comparison

As the most recent works, Vercauteren [19] and Lee et al. [14] have proposed efficient Ate pairings. Vercauteren introduced *optimal pairings*. According to [19], the basic idea is finding $\lambda = mr,\ r \nmid m$ that has p–adic expansion $\lambda = \sum c_i p^i$ with small coefficients c_i. Then, *optimal pairing* uses $f_{c_i,Q}$ with Miller's algorithm calculation. Lee et al. introduced R–ate pairing [14]. Lee's basic idea is finding $T_x = \sum c_i p^i$ with small coefficients c_i, where $T_x = (t-1)^x$. In the same, R–ate pairing uses $f_{c_i,Q}$ with Miller's algorithm calculation.

As described in **Sec.**3, the proposed method derives $\sum_j d_j \chi^i = \sum_i c_i p^i$ with small coefficients c_j and d_j, thus our approach is different from theirs [19],[14]. For example, R–ate pairing for BN curve of embedding degree 12 calculates $f_{6\chi+2,Q}$ but Xate pairing calculates $f_{\chi,Q}$ from which the difference could be understood though their calculation costs are almost the same. As previously introduced, our proposal, that is characterized by Eq.(30), Eq.(43), and Eq.(45), are efficient not only for constructing Xate pairing such as Eq.(40) but also scalar multiplication in $\mathbb{G}_2$ and exponentiation in $\mathbb{F}_{p^k}^*/(\mathbb{F}_{p^k}^*)^r$ to which the techniques [19] and [14] cannot be directly applied. For example, when one calculates sQ, $Q \in \mathbb{G}_2$ with Eq.(30), calculate the 6χ–adic representation of scalar s, then substitute 6χ by $1 + p + p^3 + p^{10}$. Then, using some Frobenius maps based on $pQ = \phi(Q)$, sQ can be efficiently calculated. *Skew Frobenius map* for $Q' \in \mathbb{G}_2'$ is also efficient as $pQ' = \psi^{-1}(\phi(\psi(Q')))$ with Eq.(11b). Efficient scalar multiplication using *skew Frobenius map* is shown in Galbraith et al.'s work [8]. Furthermore, the authors have found that Eq.(30) leads to more improvement of Twisted Ate pairing.

5 Conclusion

Using BN curve whose embedding degree is 12, this paper proposed *integer variable* χ*–based* Ate (Xate) pairing. First, the following relation was shown.

$$6\chi \equiv 1 + p + p^3 + p^{10} \pmod{r}, \tag{48}$$

where the characteristic p of BN curve was given with *integer variable* χ as

$$p(\chi) = 36\chi^4 - 36\chi^3 + 24\chi^2 - 6\chi + 1. \tag{49}$$

Let ϕ and t be Frobenius map and its trace, respectively, though plain Ate pairing calculates $f_{t-1,Q}(P)$ by using Miller's algorithm, the proposed Xate pairing calculates $f_{\chi,Q}(P)$ using χ, where $P \in E(\mathbb{F}_p)$ and $Q \in E[r] \cap \mathrm{Ker}(\phi - [p])$. Noting that $\lfloor \log_2 \chi \rfloor$ is about half of $\lfloor \log_2 (t-1) \rfloor$, it was shown that Miller's part of Xate pairing was about twice more efficient than that of plain Ate pairing. Then, the authors simulated Xate pairing on Pentium4 (3.0GHz) with C language and GMP library [9], it was shown that, when r was a 254–bit prime number, improved St–Xate pairing that included so–called *final exponentiation* was calculated within 11.0 milli–seconds.

Acknowledgements

We thank Steven D. Galbraith for a lot of suggestions and comments.

References

1. Akane, M., Kato, H., Okimoto, T., Nogami, Y., Morikawa, Y.: An Improvement of Miller's Algorithm in Ate Pairing with Barreto–Naehrig Curve. In: Proc. of Computer Security Symposium 2007 (CSS 2007), pp. 489–494 (2007)
2. Akane, M., Kato, H., Okimoto, T., Nogami, Y., Morikawa, Y.: Efficient Parameters for Ate Pairing Computation with Barreto-Naehrig Curve. In: Proc. of Computer Security Symposium 2007 (CSS 2007), pp. 495–500 (2007)
3. Barreto, P.S.L.M., Naehrig, M.: Pairing–Friendly. Elliptic Curves of Prime Order. In: Preneel, B., Tavares, S. (eds.) SAC 2005. LNCS, vol. 3897, pp. 319–331. Springer, Heidelberg (2006)
4. Boneh, D., Lynn, B., Shacham, H.: Short signatures from the Weil pairing. In: Boyd, C. (ed.) ASIACRYPT 2001. LNCS, vol. 2248, pp. 514–532. Springer, Heidelberg (2001)
5. Cohen, H., Frey, G.: Handbook of Elliptic and Hyperelliptic Curve Cryptography, Discrete Mathematics and Its Applications. Chapman & Hall CRC (2005)
6. Devegili, A.J., Scott, M., Dahab, R.: Implementing Cryptographic Pairings over Barreto-Naehrig Curves. In: Takagi, T., Okamoto, T., Okamoto, E., Okamoto, T. (eds.) Pairing 2007. LNCS, vol. 4575, pp. 197–207. Springer, Heidelberg (2007)
7. Freeman, D., Scott, M., Teske, E.: A taxonomy of pairing-friendly elliptic curves (preprint, 2006), `http://math.berkeley.edu/~dfreeman/papers/taxonomy.pdf`
8. Galbraith, S.D., Scott, M.: Exponentiation in pairing-friendly groups using homomorphisms. In: Galbraith, S.D., Paterson, K.G. (eds.) Pairing 2008. LNCS. Springer, Heidelberg (to appear, 2008)

9. GNU MP, http://gmplib.org/
10. Hess, F., Smart, N., Vercauteren, F.: The Eta Pairing Revisited. IEEE Trans. Information Theory, 4595–4602 (2006)
11. Itoh, T., Tsujii, S.: A Fast Algorithm for Computing Multiplicative Inverses in $GF(2^m)$ Using Normal Bases. Inf. and Comp. 78, 171–177 (1988)
12. Kato, H., Nogami, Y., Yoshida, T., Morikawa, Y.: Cyclic Vector Multiplication Algorithm Based on a Special Class of Gauss Period Normal Basis. ETRI Journal 29(6), 769–778 (2007),http://etrij.etri.re.kr/Cyber/servlet/BrowseAbstract?paperid=RP0702-0040
13. Knuth, D.: The Art of Computer Programming. Seminumerical Algorithms, vol. 2. Addison-Wesley, Reading (1981)
14. Lee, E., Lee, H., Park, C.: Efficient and Generalized Pairing Computation on Abelien Varieties, IACR ePrint archive, http://eprint.iacr.org/2008/040
15. Matsuda, S., Kanayama, N., Hess, F., Okamoto, E.: Optimised Versions of the Ate and Twisted Ate Pairings. In: Galbraith, S.D. (ed.) Cryptography and Coding 2007. LNCS, vol. 4887, pp. 302–312. Springer, Heidelberg (2007)
16. Nakanishi, T., Funabiki, N.: Verifier-Local Revocation Group Signature Schemes with Backward Unlinkability from Bilinear Maps. In: Roy, B. (ed.) ASIACRYPT 2005. LNCS, vol. 3788, pp. 443–454. Springer, Heidelberg (2005)
17. Nogami, Y., Morikawa, Y.: A Fast Implementation of Elliptic Curve Cryptosystem with Prime Order Defined over F_{p^8}. Memoirs of the Faculty of Engineering Okayama University 37(2), 73–88 (2003)
18. Sakemi, Y., Kato, H., Akane, M., Okimoto, T., Nogami, Y., Morikawa, Y.: An Improvement of Twisted Ate Pairing Using Integer Variable with Small Hamming Weight. In: The 2008 Symposium on Cryptography and Information Security (SCIS 2008), January 22-25 (2008)
19. Vercauteren, F.: Optimal Pairings, IACR ePrint archive, http://eprint.iacr.org/2008/096

A Bilinearity of $f_{p,Q}$

First, we have

$$f_{r,Q} = f_{p-T,Q} = f_{p,Q} \cdot g_{pQ,-TQ} \cdot f_{T,Q}^{-1} = f_{p,Q} \cdot v_{pQ} \cdot f_{T,Q}^{-1}, \tag{50}$$

where v_{pQ} denotes the vertical line that goes through $pQ = \phi(Q)$. Thus,

$$\begin{aligned} \left\{f_{p,Q}(P) \cdot v_{\phi(Q)}(P)\right\}^{(p^{12}-1)/r} &= f_{r,Q}(P)^{(p^{12}-1)/r} \cdot f_{T,Q}(P)^{(p^{12}-1)/r} \\ &= \tau(Q,P) \cdot \alpha(Q,P), \end{aligned} \tag{51}$$

According to Eq.(11b), the x–coordinate of $Q \in \mathbb{G}_2$ is given by $x_{Q'} v^{1/3}$, where $x_{Q'}$ and $v^{1/3}$ belong to $\mathbb{F}_{p^2}$ and $\mathbb{F}_{p^6}$, respectively. Therefore, $v_{\phi(Q)}(P)^{(p^{12}-1)/r}$ becomes 1. Because, the x–coordinate of P belongs to $\mathbb{F}_p$. Thus, we have

$$f_{p,Q}(P)^{(p^{12}-1)/r} = \tau(Q,P) \cdot \alpha(Q,P). \tag{52}$$

Therefore $f_{p,Q}(P)^{(p^{12}-1)/r}$ gives a bilinear map from which Eq.(35a) becomes

$$f_{p,\chi Q}^{(p^{12}-1)/r} = f_{p,Q}^{\chi(p^{12}-1)/r}. \tag{53}$$

The bilinearity of $f_{p,Q}$ has been also shown in [15].

B Proof of Eq.(41)

Eq.(14) means

$$f_{r,Q}^{uL(p^{12}-1)/r} = f_{T,Q}^{c(p^{12}-1)/r}. \tag{54}$$

From Eq.(52), we have

$$f_{r,Q}^{uL(p^{12}-1)/r} \cdot f_{T,Q}^{uL(p^{12}-1)/r} = f_{p,Q}^{uL(p^{12}-1)/r}. \tag{55}$$

Substituting Eq.(54) to Eq.(55), we have

$$f_{T,Q}^{(uL+c)(p^{12}-1)/r} = f_{p,Q}^{uL(p^{12}-1)/r}. \tag{56}$$

The $(uL+c)$–th power of the right–hand side of Eq.(39) becomes

$$\left\{f_{T,Q}^{uL+c} \cdot A^{-(uL+c)}\right\}^{(p^{12}-1)/r} = \left\{f_{p,Q}^{w}\right\}^{(p^{12}-1)/r}, \tag{57}$$

where using Eq.(56) w is given as

$$w = uL - \chi(uL+c)h'(p). \tag{58}$$

Thus, we obtain the condition as Eq.(41).

C Constructing $\mathbb{F}_{p^{12}}$ and Its Subfields $\mathbb{F}_{p^2}, \mathbb{F}_{p^4}, \mathbb{F}_{p^6}$

First, the authors prepared $\mathbb{F}_{p^4}$ with type–$\langle 1,4\rangle$ Gauss period normal basis (GNB) [5] and also $\mathbb{F}_{p^3}$ with type–$\langle 2,3\rangle$ GNB. Then, the authors prepared $\mathbb{F}_{p^{12}}$ as *tower field* $\mathbb{F}_{(p^4)^3}$ by towering $\langle 2,3\rangle$ GNB over $\mathbb{F}_{p^4}$ [17]. For multiplication with GNB, the authors implemented *cyclic vector multiplication algorithm* (CVMA) [12]. For example, CVMA calculates a multiplication in $\mathbb{F}_{(p^m)^n}$ by

$$M_{mn} = \frac{n(n+1)}{2}M_m = \frac{mn(m+1)(n+1)}{4}M_1. \tag{59}$$

For inversions in extension field and prime field, the authors implemented Itoh–Tsujii inversion algorithm [11] and *binary extended* Euclidean algorithm [13], respectively. Since GNB is a class of normal bases, one can easily prepare arithmetic operations in subfields $\mathbb{F}_{p^2}, \mathbb{F}_{p^4}, \mathbb{F}_{(p^2)^3}$.

Pairing Computation on Twisted Edwards Form Elliptic Curves

M. Prem Laxman Das and Palash Sarkar

Applied Statistics Unit
Indian Statistical Institute
203, B.T. Road
Kolkata 700108, India
{prem_r,palash}@isical.ac.in

Abstract. A new form of elliptic curve was recently discovered by Edwards and their application to cryptography was developed by Bernstein and Lange. The form was later extended to the twisted Edwards form. For cryptographic applications, Bernstein and Lange pointed out several advantages of the Edwards form in comparison to the more well known Weierstraß form. We consider the problem of pairing computation over Edwards form curves. Using a birational equivalence between twisted Edwards and Weierstraß forms, we obtain a closed form expression for the Miller function computation.

Simplification of this computation is considered for a class of supersingular curves. As part of this simplification, we obtain a distortion map similar to that obtained for Weierstraß form curves by Barreto et al and Galbraith et al. Finally, we present explicit formulae for combined doubling and Miller iteration and combined addition and Miller iteration using both inverted Edwards and projective Edwards coordinates. For the class of supersingular curves considered here, our pairing algorithm can be implemented without using any inversion.

Keywords: elliptic curve, pairings, Edwards form, Miller function, supersingular curves.

1 Introduction

BACKGROUND. Pairings on curves find many applications in cryptographic protocols. These have been used to give one-round three-party key exchange [1], identity-based encryption [2] and many other schemes. For implementing such protocols, it is essential to have curves which are pairing friendly and an efficient pairing algorithm. Construction of pairing friendly curves is itself an active area of research. See [3] for a survey.

This work concerns computing (Tate) pairing on an elliptic curve. Tate pairing was introduced in cryptology in [4]. An algorithm for finding Tate pairing on elliptic curves was first given by Miller, which was subsequently published in [5]. Tate pairing over supersingular curves was studied in [6,7]. Several techniques were described to improve the efficiency of computing the pairing.

S.D. Galbraith and K.G. Paterson (Eds.): Pairing 2008, LNCS 5209, pp. 192–210, 2008.

Edwards [8] introduced a new form of elliptic curves and gave an elegant addition rule for such curves. The work [8] considered elliptic curves over number fields. Bernstein and Lange [9] showed the usefulness of the Edwards form elliptic curves in cryptography. Among other things, they showed that, unlike the more well known Weierstraß form, the Edwards form admits a complete (and hence, unified) addition formula. This is very useful in providing resistance to side-channel attacks. Further, in [9] and [10] they developed efficient explicit formulae for doubling, addition and mixed-addition using projective and inverted coordinates. These provided the fastest methods for scalar multiplication on elliptic curves.

MOTIVATION. Pairing based cryptographic protocols use both scalar multiplications and pairing computations. In view of the advantages of Edwards form curves, a designer may wish to implement a pairing based protocol using such curves. The problem, however, is with the pairing computation. Till date, all pairing algorithms use the more well known Weierstraß form of an elliptic curve. Thus, to implement pairings, one will have to use an isomorphism to map Edwards points to points on Weierstraß form and then compute pairing on Weierstraß form curve.

This raises several questions. Is it possible to compute pairing directly on the Edwards form? How does this compare to the cost of converting to Weierstraß form and then computing the pairing? More generally, how does pairing on Edwards form compare to the cost of computing pairing on the Weierstraß form? Are there any advantages in computing pairing directly on Edwards form?

Motivated by these questions, we make a detailed investigation of pairing on Edwards form. The basic question is of course, how to perform pairing directly on Edwards form.

CONTRIBUTIONS. The following question is central to computing the Tate pairing on elliptic curves using Miller's algorithm: given points P_1 and P_2 on an elliptic curve, find a point P_3 and a rational function h such that

$$\mathrm{div}(h) = (P_1) + (P_2) - (P_3) - \mathcal{O},$$

where $\mathcal{O}$ is a distinguished rational point. This fact is emphasized in [4]. For Weierstraß form curve this is easy to do using the chord-and-tangent rule for addition. In this case, P_3 is taken to be the negative of the sum of P_1 and P_2 and one such step is called a Miller iteration.

The first contribution of this work is to work out a solution to the above problem for twisted Edwards form curve. Using the birational equivalence between twisted Edwards and Weierstraß form curves, we obtain the form of the rational function h over twisted Edwards form when P_3 is the sum of P_1 and P_2. In other words, we show how to perform Miller iteration directly on twisted Edwards form curve. Since the Miller iteration forms the basis of all pairing algorithms, including the Weil, Tate, Eta and Ate pairings, our work shows how to compute such pairings directly over twisted Edwards form curves.

In its general form, the expression for h looks a bit complicated. We show that for special curves, it is possible to simplify the computation. As examples, we consider supersingular curves over finite fields of characteristic greater than 3 (and hence having embedding degree 2). An important aspect in pairing computation over supersingular curves is the utilization of the so-called distortion map. For Weierstraß form, such a map was obtained in [6,7]. We obtain a similar distortion map for a class of Edwards form supersingular curves. Using this map and some further simplifications, we work out explicit formulae for combined doubling and Miller iteration and combined addition and Miller iteration using both inverted Edwards and projective Edwards coordinates.

The cost for doubling and Miller value computation is 9[M]+6[S] and for mixed addition and Miller value computation is 17[M]+1[S] using inverted Edwards coordinates. The corresponding values using projective Edwards coordinates are 9[M]+6[S] and 18[M]+1[S]. This is slower than the best known pairing algorithm for Weierstraß form supersingular curve $s^2 = r^3 + ar$ using Jacobian curves obtained in [11]. The corresponding values for general a, small a and $a = -3$ are (8[M]+6[S], 11[M]+3[S]), (7[M]+6[S], 11[M]+3[S]) and (8[M]+4[S], 11[M]+3[S]) respectively. (The Edwards form does not distinguish between different values of a.)

Comparison to Pairing on Weierstrass. In general, it is expected that pairing over Edwards form will be slower than pairing over Weierstraß form. To see this, consider the two ways of performing pairing over Edwards.

1. Convert the points to Weierstraß form and then perform the pairing on Weierstraß form. In this method, the total cost of pairing will also include the cost of converting points from Edwards form to Weierstraß form.
2. Perform pairing directly on twisted Edwards form using the required Miller function (obtained here). The form for this function is obtained by mapping Edwards points to Weierstraß points, obtaining the expression for Miller function on Weierstraß and then mapping back to obtain the Miller function on Edwards. So, the form for the Miller function on Edwards implicitly includes both the maps to and from Weierstraß. Consequently, it is unlikely that a Miller iteration on Edwards will be faster than a Miller iteration on Weierstraß.

The above seems to suggest that Edwards form should not be used for implementing pairing based protocols. The answer, however, is not that straightforward. Each algorithm in a protocol involves some scalar multiplications and some pairings. For the scalar multiplications, Edwards form is faster, especially if the implementation has to guard against side channel attacks. The pairing will be slower but, this may be compensated by the faster scalar multiplications. We believe that there is no general answer and a designer would have to look at the very specific details before making a proper selection of elliptic curve form.

Pairing on Edwards: Compute Pairing Directly or Via Weierstrass Form? Suppose a designer chooses to implement a protocol using the Edwards form. From Point 2 mentioned above, it seems that each Miller iteration on Edwards will be slower than that on Weierstraß. The direct method is faster if the cost of conversion to Weierstraß amortized over all the Miller iterations is more than the difference between the Miller iteration on Edwards and that of Weierstraß.

Inversion Free Pairing on Edwards Form. On the other hand, there is one advantange of the direct method. This arises in specific reference to the class of supersingular curves considered here. Suppose, a designer wants an inversion-free pairing algorithm, i.e., a pairing algorithm, which does not make any inversion. Then the implementation will not require an inversion module. For resource constrained devices this may be an important issue.

For the specific class of supersingular curves considered here, the pairing algorithm that we obtain is free from inversion. Hence, the inversion module is not required to implement this algorithm. In contrast, we show that if the pairing is computed by converting to Weierstraß, then the conversion itself will require an inversion (as otherwise the resulting algorithm will be inefficient).

2 Preliminaries and Notations

Throughout this paper p denotes a prime greater than 3 and q an odd prime power. The finite field of cardinality q will be denoted by $\mathbb{F}_q$.

An elliptic curve (over $\mathbb{F}_q$) in Weierstraß form is given by an equation $y^2 = x^3 + a_2x^2 + a_4x + a_6$, where a_2, a_4 and a_6 are from $\mathbb{F}_q$. The addition rule and other properties on this form of the curve are quite well known and hence we do not repeat these here.

An elliptic curve (over $\mathbb{F}_q$) in Edwards form is given by an equation $x^2 + y^2 = c^2(1 + dx^2y^2)$, $c, d \neq 0$. Edwards introduced this form for elliptic curves over number fields and with $d = 1$. The curve parameter d was introduced by Bernstein and Lange who also studied this equation over finite fields. The additive identity is $(0, c)$; $(0, -c)$ has order 2; $(\pm c, 0)$ have order 4. The addition rule is given by the following formula.

$$(x_1, y_1) + (x_2, y_2) \mapsto \left(\frac{x_1y_2 + y_1x_2}{c(1 + dx_1x_2y_1y_2)}, \frac{y_1y_2 - x_1x_2}{c(1 - dx_1x_2y_1y_2)} \right).$$

If E is an elliptic curve defined by a bi-variate polynomial $C(x, y)$, then the set of $\mathbb{F}_q$-rational points of E is denoted by $E(\mathbb{F}_q)$ and is defined to be the set of pairs $(\alpha, \beta) \in \mathbb{F}_q \times \mathbb{F}_q$ such that $C(\alpha, \beta) = 0$. The set $E(\mathbb{F}_q)$ forms a group under a suitably defined addition law and an additive identity. For an $\mathbb{F}_q$-rational point P, the i fold sum of P is denoted by $[i]P$.

2.1 Birational Equivalence

Rational functions on a curve are important in studying the behavior of the curve. These rational functions form a field and two (forms of) elliptic curve are said to be *birationally equivalent* if their fields of rational functions are isomorphic.

Another form of elliptic curves which is also quite well known is the Montgomery form and is given by an equation of the form $Bv^2 = u^3 + Au^2 + u$, with $B \neq 0$. Birational equivalences between Weierstraß and Edwards form use the Montgomery form as an intermediate stepping stone.

It has been observed in [9] that the form $x^2 + y^2 = 1 + dx^2y^2$ is as general as the form $X^2 + Y^2 = C^2(1 + DX^2Y^2)$ in the sense that there is an isomorphism between them. The change of variables $X = Cx$ and $Y = Cy$ transforms $x^2+y^2 = 1 + dx^2y^2$ into $X^2 + Y^2 = C^2(1 + DX^2Y^2)$ with the condition that $C^4D = d$.

An extension, called the *twisted* Edwards form has been studied in [12]. The curve equation in this case has the form $ax^2 + y^2 = 1 + dx^2y^2$ for distinct non-zero elements a and d in a finite field $\mathbb{F}$ (of characteristic not equal to 2). It has been proved in [12] that the set of twisted Edwards form curves over the field $\mathbb{F}$ is birationally equivalent to the set of Montgomery form curves over $\mathbb{F}$. Then

$$(x, y) \mapsto (u, v) = ((1 + y)/(1 - y), (1 + y)(x(1 - y))) \tag{1}$$

transforms $ax^2+y^2 = 1+dx^2y^2$ to $Bv^2 = u^3+Au^2+u$, where $A = 2(a+d)/(a-d)$ and $B = 4/(a-d)$. Since a and d are distinct and non-zero, A is not 2 or -2 and B is non-zero. The inverse map is given by $(u, v) \mapsto (x, y) = (u/v, (u-1)/(u+1))$.

The case $a = 1$ in twisted Edwards curve is the Edwards curve as considered in [9]. Theorem 3.5 of [12] shows that an elliptic curve is birationally equivalent to an Edwards form curve if and only if it has a point of order 4. Assuming the curve to be in Weierstraß form $s^2 = r^3 + a_2r^2 + a_4r$ and using a point (r_1, s_1) of order 4 on this curve, it is possible to exhibit a birational equivalence between the Weierstraß and Edwards forms. The map

$$(x, y) \mapsto (r, s) = ((r_1(1 + y))/(1 - y), (s_1(1 + y))/(x(1 - y))) \tag{2}$$

transforms $x^2 + y^2 = 1 + dx^2y^2$ to $s^2 = r^3 + a_2r^2 + a_4r$, where $a_2 = s_1^2/r_1^2 - 2r_1$; $a_4 = r_1^2$ and $d = 1 - 4r_1^3/s_1^2$. This result was essentially contained in the proof of Theorem 2.1 of [9]. The actual statement and the result were more complicated because the proof missed the fact that $r_1/(1 - d)$ equals $(s_1/(2r_1))^2$ and hence, is always a square. Instead, it was required that d is a non-square (equivalently, there is a unique point of order 2), which caused some complications.

The following observation from [9] shows how to convert from $S^2 = R^3 + A_4R + A_6$ to $s^2 = r^3 + a_2r^2 + a_4r$.

Observation 1. Let E be an elliptic curve over $\mathbb{F}$ given in the Weierstraß form $S^2 = R^3 + A_4R + A_6$ such that the group $E(\mathbb{F})$ has an element $Q = (R_1, S_1)$ of order 4. Then E can be transformed into the curve E': $s^2 = r^3+a_2r^2+a_4r$ by the change of variables $r = R-R_2$, and $s = S$. Then $a_2 = 3R_2$, $a_4 = 3R_2^2+A_4$ and R_2 is the x-coordinate of $2Q$. The point Q is transformed into a point $P = (r_1, s_1)$, where $r_1 = R_1 - R_2$ and $s_1 = S_1$ leading to $2P = (0, 0)$.

2.2 Background on Pairing

In this section, we discuss basics of Tate pairing. We first recall some fundamentals on divisors on elliptic curves. Let E be an elliptic curve over $\mathbb{F}_q$, with identity $\mathcal{O}$. Points are denoted by P, Q, etcetera, while the corresponding places are denoted by (P), (Q), etcetera. The function field of E is the quotient field of the coordinate ring of E. Elements of this field are called functions over E. Places correspond to valuation rings of the function field.

Divisors of E are formal $\mathbb{Z}$-linear combinations of places. Any non-constant function has finitely many zeros and poles at places, of some finite positive order. The collection of zeros and poles of a function, expressed as a divisor is called its *principal divisor.* For a function z, its principal divisor is denoted by $\mathrm{div}(z) = (z)_0 - (z)_\infty$. The divisor $(z)_0$ is called the *zero divisor* of z and $(z)_\infty$ its pole divisor.

The computation of Tate pairing depends on the addition rule on the elliptic curve group. Following [4], the following task forms the backbone for pairing computation:

Task 1. *Given $P_1 = (x_1, y_1)$ and $P_2 = (x_2, y_2)$, points on an elliptic curve X, find a point P_3 and a function h such that* $\mathrm{div}(h) = (P_1) + (P_2) - (P_3) - (\mathcal{O})$.

Weierstraß form is the most well-studied form of elliptic curve. The task above can be easily performed using the chord-tangent rule.

Tate pairing was first introduced in cryptography in [4]. We recall the definition of Tate pairing from [7]. Let E be an elliptic curve defined over $\mathbb{F}_q$ and r be coprime to q and $r \mid \#E(\mathbb{F}_q)$. Let k be a positive integer such that the field $\mathbb{F}_{q^k}$ contains all the rth roots of unity (that is, $r \mid (q^k - 1)$).

Definition 1. *With r as above, the smallest extension field of $\mathbb{F}_q$ which contains all the rth roots of unity is denoted by L. The extension degree $[L : \mathbb{F}_q]$ is known as embedding degree.*

Following [7], the Tate pairing is defined as follows.

Definition 2. *The choices for parameters are made as discussed above. Let $G := E(\mathbb{F}_{q^k})$. The Tate pairing is defined as*

$$e_r(\cdot,\cdot) : G[r] \times G/rG \longrightarrow \mathbb{F}_{q^k}^*/\mathbb{F}_{q^k}^{*r}$$

with $e_r(P,Q) := f_P(Q)^{\frac{q^k-1}{r}}$. The function f_P is such that $\mathrm{div}(f_P) = r(P) - r(\mathcal{O})$.

The quotient group on the right hand side is the set of equivalence classes modulo the relation "$a \equiv b$ if and only if there exists $c \in \mathbb{F}_{q^k}^*$ such that $a = bc^r$". For more properties of Tate pairing refer [4]. The pairing thus defined is well-defined, non-degenerate and bilinear.

Let $h_{P,Q}$ denote the rational function corresponding to the addition of P and Q. Let $r = (r_{l-1} \cdots r_0)$ the binary representation of r. With this setup, an algorithm for computing the Tate pairing $e_r(P,Q)$ on an elliptic curve may be given. The rational function appearing in the algorithm depends on the form of the elliptic curve. See Table 1.

The algorithm in Table 1 computes in the ith iteration a function $f_{i,P}$ having divisor $\mathrm{div}(f_{i,P}) = i(P) - ([i]P) - (i-1)(\mathcal{O})$, called Miller's functions. At each

Table 1. Miller's algorithm for computing Tate pairing

Input : Points P and Q **Output :** Tate pairing of P and Q

1. Set $f = 1$ and $P_1 = P$.
2. For $i = l - 2$ downto 0
  Set $f = f^2 \cdot h_{P,P}$ and $P_1 = 2P$.
  If $r_i = 1$ then set $f = f \cdot h_{P_1,P}$ and $P_1 = P_1 + P$.
3. Set $f = f^{\frac{q^k-1}{r}}$.
4. Return f.

step, the Miller's functions are evaluated at the second argument. After $l - 1$ iterations, the evaluation at Q of the function f having divisor $r(P) - r(\mathcal{O})$ is obtained.

3 Pairing over Twisted Edwards Form Curve

Pairing algorithms have been extensively studied. All such studies have used the Weierstraß form. Let us first consider how to implement pairings on Edwards form using pairings on Weierstraß form.

3.1 Pairing Via Weierstraß Form

Suppose we have a pairing friendly curve C in Weierstraß form having a point of order 4 and let E be the corresponding Edwards form. The birational equivalence between E and C is a group isomorphism between the corresponding group of points. Using this isomorphism, we can map points on Edwards form into Weierstraß form and compute the pairing on Weierstraß form. (Note that the output of the pairing is an element of an extension field and there is no issue of "going back" to Edwards form.) The cost of this procedure is the cost of applying the isomorphism from Edwards to Weierstraß form plus the cost of computing the pairing on Weierstraß form.

Suppose the input to the pairing are the points $P = (x_P, y_P)$ and $Q = (x_Q, y_Q)$ in Edwards form. Using (2), and recalling that (r_1, s_1) is a point of order four on Weierstraß form, we have

$$\begin{aligned}(x_P, y_P) &\mapsto \left(r_1 \times \frac{1+y_P}{1-y_P}, s_1 \times \frac{1+y_P}{x_P(1-y_P)}\right), \\ (x_Q, y_Q) &\mapsto \left(r_1 \times \frac{1+y_Q}{1-y_Q}, s_1 \times \frac{1+y_Q}{x_Q(1-y_Q)}\right).\end{aligned} \tag{3}$$

The coordinates x_P, y_P of the point P are from $\mathbb{F}_q$. However, the coordinates x_Q, y_Q of the point Q are from $\mathbb{F}_{q^k}$, where k is the embedding degree. The inverses of $(1 - y_Q)$ and x_Q are required as also the inverses of $(1 - y_P)$ and x_P. While the later is easier to obtain, depending on the embedding degree, obtaining the former inverses may be rather expensive. As example, consider the value $k = 10$ which is the focus of current research on obtaining pairing friendly curves [3]. In this case, the two inversions on $\mathbb{F}_{q^{10}}$ can be computed using one $\mathbb{F}_{q^{10}}$-inversion and three $\mathbb{F}_{q^{10}}$-multiplications. The total cost will be equivalent to a few hundred multiplications over $\mathbb{F}_q$.

The above transforms an affine representation of a Edwards point into an affine representation of a Weierstraß point. In many situations, one works with other representations such as projective or Jacobian coordinates. It is possible to convert to the desired coordinate system using a few multiplications. Suppose that the Edwards form point is given in affine coordinates as (x, y) and we want the Weistraß form point in projective coordinates. The output of (3) is equal to (r, s), where $r = a/b$ and $s = c/d$ with $a = r_1(1 + y)$, $b = (1 - y)$, $c = s_1(1 + y)$ and $d = x(1 - y)$. Then, the projective representation (R, S, T) with $r = R/T$ and $s = S/T$ is obtained by setting $R = ad$, $S = cb$ and $T = bd$. After obtaining a, b, c and d, three extra multiplications convert the point to projective coordinates. Further, the representation (RT, ST^2, T) is in Jacobian coordinates and two extra multiplications and one squaring are required for this.

The point in Edwards may not be given in affine. Projective and inverted Edwards representations have been suggested in [9,10]. The representation is (X, Y, Z), where in the former case, $x = X/Z$ and $y = Y/Z$ and in the latter case, $x = Z/X$ and $y = Z/Y$. With both coordinate systems it is possible to convert to projective (and Jacobian) Weierstraß forms. We show this for the inverted Edwards coordinates, the case for projective Edwards being similar. In this case, the affine Weierstraß form is $(r = a/b, s = c/d)$ where $a = r_1(Y + Z)$, $b = Y - Z$, $c = s_1X(Y + Z)$ and $d = Z(Y - Z)$. From this affine representation the conversion to projective or Jacobian Weierstraß is as described above.

If we use (3) to convert to affine Weierstraß then an inversion is required. Converting to projective or Jacobian can avoid inversion at the cost of several extra multiplications. There are two additional issues to consider for inversion free conversion.

1. Obtaining the point P in affine Weierstraß allows mixed addition formula to be used during Miller iteration. Obtaining P in projective and Jacobian will increase the cost of mixed addition.
2. The cost of converting the point Q will require extension field multiplications. Further, most pairing algorithms on Weierstraß form require Q in affine. If Q is given in projective, this will imply extra (extension field) multiplications when the Miller function is evaluated at Q. The last point is significant, since, even one extra extension field multiplication per Miller iteration can prove to be costly.

Thus, avoiding inversions in the conversion from Edwards to Weierstraß in general pushes up the cost for pairing computation on Weierstraß form itself. On the other hand, avoiding inversions may be required for other reasons in addition to that of computational efficiency. In resource constrained devices, it is desirable to implement the algorithm in as small hardware area or software code as possible. The ability to avoid implementing the inversion routine will be useful for such scenarios.

Based on the above discussion, we consider the problem of developing a pairing algorithm which works directly over the twisted Edwards form. The main task is to compute the Miller function at each iteration.

3.2 Miller Function for Twisted Edwards Form Curve

This section deals with efficiently performing Task 1 (of Section 2.2) on twisted Edwards form elliptic curve. As already seen, the Miller function computation forms the backbone for computing Tate pairing. The result of this section gives the Miller function corresponding to addition of P_1 and P_2.

Theorem 1. *Let $\mathbb{F}_q$ be a field of characteristic not equal to 2 and $ax^2 + y^2 = 1 + dx^2y^2$ be a twisted Edwards form curve where a and d are distinct non-zero elements of $\mathbb{F}_q$. Let $P_0 = (0, 1)$. Let $P_1 = (x_1, y_1)$ and $P_2 = (x_2, y_2)$ be two points on it. Let $P_3 = (x_3, y_3)$ be the sum of P_1 and P_2. Then the Miller function $h(x, y)$ such that*

$$\mathrm{div}(h) = (P_1) + (P_2) - (P_3) - (P_0) \tag{4}$$

is given by

$$h(x, y) = \frac{(1 - y_3)}{x(y - y_3)}((1 + y) - x(\lambda(1 + y) + \theta(1 - y))). \tag{5}$$

where $A = (2(a + d))/(a - d)$, $B = 4/(a - d)$ and

$$\lambda = \begin{cases} \frac{x_1(A(y_1^2-1)-2(1+y_1+y_1^2))}{B(y_1^2-1)} & \text{if } P_1 = P_2; \\ \frac{x_1(y_1-1)(y_2+1)-x_2(y_1+1)(y_2-1)}{2x_1x_2(y_1-y_2)} & \text{if } P_1 \neq P_2. \end{cases} \tag{6}$$

and $\theta = 2(1 + y_1)/(x(1 - y_1)) - \lambda(1 + y_1)/(1 - y_1)$ is given by

$$\theta = \begin{cases} \frac{(y_1^2-1)(Ax_1^2-B)-2x_1^2(1+y_1+y_1^2)}{Bx_1(y_1^2-1)} & \text{if } P_1 = P_2; \\ \frac{(x_1-x_2)(1+y_1)(1+y_2)}{2x_1x_2(y_1-y_2)} & \text{if } P_1 \neq P_2. \end{cases} \tag{7}$$

[**Note.** There is no assumption on the embedding degree.]

Proof. The idea of the proof is simple. In the Weierstraß form it is easy to obtain a rational function $g(x, y)$ such that a relation similar to that of Equation 4 holds. Basically $g(x, y)$ is the ratio of two lines – the line passing through P_1 and P_2 and the line passing through P_3 and $-P_3$.

Let Φ be the transformation given in (1).

$$\Phi(x, y) = (u, v) \triangleq \left(\frac{1 + y}{1 - y}, \frac{(1 + y)}{x(1 - y)}\right). \tag{8}$$

Then $x = u/v$ and $y = (u-1)/(u+1)$. This transforms the curve $ax^2 + y^2 = 1 + dx^2y^2$ into the curve $Av^2 = u^3 + Bu^2 + u$, where $A = 2(a+d)/(a-d)$ and $B = 4/(a-d)$. The later curve is in Montgomery form. But, the Miller function $g(x,y)$ for Montgomery form is still the ratio of two lines as in the case of the Weierstraß form.

The idea is to first transform points P_i on Edwards form into corresponding points Q_i on Montgomery form using Φ; compute $g(x,y)$ on Montgomery form and then use the inverse of Φ to transform $g(x,y)$ into the desired rational function $h(x,y)$. For this to work we need to note that the transformation Φ extends to several isomorphisms.

1. The map $\sum n_i(P_i) \mapsto \sum n_i(\Phi(P_i))$ is an isomorphism of the set of divisors on Edwards and Montgomery form curves.
2. The map $h(x,y) \mapsto h(\Phi(x,y))$ is an isomorphism of the function fields of the Edwards and Montgomery form curves.

Let $\mathcal{O}$ be the identity on Montgomery form curve. Then $\Phi((P_1)+(P_2)-(P_3)-(P_0)) = (Q_1)+(Q_2)-(Q_3)-(\mathcal{O})$. We use (x,y) to denote Edwards coordinates and (u,v) to denote Montgomery coordinates. Let $l_1(u,v)$ be the line through Q_1 and Q_2 and $l_2(u,v)$ be the line through Q_3 and $-Q_3$. Then $l_1(u,v) = v - \lambda u - \theta$ and $l_2(u,v) = u - u_3$ where the slope λ and the constant θ are obtained later.

Define $g(u,v) = l_1(u,v)/l_2(u,v)$ and so $g(u,v) = (v - \lambda u - \theta)/(u - u_3)$. The desired function $h(x,y)$ is $g(\Phi^{-1}(u,v)) = g((1+y)/(1-y), 2(1+y)/(x(1-y)))$. We have

$$\begin{aligned}
h(x,y) &= \frac{\frac{1+y}{x(1-y)} - \lambda\frac{1+y}{1-y} - \theta}{\frac{1+y}{1-y} - \frac{1+y_3}{1-y_3}} \\
&= \frac{(1-y_3)((1+y) - \lambda x(1+y) - \theta x(1-y))}{x((1+y)(1-y_3) - (1+y_3)(1-y))} \\
&= \frac{(1-y_3)}{2x(y-y_3)}((1+y) - x(\lambda(1+y) + \theta(1-y))).
\end{aligned}$$

It remains to obtain the expressions for λ and θ in terms of x_1, y_1, x_2 and y_2. Recall that $u_i = \frac{1+y_i}{1-y_i}$ and $v_i = \frac{(1+y_i)}{x_i(1-y_i)}$. Also, $\theta = v_1 - \lambda u_1$. The value of λ is obtained as the slope of the line through P_1 and P_2, if they are distinct; or as the slope of the tangent through P_1, if the points are equal. In the former case, $\lambda = (v_2 - v_1)/(u_2 - u_1)$. In the later case, we have to refer to the equation of the curve. The curve in question is the Montgomery form curve $Bv^2 = u^3 + Au^2 + u$. Differentiating with respect to u we have $\lambda = (3u_1^2 + 2Au_1 + 1)/(2Bv_1)$. The expressions for λ and θ in the two cases can now be obtained by substituting the values of u_i, v_i and simplifying the resulting expressions. □

4 Supersingular Curves in Edwards Form

For $p > 3$, two supersingular curves in Weierstraß form are quite well known. We provide the corresponding Edwards form. For the map given by (2) to exist, the curve must have a point of order 4. The number of $\mathbb{F}_p$-rational points on supersingular curves of characteristics greater than 3 is known to be $p+1$. So, we require $p \equiv 3 \bmod 4$ as a necessary condition for a point of order 4 to exist.

$\boldsymbol{s^2 = r^3 + a_4 r}$**.** The condition $p \equiv 3 \bmod 4$ ensures that this curve is supersingular which is compatible with the condition for a point of order 4 to exist. Let $P = (r_1, s_1)$ be a hypothesized point of order 4 on this curve. Then $a_4 = r_1^2$ and $s_1^2 = r_1(r_1^2 + a_4) = 2a_4 r_1$. The possible values of (r_1, s_1) are $\left(\sqrt{a_4}, \pm\sqrt{2a_4^{3/2}}\right)$ and $\left(-\sqrt{a_4}, \pm\sqrt{-2a_4^{3/2}}\right)$. Since $p \equiv 3 \bmod 4$, a_4 must be a square modulo p which is a necessary and sufficient condition for transforming to Edwards form.

Since $p \equiv 3 \bmod 4$, -1 is a non-square modulo p and hence exactly one of $2a_4^{3/2}$ and $-2a_4^{3/2}$ is a square modulo p. This shows that there are exactly two points of order 4.

1. If $a_4 = 1$, then $(1, \pm\sqrt{2})$ are the points of order 4 if $(p^2 - 1)/8$ is even; and $(-1, \pm\sqrt{-2})$ are the points of order 4 if $(p^2 - 1)/8$ is odd. Later we will consider pairing over this curve.
2. If $a_4 = -3$, then the curve has a point of order 4 only if 3 is a non-square modulo p, i.e., if $p \equiv \pm 5 \bmod 12$. Determining the two actual points of order 4 requires obtaining the square root of either $\sqrt{2 \times 3^{3/2}}$ or $\sqrt{-2 \times 3^{3/2}}$. We know that one of them is a square, but the exact value of the square root depends on p.

The value of d in the Edwards form curve is determined from the relation $a_2 = 0 = s_1^2/r_1^2 - 2r_1$. Then $2r_1^3 = s_1^2$ and so, $d = 1 - (4r_1^3/s_1^2) = -1$. Thus, if a_4 is a square modulo p, then the corresponding Edwards form is

$$x^2 + y^2 = 1 - x^2 y^2. \tag{9}$$

Note that d is equal to -1 irrespective of the value of a_4. Also, in (9) $a_4 = 1$ so that $A = 0$ and $B = 2$ in the Montgomery form obtained by applying (1).

Interestingly, the curve $x^2 + y^2 = 1 - x^2y^2$ was studied by Euler [13] and Edwards [8] reports that the curve was also of "great interest" to Gauss [14].

$\boldsymbol{S^2 = R^3 + \alpha}$**.** The condition $p \equiv 2 \bmod 3$ ensures that this curve is supersingular. This, along with the condition $p \equiv 3 \bmod 4$ for the point of order 4 to exist, implies that $p \equiv -1 \bmod 12$.

Here $A_4 = 0$ and $A_6 = \alpha$. Let $P = (R_1, S_1)$ be a point of order 4 on this curve and R_2 is the x-coordinate of $2P$. Since $2P$ has order 2, the y-coordinate of $2P$ must be zero and so $R_2^3 = -\alpha$. Using $2P = (R_2, 0)$, it can be shown that R_1 and S_1 are obtained by first solving $R_1^3 - 3R_2R_1 - 2\alpha = 0$ for R_1 and then solving $S_1^2 = R_1^3 + \alpha$ for S_1. So, for P to exist, first $-\alpha$ must be a cube modulo p and then these two equations should be solvable modulo p.

Once R_2 and (R_1, S_1) have been obtained, we can first apply Observation 1 followed by (2) to obtain the corresponding Edwards form.

Concrete Examples. Consider $E : y^2 = x^3 + x$ over $\mathbb{F}_p$, $p \geq 5$. In [15, Table 1], suitable values of p and r for various levels of security are given. We consider some particular values given in [15, Section 7.2]. In both cases below $p \equiv 3 \bmod 4$ and hence the curve $x^2+y^2 = 1-x^2y^2$ is supersingular over $\mathbb{F}_p$. The group $E(\mathbb{F}_p)$ has a unique element of order 2 and the points $(1, \pm\sqrt{2})$ are of order 4.

For 80-bit security level, with $k = 2$, recommended sizes of p and r are 512 and 160, respectively. A suitable set of parameters is given there as $p = 2^{520} + 2^{363} - 2^{360} - 1$, $r = 2^{160} + 2^3 - 1$.

For 128-bit security level, with $k = 2$, recommended sizes of p and r are 1536 and 256 bits respectively. A suitable set of parameters is given there as $p = 2^{1582} + 2^{1551} - 2^{1326} - 1$, $r = 2^{256} + 2^{225} - 1$.

5 Pairing Computation on $x^2 + y^2 = 1 - x^2y^2$ over $\mathbb{F}_p$, $p > 3$ and $p \equiv 3 \bmod 4$

In Section 4, we have seen that the supersingular curve $\mathcal{E} : s^2 = r^3 + ar$ over $\mathbb{F}_p$, with $p \equiv 3 \bmod 4$ transforms to $x^2 + y^2 = 1 - x^2y^2$ over $\mathbb{F}_p$, provided a is a square modulo p. Let $\mathcal{E}(\mathbb{F}_p)[r]$ be the set of all $\mathbb{F}_p$-rational r-torsion points of this curve. Let r be a prime greater than 3 and then $\langle R\rangle = \mathcal{E}(\mathbb{F}_p)[r]$. Then for any $(\alpha, \beta) \in \langle R\rangle$, $\beta \neq 0$. (If $\beta = 0$, then $\alpha = \pm 1$ and the points $(\pm 1, 0)$ are of order 4 and hence cannot be in $\langle R\rangle$; if they are, then $4|r$ which contradicts r is a prime greater than 3.)

The domain of pairing is $\mathcal{E}(\mathbb{F}_p)[r] \times \mathcal{E}(\mathbb{F}_{p^2})/r\mathcal{E}(\mathbb{F}_{p^2})$. By using a so-called "distortion map", the domain can be changed to $\mathcal{E}(\mathbb{F}_p)[r] \times \mathcal{E}(\mathbb{F}_p)[r]$. For the corresponding Weierstraß form this has been done in [6,7].

Definition 1. *[16, Section 4.2] A distortion map ϕ with respect to a cyclic group $\langle P\rangle$ of order r is an endomorphism of the curve that maps any non-zero point Q in $\langle P\rangle$ to a point $\phi(Q)$ which is independent of Q.*

The curve $s^2 = r^3 + r$ over $\mathbb{F}_p$ with $p > 3$ is supersingular for $p \equiv 3 \bmod 4$, with embedding degree $k = 2$. The map $\phi(r, s) = (-r, is)$ where $i^2 = -1$ is a distortion map for this curve. (For more details see [6].)

We obtain a distortion map for the Edwards form curve. The following result can be proved by mapping (x, y) on Edwards form curve to (r, s) on Weierstraß form; mapping (r, s) to $(-r, is)$ using the distortion map on Weierstraß form; and then mapping the resulting point back to Edwards form. The proof that we provide is more direct.

Theorem 2. *The function $\phi : \mathcal{E}(\mathbb{F}_p)[r] \to \mathcal{E}[\mathbb{F}_{p^2}]$ given by*

$$\phi(x, y) = \left(ix, \frac{1}{y}\right), \tag{10}$$

is a distortion map on the Edwards form curve $x^2 + y^2 = 1 - x^2y^2$.

Proof. First we notice that the image of ϕ is not contained in $\mathcal{E}(\mathbb{F}_p)[r]$. Next, we verify that ϕ is an endomorphism. Let $P_i = (x_i, y_i)$, for $i = 1, 2$. Let (x_3, y_3) be the sum $P_1 + P_2$. Thus, we have

$$\phi(P_1 + P_2) = \left(i\frac{x_1y_2 + x_2y_1}{1 - x_1x_2y_1y_2}, \frac{1 + x_1x_2y_1y_2}{y_1y_2 - x_1x_2} \right).$$

On the other hand,

$$\phi(P_1) + \phi(P_2) = \left(ix_1, \frac{1}{y_1} \right) + \left(ix_2, \frac{1}{y_2} \right) = \left(i\frac{x_1y_1 + x_2y_2}{x_1x_2 + y_1y_2}, \frac{1 + x_1y_1x_2y_2}{y_1y_2 - x_1x_2} \right).$$

We now verify that $(x_1y_2 + x_2y_1)(x_1x_2 + y_1y_2) = (x_1y_1 + x_2y_2)(1 - x_1x_2y_1y_2)$. Indeed, expanding the left hand side, we obtain,

$$\begin{aligned}
(x_1y_2 + x_2y_1)(x_1x_2 + y_1y_2) &= x_1^2x_2y_2 + x_1x_2^2y_1 + x_1y_1y_2^2 + x_2y_1^2y_2 \\
&= x_1y_1(x_2^2 + y_2^2) + x_2y_2(x_1^2 + y_1^2) \\
&= x_1y_1(1 - x_2^2y_2^2) + x_2y_2(1 - x_1^2y_1^2) \\
&= (x_1y_1 + x_2y_2)(1 - x_1x_2y_1y_2)
\end{aligned}$$

which proves the theorem. □

Under this distortion map, the output of $e(P, Q)$ is defined to be $e(P, \phi(Q))$. Each Miller iteration takes two points P_1 and P_2 and obtains P_3 to be the sum of P_1 and P_2 and evaluates $h(\phi(Q))$, where h is the rational function h given in Theorem 1. In other words, we have to evaluate

$$\begin{aligned}
h\left(ix_Q, \frac{1}{y_Q} \right) &= \frac{(1 - y_3)\left(\left(1 + \frac{1}{y_Q}\right) - ix_Q\left(\lambda\left(1 + \frac{1}{y_Q}\right) + \theta\left(1 - \frac{1}{y_Q}\right) \right) \right)}{ix_Q(\frac{1}{y_Q} - y_3)} \\
&= \frac{i(y_3 - 1)}{x_Q(1 - y_Qy_3)}((y_Q + 1) - ix_Q(\lambda(y_Q + 1) + \theta(y_Q - 1))) \\
&= \frac{(y_Q + 1)(y_3 - 1)}{x_Q(1 - y_Qy_3)}(x_Q\lambda + \alpha_Q\theta + i)
\end{aligned} \tag{11}$$

where $\alpha_Q = x_Q(y_Q - 1)/(y_Q + 1)$ and λ and θ are given by Equation 6 and Equation 7 respectively. Note that the expression for α_Q is the same as that of $1/v$ obtained in transforming from Edwards to Montgomery (see (1)). The value of α_Q depends only on Q and can be computed before starting the actual pairing computation.

Inversion Free Pairing. Computing α_Q, however, requires an inversion over $\mathbb{F}_p$ per pairing computation. While this cost is not severe, as discussed earlier, in resource constrained situations, it might be desirable to altogether avoid implementing the inversion module. For this, we express $h(ix_Q, 1/y_Q)$ as

$$h\left(ix_Q, \frac{1}{y_Q} \right) = \frac{(y_3 - 1)}{x_Q(1 - y_Qy_3)}(\beta_Q\lambda + \gamma_Q\theta + i\delta_Q) \tag{12}$$

where $\beta_Q = x_Q(y_Q+1)$, $\gamma_Q = x_Q(y_Q-1)$ and $\delta_Q = y_Q+1$. The quantities β_Q, γ_Q and δ_Q do not vary with Miller iteration and can be computed using two multiplications at the beginning of the pairing computation.

Observation 2. An important observation is that in Tate pairing computation, the final output of Miller loop is raised to the power $(p^2-1)/r$, where r does not divide $(p-1)$. So, $(p-1)$ divides $(p^2-1)/r$ and hence, in the computation of $h(Q)$ we can freely divide or multiply by a non-zero element of $\mathbb{F}_p$. This is because for any non-zero $\alpha \in \mathbb{F}_p$, $\alpha^{p-1}=1$. This technique has been used in [6] to speed up computation on Weierstraß form curve.

Since we can multiply and divide by non-zero elements of $\mathbb{F}_p$, we see that it is sufficient to evaluate

$$g(x_Q, \alpha_Q) = \beta_Q\lambda + \gamma_Q\theta + i\delta_Q. \tag{13}$$

In the following, we simplify this expression after substituting the values of λ and θ and using appropriate coordinates and then obtain explicit formulae for jointly computing P_3 and g.

Converting to Weierstraß and Computing the Pairing. The Weierstraß form of the supersingular curve that we are considering is $s^2 = r^3 + ar$. Explicit formulae for doubling-and-Miller and addition-and-Miller for this curve have been given in [11]. The coordinate system used was Jacobian and the pairing did not require any $\mathbb{F}_p$-inversion and still used mixed addition.

In contrast, if we use (3) to convert from Edwards to Weierstraß then an inversion is required. Due to the availability of the distortion map (for the Weierstraß form), we may assume that the coordinates of both P and Q in (3) are from $\mathbb{F}_p$. Then the four inversions can be done using 9[M] and 1[I] using Montgomery's trick ($s_1 = x_1$; $s_i = s_{i-1}x_{i-1}, 1 \le i \le 4$; $y_4 = s_4^{-1}$; $x_{i+1}^{-1} = y_{i+1}s_i, y_i = x_{i+1}y_{i+1}, 3 \ge i \ge 1$; this procedure generalizes to arbitrary number of x_is). The total operation (including multiplications by r_1 and s_1) count is 19[M]+1[I] for the conversion.

If we choose not to perform any inversion, then as discussed in Section 3, at the cost of some extra multiplications, we can put P in Jacobian and Q in either Jacobian or projective. As a result, the mixed addition on Weierstraß will be slower and the evaluation of each Miller function at Q will also be slower. The exact amount of slowdown for the Weierstraß form pairing due to these two factors is not clear and the entire pairing formulae for Weierstraß needs to be worked out to determine this. We do not do this; instead we work out the explicit formulae for performing inversion-free pairing directly on Edwards form. It does not appear that performing inversion-free pairing after converting to Weierstraß is likely to be faster.

In the following, by [M] *we will denote one* $\mathbb{F}_p$ *multiplication and by* [S] *we will denote one* $\mathbb{F}_p$ *squaring.*

5.1 Pairing Using Inverted Edwards Coordinates

The point (x, y) is said to be in affine representation. There are several other coordinate systems for representing a point. In [10], the inverted Edwards representation is used to represent the point (x, y) by (X, Y, Z), where $x = Z/X$ and $y = Z/Y$. The curve then transforms into $Z^4 = X^2Y^2 - Z^2(X^2 + Y^2)$. The addition and doubling formulae for the inverted Edwards representation have been given in [10].

Let $P_1 = (X_1, Y_1, Z_1)$, $P_2(X_2, Y_2, Z_2)$ and $P_3 = (X_3, Y_3, Z_3)$ such that P_3 is the sum of P_1 and P_2. It is possible to obtain unified formulae for X_3, Y_3 and Z_3, i.e., one which does not distinguish between $P_1 = P_2$ and $P_1 \neq P_2$. While this is useful for side channel resistance, a dedicated doubling formula is faster. We use the dedicated doubling formula, since in the current context the value of r (the order of the subgroup of $\mathcal{E}(\mathbb{F}_p)[r]$) is not a secret and the pairing computation will be computing rP for some point P.

Suppose that we want to compute the pairing value for P and Q. We assume that P is given as (X_1, Y_1, Z_1) with $Z_1 = 1$ and Q is given in affine as (x_Q, y_Q) so that $\phi(Q) = (ix_Q, 1/y_Q)$. As discussed above, for computing h, it is sufficient to compute g given in (13) or a product of g and some element of $\mathbb{F}_p$.

Doubling and Miller Iteration. Doubling a point and computing the Miller value are done together so that some computations can be shared. In Theorem 1, substituting the value of d to be -1 and using inverted Edwards coordinates, we obtain

$$\lambda = \frac{Z_1(Y_1^2 + Y_1Z_1 + Z_1^2)}{X_1(Y_1 - Z_1)(Y_1 + Z_1)}; \qquad \theta = \frac{(X_1^2(Y_1^2 - Z_1^2) - Z_1^2(Y_1^2 + Y_1Z_1 + Z_1^2))}{X_1(Y_1 - Z_1)^2 Z_1}.$$

At this point we need to substitute these values of λ and θ into (13) and simplify the resulting expression. During the simplification, we are free to multiply and divide by non-zero elements of $\mathbb{F}_p$ as done earlier. We have performed this simplification with the help of Mathematica [17] and the final expression for the Miller value turns out to be $\Psi = \beta_Q F + \gamma_Q G + 2i\delta_Q H$, where

$$\begin{aligned} F &= 4Z_1(Y_1 - Z_1)(Y_1^2 + Y_1Z_1 + Z_1^2) \\ G &= -4Y_1Z_1^2(Y_1 + Z_1) \\ H &= 2X_1(Y_1 + Z_1)(Y_1 - Z_1)^2. \end{aligned} \tag{14}$$

Explicit formulae for doubling using inverted Edwards coordinates have been given in [10] and requires 3[M]+4[S] operations over $\mathbb{F}_p$. This is shown in the column "doubling" in Table 2. Some of the expressions obtained during doubling can be used in the computation of Ψ. With one squaring, the value of $J = 2Y_1Z_1 = (Y_1 + Z_1)^2 - Y_1^2 - Z_1^2$ can be found. We also require $I = 2X_1Z_1 = (X_1 + Z_1)^2 - X_1^2 - Z_1^2$, which can be computed with one squaring. It may be easily seen that

$$F = (2Y_1Z_1 - 2Z_1^2)(2Y_1^2 + 2Y_1Z_1 + 2Z_1^2) = (J - 2M)(2B + J + 2M),$$

can be computed with one multiplication. The computation of $G = -J(J+2M)$ and

$$H = (2X_1Y_1 + 2X_1Z_1)(Y_1^2 - 2Y_1Z_1 + Z_1^2) = (E + I)(B - J + M)$$

Table 2. Combined explicit formula for doubling and Miller value computation using inverted Edwards coordinates. An alternative form for Ψ is $x_QF + \alpha_QG + 2iH$. Here, $\alpha_Q = x_Q(y_Q - 1)/(y_Q + 1)$, $\beta_Q = x_Q(y_Q + 1)$, $\gamma_Q = x_Q(y_Q - 1)$ and $\delta_Q = y_Q + 1$.

Doubling	Miller value
$A = X_1^2,\ B = Y_1^2,\ C = A + B,$ $D = A - B,\ E = (X_1 + Y_1)^2 - C = 2X_1Y_1,$ $M = Z_1^2,\ Z_3 = D \cdot E,\ X_3 = C \cdot D,$ $Y_3 = (C + 2Z_1^2)$	$J = (Y_1 + Z_1)^2 - B - M,$ $I = (X_1 + Z_1)^2 - A - M,$ $F = (J - 2M)(2B + J + 2M),$ $G = -J(J + 2M),$ $H = (E + I)(B - J + M),$ $\Psi = \beta_QF + \gamma_QG + 2i\delta_QH.$

require two multiplications. Finally, the computation of $\Psi = \beta_QF + \gamma_QG + 2i\delta_QH$ requires three additional multiplications. Thus, computing the Miller value requires an additional $6[M] + 2[S]$ operations and the combined doubling and Miller value computation require a total of $9[M] + 6[S]$ operations. The complete description is given in Table 2.

Mixed Addition and Miller Iteration. Explicit formula for computing the mixed addition of a point P_1 and a point P_2 (whose Z coordinate is 1) has been given in [9]. In the present case, the point P is taken to be P_2. (Recall that we are computing the pairing value of P and Q.) This is shown in the column "Mixed Addition" of Table 3. Proceeding as in the case of doubling, we need to compute $\Psi = \beta_QF + \gamma_QG + 2i\delta_QH$, where in this case,

$$\begin{aligned} F &= -X_2(1 + Y_2)(Y_1 - Z_1)Z_1 + X_1(-1 + Y_2)(Y_1 + Z_1) \\ G &= (1 + Y_2)(Y_1 + Z_1)(-X_1 + X_2Z_1) \\ H &= Z_1(-Y_1 + Y_2Z_1) \end{aligned} \tag{15}$$

The sequence of operations is the following. First, $J = Y_2Z_1$ and $K = X_2Z_1$ need two multiplications. This gives $J_1 = Y_1 - Y_2Z_1$, $J_2 = (Y_2 + 1)(Y_1 + Z_1)$, $J_3 = (Y_2 - 1)(Y_1 + Z_1)$, $J_4 = (Y_2 + 1)(Y_1 - Z_1)$ and $K_1 = X_2Z_1 - X_1$ without any other multiplications. Computation of $F = -X_2 \cdot J_4 + X_1 \cdot J_3$ requires two multiplications. Computations of $G = J_2 \cdot K_1$ and $H = -Z_1 \cdot J_1$ require one

Table 3. Combined explicit formula for mixed addition and Miller value computation using inverted Edwards coordinates. An alternative form for Ψ is $x_QF + \alpha_QG + 2iH$. Here, $\alpha_Q = x_Q(y_Q - 1)/(y_Q + 1)$, $\beta_Q = x_Q(y_Q + 1)$, $\gamma_Q = x_Q(y_Q - 1)$ and $\delta_Q = y_Q + 1$.

Mixed Addition	Miller Value
$B = -Z_1^2, C = X_1X_2, D = Y_1Y_2,$ $E = C \cdot D, H = C - D,$ $I = (X_1 + Y_1) \cdot (X_2 + Y_2) - C - D,$ $X_3 = (E + B) \cdot H, Y_3 = (E - B) \cdot I,$ $Z_3 = A \cdot H \cdot I$	$D = Y_1Y_2,\ J = Y_2Z_1,\ K = X_2Z_1,\ J = Y_1 - J,$ $J_2 = Y_1 + Z_1 + D + J,\ J_3 = D + J - Y_1 - Z_1,$ $J_4 = D - J + Y_1 - Z_1,\ K_1 = K - X_1,$ $F = -X_2J_4 + X_1J_3,\ G = J_2K_1,\ H = -Z_1J_1,$ $\Psi = \beta_QF + \gamma_QG + 2i\delta_QH.$

multiplication each. Thus, the value of Ψ can be computed with $9[M]$. Thus, mixed addition plus rational function computation requires $17[M] + 1[S]$ computations. The complete formula is given in Table 3.

5.2 Pairing Using Projective Edwards Coordinates

The affine point (x, y) on a Edwards form curve can be represented in projective coordinates as (X, Y, Z), where $x = X/Z$ and $y = Y/Z$. The curve equation then changes to $X^2 + Y^2 = Z^2 - X^2Y^2$. Explicit formulae for doubling and mixed addition using projective Edwards coordinates has been given in [9]. Equation 13 can be simplified using projective coordinates and formulae obtained for combined computation of double-and-Miller value and add-and-Miller value. The simplification process for doing this is similar to that done for inverted Edwards coordinates. Hence, we do not provide the details. Instead, we provide the final formulae in Tables 4 and 5. The total number of operations are 9[M]+6[S] and 18[M]+1[S] respectively.

Table 4. Doubling and computation of Miller value using projective Edwards coordinates. An alternative form for Ψ is $x_QF+\alpha_QG+2iH$. Here, $\alpha_Q = x_Q(y_Q-1)/(y_Q+1)$, $\beta_Q = x_Q(y_Q+1)$, $\gamma_Q = x_Q(y_Q-1)$ and $\delta_Q = y_Q+1$.

Doubling	Miller Value
$B = (X_1+Y_1)^2, C = X_1^2, D = Y_1^2,$ $E = C + D, M = Z_1^2, J = E - 2M,$ $X_3 = (B-E)J,\ Y_3 = E(C-D),\ Z_3 = EJ$	$B = (X_1+Y_1)^2,\ C = X_1^2,\ D = Y_1^2,$ $L = 2X_1Z_1,\ K = 2Y_1Z_1,\ Z_1^2,$ $F = (L - B + C + D)(2D + K + 2M),$ $G = -K \cdot (L + B - C - D),$ $H = (2M + K)(M + D - K),$ $\Psi = \beta_QF + \gamma_QG + 2i\delta_QH.$

Table 5. Mixed addition and computation of Miller value using projective Edwards coordinates. An alternative form for Ψ is $x_QF + \alpha_QG + 2iH$. Here, $\alpha_Q = x_Q(y_Q - 1)/(y_Q+1)$, $\beta_Q = x_Q(y_Q+1)$, $\gamma_Q = x_Q(y_Q-1)$ and $\delta_Q = y_Q+1$.

Mixed Addition	Miller Value
$B = Z_1^2, C = X_1X_2, D = Y_1Y_2, E = -CD,$ $I = B - E, J = B + E,$ $X_3 = Z_1I((X_1+Y_1)(X_2+Y_2) - C - D),$ $Y_3 = Z_1J(D-C),\ Z_3 = IJ$	$C = X_1X_2,\ K = Y_2Z_1,\ L = X_2Z_1,$ $D = Y_1Y_2,\ L_1 = X_1 - L,\ K_1 = K - Y_1,$ $K_2 = D + K + Y_1 + Y_2,$ $K_3 = D + K - Y_1 - Z_1,$ $K_4 = D - K + Y_1 - Z_1,$ $F = -X_1K_4 + LK_3,\ G = -K_2L_1,$ $H = CK_1,\ \Psi = \beta_QF + \gamma_QG + 2i\delta_QH.$

6 Concluding Remarks

In this work, we have studied pairing algorithms on Edwards form elliptic curves. A general form for the function required in a Miller iteration has been obtained. For a class of supersingular curves over fields of characteristic greater than 3, the expression for the Miller function has been simplified and explicit formulae obtained for combined doubling and Miller iteration and combined addition and Miller iteration using both inverted Edwards and projective Edwards coordinates.

Acknowledgements

The authors wish to thank Prof. Tanja Lange for her extensive comments on an earlier version of this work. The authors also wish to the thank the reviewers for their comments. The fist author is supported by Ministry of Information Technology, Govt. of India.

References

1. Joux, A.: A one round protocol for tripartite Diffie-Hellman. J. Cryptology 17(4), 263–276 (2004)
2. Boneh, D., Franklin, M.K.: Identity-based encryption from the Weil pairing. SIAM J. Comput. 32(3), 586–615 (2003)
3. Freeman, D., Scott, M., Teske, E.: A taxonomy of pairing-friendly elliptic curves. Cryptology ePrint Archive, Report 2006/372 (2006), `http://eprint.iacr.org/`
4. Frey, G., Rück, H.G.: A remark concerning m-divisibility and the discrete logarithm in the divisor class group of curves. Mathematics of Computation 62, 865–874 (1994)
5. Miller, V.S.: The Weil pairing and its efficient calculation. J. Cryptology 17(4), 235–261 (2004)
6. Barreto, P.S.L.M., Kim, H.Y., Lynn, B., Scott, M.: Efficient algorithms for pairing-based cryptosystems. In: Yung, M. (ed.) CRYPTO 2002. LNCS, vol. 2442, pp. 354–369. Springer, Heidelberg (2002)
7. Galbraith, S.D., Harrison, K., Soldera, D.: Implementing the Tate pairing. In: Fieker, C., Kohel, D.R. (eds.) ANTS 2002. LNCS, vol. 2369, pp. 324–337. Springer, Heidelberg (2002)
8. Edwards, H.M.: A normal form for elliptic curves. Bulletin of the American Mathematical Society 44, 393–422 (2007)
9. Bernstein, D.J., Lange, T.: Faster addition and doubling on elliptic curves. In: Kurosawa, K. (ed.) ASIACRYPT 2007. LNCS, vol. 4833, pp. 29–50. Springer, Heidelberg (2007)
10. Bernstein, D.J., Lange, T.: Inverted Edwards coordinates. In: Boztas, S., Lu, H.F. (eds.) AAECC 2007. LNCS, vol. 4851, pp. 20–27. Springer, Heidelberg (2007)
11. Chatterjee, S., Sarkar, P., Barua, R.: Efficient computation of Tate pairing in projective coordinate over general characteristic fields. In: Park, C.-s., Chee, S. (eds.) ICISC 2004. LNCS, vol. 3506, pp. 168–181. Springer, Heidelberg (2005)
12. Bernstein, D.J., Birkner, P., Lange, T., Peters, C.: Twisted Edwards curves. Cryptology ePrint Archive, Report 2008/013 (2008) `http://eprint.iacr.org/` (Accepted in AFRICACRYPT 2008)
13. Euler, L.: Observationes de comparatione arcuum curvarum irrectificabilium. Novi Comm. Acad. Sci. Petropolitanae 6(1761), 58–84

14. Gauss, C.F.: Werke 3, 404
15. Koblitz, N., Menezes, A.: Pairing-based cryptography at high security levels. In: Smart, N. (ed.) Cryptography and Coding 2005. LNCS, vol. 3796, pp. 13–36. Springer, Heidelberg (2005)
16. Verheul, E.R.: Evidence that XTR is more secure than supersingular elliptic curve cryptosystems. Journal of Cryptology 17, 277–296 (2004)
17. Wolfram, S.: The Mathematica Book, 5th edn. Wolfram Media (2003), `http://www.wolfram.com`

Exponentiation in Pairing-Friendly Groups Using Homomorphisms

Steven D. Galbraith* and Michael Scott**

[1] Mathematics Department,
Royal Holloway, University of London,
Egham, Surrey, TW20 0EX,
United Kingdom
steven.galbraith@rhul.ac.uk
[2] School of Computing, Dublin City University,
Ballymun, Dublin 9, Ireland
mike@computing.dcu.ie

Abstract. We present efficiently computable homomorphisms of the groups G_2 and G_T for pairings $G_1 \times G_2 \to G_T$. This allows exponentiation in G_2 and G_T to be accelerated using the Gallant-Lambert-Vanstone method.

Keywords: pairings.

1 Introduction

Let r be a prime and let G_1, G_2 and G_T be cyclic groups of order r with a bilinear pairing

$$e : G_1 \times G_2 \to G_T.$$

In practice G_1 is a set of points on some elliptic curve E over $\mathbb{F}_p$ and G_2 is a set of points on a twist E' of E over some field $\mathbb{F}_{p^e}$. The group G_T is a subgroup of $\mathbb{F}_{p^k}^*$, where k is the embedding degree, and is usually represented in a compressed form by using traces or algebraic tori.

Pairings over ordinary elliptic curves suffer in comparison to those over supersingular curves, in that a larger group G_2 is often required for one of the two parameters to the pairing. The quadratic twist is always an option if k is even, so $e = k/2$ and for the case $k = 2$ the quadratic twist is again over the base field. There is a family of pairing-friendly curves [10] of embedding degree $k = 6$ where the sextic twist applies, and again in this case $e = k/6 = 1$. However for most other cases of interest $e > 1$. For example with the BN curves [5], even though the sextic twist applies, G_2 is over the field $\mathbb{F}_{p^2}$. This suggests that manipulations of points over G_2 in some pairing-based protocols are in general likely to

* This work supported by EPSRC grant EP/D069904/1.
** This author acknowledges support from the Science Foundation Ireland under Grant No. 06/MI/006.

S.D. Galbraith and K.G. Paterson (Eds.): Pairing 2008, LNCS 5209, pp. 211–224, 2008.

be more expensive than those over G_1, and perhaps much more expensive. Here we will demonstrate that this is not necessarily the case.

Gallant, Lambert and Vanstone (GLV) [12] gave a method to speed up operations in groups when a suitable group homomorphism is available. The main result of the paper is to get such a group homomorphism from the Frobenius map in $\mathbb{F}_{p^k}$. This particularly speeds up operations in G_2, but also has implications for G_T.

The main contributions of our paper are:

1. To speed up arithmetic in G_2 and G_T using the GLV method.
2. To show that simpler GLV decompositions of an exponent are often possible for pairing friendly curves (i.e., not requiring lattice reduction as a precomputation), especially for Ate friendly curves.
3. To remark that parameters for Ate-friendly curves give rise to good parameters for XTR and torus based cryptography.
4. To note that our methods can be used to obtain larger equivalence classes for the Pollard rho method.

We now outline the paper. Sections 2 and 3 recall basic facts about pairings and the GLV method. Section 4 analyses the methods of Stam and Lenstra when applied in the target group G_T for pairing-based cryptography. Section 5 contains our main result, namely the construction of a group homomorphism on G_2. Section 6 studies some specific examples. Section 7 summarises the costs and benefits of the GLV method. Sections 8 and 9 mention some consequences for trace/torus cryptography and the difficulty of the DLP in G_2, and we conclude in Section 10.

2 Elliptic Curves and Pairings

Let E be an elliptic curve over $\mathbb{F}_p$ where p is prime. Denote by ∞ the point at infinity on E. Let $\#E(\mathbb{F}_p) = p + 1 - t$ be the number of points on the curve, where t is the trace of the Frobenius. Let $r \mid \#E(\mathbb{F}_p)$ be a large prime. The embedding degree is the smallest integer k such that $r \mid (p^k - 1)$. We assume that no proper subfield of $\mathbb{F}_{p^k}^*$ contains elements of order r.

Let $G_1 = E(\mathbb{F}_p)[r]$ and let G_T be the subgroup of $\mathbb{F}_{p^k}^*$ of elements of order r. Denote by π_p the p-power Frobenius map on E. Define G_2 to be the subgroup of $E(\mathbb{F}_{p^k})[r]$ such that π_p acts as multiplication by p. We assume we have a non-degenerate bilinear pairing (such as the Ate pairing [15])

$$e : G_1 \times G_2 \to G_T.$$

Following Section 4 of [15] we represent G_2 as a group of points on a twist E' of E. This means there is an isomorphism $\phi : E' \to E$ with field of definition $\mathbb{F}_{p^d}$. It is necessary that the automorphism group $\mathrm{Aut}(E)$ contain an element of order d. Hence the only non-trivial possibilities are $d = 2$, $d = 4$ if $j(E) = 1728$ (CM discriminant $D = -4$) and $d = 3, 6$ if $j(E) = 0$ (CM discriminant $D = -3$).

We assume $d \mid k$ and write $k = de$. Then $G_2 = E'(\mathbb{F}_{p^e})[r]$ and $\phi(G_2) \subset E(\mathbb{F}_{p^k})$. If $r > d$ then the image of $E'(\mathbb{F}_{p^e})[r]$ under ϕ does lie in the eigenspace of the q-power Frobenius on $E(\mathbb{F}_{p^k})$ with eigenvalue p.

For efficient pairing computation, much work has been done to find viable bilinear pairings, with the minimum number of iterations in Miller's algorithm. Starting with the Duursma-Lee method [9] and subsequent work by Barreto et al. [4] (in the context of supersingular curves), Hess al. extended the idea to ordinary elliptic curves with the discovery of the Ate pairing. Now the main Miller loop in the pairing computation iterates only $\lg(|t-1|)$ times, rather than $\lg(r)$ times as required by the Tate pairing. An "Ate pairing friendly curve" is defined as one where $|t-1|$ is as small as possible compared to r. It has been conjectured that the minimum possible ratio between $|t-1|$ and r is $1/\varphi(r)$ (where φ is the Euler totient function), and indeed this ideal condition is met by some pairing-friendly families of curves. Recently Lee, Lee and Park [18], Hess [16] and Vercauteren [28] have shown how to achieve the same level of loop truncation on curves, even if they are not Ate pairing friendly.

Many families of pairing-friendly curves have been found - see [10] for a survey. The most sought after curves are those with the minimum value of ρ, which is defined as the rounded fraction $\lg(p)/\lg r$. It is relatively easy to find families of curves with $\rho \approx 2$, but it is much preferred that $\rho \approx 1$, as this leads to more efficient implementations.

3 The GLV Method

Gallant, Lambert and Vanstone [12] introduced a method to speed up general point multiplication nP in $E(\mathbb{F}_p)[r]$. In its simplest form their method works if, given a point P, one can somehow have knowledge of a non-trivial multiple of P. This extra information is available if there is an efficiently computable endomorphism ψ on E defined over $\mathbb{F}_p$ such that $\psi(P) = \lambda P$. One can then compute nP efficiently by writing $n \equiv n_0 + n_1\lambda \pmod{r}$ with $|n_i| < \sqrt{r}$ and performing the double exponentiation $n_0P + n_1\psi(P)$. Decomposing n as $n_0 + n_1\lambda \pmod{r}$ is done by solving a closest vector problem in a lattice and the Euclidean algorithm can be used to compute a suitable lattice basis, see [12,23] for the details. We call this the GLV method.

Double exponentiation algorithms require precomputation and storage, but their efficiency comes from halving the number of doublings. One can simultaneously reduce the number of additions by using window methods, but this adds further precomputation and storage. Another method to reduce the number of additions is to allow signed representations for n_0 and n_1 and compute their joint sparse form (that is such that the signed expansions of n_0 and n_1 both have i-th bit equal to 0 with probability approximately $1/2$). We refer to Section 9.1.5 of [1] for further details.

The idea generalises to m-dimensional expansions $n \equiv n_0 + n_1\lambda + \cdots + n_{m-1}\lambda^{m-1} \pmod{r}$ assuming that the powers of λ are sufficiently different modulo r (the typical requirement is that the endomorphism ψ satisfies a characteristic polynomial of degree $\geq m$; see the discussion below). We call this the m-dimensional GLV method.

The task of decomposing n is again solving a closest vector problem in a lattice. This problem can be efficiently solved using Babai's rounding method [2] if an LLL-reduced lattice basis is precomputed. More precisely, define the modular lattice

$$L = \left\{ x \in \mathbb{Z}^m : \sum_{i=0}^{m-1} x_i \lambda^i \equiv 0 \pmod{r} \right\}. \tag{1}$$

The $2m$ vectors $(0, \ldots, 0, r, 0, \ldots, 0)$ and $(0, \ldots, 0, \lambda, -1, 0, \ldots, 0)$ generate the (row) lattice L if $\gcd(\lambda, r) = 1$. Run LLL on this basis to obtain a new basis. Given an exponent n use the Babai rounding technique to find a lattice vector $x = (x_0, \ldots, x_{m-1})$ close to $w = (n, 0, \ldots, 0)$. Define $u = w - x$. Then $\sum_{i=0}^{m-1} u_i \lambda^i \equiv n \pmod{r}$ by definition. If the LLL-reduced basis is sufficiently good then the coefficients u_i will be such that $|u_i| \approx r^{1/m}$. The practical performance of this approach depends on the particular parameters under consideration.

We stress that the lattice reduction is a pre-computation; the online cost in point multiplication is just the Babai rounding step. An alternative approach (when a random multiple of a point P is required) is to simply choose random coefficients $n_0, \ldots, n_{m-1}$ instead of choosing n first and then decomposing it.

We remark that there are natural boundaries on the size of m. For example, let $r \mid (p^2 - p + 1)$ and let ψ be the p-power Frobenius map in the subgroup G_T of $\mathbb{F}_{p^6}^*$ of order r. Then $\lambda \equiv p \pmod{r}$ satisfies $\lambda^6 \equiv 1 \pmod{r}$ and one might expect to be able to take $m = 6$. However, since $\lambda^2 \equiv \lambda - 1 \pmod{r}$ it follows that $n_0 + n_1\lambda + n_2\lambda^2 \equiv (n_0 - n_2) + (n_1 + n_2)\lambda \pmod{r}$. Therefore the size of the largest coefficient n_i in the 3-dimensional expansion cannot be significantly smaller than the size of the largest coefficient in the 2-dimensional case.

The original proposal of Gallant, Lambert and Vanstone specifically proposed using the automorphisms of elliptic curves E with $j(E) = 0, 1728$. Hence it is standard that the GLV method can be used to speed up point multiplication in G_1 and G_2 in the cases for which using twists gives good compression of G_2. In both cases the automorphisms satisfy a characteristic polynomial of degree 2 with coefficients in $\{0, 1\}$, so only the two-dimensional GLV method applies.

4 Using the Frobenius to Speed Up Operations in G_T

In this case much of the work has already been done by Stam and Lenstra. However here we consider their results in the context of pairings.

We call the subgroup of $\mathbb{F}_{p^k}^*$ of order $\Phi_k(p)$ (where $\Phi_k(x)$ is the k-th cyclotomic polynomial) the "cyclotomic group". The group G_T of order r is a subgroup of the cyclotomic group in $\mathbb{F}_{p^k}^*$. For the case $k = 6$, $r \mid (p^2 - p + 1)$ and G_T is a subgroup of the well studied "XTR subgroup" of $\mathbb{F}_{p^6}^*$. For the case $k = 2$, the cyclotomic group is of order $p + 1$, and was used in the LUC cryptosystem (see Stam and Lenstra [25]).

There are three approaches for efficient arithmetic in cyclotomic subgroups. The simplest approach is to perform arithmetic using a standard representation for $\mathbb{F}_{p^k}$ and to exploit tricks which arise from elements having order dividing $\Phi_k(p)$ (for example, the fact that the inverse of an element can be computed efficiently). The other approaches are based on compression of field elements using traces or algebraic tori respectively. All three methods can be applied for efficient exponentiation in G_T (for example see [13]). The latter two methods are also useful for minimising bandwidth in pairing-based cryptography.

Stam and Lenstra [26] discuss the first approach. They exploit the fact that elements in the cylotomic group have some extra properties that do not hold for general elements in $\mathbb{F}_{p^k}$. Specifically field inversion is a simple conjugation, and thus effectively free, and the field squaring operation can be significantly cheaper. Also as inversion is free, faster NAF methods of windowing are applicable [22].

For exponentiation in the XTR subgroup the most efficient method is to use traces. For XTR the trace is over $\mathbb{F}_{p^2}$, so the compression is by a factor of 3. For LUC the trace is over $\mathbb{F}_p$, and the compression is by a factor of two. However traces can only be manipulated in limited ways: for example multiplication of subgroup elements, if required by a protocol, is non-trivial. When using compression by a factor of 2 then exponentiating using a torus representation is competitive with LUC [13]. One advantage of tori is that one can efficiently multiply group elements as well as exponentiate them. In [8] the applications of higher dimensional tori are considered, and indeed it is suggested that in principle a degree 8 Frobenius automorphism can be used to split the exponent, and then use multi-exponentiation, in much the same way as suggested here.

In [25] a method for double exponentiation using traces is proposed, for both the LUC and XTR cases. This is required for example for the application of LUC/XTR to ElGamal-like digital signature verification schemes. But the authors also point out that the Frobenius endomorphism can be used to implement a single exponentiation using a variant of the GLV idea (independently discovered) with their double exponentiation algorithm, and indeed this is the fastest way to do it. In Section 4.4 of [25] it is pointed out that if $p \bmod r \approx \sqrt{r}$ then the 2-dimensional decomposition of the exponent is particularly easy, and the decomposition can be found at the cost a division and a remainder. In the sequel we will refer to such a decomposition as "natural". As we will see, in the context of pairings, natural decompositions arise quite frequently.

It is apparently non-trivial to extend the double exponentiation of traces to general multi-exponentiation [25], and so if multi-exponentiation is possibly beneficial then we must either use torus methods or else work in the full $\mathbb{F}_{p^k}^*$ (see Stam and Lenstra [26]).

Pairings evaluate as elements in G_T, often in higher degree cyclotomic fields than those considered by Stam and Lenstra. Many of the same ideas apply immediately if the embedding degree is a multiple of 2 or 6. However in the context of pairings, since we know that the pair (p, r) arise in the context of an elliptic curve, we know that $p \bmod r = t - 1$. Fortunately for us, for many pairing friendly curves $|t-1|$ is often rather small compared with r, in which case higher dimensional natural decompositions will also be possible. Application of the Frobenius to an element x of order r gives us the value $x^p \equiv x^{t-1}$, so the exponent n can be expressed to the base $(t-1)$ and multi-exponentiation applied as $x^{n_0}.(x^p)^{n_1}.(x^{p^2})^{n_3}\cdots$. See the examples below for more details.

5 A Homomorphism on G_2

As described above, the group G_2 is a subgroup of $E'(\mathbb{F}_{p^e})$ and there is a group homomorphism $\phi : E'(\mathbb{F}_{p^e}) \to E(\mathbb{F}_{p^k})$. We now explain how to use the p-power Frobenius on $E(\mathbb{F}_{p^k})$ to get an efficiently computed group homomorphism on G_2. Iijima et al [17] used essentially the same ideas to construct a homomorphism for a different application.

Lemma 1. *Let notation be as above. Denote by π_p the p-power Frobenius map on E. Then $\psi = \phi^{-1}\pi_p\phi$ is an endomorphism of E' such that $\psi : G_2 \to G_2$. Further, for $Q \in G_2$ we have $\psi^k(Q) = Q$, $\psi(Q) = pQ$ and $\Phi_k(\psi)(Q) = \infty$ where $\Phi_k(x)$ is the k-th cyclotomic polynomial.*

Proof. Clearly ψ is a morphism from E' to E' which fixes the point at infinity. Hence ψ is an endomorphism of E'.

Let $Q \in E'(\mathbb{F}_{p^e})[r]$. Then $\phi(Q) \in E(\mathbb{F}_{p^k})$ and, as mentioned in Section 2, we have $\pi_p(\phi(Q)) = p\phi(Q)$. Hence $Q' = \pi_p(\phi(Q))$ lies in the image of $E'(\mathbb{F}_{p^e})$ under ϕ and so $Q'' = \phi^{-1}(Q') \in E'(\mathbb{F}_{p^e})$.

Clearly $\psi^k = \phi^{-1}\pi_p^k\phi = \phi^{-1}\pi_{p^k}\phi$. Since $\pi_p^k = 1$ on $E(\mathbb{F}_{p^k})$ it follows that $\psi^k(Q) = Q$. Further, as noted above, $\pi_p(\phi(Q)) = p\phi(Q)$ and so

$$\psi(Q) = \phi^{-1}\pi_p\phi(Q) = \phi^{-1}p\phi(Q) = pQ.$$

Finally, since Q has order r and $r \mid \Phi_k(p)$ it follows that $\Phi_k(\psi)(Q) = \Phi_k(p)Q = \infty$. This completes the proof. □

The group homomorphism ψ can be computed efficiently and so is potentially useful for the GLV method. However, there are cases when this map is just a familiar homomorphism arising in an unfamiliar way. Our main interest is when the construction gives something which was not previously used for efficient computation. The following result shows that if $e = 1$ then we are just recovering elements of the automorphism group of the curve.

Lemma 2. *If $e = 1$ then ψ is equal to $\rho\pi'_p$ where π'_p is the p-power Frobenius on E' and where ρ is an element of Aut(E').*

Proof. By Corollary 2.12 of [24] ψ can be written as $\rho\pi_p$ where $\pi_p : E' \to E'^{(p)}$ is the p-power Frobenius to a Galois conjugate of E' and $\rho : E'^{(p)} \to E'$ is an isomorphism. In the case $e = 1$ we have $E' = E'^{(p)}$ and so $\rho \in \mathrm{Aut}(E')$. □

This result shows that our methods give no new result in the case $e = 1$ (although decomposition of a random exponent is always simpler than the general case of

GLV). The case $e > 1$ is interesting as it gives potential for new and improved applications of the GLV method. In particular, we have homomorphisms which do not come from $\mathrm{Aut}(E')$.

We mention that a similar optimisation for G_1 was proposed by Granger, Page and Stam in Section 4 of [13]. They considered a supersingular elliptic curve $E(\mathbb{F}_{3^m})$ and used the fact that multiplication by 3^m on E is given by a simple and easy to compute formula. Since $3^m \equiv \pm 3^{(m+1)/2} - 1 \pmod{r}$ they remarked that it is easy to obtain a GLV decomposition in this case.

6 Examples

Pairing friendly families vary significantly in detail, so the benefits of our methods are best considered on a case-by-case basis. The first two examples correspond to the case $e = 1$ and, as explained earlier, our methods give nothing new in this case. However, it is useful to demonstrate how simple the GLV decomposition is in these cases.

Example 1. Consider the pairing-friendly family of $k = 6, \rho = 2$ curves (see Section 6.7 of [10]), with $D = -3$ and $j(E) = 0$

$$p = 27x^4 + 9x^3 + 3x^2 + 3x + 1 \qquad r = 9x^2 + 3x + 1 \qquad t = 3x + 2$$

One can construct an elliptic curve $E : Y^2 = X^3 + B$ over $\mathbb{F}_p$ having $r \mid \#E(\mathbb{F}_p)$. The embedding degree is 6 and one can identify G_2 with $E'(\mathbb{F}_p)$ where E' is the sextic twist of E defined over $\mathbb{F}_p$ (in other words, $e = 1$).

Since $j(E) = 0$ the standard GLV method applies immediately to G_1. However observe that r is of the form $\lambda^2 + \lambda + 1$, with $\lambda = 3x$. Therefore the standard automorphism $\rho(x, y) = (\zeta_3 x, y)$ applied to a point $P = (x, y)$, gives us the point λP, and presents us with a natural 2-dimensional decomposition of a point multiplier into its quotient and remainder modulo $3x$.

Now consider the homomorphism ψ of Lemma 1. For $Q \in G_2$ we have $\psi(Q) = TQ$ where $T = t-1 = 3x+1$. Hence $\psi = \rho+1$, which can naturally be interpreted as $-\rho^2$. The point multiplication by $n < r$ can be written as $n_0 Q + n_1 \psi(Q)$ by taking the base T representation of n.

Exponentiation in G_T can use the fast trace methods of [25]. However the decomposition is again simple to obtain, as $p \bmod r = 3x + 1 \approx \sqrt{r}$. Note that the fast squaring operations of [25,26] do not apply since $p \not\equiv 2 \pmod{3}$, but one can still obtain very efficient field arithmetic in this case.

Example 2. Miyaji, Nakabayashi and Takano [21] gave parameters for curves of prime order r over $\mathbb{F}_p$ with embedding degree 6. These curves are ideal, in the sense that $\rho = 1$.

$$p = x^2 + 1 \qquad r = x^2 - x + 1 \qquad t = x + 1$$

One major drawback of the MNT method is the necessity of solving Pell equations to generate the curves. Furthermore, certain CM discriminants cannot be used. Indeed, it is not possible to generate a suitable curve with $j(E) = 0$. In the more general setting we have $\mathrm{Aut}(E) = \{1, -1\}$ and the GLV method cannot usefully be applied. The best representation for G_2 is then as a subgroup of $E'(\mathbb{F}_{p^3})$ where E' is a quadratic twist of E which is defined over $\mathbb{F}_p$.

In this case nothing can be done for G_1, but E' is now a "subfield curve" so it is natural to use the Frobenius map π'_p on E' to speed up arithmetic on $E'(\mathbb{F}_{p^3})$. For the subgroup of relevance π_p satisfies $\pi_p'^2 + \pi'_p + 1 = 0$ and so a 2-dimensional GLV method is the best on can hope for.

As with the previous example, our approach gives the same performance with simpler decomposition of the exponents. The group homomorphism ψ on G_2 defined above satisfies $\psi^2 - \psi + 1 = 0$ and acts as multiplication by $t - 1$.

Example 3. Consider this family of Ate pairing-friendly curves [3], with $k = 12, D = -3, \rho = 3/2$.

$$p = (x^6 - 2x^5 + 2x^3 + x + 1)/3 \qquad r = x^4 - x^2 + 1 \qquad t = x + 1$$

In this case standard GLV applies to G_1, and again a natural 2-dimensional decomposition is possible with the standard automorphism, given the special form of r. The group G_2 is a subgroup of the sextic twist $E'(\mathbb{F}_{p^2})$. Since $j(E') = 0$ we could use the standard GLV method, but in this case $e = 2$ so it is possible to do better. In this case for G_2 and G_T we get a natural 4-dimensional decomposition, as any multiplier in G_2 or exponent in G_T can be written as a degree 4 polynomial in $T = t - 1 = x$. For G_T trace methods are probably not practical for a degree 4 multi-exponentiation, so fast non-trace based methods should be used here instead.

Example 4. Consider this family of Ate pairing-friendly curves [3], with $k = 24, D = -3, \rho = 5/4$. This curve might be appropriate at the highest levels of security.

$$p = (x^{10} - 2x^9 + x^8 - x^6 + 2x^5 - x^4 + x^2 + x + 1)/3 \quad r = x^8 - x^4 + 1 \quad t = x + 1$$

As before standard GLV applies to G_1, again with a natural 2-dimensional decomposition. G_2 is a subgroup of the sextic twist $E'(\mathbb{F}_{p^4})$. In this case for G_2 and G_T we get a natural 8-dimensional decomposition, as any multiplier in G_2 or exponent in G_T can be written as a degree 8 polynomial in $T = t - 1 = x$. Again for G_T fast non-trace-based methods should be used.

Example 5. (BN curves [5]) Consider the BN parameters

$$t = 6x^2 + 1, \qquad p = 36x^4 + 36x^3 + 24x^2 + 6x + 1, \qquad r = p + 1 - t.$$

One can construct an elliptic curve $E : Y^2 = X^3 + a$ over $\mathbb{F}_p$ having r points. The embedding degree is 12 and one can identify G_2 with a subgroup of $E'(\mathbb{F}_{p^2})$ where E' is a twist of E defined over $\mathbb{F}_{p^2}$.

Taking $\phi^{-1}\pi_p^2\phi$ gives the usual automorphism $\zeta_6(x, y) = (\zeta_3 x, -y)$ which satisfies the characteristic polynomial $\zeta_6^2 - \zeta_6 + 1 = 0$. It is standard that the GLV method using this automorphism speeds up point multiplication on E'.

Now consider $\psi = \phi^{-1}\pi_p\phi$, which satisfies $\psi^4 - \psi^2 + 1 = 0$ and so behaves as ζ_{12}. Note that $\mathrm{Aut}(E')$ does not contain an element of order 12. Since ψ acts as multiplication by p and $p \equiv (t-1) \pmod r$ one can naturally decompose n as $n_0 + n_1(t-1)$ such that $|n_0| < |t-1|$ and n_1 is a similar size. Hence one gets the 2-dimensional GLV method with natural decomposition.

Unlike the previous two examples, $|t-1| \not\approx r^{1/m}$ and so obtaining the GLV expansion is not as simple as writing the exponent n in base $(t-1)$. In this case it is necessary to use lattice reduction. Let x be the parameter in the BN polynomial family. Then a reduced basis for the lattice L of equation (1) with $\lambda = T = 6x^2$ is

$$B = \begin{pmatrix} x+1 & x & x & -2x \\ 2x+1 & -x & -(x+1) & -x \\ 2x & 2x+1 & 2x+1 & 2x+1 \\ x-1 & 4x+2 & -(2x-1) & x-1 \end{pmatrix}.$$

The determinant of B is $-3r(x)$.

To decompose an integer n one needs to find a vector x close to $w = (n, 0, 0, 0)$ in the lattice L. One first computes a vector $v \approx wB^{-1}$. As pointed out to us by Barreto, for the above choice of B one has

$$wB^{-1} = \left(\frac{n(2x^2+3x+1)}{r}, \frac{n(12x^3+8x^2+x)}{r}, \frac{n(6x^3+4x^2+x)}{r}, \frac{n(-2x^2-x)}{r} \right)$$

and so computing v can be done using integer multiplication and division by r. One then computes the vector $u = w - vB$ whose entries are the coefficients n_i for the decomposition of n.

We illustrate the method with a toy example. Let $x = 10267$ and choose the "random" exponent $n = 123456789123456789$. The first step is to decompose the vector $(n, 0, 0, 0)$ with respect to the basis formed by the rows of B. This gives

$$\begin{aligned}(n,0,0,0)B^{-1} = (&26031281270628101244596820/r, 160344884510280497561435613 2115/r, \\ &801724423185167914772443492389/r, -26028746085463451059434705/r)\end{aligned}$$

Rounding these coefficients to the nearest integer gives a vector v such that vB is a close vector in the lattice to $(n, 0, 0, 0)$. Finally, compute

$$u = (n, 0, 0, 0) - vB = (-11418, -5569, -4753, -8683)$$

and one can check that $n \equiv \sum_{i=0}^{3} u_i T^i \pmod r$ as required. Note that all the entries in the vector u satisfy $|u_i| < r^{1/4}$. Experiments with 64-bit x (i.e., 258-bit p) always had coefficients u_i satisfying $|u_i| < 2^{65}$ as desired.

Example 6. Pairing friendly elliptic curves with $k = 9$ were considered by [19]. Since $\varphi(9) = 6$ one would get a 6-dimensional GLV method in this case.

7 Multi-exponentiation

As high-dimensional exponent decompositions are now possible, it is a useful exercise to see just how much improvement can be expected from using them. Here we follow the analysis and methods of Möller [22]. In particular we consider the wNAF-based interleaving windowed exponentiation method, which applies both for G_2 and for G_T. NAF methods apply when inversion is easy in the group. It is well known that inversion is easy for points on an elliptic curve, but perhaps not as well known that this also applies to elements in G_T. Indeed as part of the final exponentiation of the pairing, there is a component in that exponentiation of $p^{k/2}-1$. After this exponentiation elements become "unitary" (i.e., norm 1), and with this property inversion becomes a simple conjugation, and field squaring becomes significantly cheaper [26].

We stress that we are considering exponentiation for a variable base. Hence our estimates and timings include the cost of any "precomputation" required. If the base in exponentiation is fixed then there are all sorts of different optimisations based on precomputation which can be adopted.

Here for simplicity, we do not further consider trace-based methods, as they are limited by the extent to which they can exploit multi-exponentiation. But we will of course exploit the "unitary" property of elements in G_T.

When estimating the cost of multi-exponentiation, it is important to estimate the relative costs of field multiplication and squaring in G_T, and of point doubling and addition in G_2. So we make the assumption that a point addition/field multiplication is c times the cost of a point doubling/field squaring, where we will keep c as a variable.

In fact the relative costs of these operations for an elliptic curve over a prime field is the subject of much debate, and improved formulae for both doubling and addition are still being found, often using novel coordinate systems [6]. On the other hand, for curves over larger extension fields the subject has not received much attention. Indeed it seems likely that affine coordinates may be faster than projective coordinates for higher extensions. In G_T the matter is also not so clear cut - but the fast methods for field squaring of unitary elements [26] are certainly relevant. (Even just exploiting their quadratic extension formulae leads to significant improvements when applied over large even extension fields; see [14].)

Assuming that the same window size is used for all exponents, the cost of multi-exponentiation [22] is approximated by

$$(mc(2^{w-1}-1+b/(w+2))+b)$$

point doublings/field squarings for an m-dimensional decomposition, using a window size of w, and exponents of constant size b bits. Here w is simply chosen to minimise this cost – we ignore the space required for the precomputation. Clearly we have a choice as to the extent to exploit the possible decomposition, so we might double m (which will halve the size of b) to see how this effects the cost.

Table 1. Cost of multi-exponentiation (Optimal w in brackets)

m	c=1.0	c=1.33	c=1.66	c=2.0	c=3.0
1	306 (4)	322 (4)	338 (4)	355 (4)	405 (4)
2	185 (4)	203 (4)	222 (4)	241 (4)	298 (4)
4	127 (3)	148 (3)	169 (3)	190 (3)	254 (3)

Our estimates (based on the above formula) are given in Table 1, for a group whose order r is 256-bits, assuming that a 1, 2, or 4 dimensional decomposition is possible (as is the case for the BN curve). We conclude that it is beneficial to decompose to the maximum extent possible, assuming that space for precomputation is not an issue.

8 Hashing to G_2

Some pairing based protocols, for example the original Boneh and Franklin IBE scheme [7], require hashing of identities to G_1 or G_2. In the latter case this might be considered inefficient, as a large co-factor multiplication would be required. For example consider hashing an identity to the group $G_2 \subset E'(\mathbb{F}_{p^2})$ on a BN curve. The number of points on $E'(\mathbb{F}_{p^2})$ is $r(p-1+t)$ (see [5]) where t is the trace of Frobenius of $E(\mathbb{F}_p)$. To hash-and-map an identity to a point of order r, the approach might be to hash the identity to an x coordinate, solve the quadratic curve equation to find a y coordinate (and iterate on x if one should not exist), and finally multiply this point by the co-factor $p-1+t$.

However, in this case the homomorphism ψ of Lemma 1 can be exploited to advantage. As we have seen, ψ satisfies the equation

$$\psi^2(P) - [t]\psi(P) + [p]P = 0$$

for $P \in E'(\mathbb{F}_{p^2})$. Therefore by simple substitution

$$[p-1+t]P = [t](\pi(P) + P) - \pi^2(P) - P.$$

The major cost of the cofactor calculation is therefore a multiplication by t, which is "half-sized" compared to a full multiplication by $p-1+t$.

9 Application to XTR and Torus-Based Cryptography

As mentioned in Section 4.7 of Stam's thesis [27], a natural problem is to develop the XTR cryptosystem in $\mathbb{F}_{p^{6m}}$. The main obstacle is efficient key generation. A key generation algorithm was given in [20] but it requires factoring integers so is not very practical for large security levels.

A fact (which does not seem to have been noted before) is that polynomial families of parameters for pairing-friendly curves give efficient key generation algorithms for XTR or torus based cryptography over extension fields. Once such parameters are available then one can immediately apply the GLV method to speed up exponentiation (see [13,25,26]).

Furthermore, if one works in a subgroup of order r where $r = p - T$ is "super Ate friendly" then one can also benefit from the easy decomposition of exponents using the base-$|T|$ expansion and hence get very efficient multi-exponentiation in dimension > 2.

10 Security Implications

Gallant, Lambert and Vanstone [11] and Wiener and Zuccherato [29] showed how to speed up the parallel Pollard rho algorithm by using equivalence classes coming from efficiently computable endomorphisms on elliptic curves. One can always work with equivalence classes of size $\#\mathrm{Aut}(E)$.

Such methods can also exploit our homomorphism, giving a slight lowering of security for the group G_2 compared with what was previously believed. As shown in Lemma 1, the homomorphism ψ on G_2 has order k and so we can partition $G_2 - \{0\}$ into equivalence classes of size k. Similarly $G_T - \{1\}$ can be partitioned into equivalence classes of size k.

The size of equivalence classes for G_2 and G_T is therefore k, while the size of equivalence classes for G_1 is $\mathrm{Aut}(E)$. When $e = 1$ then $k = \#\mathrm{Aut}(E)$ and so our result is not new, but when $e > 1$ then $k > \#\mathrm{Aut}(E)$. For example, with BN curves the size of equivalence classes is 6 for G_1 and 12 for G_2 and G_T. This does not imply that the DLP is easier by a factor $\sqrt{2}$ in G_2 and G_T than G_1, since those groups are defined over larger fields; in practice it will still be quicker to solve the DLP in $G_1 = E(\mathbb{F}_p)[r]$ than in G_2 or G_T.

11 Conclusion

In the deployment of pairing-based cryptography there has been much emphasis on the efficiency of the pairing itself. But in real protocols the efficiency of operations in the groups G_1, G_2 and G_T are also of significance, but have been rather overlooked. In this paper we address this imbalance by suggesting faster algorithms for group operations in G_T, and particularly in G_2. The latter is of particular significance for pairing-friendly ordinary elliptic curves, where G_2 may be defined over an extension field. Further work is required to determine more precisely the speed-up that can be achieved in practise.

Acknowledgements

The authors thank Paulo Barreto, Rob Granger, Xibin Lin, Nigel Smart, Martijn Stam and the anonymous referees for comments and suggestions.

References

1. Avanzi, R., Cohen, H., Doche, C., Frey, G., Lange, T., Nguyen, K., Vercauteren, F.: Handbook of elliptic and hyperelliptic cryptography. Chapman and Hall/CRC (2006)
2. Babai, L.: On Lovasz lattice reduction and the nearest lattice point problem. Combinatorica 6(1), 1–13 (1986)
3. Barreto, P.S.L.M., Lynn, B., Scott, M.: Constructing Elliptic Curves with Prescribed Embedding Degrees. In: Cimato, S., Galdi, C., Persiano, G. (eds.) SCN 2002. LNCS, vol. 2576, pp. 263–273. Springer, Heidelberg (2003)
4. Barreto, P.S.L.M., Galbraith, S., OhEigeartaigh, C., Scott, M.: Efficient Pairing Computation on Supersingular Abelian Varieties. Designs, Codes and Cryptography 42, 239–271 (2007)
5. Barreto, P.S.L.M., Naehrig, M.: Pairing-friendly elliptic curves of prime order. In: Preneel, B., Tavares, S. (eds.) SAC 2005. LNCS, vol. 3897, pp. 319–331. Springer, Heidelberg (2006)
6. Bernstein, D.J., Lange, T.: Inverted Edwards coordinates. In: Boztas, S., Lu, H.-F. (eds.) AAECC 2007. LNCS, vol. 4851, pp. 20–27. Springer, Heidelberg (2007)
7. Boneh, D., Franklin, M.: Identity-based encryption from the Weil pairing. SIAM Journal of Computing 32(3), 586–615 (2003)
8. van Dijk, M., Granger, R., Page, D., Rubin, K., Silverberg, A., Stam, M., Woodruff, D.: Practical Cryptography in High Dimensional Tori. In: Cramer, R. (ed.) EUROCRYPT 2005. LNCS, vol. 3494, pp. 234–250. Springer, Heidelberg (2005)
9. Duursma, I., Lee, H.-S.: Tate pairing implementation for hyperelliptic curves $y^2 = x^p - x + d$. In: Laih, C.-S. (ed.) ASIACRYPT 2003. LNCS, vol. 2894, pp. 111–123. Springer, Heidelberg (2003)
10. Freeman, D., Scott, M., Teske, E.: A taxonomy of pairing-friendly elliptic curves. Cryptology ePrint Archive, Report 2006/372 (2006)
11. Gallant, R.P., Lambert, R.J., Vanstone, S.A.: Improving the parallelized Pollard lambda search on anomalous binary curves. Math. Comp. 69, 1699–1705 (2000)
12. Gallant, R.P., Lambert, R.J., Vanstone, S.A.: Faster Point Multiplication on Elliptic Curves with Efficient Endomorphisms. In: Kilian, J. (ed.) CRYPTO 2001. LNCS, vol. 2139, pp. 190–200. Springer, Heidelberg (2001)
13. Granger, R., Page, D., Stam, M.: On small characteristic algebraic tori in pairing-based cryptography. LMS Journal of Computation and Mathematics 9, 64–85 (2006)
14. Hankerson, D., Menezes, A., Scott, M.: Software Implementation of Pairings. University of Waterloo, Centre for Applied Cryptographic Research, Technical report CACR 2008-08
15. Hess, F., Smart, N.P., Vercauteren, F.: The Eta Pairing Revisited. IEEE Trans. Information Theory 52(10), 4595–4602 (2006)
16. Hess, F.: Pairing lattices. In: Galbraith, S.D., Paterson, K.G. (eds.) Pairing 2008, LNCS, vol. 5209, Springer, Heidelberg (2008)
17. Iijima, T., Matsuo, K., Chao, J., Tsujii, S.: Construction of Frobenius maps of twists elliptic curves and its application to elliptic scalar multiplication, SCIS 2002 (2002)
18. Lee, E., Lee, H.-S., Park, C.-M.: Efficient and Generalized Pairing Computation on Abelian Varieties. Cryptology ePrint Archive, Report 2008/040 (2008)
19. Lin, X., Zhao, C.-A., Zhang, F., Wang, Y.: Computing the Ate Pairing on Elliptic Curves with Embedding Degree $k = 9$. IEICE transactions A E91-A(9) (to appear, 2008)

20. Lim, S.-G., Kim, S.-J., Yie, I.-W., Kim, J.-M., Lee, H.-S.: XTR Extended to GF(p^{6m}). In: Vaudenay, S., Youssef, A.M. (eds.) SAC 2001. LNCS, vol. 2259, pp. 301–312. Springer, Heidelberg (2001)
21. Miyaji, A., Nakabayashi, M., Takano, S.: New explicit conditions of elliptic curve traces for FR-reduction. IEICE Trans. Fundamentals E84, 1234–1243 (2001)
22. Möller, B.: Algorithms for multi-exponentiation. In: Vaudenay, S., Youssef, A.M. (eds.) SAC 2001. LNCS, vol. 2259, pp. 165–180. Springer, Heidelberg (2001)
23. Sica, F., Ciet, M., Quisquater, J.-J.: Analysis of the Gallant-Lambert-Vanstone method based on efficient endomorphisms: Elliptic and hyperelliptic curves. In: Nyberg, K., Heys, H.M. (eds.) SAC 2002. LNCS, vol. 2595, pp. 21–36. Springer, Heidelberg (2003)
24. Silverman, J.H.: The Arithmetic of Elliptic Curves. Graduate Texts in Mathematics 106. Springer, Heidelberg (1986)
25. Stam, M., Lenstra, A.K.: Speeding Up XTR. In: Boyd, C. (ed.) ASIACRYPT 2001. LNCS, vol. 2248, pp. 125–143. Springer, Heidelberg (2001)
26. Stam, M., Lenstra, A.K.: Efficient Subgroup Exponentiation in Quadratic and Sixth Degree Extensions. In: Kaliski Jr., B.S., Koç, Ç.K., Paar, C. (eds.) CHES 2002. LNCS, vol. 2523, pp. 318–332. Springer, Heidelberg (2003)
27. Stam, M.: Speeding up Subgroup Cryptosystems. PhD thesis (2003), `http://www.cs.bris.ac.uk/Publications/Papers/2000036.pdf`
28. Vercauteren, F.: Optimal pairings. Cryptology ePrint Archive, Report 2008/096 (2008)
29. Wiener, M.J., Zuccherato, R.J.: Faster Attacks on Elliptic Curve Cryptosystems. In: Tavares, S., Meijer, H. (eds.) SAC 1998. LNCS, vol. 1556, pp. 190–200. Springer, Heidelberg (1999)

Generators for the ℓ-Torsion Subgroup of Jacobians of Genus Two Curves

Christian Robenhagen Ravnshøj

Department of Mathematical Sciences
University of Aarhus
Ny Munkegade
Building 1530
DK-8000 Aarhus C
cr@imf.au.dk

Abstract. We give an explicit description of the matrix representation of the Frobenius endomorphism on the Jacobian of a genus two curve on the subgroup of ℓ-torsion points. By using this description, we can describe the matrix representation of the Weil-pairing on the subgroup of ℓ-torsion points explicitly. Finally, the explicit description of the Weil-pairing provides us with an efficient, probabilistic algorithm to find generators of the subgroup of ℓ-torsion points on the Jacobian of a genus two curve.

1 Introduction

In [13], Koblitz described how to use elliptic curves to construct a public key cryptosystem. To get a more general class of curves, and possibly larger group orders, Koblitz [14] then proposed using Jacobians of hyperelliptic curves. After Boneh and Franklin [1] proposed an identity based cryptosystem by using the Weil-pairing on an elliptic curve, pairings have been of great interest to cryptography [8]. The next natural step was to consider pairings on Jacobians of hyperelliptic curves.

Galbraith *et al.* [9] survey the recent research on pairings on Jacobians of hyperelliptic curves. Their conclusion is that, for most applications, elliptic curves provide more efficient solutions than hyperelliptic curves. One way of making pairing based cryptography on Jacobians of hyperelliptic curves interesting is to exploit the full torsion subgroup of the Jacobian of a hyperelliptic curve. In particular, cryptographic applications of pairings on groups which require three or more generators will be interesting. If such applications are found, the next natural problem will be to give efficient methods to choose points in the particular subgroups. The present paper addresses this problem.

Let $\mathcal{J}_C$ be the Jacobian of a genus two curve defined over $\mathbb{F}_q$. In [5, Algorithm 4.3], Freeman and Lauter describe a probabilistic algorithm to determine generators of the subgroup $\mathcal{J}_C[\ell]$ of points of order ℓ, but the algorithm is incomplete in the sense that the output only *probably* is a generating set - it is not

S.D. Galbraith and K.G. Paterson (Eds.): Pairing 2008, LNCS 5209, pp. 225–242, 2008.

tested whether the output in fact *is* a generating set. Furthermore, if the output happens to be a generating set, it still may not be a *basis* of $\mathfrak{J}_C[\ell]$.

In [21], the author describes an algorithm based on the Tate-pairing to determine a basis of the subgroup $\mathfrak{J}_C(\mathbb{F}_q)[m]$ of points of order m on the Jacobian, where m is a number dividing $q-1$. The key ingredient of the algorithm is a "diagonalization" of a set of randomly chosen points $\{P_1,\dots,P_4,Q_1,\dots,Q_4\}$ on the Jacobian with respect to the (reduced) Tate-pairing ε; i.e. a modification of the set such that $\varepsilon(P_i,Q_j)\neq 1$ if and only if $i=j$. This procedure is based on solving the discrete logarithm problem in $\mathfrak{J}_C(\mathbb{F}_q)[m]$. Contrary to the special case where m divides $q-1$, it is in general infeasible to solve the discrete logarithm problem in $\mathfrak{J}_C(\mathbb{F}_q)[m]$. Hence, in general the algorithm in [21] does not apply.

Results

In the present paper, we generalize the algorithm in [21] to subgroups of points of prime order ℓ, where ℓ does not divide $q-1$. In order to do so, we must somehow alter the diagonalization step. We show and exploit the fact that the matrix representation on $\mathfrak{J}_C[\ell]$ of the q-power Frobenius endomorphism on $\mathfrak{J}_C$ can be described explicitly. This description enables us to describe the matrix representation of the Weil pairing on $\mathfrak{J}_C[\ell]$ explicitly. Miller [18] uses the Weil pairing to determine generators of $E(\mathbb{F}_{q^a})$, where E is an elliptic curve defined over a finite field $\mathbb{F}_q$ and $a\in\mathbb{N}$. The basic idea of his algorithm is to decide whether points on the curve are independent by means of calculating pairing values. The explicit description of the matrix representation of the Weil pairing lets us transfer this idea to Jacobians of genus two curves. Hereby, computations of discrete logarithms are avoided, yielding the desired altering of the diagonalization step.

Setup. Consider the Jacobian $\mathfrak{J}_C$ of a genus two curve C defined over a finite field $\mathbb{F}_q$. Let ℓ be an odd prime number dividing the number of $\mathbb{F}_q$-rational points on $\mathfrak{J}_C$, and with ℓ dividing neither q nor $q-1$. Assume that the $\mathbb{F}_q$-rational subgroup $\mathfrak{J}_C(\mathbb{F}_q)[\ell]$ of points on the Jacobian of order ℓ is cyclic. Let k be the multiplicative order of q modulo ℓ, and let k_0 be the least number, such that $\mathfrak{J}_C[\ell]\subseteq\mathfrak{J}_C(\mathbb{F}_{q^{k_0}})$. (Obviously, in applications k_0 must be small enough for representation of and computations with points on $\mathfrak{J}_C(\mathbb{F}_{q^{k_0}})$ to be feasible. Hence, the algorithms presented are only efficient if k_0 is "small"). Write the characteristic polynomial of the q^k-power Frobenius endomorphism on $\mathfrak{J}_C$ as $P_k(X)=X^4+sX^3+(2q^k+(s^2-\tau_k)/4)X^2+sq^kX+q^{2k}$. Let $\omega_k\in\mathbb{C}$ be a root of $P_k(X)$. Finally, if ℓ divides τ_k, we assume that ℓ is unramified in $\mathbb{Q}(\omega_k)$.

Remark 1. Notice that most likely, in cases relevant to pairing based cryptography the considered Jacobian of a genus two curve fulfills these assumptions. Cf. Remark 13 and 21.

The Algorithm. Let $\mathfrak{J}_C$, ℓ, q, k, k_0 and τ_k be given as in the above setup. Note that the numbers k and k_0 are *computed* from $\mathfrak{J}_C$, ℓ and q - they are *not* chosen.

Since ℓ divides the number of $\mathbb{F}_q$-rational points on $\mathcal{J}_C$, it is implicitly assumed that $\mathcal{J}_C$ contains points of order ℓ defined over $\mathbb{F}_q$, i.e. that $\mathcal{J}_C(\mathbb{F}_q)[\ell]$ is non-trivial. Notice also that we assume to know the Weil polynomial (see Section 3) of $\mathcal{J}_C$ already - it is *not* computed in the algorithm. In particular, we know τ_k.

Now, first of all we notice that in the above setup the q-power Frobenius endomorphism φ on $\mathcal{J}_C$ can be represented on $\mathcal{J}_C[\ell]$ by either a diagonal matrix or a matrix of a particular form with respect to an appropriate basis $\mathcal{B}$ of $\mathcal{J}_C[\ell]$; cf. Theorem 14. (In fact, to show this we do not need the $\mathbb{F}_q$-rational subgroup $\mathcal{J}_C(\mathbb{F}_q)[\ell]$ of points on the Jacobian of order ℓ to be cyclic). From this observation it follows that all non-degenerate, bilinear, anti-symmetric and Galois-invariant pairings on $\mathcal{J}_C[\ell]$ are given by the matrices

$$\mathcal{E}_{a,b} = \begin{bmatrix} 0 & a & 0 & 0 \\ -a & 0 & 0 & 0 \\ 0 & 0 & 0 & b \\ 0 & 0 & -b & 0 \end{bmatrix}, \qquad a, b \in (\mathbb{Z}/\ell\mathbb{Z})^\times$$

with respect to $\mathcal{B}$; cf. Theorem 19. By using this description of the pairings, the desired algorithm is given as follows.

Algorithm 16. *Let the notation and assumptions be as in the above setup. On input the Jacobian $\mathcal{J}_C$, the numbers ℓ, q, k, k_0, τ_k and a number $n \in \mathbb{N}$, the following algorithm outputs a basis of $\mathcal{J}_C[\ell]$ or "failure".*

1. *If ℓ does not divide τ_k, then do the following:*
   (a) *Choose points $\mathcal{O} \neq x_1 \in \mathcal{J}_C(\mathbb{F}_q)[\ell]$, $x_2 \in \mathcal{J}_C(\mathbb{F}_{q^k})[\ell]$ and $x_3' \in \mathcal{J}_C(\mathbb{F}_{q^{k_0}})[\ell]$ (cf. Section 8 for details on how to choose points); compute $x_3 = q(x_3' - \varphi(x_3')) - \varphi(x_3' - \varphi(x_3'))$. If $\varepsilon(x_3, \varphi(x_3)) \neq 1$, then output $\{x_1, x_2, x_3, \varphi(x_3)\}$ and stop.*
   (b) *Let $i = j = 0$. While $i < n$ do the following:*
      i. *Choose a random point $x_4 \in \mathcal{J}_C(\mathbb{F}_{q^{k_0}})[\ell]$.*
      ii. *If $\varepsilon(x_3, x_4) = 1$, then $i := i + 1$. Else $i := n$ and $j := 1$.*
   (c) *If $j = 0$, then output "failure". Else output $\{x_1, x_2, x_3, x_4\}$.*
2. *If ℓ divides τ_k, then do the following:*
   (a) *Choose a random point $\mathcal{O} \neq x_1 \in \mathcal{J}_C(\mathbb{F}_q)[\ell]$.*
   (b) *Let $i = j = 0$. While $i < n$ do the following:*
      i. *Choose a random point $x_2 \in \mathcal{J}_C(\mathbb{F}_{q^{k_0}})[\ell]$.*
      ii. *If $\varepsilon(x_1, x_2) = 1$, then $i := i + 1$. Else $i := n$ and $j := 1$.*
   (c) *If $j = 0$, then output "failure" and stop.*
   (d) *Let $i = j = 0$. While $i < n$ do the following:*
      i. *Choose random points $y_3, y_4 \in \mathcal{J}_C(\mathbb{F}_{q^{k_0}})[\ell]$; compute $x_\nu := q(y_\nu - \varphi(y_\nu)) - \varphi(y_\nu - \varphi(y_\nu))$ for $\nu = 3, 4$.*
      ii. *If $\varepsilon(x_3, x_4) = 1$, then $i := i + 1$. Else $i := n$ and $j := 1$.*
   (e) *If $j = 0$, then output "failure". Else output $\{x_1, x_2, x_3, x_4\}$.*

Algorithm 24 finds generators of $\mathcal{J}_C[\ell]$ with probability at least $(1 - 1/\ell^n)^2$ and in expected running time $O\left(\log \ell \log \frac{q^{k_0}-1}{\ell} {k_0}^3 \log k_0 \log q\right)$ field operations in

$\mathbb{F}_q$ (ignoring $\log\log q$ factors); this is contained in Theorem 25. The algorithm [5, Algorithm 4.3] runs in expected time $O(k^2 \log k (\log p)^2 \ell^{s-4}\sqrt{-\log \epsilon})$, where the number s is given by $|\mathcal{J}_C(\mathbb{F}_{q^{k_0}})| = m\ell^s$ and $\ell \nmid m$, and ϵ is the rate of failure. Hence, if $s > 4$, then Algorithm 24 is by far more efficient than [5, Algorithm 4.3]. [5, Algorithm 4.3] is used in [5] to compute endomorphism rings of Jacobians of genus two curves, and this in turn has applications for generating Jacobians of genus two curves using the CRT version of the CM method [4]. Hence, Algorithm 24 also has applications for generating Jacobians of genus two curves.

If the Weil polynomial splits in distinct factors modulo ℓ, then the problem of determining a basis of the ℓ-torsion subgroup is trivially solved: the ℓ-torsion subgroup decomposes in four eigenspaces of the q-power Frobenius endomorphism, so to find a basis, simply choose an ℓ-torsion point and project it to the eigenspaces. A standard example is the Jacobian $\mathcal{J}_C$ of the curve over $\mathbb{F}_3$ given by $y^2 = x^5 + 1$. The Weil polynomial of $\mathcal{J}_C$ is given by $P(X) = X^4 + 9$, the number of $\mathbb{F}_3$-rational points on $\mathcal{J}_C$ is $|\mathcal{J}_C(\mathbb{F}_3)| = P(1) = 10$, and $P(X)$ factors modulo 5 as $P(X) \equiv (X-1)(X-2)(X-3)(X-4) \pmod 5$. But there *are* cases where the Weil polynomial does not split in distinct factors; cf. the following example.

Example 1. Consider the Jacobian $\mathcal{J}_C$ of the curve over $\mathbb{F}_3$ given by

$$y^2 = x^5 + 2x^2 + x + 1 \ .$$

The Weil polynomial of $\mathcal{J}_C$ is given by $P(X) = X^4 + X^3 - X^2 + 3X + 9$, the number of $\mathbb{F}_3$-rational points on $\mathcal{J}_C$ is $|\mathcal{J}_C(\mathbb{F}_3)| = P(1) = 13$, and $P(X)$ factors modulo 13 as $P(X) \equiv (X-1)(X-3)(X-4)^2 \pmod{13}$.

Remark 2. To implement Algorithm 24, we need to find the *Weil polynomial* of the Jacobian. On Jacobians generated by the *complex multiplication method* [23, 10, 4], we know the Weil polynomial in advance. Hence, Algorithm 24 is particularly well suited for such Jacobians.

Assumption

In this paper, a *curve* is an irreducible nonsingular projective variety of dimension one.

2 Genus Two Curves

A hyperelliptic curve is a projective curve $C \subseteq \mathbb{P}^n$ of genus at least two with a separable, degree two morphism $\phi : C \to \mathbb{P}^1$. It is well known, that any genus two curve is hyperelliptic. Throughout this paper, let C be a curve of genus two defined over a finite field $\mathbb{F}_q$ of characteristic p. By the Riemann-Roch Theorem there exists a birational map $\psi : C \to \mathbb{P}^2$, mapping C to a curve given by an equation of the form

$$y^2 + g(x)y = h(x) \ ,$$

where $g, h \in \mathbb{F}_q[x]$ are of degree $\deg(g) \leq 3$ and $\deg(h) \leq 6$; cf. [2, chapter 1].

The set of principal divisors $\mathcal{P}(C)$ on C constitutes a subgroup of the degree zero divisors $\mathrm{Div}_0(C)$. The Jacobian $\mathcal{J}_C$ of C is defined as the quotient

$$\mathcal{J}_C = \mathrm{Div}_0(C)/\mathcal{P}(C) \ .$$

The Jacobian is an abelian group. We write the group law additively, and denote the zero element of the Jacobian by $\mathcal{O}$.

Let $\ell \neq p$ be a prime number. The ℓ^n-torsion subgroup $\mathcal{J}_C[\ell^n] \subseteq \mathcal{J}_C$ of points of order dividing ℓ^n is a $\mathbb{Z}/\ell^n\mathbb{Z}$-module of rank four, i.e.

$$\mathcal{J}_C[\ell^n] \simeq \mathbb{Z}/\ell^n\mathbb{Z} \times \mathbb{Z}/\ell^n\mathbb{Z} \times \mathbb{Z}/\ell^n\mathbb{Z} \times \mathbb{Z}/\ell^n\mathbb{Z} \ ;$$

cf. [15, Theorem 6, p. 109].

The multiplicative order k of q modulo ℓ plays an important role in cryptography, since the (reduced) Tate-pairing is non-degenerate over $\mathbb{F}_{q^k}$; cf. [11].

Definition 3 (Embedding degree). *Consider a prime number $\ell \neq p$ dividing the number of $\mathbb{F}_q$-rational points on the Jacobian $\mathcal{J}_C$. The embedding degree of $\mathcal{J}_C(\mathbb{F}_q)$ with respect to ℓ is the least number k, such that $q^k \equiv 1 \pmod{\ell}$.*

Closely related to the embedding degree, we have the *full* embedding degree.

Definition 4 (Full embedding degree). *Consider a prime number $\ell \neq p$ dividing the number of $\mathbb{F}_q$-rational points on the Jacobian $\mathcal{J}_C$. The full embedding degree of $\mathcal{J}_C(\mathbb{F}_q)$ with respect to ℓ is the least number k_0, such that $\mathcal{J}_C[\ell] \subseteq \mathcal{J}_C(\mathbb{F}_{q^{k_0}})$.*

Remark 5. If $\mathcal{J}_C[\ell] \subseteq \mathcal{J}_C(\mathbb{F}_{q^{k_0}})$, then $\ell \mid q^{k_0} - 1$; cf. [15, Theorem 6, p. 109] and [6, Proposition 5.78, p. 111]. Hence, the full embedding degree is a multiple of the embedding degree.

3 The Frobenius Endomorphism

Since C is defined over $\mathbb{F}_q$, the mapping $(x, y) \mapsto (x^q, y^q)$ is a morphism on C. This morphism induces the q-power Frobenius endomorphism φ on the Jacobian $\mathcal{J}_C$. Let $P(X)$ be the characteristic polynomial of φ; cf. [15, pp. 109–110]. $P(X)$ is called the *Weil polynomial* of $\mathcal{J}_C$, and

$$|\mathcal{J}_C(\mathbb{F}_q)| = P(1)$$

by the definition of $P(X)$ (see [15, pp. 109–110]); i.e. the number of $\mathbb{F}_q$-rational points on the Jacobian is $P(1)$.

Definition 6 (Weil number). *Let notation be as above. Let $P_k(X)$ be the characteristic polynomial of the q^m-power Frobenius endomorphism φ_m on $\mathfrak{J}_C$. A complex number $\omega_m \in \mathbb{C}$ with $P_m(\omega_m) = 0$ is called a q^m-*Weil number *of $\mathfrak{J}_C$.*

Remark 7. Note that $\mathfrak{J}_C$ has four q^m-Weil numbers. If $P_1(X) = \prod_i (X - \omega_i)$, then $P_m(X) = \prod_i (X - \omega_i^m)$. Hence, if ω is a q-Weil number of $\mathfrak{J}_C$, then ω^m is a q^m-Weil number of $\mathfrak{J}_C$.

4 Non-cyclic Subgroups

Consider a genus two curve C defined over a finite field $\mathbb{F}_q$. Let $P_m(X)$ be the characteristic polynomial of the q^m-power Frobenius endomorphism φ_m on the Jacobian $\mathfrak{J}_C$. $P_m(X)$ is of the form $P_m(X) = X^4 + sX^3 + tX^2 + sq^m X + q^{2m}$, where $s, t \in \mathbb{Z}$. Let $\tau = 8q^m + s^2 - 4t$. Then $P_m(X) = X^4 + sX^3 + (2q^m + (s^2 - \tau)/4)X^2 + sq^m X + q^{2m}$. In [22], the author proves the following Theorem 8 and Theorem 9.

Theorem 8. *Consider the Jacobian $\mathfrak{J}_C$ of a genus two curve C defined over a finite field $\mathbb{F}_q$. Write the characteristic polynomial of the q^m-power Frobenius endomorphism on $\mathfrak{J}_C$ as $P_m(X) = X^4 + sX^3 + (2q^m + (s^2 - \tau)/4)X^2 + sq^m X + q^{2m}$. Let ℓ be an odd prime number dividing the number of $\mathbb{F}_q$-rational points on $\mathfrak{J}_C$, and with $\ell \nmid q$ and $\ell \nmid q - 1$. If $\ell \nmid \tau$, then*

1. *$\mathfrak{J}_C(\mathbb{F}_{q^m})[\ell]$ is of rank at most two as a $\mathbb{Z}/\ell\mathbb{Z}$-module, and*
2. *$\mathfrak{J}_C(\mathbb{F}_{q^m})[\ell]$ is bicyclic if and only if ℓ divides $q^m - 1$.*

Theorem 9. *Let notation be as in Theorem 8. Furthermore, let ω_m be a q^m-Weil number of $\mathfrak{J}_C$, and assume that ℓ is unramified in $\mathbb{Q}(\omega_m)$. Now assume that $\ell \mid \tau$. Then the following holds.*

1. *If $\omega_m \in \mathbb{Z}$, then $\ell \mid q^m - 1$ and $\mathfrak{J}_C[\ell] \subseteq \mathfrak{J}_C(\mathbb{F}_{q^m})$.*
2. *If $\omega_m \notin \mathbb{Z}$, then $\ell \nmid q^m - 1$, $\mathfrak{J}_C(\mathbb{F}_{q^m})[\ell] \simeq (\mathbb{Z}/\ell\mathbb{Z})^2$ and $\mathfrak{J}_C[\ell] \subseteq \mathfrak{J}_C(\mathbb{F}_{q^{mk}})$ if and only if $\ell \mid q^{mk} - 1$.*

Example 10 (The case $\ell \nmid \tau_k$). Let $P(X) = X^4 + X^3 - X^2 + 3X + 9 \in \mathbb{Q}[X]$. By [16] and [12] it follows that $P(X)$ is the Weil polynomial of the Jacobian of a genus two curve C defined over $\mathbb{F}_3$. The number of $\mathbb{F}_3$-rational points on the Jacobian is $P(1) = 13$, and the embedding degree of $\mathfrak{J}_C(\mathbb{F}_3)$ with respect to $\ell = 13$ is $k = 3$. The characteristic polynomial of the 3^3-power Frobenius endomorphisms is given by $P_3(X) = X^4 + 13X^3 + 89X^2 + 351X + 729$. Hence, $\mathfrak{J}_C(\mathbb{F}_{27})[13]$ is bicyclic by Theorem 8.

Example 11 (The case $\ell \mid \tau_k$). Let $P(X) = (X^2 - 5X + 9)^2 \in \mathbb{Q}[X]$. By [16] and [12] it follows that $P(X)$ is the Weil polynomial of the Jacobian of a genus two curve C defined over $\mathbb{F}_9$. The number of $\mathbb{F}_9$-rational points on the Jacobian is $P(1) = 25$, so $\ell = 5$ is an odd prime divisor of $|\mathfrak{J}_C(\mathbb{F}_9)|$ not dividing $q = 9$. Notice that $P(X) \equiv X^4 + 2qX^2 + q^2 \pmod 5$. The complex roots of $P(X)$ are given by $\omega = \frac{5+\sqrt{-11}}{2}$ and $\bar{\omega}$, and 5 is unramified in $\mathbb{Q}(\omega)$. Since $9^2 \equiv 1 \pmod 5$, it follows by Theorem 9 that $\mathfrak{J}_C(\mathbb{F}_9)[5] \simeq \mathbb{Z}/5\mathbb{Z} \oplus \mathbb{Z}/5\mathbb{Z}$ and $\mathfrak{J}_C[5] \subseteq \mathfrak{J}_C(\mathbb{F}_{81})$.

Inspired by Theorem 8 and Theorem 9 we introduce the following notation.

Definition 12. *Consider the Jacobian $\mathcal{J}_C$ of a genus two curve C defined over a finite field $\mathbb{F}_q$. We say that the Jacobian is a $\mathbb{J}(\ell, q, k, \tau_k)$-Jacobian or is of type $\mathbb{J}(\ell, q, k, \tau_k)$, and write $\mathcal{J}_C \in \mathbb{J}(\ell, q, k, \tau_k)$, if the following holds.*

1. *The number ℓ is an odd prime number dividing the number of $\mathbb{F}_q$-rational points on $\mathcal{J}_C$, ℓ divides neither q nor $q-1$, and $\mathcal{J}_C(\mathbb{F}_q)$ is of embedding degree k with respect to ℓ.*
2. *The characteristic polynomial of the q^k-power Frobenius endomorphism on $\mathcal{J}_C$ is given by $P_k(X) = X^4 + sX^3 + (2q^k + (s^2 - \tau_k)/4)X^2 + sq^kX + q^{2k}$.*
3. *Let ω_k be a q^k-Weil number of $\mathcal{J}_C$. If ℓ divides τ_k, then ℓ is unramified in $\mathbb{Q}(\omega_k)$.*

Remark 13. Since ℓ is ramified in $\mathbb{Q}(\omega_k)$ if and only if ℓ divides the discriminant of $\mathbb{Q}(\omega_k)$ (see [20, Theorem 2.6, p. 199]), ℓ is unramified in $\mathbb{Q}(\omega_k)$ with probability approximately $1 - 1/\ell$. Hence, most likely, in cases relevant to pairing based cryptography the considered Jacobian is a $\mathbb{J}(\ell, q, k, \tau_k)$-Jacobian.

5 Matrix Representation of the Frobenius Endomorphism

An endomorphism $\psi : \mathcal{J}_C \to \mathcal{J}_C$ induces a linear map $\bar{\psi} : \mathcal{J}_C[\ell] \to \mathcal{J}_C[\ell]$ by restriction. Hence, ψ is represented by a matrix $M \in \mathrm{Mat}_4(\mathbb{Z}/\ell\mathbb{Z})$ on $\mathcal{J}_C[\ell]$. If ψ can be represented on $\mathcal{J}_C[\ell]$ by a diagonal matrix with respect to an appropriate basis of $\mathcal{J}_C[\ell]$, then we say that ψ is *diagonalizable* or has a *diagonal representation* on $\mathcal{J}_C[\ell]$.

Let $f \in \mathbb{Z}[X]$ be the characteristic polynomial of ψ (see [15, pp. 109–110]), and let $\bar{f} \in (\mathbb{Z}/\ell\mathbb{Z})[X]$ be the characteristic polynomial of $\bar{\psi}$. Then f is a monic polynomial of degree four, and by [15, Theorem 3, p. 186],

$$f(X) \equiv \bar{f}(X) \pmod{\ell} .$$

By Theorem 8 and Theorem 9 we get the following explicit description of the matrix representation of the Frobenius endomorphism on the Jacobian of a genus two curve.

Theorem 14. *Consider a Jacobian $\mathcal{J}_C \in \mathbb{J}(\ell, q, k, \tau_k)$. Let φ be the q-power Frobenius endomorphism of $\mathcal{J}_C$. If φ is not diagonalizable on $\mathcal{J}_C[\ell]$, then φ is represented on $\mathcal{J}_C[\ell]$ by a matrix of the form*

$$M = \begin{bmatrix} 1 & 0 & 0 & 0 \\ 0 & q & 0 & 0 \\ 0 & 0 & 0 & -q \\ 0 & 0 & 1 & c \end{bmatrix} \tag{1}$$

with respect to an appropriate basis of $\mathcal{J}_C[\ell]$. In particular, $c \not\equiv q+1 \pmod{\ell}$.

Proof. Assume at first that ℓ does not divide τ_k. Then we know that $\mathcal{J}_C(\mathbb{F}_q)[\ell]$ is cyclic and that $\mathcal{J}_C(\mathbb{F}_{q^k})[\ell]$ is bicyclic; cf. Theorem 8. Choose points $x_1, x_2 \in$

$\mathcal{J}_C[\ell]$, such that $\varphi(x_1) = x_1$ and $\varphi(x_2) = qx_2$. Then the set $\{x_1, x_2\}$ is a basis of $\mathcal{J}_C(\mathbb{F}_{q^k})[\ell]$. Now, extend $\{x_1, x_2\}$ to a basis $\mathcal{B} = \{x_1, x_2, x_3, x_4\}$ of $\mathcal{J}_C[\ell]$. If x_3 and x_4 are eigenvectors of φ, then φ is represented by a diagonal matrix on $\mathcal{J}_C[\ell]$ with respect to $\mathcal{B}$. Assume x_3 is not an eigenvector of φ. Then $\mathcal{B}' = \{x_1, x_2, x_3, \varphi(x_3)\}$ is a basis of $\mathcal{J}_C[\ell]$, and φ is represented by a matrix of the form (1) with respect to $\mathcal{B}'$.

Now, assume ℓ divides τ_k. Since ℓ divides $q^k - 1$, it follows that $\mathcal{J}_C[\ell] \subseteq \mathcal{J}_C(\mathbb{F}_{q^k})$; cf. Theorem 9. Since ℓ divides the number of $\mathbb{F}_q$-rational points on $\mathcal{J}_C$, 1 is a root of the Weil polynomial $P(X)$ modulo ℓ. Assume that 1 is an root of $P(X)$ modulo ℓ of multiplicity ν. Since the roots of $P(X)$ occur in pairs of the form $(\alpha, q/\alpha)$, it follows that

$$P(X) \equiv (X-1)^\nu (X-q)^\nu Q(X) \pmod{\ell} ,$$

where $Q \in \mathbb{Z}[X]$ is a polynomial of degree $4-2\nu$, $Q(1) \not\equiv 0 \pmod{\ell}$ and $Q(q) \not\equiv 0 \pmod{\ell}$. Let $U = \ker(\varphi-1)^\nu$, $V = \ker(\varphi-q)^\nu$ and $W = \ker(Q(\varphi))$. Then U, V and W are φ-invariant submodules of the $\mathbb{Z}/\ell\mathbb{Z}$-module $\mathcal{J}_C[\ell]$, $\operatorname{rank}_{\mathbb{Z}/\ell\mathbb{Z}}(U) = \operatorname{rank}_{\mathbb{Z}/\ell\mathbb{Z}}(V) = \nu$, and $\mathcal{J}_C[\ell] \simeq U \oplus V \oplus W$. If $\nu = 1$, then it follows as above that φ is either diagonalizable on $\mathcal{J}_C[\ell]$ or represented by a matrix of the form (1) with respect to some basis of $\mathcal{J}_C[\ell]$. Hence, we may assume that $\nu = 2$. Now, choose $x_1 \in U$ such that $\varphi(x_1) = x_1$, and extend $\{x_1\}$ to a basis $\{x_1, x_2\}$ of U. Similarly, choose a basis $\{x_3, x_4\}$ of V with $\varphi(x_3) = qx_3$. With respect to the basis $\mathcal{B} = \{x_1, x_2, x_3, x_4\}$, φ is represented by a matrix of the form

$$M = \begin{bmatrix} 1 & \alpha & 0 & 0 \\ 0 & 1 & 0 & 0 \\ 0 & 0 & q & \beta \\ 0 & 0 & 0 & q \end{bmatrix} .$$

Notice that

$$M^k = \begin{bmatrix} 1 & k\alpha & 0 & 0 \\ 0 & 1 & 0 & 0 \\ 0 & 0 & 1 & kq^{k-1}\beta \\ 0 & 0 & 0 & 1 \end{bmatrix} .$$

Since $\mathcal{J}_C[\ell] \subseteq \mathcal{J}_C(\mathbb{F}_{q^k})$, we know that $\varphi^k = \varphi_k$ is the identity on $\mathcal{J}_C[\ell]$. Hence, $M^k = I$. So $\alpha \equiv \beta \equiv 0 \pmod{\ell}$, i.e. φ is represented by a diagonal matrix with respect to $\mathcal{B}$.

Finally, if $c \equiv q+1 \pmod{\ell}$, then M is diagonalizable. The theorem is proved. □

6 Determining Fields of Definition

In [5], Freeman and Lauter consider the problem of determining the field of definition of the ℓ-torsion points on the Jacobian of a genus two curve, i.e. the problem of determining the full embedding degree k_0. They describe a probabilistic algorithm to determine if $\mathcal{J}_C[\ell] \subseteq \mathcal{J}_C(\mathbb{F}_{q^\kappa})$; see [5, Algorithm 4.3]. (Notice that Freeman and Lauter consider a Jacobian defined over a prime field $\mathbb{F}_p$, and

[5, Algorithm 4.3] determines if $\mathfrak{J}_C[\ell^d] \subseteq \mathfrak{J}_C(\mathbb{F}_q)$, where $q = p^k$ and $d \in \mathbb{N}$. This algorithm is easily generalized to determine if $\mathfrak{J}_C[\ell] \subseteq \mathfrak{J}_C(\mathbb{F}_{q^\kappa})$ for Jacobians defined over $\mathbb{F}_q$, $q = p^a$).

In most applications, a probabilistic algorithm to determine k_0 is sufficient. But we may have to compute k_0. To this end, consider a $\mathbb{J}(\ell, q, k, \tau_k)$-Jacobian $\mathfrak{J}_C$. Let ω be a q-Weil number of $\mathfrak{J}_C$. In cases relevant to pairing based cryptography, ℓ is most likely unramified in $\mathbb{Q}(\omega)$; cf. Remark 13. But then the full embedding degree of $\mathfrak{J}_C$ with respect to ℓ can be computed directly by the following Algorithm 15.

Algorithm 15. *Consider a Jacobian $\mathfrak{J}_C \in \mathbb{J}(\ell, q, k, \tau_k)$. Let ω be a q-Weil number of $\mathfrak{J}_C$. Assume that ℓ is unramified in $\mathbb{Q}(\omega)$. Choose an upper bound $N \in \mathbb{N}$ of the full embedding degree k_0 of $\mathfrak{J}_C$ with respect to ℓ. If $k_0 \leq N$, then the following algorithm outputs k_0. If $k_0 > N$, then the algorithm outputs "$k_0 > N$".*

1. *Let $j = 1$.*
2. *If the Weil polynomial $P(X)$ of $\mathfrak{J}_C$ does not split in linear factors modulo ℓ, then φ is represented by a matrix M of the form (1) on $\mathfrak{J}_C[\ell]$. In this case, let $k_0 = \min\{\kappa \in k\mathbb{N}, \kappa \leq N, M^\kappa \equiv I \pmod{\ell}\}$, if the minimum exists. Else let $j = 0$.*
3. *If $P(X) \equiv (X-1)(X-q)(X-\alpha)(X-q/\alpha) \pmod{\ell}$, then do the following:*
   (a) If $\alpha \not\equiv 1, q, q/\alpha \pmod{\ell}$, then let $k_0 = \min\{\kappa \in k\mathbb{N}, \kappa \leq N, \alpha^\kappa \equiv 1 \pmod{\ell}\}$, if the minimum exists. Else let $j = 0$.
   (b) If $\alpha \equiv 1, q \pmod{\ell}$, then let $k_0 = k$.
   (c) If $\alpha \equiv q/\alpha \pmod{\ell}$, then let $k_0 = 2k$.
4. *If $j = 0$ then output "$k_0 > N$". Else output k_0.*

Proof. First of all, recall that $k_0 \in k\mathbb{N}$; cf. Remark 5. As usual, let φ be the q-power Frobenius endomorphism of $\mathfrak{J}_C$.

Assume at first that the Weil polynomial of $\mathfrak{J}_C$ does not split in linear factors modulo ℓ. Then φ is not diagonalizable on $\mathfrak{J}_C[\ell]$. Thus, φ is represented by a matrix M of the form (1) on $\mathfrak{J}_C[\ell]$. Since φ^{k_0} is the identity on $\mathfrak{J}_C[\ell]$, it is represented by the identity matrix I on $\mathfrak{J}_C[\ell]$. But φ^{k_0} is also represented by M^{k_0} on $\mathfrak{J}_C[\ell]$. So $M^{k_0} \equiv I \pmod{\ell}$. On the other hand, if $M^\kappa \equiv I \pmod{\ell}$ for some number $\kappa \leq k_0$, then φ^κ is the identity on $\mathfrak{J}_C[\ell]$, i.e. $\mathfrak{J}_C[\ell] \subseteq \mathfrak{J}_C(\mathbb{F}_{q^\kappa})$. But then $\kappa = k_0$ by the definition of k_0. Hence, k_0 is the least number, such that $M^{k_0} \equiv I \pmod{\ell}$.

Now, assume the Weil polynomial factors modulo ℓ as

$$P(X) \equiv (X-1)(X-q)(X-\alpha)(X-q/\alpha) \pmod{\ell} .$$

The case $\alpha \not\equiv 1, q, q/\alpha \pmod{\ell}$ is obvious. If $\alpha \equiv 1, q \pmod{\ell}$, then

$$P(X) \equiv (X-1)^2(X-q)^2 \equiv X^4 + 2\sigma X^3 + (2q + \sigma^2 - \tau)X^2 + 2\sigma q X + q^2 \pmod{\ell} ,$$

where $\sigma \equiv -(q+1) \pmod{\ell}$ and $\tau \equiv 0 \pmod{\ell}$. By Theorem 9 it follows that $\mathfrak{J}_C[\ell] \subseteq \mathfrak{J}_C(\mathbb{F}_{q^k})$; i.e. $k_0 = k$ in this case. Finally, assume that $\alpha \equiv q/\alpha$

(mod ℓ), i.e. that $\alpha^2 \equiv q \pmod{\ell}$. Then the q-power Frobenius endomorphism is represented on $\mathfrak{J}_C[\ell]$ by a matrix of the form

$$M = \begin{bmatrix} 1 & 0 & 0 & 0 \\ 0 & q & 0 & 0 \\ 0 & 0 & \alpha & \beta \\ 0 & 0 & 0 & \alpha \end{bmatrix}$$

with respect to an appropriate basis of $\mathfrak{J}_C[\ell]$. Notice that

$$M^{2k} = \begin{bmatrix} 1 & 0 & 0 & 0 \\ 0 & 1 & 0 & 0 \\ 0 & 0 & 1 & 2k\alpha^{2k-1}\beta \\ 0 & 0 & 0 & 1 \end{bmatrix} .$$

Thus, $P_{2k}(X) \equiv (X-1)^4 \pmod{\ell}$. By Theorem 9 it follows that $\mathfrak{J}_C[\ell] \subseteq \mathfrak{J}_C(\mathbb{F}_{q^{2k}})$, i.e. $k_0 = 2k$. □

Theorem 16. *Let the notation and assumptions be as in Algorithm 15. On input $\mathfrak{J}_C$, the Weil polynomial modulo ℓ and a number $N \in \mathbb{N}$, Algorithm 15 outputs either "$k_0 > N$" or the full embedding degree of $\mathfrak{J}_C$ with respect to ℓ in at most $O(N)$ number of operations in $\mathbb{F}_\ell$.*

Proof. If the Weil polynomial of $\mathfrak{J}_C$ does not split in linear factors modulo ℓ, then powers $\{M^k, (M^k)^2, \ldots, (M^k)^{\lfloor N/k \rfloor}\}$ of M modulo ℓ are computed; here, M is the matrix representation of the q-power Frobenius endomorphism on $\mathfrak{J}_C[\ell]$. M is of the form

$$M = \begin{bmatrix} 1 & 0 & 0 & 0 \\ 0 & q & 0 & 0 \\ 0 & 0 & 0 & -q \\ 0 & 0 & 1 & c \end{bmatrix} .$$

Hence, computing powers of M is equivalent to computing powers of $M' = \left[\begin{smallmatrix} 0 & -q \\ 1 & c \end{smallmatrix}\right]$ and powers of q. Computation of the product of two matrices $A, B \in \mathrm{Mat}_2(\mathbb{F}_\ell)$ takes 12 operations in $\mathbb{F}_\ell$, so computing the powers of M modulo ℓ takes $O(N)$ operations in $\mathbb{F}_\ell$.

Assume the Weil polynomial factors as $(X-1)(X-q)(X-\alpha)(X-q/\alpha)$ modulo ℓ. If $\alpha \equiv 1, q, q/\alpha \pmod{\ell}$, then no computations are needed. If $\alpha \not\equiv 1, q, q/\alpha \pmod{\ell}$, then powers $\{\alpha^k, (\alpha^k)^2, \ldots, (\alpha^k)^{\lfloor N/k \rfloor}\}$ of α modulo ℓ are computed; this takes $O(N)$ operations in $\mathbb{F}_\ell$. □

Remark 17. Recall that $q = p^a$ for some power $a \in \mathbb{N}$. Assume ℓ and p are of the same size. For small N (e.g. $N < 200$), a limit of $O(N)$ number of operations in $\mathbb{F}_\ell$ is a better result than the expected number of operations in $\mathbb{F}_p$ of [5, Algorithm 4.3] given by [5, Proposition 4.6]. Furthermore, the algorithm of [5] only checks if a given number $\kappa \in \mathbb{N}$ is the full embedding degree k_0 of the Jacobian. Hence, to find k_0 using [5, Algorithm 4.3], we must apply it to every number in the set $\{\kappa \in k\mathbb{N} | \kappa \leq N\}$. Thus, we must multiply the number of expected operations in $\mathbb{F}_p$ with a factor $O(\lfloor N/k \rfloor)$. So if ℓ and p are of the same size, then Algorithm 15 is more efficient than [5, Algorithm 4.3]. On the other hand, if $\ell \gg p$, then field operations in $\mathbb{F}_p$ is faster than field operations in $\mathbb{F}_\ell$, and [5, Algorithm 4.3] may be the more efficient one. Hence, the choice of algorithm to compute the full embedding degree depends strongly on the values of ℓ and p in the implementation.

7 Anti-symmetric Pairings on the Jacobian

On $\mathcal{J}_C[\ell]$, a non-degenerate, bilinear, anti-symmetric and Galois-invariant pairing

$$\varepsilon : \mathcal{J}_C[\ell] \times \mathcal{J}_C[\ell] \to \mu_\ell = \langle \zeta \rangle \subseteq \mathbb{F}_{q^k}^\times$$

exists, e.g. the Weil pairing; cf. e.g. [19, chapter 12]. Here, μ_ℓ is the group of ℓth roots of unity. A fast algorithm for computing the Weil pairing is given in [3]. Since ε is bilinear, it is given by

$$\varepsilon(x, y) = \zeta^{x^T \mathcal{E} y} , \tag{2}$$

for some matrix $\mathcal{E} \in \mathrm{Mat}_4(\mathbb{Z}/\ell\mathbb{Z})$ with respect to a basis $\mathcal{B} = \{x_1, x_2, x_3, x_4\}$ of $\mathcal{J}_C[\ell]$.

Remark 18. To be more precise, the points x and y on the right hand of equation (2) should be replaced by their column vectors $[x]_\mathcal{B}$ and $[y]_\mathcal{B}$ with respect to $\mathcal{B}$. To ease notation, this has been omitted.

Let φ denote the q-power Frobenius endomorphism on $\mathcal{J}_C$. Since ε is Galois-invariant,

$$\forall x, y \in \mathcal{J}_C[\ell] : \varepsilon(x, y)^q = \varepsilon(\varphi(x), \varphi(y)) \ .$$

This is equivalent to

$$\forall x, y \in \mathcal{J}_C[\ell] : q(x^T \mathcal{E} y) = (Mx)^T \mathcal{E} (My) \ ,$$

where M is the matrix representation of φ on $\mathcal{J}_C[\ell]$ with respect to $\mathcal{B}$. Since $(Mx)^T \mathcal{E}(My) = x^T M^T \mathcal{E} M y$, it follows that

$$\forall x, y \in \mathcal{J}_C[\ell] : x^T q \mathcal{E} y = x^T M^T \mathcal{E} M y \ ,$$

or equivalently, that $q\mathcal{E} = M^T \mathcal{E} M$.

Now, let $\varepsilon(x_i, x_j) = \zeta^{a_{ij}}$. By anti-symmetry,

$$\mathcal{E} = \begin{bmatrix} 0 & a_{12} & a_{13} & a_{14} \\ -a_{12} & 0 & a_{23} & a_{24} \\ -a_{13} & -a_{23} & 0 & a_{34} \\ -a_{14} & -a_{24} & -a_{34} & 0 \end{bmatrix} .$$

At first, assume that φ is represented by a matrix of the form (1) with respect to $\mathcal{B}$. Since $M^T \mathcal{E} M = q\mathcal{E}$, it follows that

$$a_{14} - qa_{13} \equiv a_{23} - a_{24} \equiv a_{14}(c-(1+q)) \equiv a_{24}(c-(1+q)) \equiv 0 \pmod{\ell} .$$

Thus, $a_{13} \equiv a_{14} \equiv a_{23} \equiv a_{24} \equiv 0 \pmod{\ell}$, cf. Theorem 14. So

$$\mathcal{E} = \begin{bmatrix} 0 & a_{12} & 0 & 0 \\ -a_{12} & 0 & 0 & 0 \\ 0 & 0 & 0 & a_{34} \\ 0 & 0 & -a_{34} & 0 \end{bmatrix} .$$

Since ε is non-degenerate, $a_{12}^2 a_{34}^2 = \det \mathcal{E} \not\equiv 0 \pmod{\ell}$.

Finally, assume that φ is represented by a diagonal matrix $\mathrm{diag}(1, q, \alpha, q/\alpha)$ with respect to $\mathcal{B}$. Then it follows from $M^T \mathcal{E} M = q\mathcal{E}$, that

$$a_{13}(\alpha - q) \equiv a_{14}(\alpha - 1) \equiv a_{23}(\alpha - 1) \equiv a_{24}(\alpha - q) \equiv 0 \pmod{\ell} .$$

If $\alpha \equiv 1, q \pmod{\ell}$, then $\mathcal{J}_C(\mathbb{F}_q)[\ell]$ is bi-cyclic. Hence the following theorem holds.

Theorem 19. *Consider a Jacobian $\mathcal{J}_C \in \mathbb{J}(\ell, q, k, \tau_k)$. Let φ be the q-power Frobenius endomorphism on $\mathcal{J}_C$. Choose a basis $\mathcal{B}$ of $\mathcal{J}_C[\ell]$, such that φ is represented by either a diagonal matrix* $\mathrm{diag}(1, q, \alpha, q/\alpha)$ *or a matrix of the form* (1) *with respect to $\mathcal{B}$. If the $\mathbb{F}_q$-rational subgroup $\mathcal{J}_C(\mathbb{F}_q)[\ell]$ of ℓ-torsion points on the Jacobian is cyclic, then all non-degenerate, bilinear, anti-symmetric and Galois-invariant pairings on $\mathcal{J}_C[\ell]$ are given by the matrices*

$$\mathcal{E}_{a,b} = \begin{bmatrix} 0 & a & 0 & 0 \\ -a & 0 & 0 & 0 \\ 0 & 0 & 0 & b \\ 0 & 0 & -b & 0 \end{bmatrix}, \qquad a, b \in (\mathbb{Z}/\ell\mathbb{Z})^\times$$

with respect to $\mathcal{B}$.

Remark 20. Let notation and assumptions be as in Theorem 19. Let ε be a non-degenerate, bilinear, anti-symmetric and Galois-invariant pairing on $\mathcal{J}_C[\ell]$, and let ε be given by $\mathcal{E}_{a,b}$ with respect to a basis $\{x_1, x_2, x_3, x_4\}$ of $\mathcal{J}_C[\ell]$. Then ε is given by $\mathcal{E}_{1,1}$ with respect to $\{a^{-1}x_1, x_2, b^{-1}x_3, x_4\}$.

Remark 21. In cases relevant to pairing based cryptography, we consider a prime divisor ℓ of size q^2. Assume ℓ is of size q^2. Then ℓ divides neither q nor $q-1$. The number of $\mathbb{F}_q$-rational points on the Jacobian is approximately q^2. Thus, $\mathcal{J}_C(\mathbb{F}_q)[\ell]$ is cyclic in cases relevant to pairing based cryptography.

8 Generators of $\mathcal{J}_C[\ell]$

Consider a Jacobian $\mathcal{J}_C \in \mathbb{J}(\ell, q, k, \tau_k)$. Assume the $\mathbb{F}_q$-rational subgroup of ℓ-torsion points $\mathcal{J}_C(\mathbb{F}_q)[\ell]$ is cyclic. Let φ be the q-power Frobenius endomorphism of $\mathcal{J}_C$. Let ε be a non-degenerate, bilinear, anti-symmetric and Galois-invariant pairing

$$\varepsilon : \mathcal{J}_C[\ell] \times \mathcal{J}_C[\ell] \to \mu_\ell = \langle \zeta \rangle \subseteq \mathbb{F}_{q^k}^\times \ .$$

In the following, frequently we will choose a random point $P \in \mathcal{J}_C(\mathbb{F}_{q^a})[\ell]$ for some power $a \in \mathbb{N}$. This is done as follows: (1) Choose a random point $P \in \mathcal{J}_C(\mathbb{F}_{q^a})$. (2) Compute $P := [m](P)$, where $|\mathcal{J}_C(\mathbb{F}_{q^a})| = m\ell^s$ and $\ell \nmid m$. (3) Compute the order $|P| = \ell^{t(P)}$ of P. (4) If $t(P) > 0$, then let $P := [\ell^{t(P)-1}](P)$. Since the power $t(P)$ will be different for each point P, this procedure does not define a group homomorphism from $\mathcal{J}_C(\mathbb{F}_{q^a})$ to $\mathcal{J}_C(\mathbb{F}_{q^a})[\ell]$. Thus, the image of points uniformly distributed in $\mathcal{J}_C(\mathbb{F}_{q^a})$ will not necessarily be uniformly distributed in $\mathcal{J}_C(\mathbb{F}_{q^a})[\ell]$. A method of choosing points uniformly at random is given in [5, Section 5.3], but it leads to a significant extra cost. In practice we believe it is better to not use the method in [5], even though this means one might need to sample a few extra points.

We consider the cases where $\ell \nmid \tau_k$ and where $\ell \mid \tau_k$ separately.

8.1 The Case $\ell \nmid \tau_k$

If ℓ does not divide τ_k, then $\mathcal{J}_C(\mathbb{F}_{q^k})[\ell]$ is bicyclic; cf. Theorem 8. Choose a random point $\mathcal{O} \neq x_1 \in \mathcal{J}_C(\mathbb{F}_q)[\ell]$, and extend $\{x_1\}$ to a basis $\{x_1, y_2\}$ of $\mathcal{J}_C(\mathbb{F}_{q^k})[\ell]$, where $\varphi(y_2) = qy_2$. Let $x_2' \in \mathcal{J}_C(\mathbb{F}_{q^k})[\ell]$ be a random point. If $x_2' \in \mathcal{J}_C(\mathbb{F}_q)[\ell]$, then choose another random point $x_2' \in \mathcal{J}_C(\mathbb{F}_{q^k})[\ell]$. After two trials, $x_2' \notin \mathcal{J}_C(\mathbb{F}_q)[\ell]$ with probability $1 - {}^1/_{\ell^2}$. Hence, we may ignore the case where $x_2' \in \mathcal{J}_C(\mathbb{F}_q)[\ell]$. Write $x_2' = \alpha_1 x_1 + \alpha_2 y_2$. Then

$$\mathcal{O} \neq x_2 = x_2' - \varphi(x_2') = \alpha_2(1-q)y_2 \in \langle y_2 \rangle \ ,$$

i.e. $\varphi(x_2) = qx_2$. Now, let $\mathcal{J}_C[\ell] \simeq \mathcal{J}_C(\mathbb{F}_{q^k})[\ell] \oplus W$, where W is a φ-invariant submodule of rank two. Choose a random point $x_3' \in \mathcal{J}_C[\ell]$. Since $x_3' - \varphi(x_3') \in \langle y_2 \rangle \oplus W$, we may assume that $x_3' \in \langle y_2 \rangle \oplus W$. But then

$$x_3 = qx_3' - \varphi(x_3') \in W$$

as above. If $\varphi(x_3') = qx_3'$, then $x_3' \in \mathcal{J}_C(\mathbb{F}_{q^k})[\ell]$. This will only happen with probability ${}^1/_{\ell^2}$. Hence, we may ignore this case. Notice that

$$\mathcal{J}_C[\ell] = \langle x_1, x_2, x_3, \varphi(x_3) \rangle \text{ if and only if } \varepsilon(x_3, \varphi(x_3)) \neq 1;$$

cf. Theorem 19.

Assume $\varepsilon(x_3, \varphi(x_3)) = 1$. Then x_3 is an eigenvector of φ. Let $\varphi(x_3) = \alpha x_3$. Then the Weil polynomial of $\mathcal{J}_C$ is given by

$$P(X) \equiv (X-1)(X-q)(X-\alpha)(X-q/\alpha) \pmod{\ell}$$

modulo ℓ. Assume $\alpha \equiv q/\alpha \pmod{\ell}$. Then $\alpha^2 \equiv q \pmod{\ell}$, and it follows that the characteristic polynomial of φ^k is given by

$$P_k(X) \equiv (X-1)^2(X+1)^2 \equiv X^4 - 2q^kX^2 + q^{2k} \pmod{\ell}$$

modulo ℓ. But then $\ell \mid \tau_k$. This is a contradiction. So $\alpha \not\equiv q/\alpha \pmod{\ell}$. Therefore, we can extend $\{x_1, x_2, x_3\}$ to a basis $\mathcal{B} = \{x_1, x_2, x_3, x_4\}$ of $\mathfrak{J}_C[\ell]$, such that φ is represented by a diagonal matrix on $\mathfrak{J}_C[\ell]$ with respect to $\mathcal{B}$. We may assume that ε is given by $\mathcal{E}_{1,1}$ with respect to $\mathcal{B}$; cf. Remark 20.

Now, choose a random point $x \in \mathfrak{J}_C[\ell]$. Write $x = \alpha_1x_1 + \alpha_2x_2 + \alpha_3x_3 + \alpha_4x_4$. Then $\varepsilon(x_3, x) = \zeta^{\alpha_4}$. So $\varepsilon(x_3, x) \neq 1$ if and only if ℓ does not divide α_4. On the other hand, $\{x_1, x_2, x_3, x\}$ is a basis of $\mathfrak{J}_C[\ell]$ if and only ℓ does not divide α_4. Thus, if ℓ does not divide τ_k, then the following Algorithm 22 outputs generators of $\mathfrak{J}_C[\ell]$ with probability at least $1 - 1/\ell^n$.

Algorithm 22. *On input a Jacobian $\mathfrak{J}_C \in \mathbb{J}(\ell, q, k, \tau_k)$, the numbers ℓ, q, k and τ_k, the full embedding degree k_0 of $\mathfrak{J}_C$ with respect to ℓ and a number $n \in \mathbb{N}$, if ℓ does not divide τ_k, then the following algorithm outputs a basis of $\mathfrak{J}_C[\ell]$ or "failure".*

1. *Choose points $\mathcal{O} \neq x_1 \in \mathfrak{J}_C(\mathbb{F}_q)[\ell]$, $x_2 \in \mathfrak{J}_C(\mathbb{F}_{q^k})[\ell]$ and $x_3' \in \mathfrak{J}_C(\mathbb{F}_{q^{k_0}})[\ell]$; compute $x_3 = q(x_3' - \varphi(x_3')) - \varphi(x_3' - \varphi(x_3'))$. If $\varepsilon(x_3, \varphi(x_3)) \neq 1$, then output $\{x_1, x_2, x_3, \varphi(x_3)\}$ and stop.*
2. *Let $i = j = 0$. While $i < n$ do the following:*
   (a) Choose a random point $x_4 \in \mathfrak{J}_C(\mathbb{F}_{q^{k_0}})[\ell]$.
   (b) If $\varepsilon(x_3, x_4) = 1$, then $i := i + 1$. Else $i := n$ and $j := 1$.
3. *If $j = 0$, then output "failure". Else output $\{x_1, x_2, x_3, x_4\}$.*

8.2 The Case $\ell \mid \tau_k$

Assume ℓ divides τ_k. Then $\mathfrak{J}_C[\ell] \subseteq \mathfrak{J}_C(\mathbb{F}_{q^k})$; cf. Theorem 9. Choose a random point $\mathcal{O} \neq x_1 \in \mathfrak{J}_C(\mathbb{F}_q)[\ell]$, and let $y_2 \in \mathfrak{J}_C[\ell]$ be a point with $\varphi(y_2) = qy_2$. Write $\mathfrak{J}_C[\ell] = \langle x_1, y_2 \rangle \oplus W$, where W is a φ-invariant submodule of rank two; cf. the proof of Theorem 14. Let $\{y_3, y_4\}$ be a basis of W, such that φ is represented on $\mathfrak{J}_C[\ell]$ with respect to the basis $\mathcal{B} = \{x_1, y_2, y_3, y_4\}$ by either a diagonal matrix

$$M_1 = \mathrm{diag}(1, q, \alpha, q/\alpha) \ ,$$

or a matrix of the form

$$M_2 = \begin{bmatrix} 1 & 0 & 0 & 0 \\ 0 & q & 0 & 0 \\ 0 & 0 & 0 & -q \\ 0 & 0 & 1 & c \end{bmatrix} \ ,$$

where $c \not\equiv q + 1 \pmod{\ell}$; cf. Theorem 14.

Now, choose a random point $z \in \mathcal{J}_C[\ell]$. Since $z - \varphi(z) \in \langle y_2, y_3, y_4 \rangle$, we may assume that $z \in \langle y_2, y_3, y_4 \rangle$. Write $z = \alpha_2 y_2 + \alpha_3 y_3 + \alpha_4 y_4$. Assume at first that φ is represented on $\mathcal{J}_C[\ell]$ by M_1 with respect to $\mathcal{B}$. Then

$$\begin{aligned} qz - \varphi(z) &= \alpha_2 q y_2 + \alpha_3 q y_3 + \alpha_4 q y_4 - (\alpha_2 q y_2 + \alpha_3 \alpha y_3 + \alpha_4 (q/\alpha) y_4) \\ &= \alpha_3 (q - \alpha) y_3 + \alpha_4 (q - q/\alpha) y_4; \end{aligned}$$

so $qz - \varphi(z) \in \langle y_3, y_4 \rangle$. If $qz - \varphi(z) = 0$, then it follows that $q \equiv 1 \pmod{\ell}$. This contradicts the choice of the Jacobian $\mathcal{J}_C \in \mathbb{J}(\ell, q, k, \tau_k)$. Hence, we have a procedure to choose a point $\mathcal{O} \neq w \in W$ in this case. Now assume that φ is represented on $\mathcal{J}_C[\ell]$ by M_2 with respect to $\mathcal{B}$. Then

$$\begin{aligned} qz - \varphi(z) &= \alpha_2 q y_2 + \alpha_3 q y_3 + \alpha_4 q y_4 - (\alpha_2 q y_2 + \alpha_3 y_4 + \alpha_4 (-q y_3 + c y_4)) \\ &= q(\alpha_3 + \alpha_4) y_3 + (\alpha_4 q - \alpha_3 - \alpha_4 c) y_4; \end{aligned}$$

so again $qz - \varphi(z) \in \langle y_3, y_4 \rangle$. If $qz - \varphi(z) = 0$, then it follows that $c \equiv q + 1 \pmod{\ell}$. This is a contradiction. Hence, we have a procedure to choose a point $\mathcal{O} \neq w \in W$ also in this case.

Choose random points $x_3, x_4 \in W$. Write $x_i = \alpha_{i3} y_3 + \alpha_{i4} y_4$ for $i = 3, 4$. We may assume that ε is given by $\mathcal{E}_{1,1}$ with respect to $\mathcal{B}$; cf. Remark 20. But then $\varepsilon(x_3, x_4) = \zeta^{\alpha_{33}\alpha_{44} - \alpha_{34}\alpha_{43}}$. Hence, $\varepsilon(x_3, x_4) = 1$ if and only if $\alpha_{33}\alpha_{44} \equiv \alpha_{34}\alpha_{43} \pmod{\ell}$. So $\varepsilon(x_3, x_4) \neq 1$ with probability $1 - 1/\ell$. Hence, we have a procedure to find a basis of W.

Until now, we have found points $x_1 \in \mathcal{J}_C(\mathbb{F}_q)[\ell]$ and $x_3, x_4 \in W$, such that $W = \langle x_3, x_4 \rangle$. Now, choose a random point $x_2 \in \mathcal{J}_C[\ell]$. Write $x_2 = \alpha_1 x_1 + \alpha_2 y_2 + \alpha_3 y_3 + \alpha_4 y_4$. Then $\varepsilon(x_1, x_2) = \zeta^{\alpha_2}$, i.e. $\varepsilon(x_1, x_2) = 1$ if and only if $\alpha_2 \equiv 0 \pmod{\ell}$. Thus, with probability $1 - 1/\ell$, the set $\{x_1, x_2, x_3, x_4\}$ is a basis of $\mathcal{J}_C[\ell]$.

Summing up, if ℓ divides τ_k, then the following Algorithm 23 outputs generators of $\mathcal{J}_C[\ell]$ with probability at least $(1 - 1/\ell^n)^2$.

Algorithm 23. *On input a Jacobian $\mathcal{J}_C \in \mathbb{J}(\ell, q, k, \tau_k)$, the numbers ℓ, q, k and τ_k, the full embedding degree k_0 of $\mathcal{J}_C$ with respect to ℓ and a number $n \in \mathbb{N}$, if ℓ divides τ_k, then the following algorithm outputs a basis of $\mathcal{J}_C[\ell]$ or "failure".*

1. *Choose a random point $\mathcal{O} \neq x_1 \in \mathcal{J}_C(\mathbb{F}_q)[\ell]$.*
2. *Let $i = j = 0$. While $i < n$ do the following:*
   (a) *Choose a random point $x_2 \in \mathcal{J}_C(\mathbb{F}_{q^{k_0}})[\ell]$.*
   (b) *If $\varepsilon(x_1, x_2) = 1$, then $i := i + 1$. Else $i := n$ and $j := 1$.*
3. *If $j = 0$, then output "failure" and stop.*
4. *Let $i = j = 0$. While $i < n$ do the following:*
   (a) *Choose random points $y_3, y_4 \in \mathcal{J}_C(\mathbb{F}_{q^{k_0}})[\ell]$; compute $x_\nu := q(y_\nu - \varphi(y_\nu)) - \varphi(y_\nu - \varphi(y_\nu))$ for $\nu = 3, 4$.*
   (b) *If $\varepsilon(x_3, x_4) = 1$, then $i := i + 1$. Else $i := n$ and $j := 1$.*
5. *If $j = 0$, then output "failure". Else output $\{x_1, x_2, x_3, x_4\}$.*

8.3 The Complete Algorithm

Combining Algorithm 22 and 23, we obtain the desired algorithm to find generators of $\mathfrak{J}_C[\ell]$.

Algorithm 24. *On input a Jacobian $\mathfrak{J}_C \in \mathbb{J}(\ell, q, k, \tau_k)$, the numbers ℓ, q, k and τ_k, the full embedding degree k_0 of $\mathfrak{J}_C$ with respect to ℓ and a number $n \in \mathbb{N}$, the following algorithm outputs a basis of $\mathfrak{J}_C[\ell]$ or "failure".*

1. *If $\ell \nmid \tau_k$, run Algorithm 22 on input $(\mathfrak{J}_C, \ell, q, k, \tau_k, k_0, n)$.*
2. *If $\ell \mid \tau_k$, run Algorithm 23 on input $(\mathfrak{J}_C, \ell, q, k, \tau_k, k_0, n)$.*

Theorem 25. *Let $\mathfrak{J}_C$ be a $\mathbb{J}(\ell, q, k, \tau_k)$-Jacobian of full embedding degree k_0 with respect to ℓ. On input $(\mathfrak{J}_C, \ell, q, k, \tau_k, k_0, n)$, Algorithm 24 outputs generators of $\mathfrak{J}_C[\ell]$ with probability at least $(1 - 1/\ell^n)^2$. We expect Algorithm 24 to run in*

$$O\left(\log \ell \log \frac{q^{k_0} - 1}{\ell} {k_0}^3 \log k_0 \log q\right)$$

field operations in $\mathbb{F}_q$ (ignoring $\log \log q$ factors).

Proof. We must compare the cost of the steps in Algorithm 24. From [5, proof of Proposition 4.6], [7, proof of Corollary 1] and [17] we get the following estimates: (1) Choosing a random point on $\mathfrak{J}_C(\mathbb{F}_{q^a})$ for some power $a \in \mathbb{N}$ takes $O(a \log q)$ field operations in $\mathbb{F}_{q^a}$, and computing a multiple $[m](P)$ of a point $P \in \mathfrak{J}_C(\mathbb{F}_{q^a})$ takes $O(a \log q)$ field operations in $\mathbb{F}_{q^a}$. (2) Evaluating the q^a-power Frobenius endomorphism of the Jacobian on a point $P \in \mathfrak{J}_C[\ell]$ takes $O(a \log q)$ field operations in $\mathbb{F}_{q^a}$. (3) Evaluating the Tate pairing on two point of $\mathfrak{J}_C(\mathbb{F}_{q^{k_0}})[\ell]$ takes $O(\log \ell)$ field operations in $\mathbb{F}_{q^{k_0}}$. The Weil pairing can be computed by computing two Tate pairings, raising the results to the power $\frac{q^{k_0}-1}{\ell}$ and finally computing the quotient of these numbers; see [8]. The exponentiation takes $O(\log \frac{q^{k_0}-1}{\ell})$ field operations in $\mathbb{F}_{q^{k_0}}$, and a division takes $O({k_0}^2)$ field operations in $\mathbb{F}_{q^{k_0}}$. Hence, evaluating the Weil pairing on two point of $\mathfrak{J}_C(\mathbb{F}_{q^{k_0}})[\ell]$ takes $O(\log \ell) O(\log \frac{q^{k_0}-1}{\ell}) O({k_0}^2)$ field operations in $\mathbb{F}_{q^{k_0}}$. (4) By using fast multiplication techniques, one field operation in $\mathbb{F}_{q^a}$ takes $O(\log q^a \log \log q^a) = O(a \log a \log q)$ field operations in $\mathbb{F}_q$ (ignoring $\log \log q$ factors).

We see that the pairing computation is the most expensive step in Algorithm 24. Thus, Algorithm 24 runs in $O(\log \ell \log \frac{q^{k_0}-1}{\ell} {k_0}^3 \log k_0 \log q)$ field operations in $\mathbb{F}_q$ (ignoring $\log \log q$ factors). □

9 Implementation Issues

To check if ℓ ramifies in $\mathbb{Q}(\omega_k)$ in the case where ℓ divides τ_k, a priori we need to find a q^k-Weil number ω_k of the Jacobian $\mathfrak{J}_C$. On Jacobians generated by the *complex multiplication method* [23, 10, 4], we know the Weil numbers in advance. Hence, Algorithm 24 is particularly well suited for such Jacobians.

Fortunately, most likely ℓ does not divide τ_k, and then we do not have to find a q^k-Weil number (ℓ divides a random number $n \in \mathbb{Z}$ with vanishing probability $1/\ell$). And if the Weil polynomial splits in distinct linear factors modulo ℓ, then we do not even have to compute τ_k. To see this, assume that the Weil polynomial of $\mathcal{J}_C$ splits as

$$P(X) \equiv (X-1)(X-q)(X-\alpha)(X-q/\alpha) \pmod{\ell} ,$$

where $\alpha \not\equiv 1, q, q/\alpha \pmod{\ell}$. Let φ be the q-power Frobenius endomorphism of $\mathcal{J}_C$, and let $P_k(X)$ be the characteristic polynomial of φ^k. Then

$$P_k(X) \equiv (X-1)^2(X-\alpha^k)(X-1/\alpha^k) \pmod{\ell} .$$

If ℓ divides τ_k, then $\mathcal{J}_C[\ell] \subseteq \mathcal{J}_C(\mathbb{F}_{q^k})$; cf. Theorem 9. But then $P_k(X) \equiv (X-1)^4 \pmod{\ell}$. Hence,

$$\ell \text{ divides } \tau_k \text{ if and only if } \alpha^k \equiv 1 \pmod{\ell}. \tag{3}$$

Assume $\alpha^k \equiv 1 \pmod{\ell}$. Then $P_k(X) \equiv (X-1)^4 \pmod{\ell}$. Hence,

$$\ell \text{ ramifies in } \mathbb{Q}(\omega^k) \text{ if and only if } \omega^k \notin \mathbb{Z}. \tag{4}$$

See [20, Proposition 8.3, p. 47]. Here, ω is a q-Weil number of $\mathcal{J}_C$.

Consider the case where $\alpha^k \equiv 1 \pmod{\ell}$ and $\omega^k \in \mathbb{Z}$. Then $\omega = \sqrt{q}e^{in\pi/k}$ for some $n \in \mathbb{Z}$ with $0 < n < k$. Assume k divides mn for some $m < k$. Then $\omega^{2m} = q^m \in \mathbb{Z}$. Since the q-power Frobenius endomorphism is the identity on the $\mathbb{F}_q$-rational points on the Jacobian, it follows that $\omega^{2m} \equiv 1 \pmod{\ell}$. Hence, $q^m \equiv 1 \pmod{\ell}$, i.e. k divides m. This is a contradiction. So n and k has no common divisors. Let $\xi = \omega^2/q = e^{in2\pi/k}$. Then ξ is a primitive k^{th} root of unity, and $\mathbb{Q}(\xi) \subseteq \mathbb{Q}(\omega)$. Since $[\mathbb{Q}(\omega):\mathbb{Q}] \leq 4$ and $[\mathbb{Q}(\xi):\mathbb{Q}] = \phi(k)$, where ϕ is the Euler phi function, it follows that $k \leq 12$. Hence,

$$\text{if } \alpha^k \equiv 1 \pmod{\ell}, \text{ then } \omega^k \in \mathbb{Z} \text{ if and only if } k \leq 12. \tag{5}$$

The criteria (3), (4) and (5) provides the following efficient algorithm to check whether a given Jacobian is of type $\mathbb{J}(\ell, q, k, \tau_k)$, and whether ℓ divides τ_k.

Algorithm 26. *Let $\mathcal{J}_C$ be the Jacobian of a genus two curve C. Assume that the odd prime number ℓ divides the number of $\mathbb{F}_q$-rational points on $\mathcal{J}_C$, and that ℓ divides neither q nor $q-1$. Let k be the multiplicative order of q modulo ℓ.*

1. *Compute the Weil polynomial $P(X)$ of $\mathcal{J}_C$. Let $P(X) \equiv \prod_{i=1}^{4}(X-\alpha_i) \pmod{\ell}$.*
2. *If $\alpha_i^k \not\equiv 1 \pmod{\ell}$ for an $i \in \{1,2,3,4\}$, then output "$\mathcal{J}_C \in \mathbb{J}(\ell, q, k, \tau_k)$ and ℓ does not divide τ_k" and stop.*
3. *If $k > 12$ then output "$\mathcal{J}_C \notin \mathbb{J}(\ell, q, k, \tau_k)$" and stop.*
4. *Output "$\mathcal{J}_C \in \mathbb{J}(\ell, q, k, \tau_k)$ and ℓ divides τ_k" and stop.*

References

[1] Boneh, D., Franklin, M.: Identity-based encryption from the weil pairing. SIAM J. Computing 32(3), 586–615 (2003)

[2] Cassels, J.W.S., Flynn, E.V.: Prolegomena to a Middlebrow Arithmetic of Curves of Genus 2. London Mathematical Society Lecture Note Series. Cambridge University Press, Cambridge (1996)

[3] Duursma, I., Lee, H.-S.: Tate pairing implementation for hyperelliptic curves $y^2 = x^p - x + d$. In: Laih, C.-S. (ed.) ASIACRYPT 2003. LNCS, vol. 2894, pp. 111–123. Springer, Heidelberg (2003)

[4] Eisenträger, K., Lauter, K.: A CRT algorithm for constructing genus 2 curves over finite fields, arXiv:math/0405305, to appear in Proceedings of AGCT-10 (preprint, 2007)

[5] Freeman, D., Lauter, K.: Computing endomorphism rings of jacobians of genus 2 curves over finite fields. In: Hirschfeld, J., Chaumine, J., Rolland, R. (eds.) Algebraic geometry and its applications, Proceedings of the First SAGA conference, Papeete, May 7–11, 2007. Number Theory and Its Applications, vol. 5, pp. 29–66. World Scientific, Singapore (2008)

[6] Frey, G., Lange, T.: Varieties over special fields. In: Cohen, H., Frey, G. (eds.) Handbook of Elliptic and Hyperelliptic Curve Cryptography, pp. 87–113. Chapman & Hall/CRC (2006)

[7] Frey, G., Rück, H.-G.: A remark concerning m-divisibility and the discrete logarithm in the divisor class group of curves. Math. Comp. 62, 865–874 (1994)

[8] Galbraith, S.D.: Pairings. In: Blake, I.F., Seroussi, G., Smart, N.P. (eds.) Advances in Elliptic Curve Cryptography. London Mathematical Society Lecture Note Series, vol. 317, pp. 183–213. Cambridge University Press, Cambridge (2005)

[9] Galbraith, S.D., Hess, F., Vercauteren, F.: Hyperelliptic pairings. In: Takagi, T., Okamoto, T., Okamoto, E., Okamoto, T. (eds.) Pairing 2007. LNCS, vol. 4575, pp. 108–131. Springer, Heidelberg (2007)

[10] Gaudry, P., Houtmann, T., Kohel, D., Ritzenthaler, C., Weng, A.: The p-adic CM-method for genus 2 (preprint, 2005) arXiv:math/0503148

[11] Hess, F.: A note on the tate pairing of curves over finite fields. Arch. Math. 82, 28–32 (2004)

[12] Howe, E.W., Nart, E., Ritzenthaler, C.: Jacobians in isogeny classes of abelian surfaces over finite fields (preprint, 2007) arXiv:math/0607515

[13] Koblitz, N.: Elliptic curve cryptosystems. Math. Comp. 48, 203–209 (1987)

[14] Koblitz, N.: Hyperelliptic cryptosystems. J. Cryptology 1, 139–150 (1989)

[15] Lang, S.: Abelian Varieties. Interscience (1959)

[16] Maisner, D., Nart, E., Howe, E.W.: Abelian surfaces over finite fields as jacobians. Experimental Mathematics 11(3), 321–337 (2002)

[17] Menezes, A., van Oorschot, P., Vanstone, S.: Handbook of Applied Cryptography. CRC Press, Boca Raton (1997)

[18] Miller, V.S.: The weil pairing, and its efficient calculation. J. Cryptology 17, 235–261 (2004)

[19] Milne, J.S.: Abelian varieties (1998), `http://www.jmilne.org`

[20] Neukirch, J.: Algebraic Number Theory. Springer, Heidelberg (1999)

[21] Ravnshøj, C.R.: Generators of Jacobians of hyperelliptic curves, (preprint, 2007) arXiv:0704.3339

[22] Ravnshøj, C.R.: Non-cyclic subgroups of Jacobians of genus two curves (preprint, 2008) arXiv:0801.2835

[23] Weng, A.: Constructing hyperelliptic curves of genus 2 suitable for cryptography. Math. Comp. 72, 435–458 (2003)

Speeding Up Pairing Computations on Genus 2 Hyperelliptic Curves with Efficiently Computable Automorphisms

Xinxin Fan[1,⋆], Guang Gong[1,⋆], and David Jao[2,⋆⋆]

[1] Department of Electrical and Computer Engineering
[2] Department of Combinatorics and Optimization
University of Waterloo
Waterloo, Ontario, N2L 3G1, CANADA
{x5fan@engmail,ggong@calliope,djao@math}.uwaterloo.ca

Abstract. Pairings on the Jacobians of (hyper-)elliptic curves have received considerable attention not only as a tool to attack curve based cryptosystems but also as a building block for constructing cryptographic schemes with new and novel properties. Motivated by the work of Scott, we investigate how to use efficiently computable automorphisms to speed up pairing computations on two families of non-supersingular genus 2 hyperelliptic curves over prime fields. Our findings lead to new variants of Miller's algorithm in which the length of the main loop can be up to 4 times shorter than that of the original Miller's algorithm in the best case. We also implement the calculation of the Tate pairing on both a supersingular and a non-supersingular genus 2 curve with the same embedding degree of $k = 4$. Combining the new algorithm with known optimization techniques, we show that pairing computations on non-supersingular genus 2 curves over prime fields use up to 55.8% fewer field operations and run about 10% faster than supersingular genus 2 curves for the same security level.

Keywords: Genus 2 non-supersingular hyperelliptic curves, Tate pairing, Miller's algorithm, Automorphism, Efficient implementation.

1 Introduction

Pairing based cryptography is a relatively new area of cryptography centered around particular functions with interesting properties. Initially, bilinear pairings were introduced to cryptography for attacking instances of the discrete logarithm problem (DLP) on elliptic curves and hyperelliptic curves [14,28]. With the advent of non-interactive key distribution [33], tripartite key exchange [24], and identity based encryption [5], the design of pairing based cryptographic protocols has received a lot of attention from the research community. The implementation of pairing based protocols requires the efficient computation of

⋆ Supported by NSERC Strategic Project Grants (SPG).
⋆⋆ Partially supported by NSERC.

S.D. Galbraith and K.G. Paterson (Eds.): Pairing 2008, LNCS 5209, pp. 243–264, 2008.

pairings. To date, the Weil and Tate pairings and their variants such as the Eta and Ate pairings on Jacobians of (hyper-)elliptic curves are the only efficient instantiations of cryptographically useful bilinear maps.

There has been a lot of work on efficient implementation of pairings on elliptic curves, and many important techniques have been proposed to accelerate the computation of the Tate pairing and its variants on elliptic curves [2,3,4,22]. Furthermore, the subject of pairing computations on hyperelliptic curves is also receiving an increasing amount of attention. Choie and Lee [6] investigated the implementation of the Tate pairing on supersingular genus 2 hyperelliptic curves over prime fields. Later on, hÉigeartaigh and Scott [21] improved the implementation of [6] significantly by using a new variant of Miller's algorithm combined with various optimization techniques. Duursma and Lee [10] presented a closed formula for the Tate pairing computation on a very special family of supersingular hyperelliptic curves. Barreto *et. al.* [2] generalized the results of Duursma and Lee and proposed the Eta pairing approach for efficiently computing the Tate pairing on supersingular genus 2 curves over binary fields. In particular, their algorithm leads to the fastest pairing implementation in the literature. In [27], Lee *et. al.* considered the Eta pairing computation on general divisors on supersingular genus 3 hyperelliptic curves of the form of $y^2 = x^7 - x \pm 1$. Recently, the Ate pairing, originally defined for elliptic curves, has been generalized to hyperelliptic curves [18] as well. Although the Eta and Ate pairings hold the record for speed at the present time, we will focus on the Tate pairing in this paper. The main reason is that the Tate pairing is uniformly available across a wide range of hyperelliptic curves and subgroups, whereas the Eta pairing is only defined for supersingular curves and the Ate pairing incurs a huge performance penalty in the context of ordinary genus 2 curves [18, Table 6].

Motivated by previous work in [34,38,39], we consider pairing computations on two families of non-supersingular genus 2 hyperelliptic curves over prime fields. We first explicitly give efficiently computable automorphisms (also isogenies) and the dual isogenies on the divisor class group of these curves. We then exploit the automorphism in lieu of the order of the torsion group to construct the rational functions required in Miller's algorithm, and shorten the length of the main loop in Miller's algorithm as a result. Based on the new construction of the rational functions, we propose new variants of Miller's algorithm for computing the Tate pairing on certain non-supersingular genus 2 curves over prime fields. In the best case, the length of the loop in our new variant can be up to 4 times shorter than that of the original Miller's algorithm. Furthermore, we generate a non-supersingular pairing-friendly genus 2 curve with embedding degree 4 and compare the performance of our new algorithm with that of the variant proposed by hÉigeartaigh and Scott [21] for supersingular genus 2 curves. Theoretical analysis shows that our new algorithm uses 55.8% fewer field operations than that of [21] for the same security level. However, the size of the field where the non-supersingular curve is defined is 1.285 times larger than that of the field used for supersingular curves, which somewhat offsets these gains. Nevertheless, our experimental results show that using the non-supersingular genus 2 curve one can still obtain a 10% performance improvement over the supersingular curve.

The rest of this paper is organized as follows. Section 2 gives a short introduction to the Tate pairing on hyperelliptic curves and Miller's algorithm for computing the pairing. In Section 3 we recall supersingular genus 2 curves over prime fields which have been used for pairing computations, and introduce two families of non-supersingular genus 2 curves with efficiently computable automorphisms. In Section 4 we prove the main results of our contribution and propose new variants of Miller's algorithm. Section 5 details the various techniques for efficiently implementing the Tate pairing on a non-supersingular genus 2 curve with embedding degree 4, analyzes the computational cost of our new algorithm and gives experimental results. Finally, Section 6 concludes this paper.

2 Mathematical Background

In this section, we present a brief introduction to the definition of the Tate pairing on hyperelliptic curves and also review Miller's algorithm for computing pairings. For more details, the reader is referred to [1].

2.1 Tate Pairing on Hyperelliptic Curves

Let $\mathbb{F}_q$ be a finite field with q elements, and $\overline{\mathbb{F}}_q$ be its algebraic closure. Let C be a hyperelliptic curve of genus g over $\mathbb{F}_q$, and let $\mathcal{J}_C$ denote the degree zero divisor class group of C. We say that a subgroup of the divisor class group $\mathcal{J}_C(\mathbb{F}_q)$ has *embedding degree* k if the order n of the subgroup divides $q^k - 1$, but does not divide $q^i - 1$ for any $0 < i < k$. For our purpose, n should be a (large) prime with $n \mid \#\mathcal{J}_C(\mathbb{F}_q)$ and $\gcd(n, q) = 1$. Let $\mathcal{J}_C(\mathbb{F}_{q^k})[n]$ be the n-torsion group and $\mathcal{J}_C(\mathbb{F}_{q^k})/n\mathcal{J}_C(\mathbb{F}_{q^k})$ be the quotient group. Then the Tate pairing is a well defined, non-degenerate, bilinear map [14]:

$$\langle \cdot, \cdot \rangle_n : \mathcal{J}_C(\mathbb{F}_{q^k})[n] \times \mathcal{J}_C(\mathbb{F}_{q^k})/n\mathcal{J}_C(\mathbb{F}_{q^k}) \to \mathbb{F}_{q^k}^*/(\mathbb{F}_{q^k}^*)^n,$$

defined as follows: let $D_1 \in \mathcal{J}_C(\mathbb{F}_{q^k})[n]$, with $\mathrm{div}(f_{n,D_1}) = nD_1$ for some rational function $f_{n,D_1} \in \mathbb{F}_{q^k}(C)^*$. Let $D_2 \in \mathcal{J}_C(\mathbb{F}_{q^k})/n\mathcal{J}_C(\mathbb{F}_{q^k})$ with $\mathrm{supp}(D_1) \cap \mathrm{supp}(D_2) = \emptyset$ (to ensure a non-trivial pairing value). The Tate pairing of two divisor classes $\overline{D}_1$ and $\overline{D}_2$ is then defined by

$$\langle \overline{D}_1, \overline{D}_2 \rangle_n = f_{n,D_1}(D_2) = \prod_{P \in C(\overline{\mathbb{F}}_q)} f_{n,D_1}(P)^{\mathrm{ord}_P(D_2)}.$$

Note that the Tate pairing as detailed above is only defined up to n-th powers. One can show that if the function f_{n,D_1} is properly normalized, we only need to evaluate the rational function f_{n,D_1} at the effective part of the reduced divisor D_2 in order to compute the Tate pairing [3,18].

In practice, the fact that the Tate pairing is only defined up to n-th power is usually undesirable, and many pairing-based protocols require a unique pairing value. Hence one defines the *reduced* pairing as

$$\langle \overline{D}_1, \overline{D}_2 \rangle_n^{(q^k-1)/n} = f_{n,D_1}(D_2)^{(q^k-1)/n} \in \mu_n \subset \mathbb{F}_{q^k}^*,$$

where $\mu_n = \{u \in \mathbb{F}_{q^k}^* \mid u^n = 1\}$ is the group of n-th roots of unity. In the rest of this paper we will refer to the extra powering required to compute the reduced pairing as the *final exponentiation.* One of the important properties of the reduced pairing we will use in this paper is that for any positive integer N satisfying $n \mid N$ and $N \mid q^k - 1$ we have

$$\langle D_1, D_2 \rangle_n^{(q^k-1)/n} = \langle D_1, D_2 \rangle_N^{(q^k-1)/N}. \tag{1}$$

2.2 Miller's Algorithm

The main task involved in the computation of the Tate pairing $\langle \overline{D}_1, \overline{D}_2 \rangle_n$ is to construct a rational function f_{n,D_1} such that $\mathrm{div}(f_{n,D_1}) = nD_1$. In [29] (see also [30]), Miller described a polynomial time algorithm, known universally as Miller's algorithm, to construct the function f_{n,D_1} and compute the Weil pairing on elliptic curves. However, the algorithm can be easily adapted to compute the Tate pairing on hyperelliptic curves.

Let $G_{iD_1,jD_1} \in \mathbb{F}_{q^k}(C)^*$ be a rational function with $\mathrm{div}(G_{iD_1,jD_1}) = iD_1 + jD_1 - (iD_1 \oplus jD_1)$ where $\oplus$ is the group law on $\mathcal{J}_C$ and $(iD_1 \oplus jD_1)$ is reduced. Miller's algorithm constructs the rational function f_{n,D_1} based on the following iterative formula:

$$f_{i+j,D_1} = f_{i,D_1} f_{j,D_1} G_{iD_1,jD_1} \,.$$

The following Algorithm 1 shows the basic version of Miller's algorithm for computing the reduced Tate pairing on hyperelliptic curves according to the above iterative relation. A more detailed version of Miller's algorithm for hyperelliptic curves can be found in [18].

Choie and Lee [6] described how to explicitly find the rational function $G(x, y)$ in the Algorithm 1 for the case of genus 2 hyperelliptic curves. Their results can be summarized as follows: Let $D_1 = [u_1, v_1]$ and $D_2 = [u_2, v_2]$ be the two reduced divisors in $\mathcal{J}_C(\mathbb{F}_{q^k})$ that are being added. In the composition stage of Cantor's algorithm, we compute the polynomial $\delta(x)$ which is the greatest common divisor of u_1, u_2 and $v_1 + v_2 + h$ and a divisor $D_3' = [u_3', v_3']$, which is in the same divisor class as $D_3 = [u_3, v_3] = \overline{D}_1 + \overline{D}_2$. At this point, two cases may occur:

1. If the divisor D_3' is already reduced following the composition stage, then the rational function $G(x, y) = \delta(x)$.
2. If this is not the case, then the rational function $G(x, y) = \delta(x) \cdot \frac{y - v_3'(x)}{u_3(x)}$.

In the most frequent cases[1] $\delta = 1$ and thus $G(x, y) = \frac{y - v_3'(x)}{u_3(x)}$.

[1] For addition, the inputs are two co-prime polynomials of degree 2, and for doubling the input is a square free polynomial of degree 2.

Algorithm 1. Miller's Algorithm for Hyperelliptic Curves (basic version)

IN: $\overline{D}_1 \in \mathcal{J}_C(\mathbb{F}_{q^k})[n], \overline{D}_2 \in \mathcal{J}_C(\mathbb{F}_{q^k})$, represented by D_1 and D_2
with $\mathrm{supp}(D_1) \cap \mathrm{supp}(D_2) = \emptyset$

OUT: $\langle D_1, D_2 \rangle_n^{(q^k-1)/n}$

1. $f \leftarrow 1, T \leftarrow D_1$

2. for $i \leftarrow \lfloor \log_2(n) \rfloor - 1$ **downto 0 do**

3. $\triangleright$ Compute T' and $G_{T,T}(x,y)$ such that $T' = 2T - \mathrm{div}(G_{T,T})$

4. $f \leftarrow f^2 \cdot G_{T,T}(D_2), \overline{T} \leftarrow [2]\overline{T}$

5. if $n_i = 1$ **then**

6. $\triangleright$ Compute T' and $G_{T,D_1}(x,y)$ such that $T' = T + D_1 - \mathrm{div}(G_{T,D_1})$

7. $f \leftarrow f \cdot G_{T,D_1}(D_2), \overline{T} \leftarrow \overline{T} \oplus \overline{D}_1$

8. Return $f^{(q^k-1)/n}$

3 Supersingular Curves and Non-supersingular Curves

In this section, we first recall the supersingular genus 2 curves over $\mathbb{F}_p$ which have been used to implement the Tate pairing. Then, by making a small change to the definition of these curves, we produce two families of non-supersingular genus 2 curves over $\mathbb{F}_p$ with efficiently computable automorphisms which provide potential advantages for pairing computations.

3.1 Supersingular Genus 2 Curves over $\mathbb{F}_p$

Theoretically, supersingular genus 2 hyperelliptic curves exist over $\mathbb{F}_p$ with an embedding degree of $k = 6$ [32]. However, only supersingular genus 2 curves with an embedding degree of $k = 4$ are known to the cryptographic community at the present time [7]. In [6,21], the authors investigated the efficient implementation of the Tate pairing on supersingular genus 2 curves with embedding degree 4. The curve used in their implementation is defined by the following equation:

$$C_1 : y^2 = x^5 + a, \quad a \in \mathbb{F}_p^* \text{ and } p \equiv 2, 3 \pmod 5.$$

On these supersingular curves a *modified pairing* $\langle D_1, \psi(D_1) \rangle_n^{(p^k-1)/n}$ is computed, where the map $\psi_1(\cdot)$ is a *distortion map* that maps elements in $C_1(\mathbb{F}_p)$ to $C_1(\mathbb{F}_{p^4})$. The distortion map ψ_1 is given by $\psi_1(x,y) = (\xi_5 x, y)$, where ξ_5 is a primitive 5-th root of unity in $\mathbb{F}_{p^4}$. We also note that another family of supersingular genus 2 curves over $\mathbb{F}_p$ with embedding degree 4 [7] is also suitable for implementing pairings. Such curves are given by an equation of the form

$$C_2 : y^2 = x^5 + ax, \quad a \in \mathbb{F}_p^* \cap \overline{QR}_p \text{ and } p \equiv 5 \pmod 8,$$

where $\overline{QR}_p$ denotes the set of quadratic non-residues modulo p. The distortion map for the curve C_2 is defined by $\psi_2(x,y) = (\xi_8^2 x, \xi_8 y)$, where ξ_8 is a primitive 8-th root of unity in $\mathbb{F}_{p^4}$.

3.2 Non-supersingular Genus 2 Curves over $\mathbb{F}_p$

Motivated by the work in [34,38,39], we consider now the following two families of non-supersingular genus 2 hyperelliptic curves over $\mathbb{F}_p$:

$$C_1' : y^2 = x^5 + a, \quad a \in \mathbb{F}_p^* \text{ and } p \equiv 1 \pmod 5,$$
$$C_2' : y^2 = x^5 + ax, \quad a \in \mathbb{F}_p^* \text{ and } p \equiv 1 \pmod 8.$$

Curves of this form exist which are pairing friendly (see Section 4). Note that the only difference between the curves C_i and C_i' $(i = 1, 2)$ is the congruence condition applied to the characteristic p. Although distortion maps do not exist on these non-supersingular curves, both C_1' and C_2' have efficiently-computable *endomorphisms*. In fact, these endomorphisms also induce efficient *automorphisms* on the divisor class groups of C_1' and C_2', which have been used to accelerate the scalar multiplication for hyperelliptic curve cryptosystems [31] and to attack the discrete log problems on the Jacobians [9,17]. In the next section, we will show how to speed up the computation of the Tate pairing using these efficiently computable automorphisms on the curves C_1' and C_2'. We first recall some basic facts which will be used in this work.

For the curve C_1', the morphism ψ_1 defined by $P = (x, y) \mapsto \psi_1(P) = (\xi_5 x, y)$ (see Section 3.1 and notice $\xi_5 \in \mathbb{F}_p$ now) induces an efficient non-trivial automorphism of order 5 on the divisor class group $\mathcal{J}_{C_1'}(\mathbb{F}_p)$ [31]. The formulae for ψ_1 on the Jacobian are given by

$$\begin{aligned} \psi_1 : [x^2 + u_1 x + u_0, v_1 x + v_0] &\mapsto [x^2 + \xi_5 u_1 x + \xi_5^2 u_0, \xi_5^{-1} v_1 x + v_0] \\ [x + u_0, v_0] &\mapsto [x + \xi_5 u_0, v_0] \\ \mathcal{O} &\mapsto \mathcal{O}. \end{aligned}$$

The characteristic polynomial of ψ_1 is given by $P(t) = t^4 + t^3 + t^2 + t + 1$. Since the automorphism ψ_1 is also an *isogeny*, we can construct its *dual isogeny* as follows:

$$\begin{aligned} \widehat{\psi_1} : [x^2 + u_1 x + u_0, v_1 x + v_0] &\mapsto [x^2 + \xi_5^{-1} u_1 x + \xi_5^{-2} u_0, \xi_5 v_1 x + v_0] \\ [x + u_0, v_0] &\mapsto [x + \xi_5^{-1} u_0, v_0] \\ \mathcal{O} &\mapsto \mathcal{O}. \end{aligned}$$

Note that $\psi_1 \circ \widehat{\psi_1} = [1]$ and $\#\,\mathrm{Ker}\ \psi_1 = \deg[1] = 1$, and $\widehat{\psi_1}$ is also a non-trivial automorphism on the curve C_1'.

Let $D \in \mathcal{J}_{C_1'}(\mathbb{F}_p)$ be a reduced divisor of a large prime order n. From [31], we know that the automorphisms ψ_1 and $\widehat{\psi_1}$ act respectively as multiplication maps by $[\lambda_1]$ and $[\widehat{\lambda_1}]$ on the subgroup $\langle D \rangle$ of $\mathcal{J}_{C_1'}(\mathbb{F}_p)$, where λ_1 and $\widehat{\lambda_1}$ are the two roots of the equation $t^4 + t^3 + t^2 + t + 1 \equiv 0 \pmod n$. Furthermore, it is easily seen that $[\lambda_1]D = \psi_1(D)$ can be obtained with only three or one field multiplications in $\mathbb{F}_p$ for a *general divisor* and a *degenerate divisor*, respectively. In the genus 2 context, a general divisor has two finite points in the support, whereas a degenerate divisor has only a single finite point in its support.

Similarly, for the curve C'_2, the morphism ψ_2 defined by $P = (x, y) \mapsto \psi_2(P) = (\xi_8^2 x, \xi_8 y)$ (see Section 3.1 and notice $\xi_8 \in \mathbb{F}_p$ now) induces an efficient non-trivial automorphism of order 8 on the divisor class group $\mathcal{J}_C(\mathbb{F}_p)$ as follows [31].

$$\begin{aligned}\psi_2 : [x^2 + u_1 x + u_0, v_1 x + v_0] &\mapsto [x^2 + \xi_8^2 u_1 x + \xi_8^4 u_0, \xi_8^{-1} v_1 x + \xi_8 v_0] \\ [x + u_0, v_0] &\mapsto [x + \xi_8^2 u_0, \xi_8 v_0] \\ \mathcal{O} &\mapsto \mathcal{O}.\end{aligned}$$

The characteristic polynomial of ψ_2 is given by $P(t) = t^4 + 1$ and the dual isogeny of ψ_2 is defined as follows

$$\begin{aligned}\widehat{\psi_2} : [x^2 + u_1 x + u_0, v_1 x + v_0] &\mapsto [x^2 + \xi_8^{-2} u_1 x + \xi_8^4 u_0, \xi_8 v_1 x + \xi_8^{-1} v_0] \\ [x + u_0, v_0] &\mapsto [x + \xi_8^{-2} u_0, \xi_8^{-1} v_0] \\ \mathcal{O} &\mapsto \mathcal{O}.\end{aligned}$$

It is not difficult to show that $\psi_2 \circ \widehat{\psi_2} = [1]$ and $\#\operatorname{Ker} \psi_2 = \deg[1] = 1$, and $\widehat{\psi_2}$ is also a non-trivial automorphism on the curve C'_2. Let $D \in \mathcal{J}_{C'_2}(\mathbb{F}_p)$ be a reduced divisor of a large prime order n. Then the automorphism ψ_2 acts as a multiplication map by λ_2 on the subgroup $\langle D \rangle$ of $\mathcal{J}_{C'_2}(\mathbb{F}_p)$, where λ_2 is a root of the equation $t^4 + 1 \equiv 0 \pmod{n}$. Moreover, $[\lambda_2]D = \psi_2(D)$ can be computed with four or two field multiplications in $\mathbb{F}_p$ for a general divisor and a degenerate divisor, respectively.

4 Efficient Pairings on Non-supersingular Genus 2 Curves

In this section we investigate efficient algorithms for computing the Tate pairing on the two families of genus 2 hyperelliptic curves C'_1 and C'_2 defined in Section 3.2. We show how to use the efficiently-computable automorphisms on these curves to shorten the length of the loop in Miller's algorithm. As a result, we propose new variants of Miller's algorithm for certain non-supersingular genus 2 curves over large prime fields.

4.1 Pairing Computation with Efficient Automorphisms

In this subsection, we present the main results of this paper in the following theorems and show their correctness. The pairing computation on the curve C'_1 is first examined.

Theorem 1. *Let C'_1 be a non-supersingular genus 2 hyperelliptic curve over $\mathbb{F}_p$ as above, with embedding degree $k > 1$ and automorphisms ψ_1 and $\widehat{\psi_1}$ defined as above. Let $D_1 = [u_1(x), v_1(x)] \in \mathcal{J}_{C'_1}(\mathbb{F}_p)$ be a reduced divisor of prime order n, where $n^2 \nmid \#\mathcal{J}_{C'_1}(\mathbb{F}_p)$. Let $[\lambda_1]$ act as the multiplication map on the subgroup $\langle D_1 \rangle$ defined as above such that $[\lambda_1]D_1 = \psi_1(D_1)$. Suppose $m \in \mathbb{Z}$ is such that*

$mn = \lambda_1^4+\lambda_1^3+\lambda_1^2+\lambda_1+1$. *Let rational functions* $\frac{c_1(x,y)}{d_1(x,y)}, \frac{c_2(x,y)}{d_2(x,y)}, \frac{c_3(x,y)}{d_3(x,y)} \in \mathbb{F}_p(C_1')^*$ *respectively satisfy the following three relations:*

$$[\lambda_1]D_1 + [\lambda_1^2]D_1 - ([\lambda_1]D_1 \oplus [\lambda_1^2]D_1) = \operatorname{div}\left(\frac{c_1(x,y)}{d_1(x,y)}\right),$$

$$\left[\lambda_1^3\right] D_1 + [\lambda_1^4]D_1 - ([\lambda_1^3]D_1 \oplus [\lambda_1^4]D_1) = \operatorname{div}\left(\frac{c_2(x,y)}{d_2(x,y)}\right),$$

$$\left[\lambda_1 + \lambda_1^2\right] D_1 + [\lambda_1^3 + \lambda_1^4]D_1 - ([\lambda_1 + \lambda_1^2]D_1 \oplus [\lambda_1^3 + \lambda_1^4]D_1) = \operatorname{div}\left(\frac{c_3(x,y)}{d_3(x,y)}\right).$$

Let $g(x,y) = \frac{c_1(x,y)\cdot c_2(x,y)\cdot c_3(x,y)}{d_1(x,y)\cdot d_2(x,y)}$. *Then for* $D_2 \in \mathcal{J}_{C_1'}(\mathbb{F}_{p^k})$, *we have*

$$\langle D_1, D_2\rangle_n^{\frac{m(p^k-1)}{n}} = \Big[f_{\lambda_1,D_1}^{\lambda_1^3+\lambda_1^2+\lambda_1+1}(D_2)\cdot f_{\lambda_1,D_1}^{\lambda_1^2+\lambda_1+1}\left(\widehat{\psi}_1(D_2)\right)\cdot f_{\lambda_1,D_1}^{\lambda_1+1}\left(\widehat{\psi}_1^2(D_2)\right)\cdot$$
$$f_{\lambda_1,D_1}\left(\psi_1^2(D_2)\right)\cdot g(D_2)\Big]^{\frac{p^k-1}{n}}.$$

Note that the condition that λ_1 satisfies $\lambda_1^4 + \lambda_1^3 + \lambda_1^2 + \lambda_1 + 1 \equiv 0 \pmod n$ guarantees the existence of the integer m. Moreover, the pairing will be non-degenerate if $n \nmid m$ and $\operatorname{supp}(D_1) \cap \operatorname{supp}(D_2) = \emptyset$. We split the proof of the Theorem 1 into the following lemmas.

Lemma 1. *With notation as above, we have*

$$\langle D_1, D_2\rangle_n^{\frac{m(p^k-1)}{n}} = \left(f_{\lambda_1^4+\lambda_1^3+\lambda_1^2+\lambda_1,D_1}(D_2)\cdot u_1(D_2)\right)^{\frac{p^k-1}{n}}.$$

Proof. From the important property of the reduced pairing (see equation (1)), we have

$$\langle D_1, D_2\rangle_n^{\frac{m(p^k-1)}{n}} = \langle D_1, D_2\rangle_{mn}^{\frac{p^k-1}{n}} = f_{mn,D_1}(D_2)^{\frac{p^k-1}{n}}.$$

From the condition that $mn = \lambda_1^4 + \lambda_1^3 + \lambda_1^2 + \lambda_1 + 1$, we get

$$\langle D_1, D_2\rangle_n^{\frac{m(p^k-1)}{n}} = f_{mn,D_1}(D_2)^{\frac{p^k-1}{n}} = f_{\lambda_1^4+\lambda_1^3+\lambda_1^2+\lambda_1+1,D_1}(D_2)^{\frac{p^k-1}{n}}.$$

Since $[\lambda_1^4 + \lambda_1^3 + \lambda_1^2 + \lambda_1]D_1 = -D_1$, we obtain the following relation

$$D_1 + [\lambda_1 + \lambda_1^2 + \lambda_1^3 + \lambda_1^4]D_1 = D_1 + (-D_1) = \operatorname{div}(u_1(x)).$$

Therefore, we have

$$\begin{aligned}\operatorname{div}\left(f_{\lambda_1^4+\lambda_1^3+\lambda_1^2+\lambda_1+1,D_1}\right) &= (\lambda_1^4 + \lambda_1^3 + \lambda_1^2 + \lambda_1)D_1 + D_1\\ &= \operatorname{div}\left(f_{\lambda_1^4+\lambda_1^3+\lambda_1^2+\lambda_1,D_1}\right) + D_1 + [\lambda_1 + \lambda_1^2 + \lambda_1^3 + \lambda_1^4]D_1\\ &= \operatorname{div}\left(f_{\lambda_1^4+\lambda_1^3+\lambda_1^2+\lambda_1,D_1}\cdot u_1(x)\right),\end{aligned}$$

which proves the result. □

The next lemma shows the relation between div $\left(f_{\lambda_1^4+\lambda_1^3+\lambda_1^2+\lambda_1,D_1}\cdot u_1(x)\right)$ and the divisors div $\left(f_{\lambda_1,[\lambda_1^i]D_1}\right)$ for $i=0,1,2$, and 3.

Lemma 2. *With notation as above, we have*

$$\operatorname{div}\left(f_{\lambda_1^4+\lambda_1^3+\lambda_1^2+\lambda_1,D_1}\cdot u_1(x)\right)=$$
$$\operatorname{div}\left(f_{\lambda_1,D_1}^{\lambda_1^3+\lambda_1^2+\lambda_1+1}\cdot f_{\lambda_1,[\lambda_1]D_1}^{\lambda_1^2+\lambda_1+1}\cdot f_{\lambda_1,[\lambda_1^2]D_1}^{\lambda_1+1}\cdot f_{\lambda_1,[\lambda_1^3]D_1}\cdot g(x,y)\right).$$

Proof. We first note the following relation

$$\begin{aligned}\operatorname{div}\left(f_{\lambda_1^4+\lambda_1^3+\lambda_1^2+\lambda_1,D_1}\right)&=(\lambda_1^4+\lambda_1^3+\lambda_1^2+\lambda_1)D_1-[\lambda_1^4+\lambda_1^3+\lambda_1^2+\lambda_1]D_1\\&=\operatorname{div}\left(f_{\lambda_1^4+\lambda_1^3,D_1}\right)+\operatorname{div}\left(f_{\lambda_1^2+\lambda_1,D_1}\right)+[\lambda_1+\lambda_1^2]D_1+\\&\quad[\lambda_1^3+\lambda_1^4]D_1-([\lambda_1+\lambda_1^2]D_1\oplus[\lambda_1^3+\lambda_1^4]D_1)\\&=\operatorname{div}\left(f_{\lambda_1^4+\lambda_1^3,D_1}\right)+\operatorname{div}\left(f_{\lambda_1^2+\lambda_1,D_1}\right)+\operatorname{div}\left(\frac{c_3(x,y)}{d_3(x,y)}\right)\\&=\operatorname{div}\left(f_{\lambda_1^4+\lambda_1^3,D_1}\cdot f_{\lambda_1^2+\lambda_1,D_1}\cdot\frac{c_3(x,y)}{d_3(x,y)}\right)\end{aligned}$$

Since $[\lambda_1^4+\lambda_1^3+\lambda_1^2+\lambda_1]D_1=-D_1$, we get $d_3(x,y)=u_1(x)$. Therefore, we have

$$\operatorname{div}\left(f_{\lambda_1^4+\lambda_1^3+\lambda_1^2+\lambda_1,D_1}\cdot u_1(x)\right)=\operatorname{div}\left(f_{\lambda_1^4+\lambda_1^3,D_1}\cdot f_{\lambda_1^2+\lambda_1,D_1}\cdot c_3(x,y)\right).\qquad(2)$$

Similarly, we can obtain the following two equalities

$$\begin{aligned}&\operatorname{div}\left(f_{\lambda_1^4+\lambda_1^3,D_1}\right)=(\lambda_1^4+\lambda_1^3)D_1-[\lambda_1^4+\lambda_1^3]D_1\\&=\operatorname{div}\left(f_{\lambda_1^4,D_1}\right)+\operatorname{div}\left(f_{\lambda_1^3,D_1}\right)+[\lambda_1^4]D_1+[\lambda_1^3]D_1-([\lambda_1^3]D_1\oplus[\lambda_1^4]D_1)\\&=\operatorname{div}\left(f_{\lambda_1^4,D_1}\right)+\operatorname{div}\left(f_{\lambda_1^3,D_1}\right)+\operatorname{div}\left(\frac{c_2(x,y)}{d_2(x,y)}\right)\\&=\operatorname{div}\left(f_{\lambda_1^4,D_1}\cdot f_{\lambda_1^3,D_1}\cdot\frac{c_2(x,y)}{d_2(x,y)}\right)\end{aligned}$$

and

$$\begin{aligned}&\operatorname{div}\left(f_{\lambda_1^2+\lambda_1,D_1}\right)=(\lambda_1^2+\lambda_1)D_1-[\lambda_1^2+\lambda_1]D_1\\&=\operatorname{div}\left(f_{\lambda_1^2,D_1}\right)+\operatorname{div}\left(f_{\lambda_1,D_1}\right)+[\lambda_1^2]D_1+[\lambda_1]D_1-([\lambda_1]D_1\oplus[\lambda_1^2]D_1)\\&=\operatorname{div}\left(f_{\lambda_1^2,D_1}\right)+\operatorname{div}\left(f_{\lambda_1,D_1}\right)+\operatorname{div}\left(\frac{c_1(x,y)}{d_1(x,y)}\right)\\&=\operatorname{div}\left(f_{\lambda_1^2,D_1}\cdot f_{\lambda_1,D_1}\cdot\frac{c_1(x,y)}{d_1(x,y)}\right)\end{aligned}$$

Some easy calculations (see Lemma 2 in [2]) give us

$$\operatorname{div}\left(f_{\lambda_1^4,D_1}\right) = \operatorname{div}\left(f_{\lambda_1,D_1}^{\lambda_1^3} \cdot f_{\lambda_1,[\lambda_1]D_1}^{\lambda_1^2} \cdot f_{\lambda_1,[\lambda_1^2]D_1}^{\lambda_1} \cdot f_{\lambda_1,[\lambda_1^3]D_1}\right) \quad (3)$$

$$\operatorname{div}\left(f_{\lambda_1^3,D_1}\right) = \operatorname{div}\left(f_{\lambda_1,D_1}^{\lambda_1^2} \cdot f_{\lambda_1,[\lambda_1]D_1}^{\lambda_1} \cdot f_{\lambda_1,[\lambda_1^2]D_1}\right) \quad (4)$$

$$\operatorname{div}\left(f_{\lambda_1^2,D_1}\right) = \operatorname{div}\left(f_{\lambda_1,D_1}^{\lambda_1} \cdot f_{\lambda_1,[\lambda_1]D_1}\right) \quad (5)$$

Combining the equations (2)–(7) proves the result. □

The following lemma shows how to evaluate functions $f_{\lambda_1,[\lambda_1^i]D_1}$ $(i = 1, 2, 3)$ at the image divisor D_2 by using the function f_{λ_1,D_1}.

Lemma 3. *With notation as above, we have (up to a scalar multiple in* $\mathbb{F}_p^*$*)*

$$\begin{aligned} f_{\lambda_1,[\lambda_1]D_1}(D_2) &= f_{\lambda_1,D_1}(\widehat{\psi_1}(D_2)), \\ f_{\lambda_1,[\lambda_1^2]D_1}(D_2) &= f_{\lambda_1,D_1}(\widehat{\psi_1^2}(D_2)), \\ f_{\lambda_1,[\lambda_1^3]D_1}(D_2) &= f_{\lambda_1,D_1}(\psi_1^2(D_2)). \end{aligned}$$

Proof. Using the pullback property (see Silverman [36] p. 33) and following the same proof as the Lemma 1 in [2], we obtain (up to a scalar multiple in $\mathbb{F}_p^*$)

$$\begin{aligned} f_{\lambda_1,[\lambda_1]D_1} \circ \psi_1 &= f_{\lambda_1,D_1}, \\ f_{\lambda_1,[\lambda_1^2]D_1} \circ \psi_1^2 &= f_{\lambda_1,D_1}, \\ f_{\lambda_1,[\lambda_1^3]D_1} \circ \psi_1^3 &= f_{\lambda_1,D_1}. \end{aligned}$$

Using the relations between the isogeny ψ_1 and its dual isogeny $\widehat{\psi_1}$ (see Section 3.2), we have

$$\begin{aligned} f_{\lambda_1,[\lambda_1]D_1} \circ \psi_1 \circ \widehat{\psi_1} &= f_{\lambda_1,[\lambda_1]D_1} = f_{\lambda_1,D_1} \circ \widehat{\psi_1}, \\ f_{\lambda_1,[\lambda_1^2]D_1} \circ \psi_1^2 \circ \widehat{\psi_1^2} &= f_{\lambda_1,[\lambda_1^2]D_1} = f_{\lambda_1,D_1} \circ \widehat{\psi_1^2}, \\ f_{\lambda_1,[\lambda_1^3]D_1} \circ \psi_1^3 \circ \widehat{\psi_1^3} &= f_{\lambda_1,[\lambda_1^3]D_1} = f_{\lambda_1,D_1} \circ \widehat{\psi_1^3} = f_{\lambda_1,D_1} \circ \psi_1^2, \end{aligned}$$

which proves the results. □

With the above three lemmas, we can now prove the statement of Theorem 1 as follows:

Proof (of Theorem 1). For $D_1 \in \mathcal{J}_{C_1'}(\mathbb{F}_p)[n]$ and $D_2 \in \mathcal{J}_{C_1'}(\mathbb{F}_{p^k})$, Lemma 3 shows that up to a scalar multiple in $\mathbb{F}_p^*$ we have

$$\begin{aligned} f_{\lambda_1,[\lambda_1]D_1}(D_2) &= f_{\lambda_1,D_1}(\widehat{\psi_1}(D_2)), \\ f_{\lambda_1,[\lambda_1^2]D_1}(D_2) &= f_{\lambda_1,D_1}(\widehat{\psi_1^2}(D_2)), \\ f_{\lambda_1,[\lambda_1^3]D_1}(D_2) &= f_{\lambda_1,D_1}(\psi_1^2(D_2)). \end{aligned}$$

Now, substituting the above three equalities into Lemma 2 implies that

$$f_{\lambda_1^4+\lambda_1^3+\lambda_1^2+\lambda_1,D_1}(D_2)\cdot u_1(D_2) = f_{\lambda_1,D_1}^{\lambda_1^3+\lambda_1^2+\lambda_1+1}(D_2)\cdot f_{\lambda_1,D_1}^{\lambda_1^2+\lambda_1+1}\left(\widehat{\psi}_1(D_2)\right)\cdot f_{\lambda_1,D_1}^{\lambda_1+1}\left(\widehat{\psi}_1^2(D_2)\right)\cdot f_{\lambda_1,D_1}\left(\psi_1^2(D_2)\right)\cdot g(D_2).$$

Finally, substituting the above equation into Lemma 1 gives the result of Theorem 1. □

Next, we consider how to use the efficiently-computable automorphism ψ_2 to accelerate the computation of the Tate pairing on the curve C_2'. The following Theorem 2 describes our result.

Theorem 2. *Let C_2' be a non-supersingular genus 2 hyperelliptic curve over $\mathbb{F}_p$ as above, with embedding degree $k>1$ and automorphisms ψ_2 and $\widehat{\psi}_2$ defined as above. Let $D_1=[u_1(x),v_1(x)]\in\mathcal{J}_{C_2'}(\mathbb{F}_p)$ be a reduced divisor of prime order n, where $n^2\nmid\#\mathcal{J}_{C_2'}(\mathbb{F}_p)$. Let $[\lambda_2]$ act as the multiplication map on the subgroup $\langle D_1\rangle$ defined as above such that $[\lambda_2]D_1=\psi_2(D_1)$. Suppose $m\in\mathbb{Z}$ is such that $mn=\lambda_2^4+1$. Then for $D_2\in\mathcal{J}_{C_2'}(\mathbb{F}_{p^k})$, we have*

$$\langle D_1,D_2\rangle_n^{\frac{m(p^k-1)}{n}} = \left[f_{\lambda_2,D_1}^{\lambda_2^3}(D_2)\cdot f_{\lambda_2,D_1}^{\lambda_2^2}\left(\widehat{\psi}_2(D_2)\right)\cdot f_{\lambda_2,D_1}^{\lambda_2}\left(\widehat{\psi}_2^2(D_2)\right)\cdot f_{\lambda_2,D_1}\left(\widehat{\psi}_2^3(D_2)\right)\cdot u_1(D_2)\right]^{\frac{p^k-1}{n}}.$$

Proof. The proof is similar to that of Theorem 1. Therefore, we omit it here. □

From Theorem 1 and Theorem 2, we note that the pairing computation on curve C_2' is more efficient than that on curve C_1'. Hence, the following discussions only focus on the curve C_2'.

4.2 A New Variant of Miller's Algorithm

In this subsection, we propose a new variant of Miller's algorithm for the family of genus 2 hyperelliptic curves C_2' over $\mathbb{F}_p$ with efficiently computable automorphisms ψ_2 and $\widehat{\psi}_2$. From Theorem 2, we obtain the following Algorithm 2 for computing the Tate pairing on such curves C_2', which is a variant of Miller's Algorithm (see Algorithm 1 in Section 2.2). For the curve C_1', we can also obtain a similar variant of Miller's algorithm as in Algorithm 2, based on Theorem 1.

5 Implementing the Tate Pairing with Efficient Automorphisms

In this section, we describe various techniques that enable the efficient implementation of the Tate pairing on a non-supersingular genus 2 curve of type C_2' over $\mathbb{F}_p$. Furthermore, we also analyze the computational cost of our new algorithm in detail and give timings for our implementation.

Algorithm 2. Computing the Tate Pairing with Efficient Automorphisms

IN: $\overline{D}_1 = [u_1, v_1] \in \mathcal{J}_{C_2'}(\mathbb{F}_p)[n], \overline{D}_2 \in \mathcal{J}_{C_2'}(\mathbb{F}_{p^k})$, represented by D_1 and D_2 with $\text{supp}(D_1) \cap \text{supp}(D_2) = \emptyset$, $\lambda_2 = (e_r, e_{r-1}, \ldots, e_0)_2$, where $e_i \in \{0, 1\}$ for $i = 0, \ldots, r-1$ and $e_r = 1$, and $mn = \lambda_2^4 + 1$.

OUT: $\langle D_1, D_2 \rangle_n^{m(p^k-1)/n}$

1. $T \leftarrow D_1, f_1 \leftarrow 1, f_2 \leftarrow 1, f_3 \leftarrow 1, f_4 \leftarrow 1, f_5 \leftarrow u_1(D_2)$

2. for i **from** $r-1$ **downto** 0 **do**

3. $\triangleright$ Compute T' and $G_{T,T}(x, y)$ such that $T' = 2T - \text{div}(G_{T,T})$

4. $\overline{T} \leftarrow [2]\overline{T}, f_1 \leftarrow f_1^2 \cdot G_{T,T}(D_2), f_2 \leftarrow f_2^2 \cdot G_{T,T}(\widehat{\psi_2}(D_2))$

5. $f_3 \leftarrow f_3^2 \cdot G_{T,T}(\widehat{\psi_2^2}(D_2)), f_4 \leftarrow f_4^2 \cdot G_{T,T}(\widehat{\psi_2^3}(D_2))$

6. if $e_i = 1$ **then**

7. $\triangleright$ Compute T' and $G_{T,D_1}(x, y)$ such that $T' = T + D_1 - \text{div}(G_{T,D_1})$

8. $\overline{T} \leftarrow \overline{T} \oplus \overline{D}_1, f_1 \leftarrow f_1 \cdot G_{T,D_1}(D_2), f_2 \leftarrow f_2 \cdot G_{T,D_1}(\widehat{\psi_2}(D_2))$

9. $f_3 \leftarrow f_3 \cdot G_{T,D_1}(\widehat{\psi_2^2}(D_2)), f_4 \leftarrow f_4 \cdot G_{T,D_1}(\widehat{\psi_2^3}(D_2))$

10. $f \leftarrow ((f_1^{\lambda_2} \cdot f_2)^{\lambda_2} \cdot f_3)^{\lambda_2} \cdot f_4 \cdot f_5$

11. $f \leftarrow f^{(p^k-1)/n}$

12. Return f

5.1 Curve Generation

While generating suitable parameters for supersingular genus 2 hyperelliptic curves over prime fields is easy, it seems to be more difficult to generate pairing-friendly non-supersingular genus 2 curves over $\mathbb{F}_p$ because of the complicated algebraic structure of hyperelliptic curves. Only a few results have appeared in the literature to address this issue [12,16,23,25] and there is ongoing research in this direction. In [12], Freeman proposed the first explicit construction of pairing-friendly genus 2 hyperelliptic curves over prime fields with ordinary Jacobians by modeling on the Cocks-Pinch method for the elliptic curve case [8]. In the appendix of [12], we find two examples which belong to the curve family C_1' considered in this paper. Unfortunately, the curve parameters in the two examples are too large to be optimal for efficient implementation. In a recent paper [25], Kawazoe and Takahashi presented two different approaches for explicitly constructing pairing-friendly genus 2 curves of the type $y^2 = x^5 + ax$ over $\mathbb{F}_p$. One is an analogue of the Cocks-Pinch method and the other is a cyclotomic method. Their findings are based on the closed formulae [15,19] for the order of the Jacobian of hyperelliptic curves of the above type. In this paper we will restrict to the case $p \equiv 1 \pmod 8$ and generate a suitable non-supersingular pairing-friendly genus 2 hyperelliptic curves C_2' with embedding degree 4 using the theorems in [25]. The reason that we only consider curves with embedding degree 4 in this section is to facilitate performance comparisons between supersingular and non-supersingular genus 2 curves. However, we would like to point out that non-supersingular curves with higher embedding degree are available from [25] and that our method is also applicable to such curves.

To find good curve parameters which are suitable for applying our new algorithm, we use the following searching strategies. From Theorem 2 we note that the subgroup order n should satisfy $mn = \lambda_2^4+1$ for an integer m. Assume that we require the (160/1024) bit security level. Then n is a prime around 160 bits and λ_2 is at least 40 bits. Furthermore, since the bit length of λ_2 determines the length of the loop in Algorithm 2, λ_2 should be taken as small as possible. Based on these observations, we first check all λ_2's of the form $\lambda_2 = 2^a, a \in \{41, 42, \ldots, 60\}$. We found two λ_2's, namely $\lambda_2 = 2^{58}$ and 2^{59}, for which $\lambda_2^4 + 1$ has a prime factor of 164 bits and 162 bits, respectively. However, using the above two primes as the subgroup order n and running the algorithms of [25], we cannot find any curve. Therefore, we consider the slightly more complicated choice of $\lambda_2 = 2^a + 2^b$, where $a, b \in \{41, 42, \ldots, 50\}$ and $a > b$. After a couple of trials, we found that choosing $\lambda_2 = 2^{43} + 2^{10}$ generates a non-supersingular pairing-friendly genus 2 hyperelliptic curve whose Jacobian has embedding degree 4 with respect to a 163-bit prime n. The curve is given by the equation

$$C_2^* : y^2 = x^5 + 9x$$

over $\mathbb{F}_p$, for a 329-bit prime p, where the hexadecimal representations of n and p are as follows:

$n =$ `00000006 a37991af 81ddfa3a ead6ec83 1ca0fc44 75d5add9 (163 bits)`

$p =$ `0000016b 953ca333 acf202b3 0476f30f ff085473 6d0a0be4`
`c542fa48 66e5afba 7bc6cd6d 21ca9fad eef796f1 (329 bits)`

In the following five subsections, we will detail various techniques required to efficiently implement the calculation of the Tate pairing on the curve C_2^*.

5.2 Finite Field Arithmetic

As the embedding degree of the curve C_2^* in our implementation is $k = 4$, we first discuss how to construct the finite field $\mathbb{F}_{p^4}$. Rather than construct $\mathbb{F}_{p^4}$ as a direct quartic extension of $\mathbb{F}_p$, the best way is to define the field $\mathbb{F}_{p^4}$ as a quadratic extension of $\mathbb{F}_{p^2}$, which is in turn a quadratic extension of $\mathbb{F}_p$. Since the p is congruent to 5 modulo 12 in our implementation, the field $\mathbb{F}_{p^2}$ can be constructed by the irreducible binomial $x^2 + 3$ and the field $\mathbb{F}_{p^4}$ can be constructed as a quadratic extension of $\mathbb{F}_{p^2}$ by the irreducible binomial $x^2 - \sqrt{-3}$. Letting $\beta = -3$, elements of the field $\mathbb{F}_{p^2}$ can be represented as $a + b\sqrt{\beta}$, where $a, b \in \mathbb{F}_p$, whereas elements of the field $\mathbb{F}_{p^4}$ can be represented as $c + d\sqrt[4]{\beta}$, where $c, d \in \mathbb{F}_{p^2}$. Under this tower construction, a multiplication of two elements and a squaring of one element in $\mathbb{F}_{p^4}$ cost $9M$ and $6M$ in $\mathbb{F}_p$, respectively [21].

5.3 Encapsulated Group Operations

In [11], Fan *et. al.* proposed a method to encapsulate the computation of the line function with the group operations for genus 2 hyperelliptic curves over prime fields, and derived new mix-addition and doubling formulae in projective and new (weighted projective) coordinates, respectively. Applying their explicit formulae to the curve C_2^* defined above, we can respectively calculate the divisor class addition and doubling with $36M + 5S$ and $32M + 6S$ in $\mathbb{F}_p$ in new coordinates. We also include their explicit formulae in the appendix with some modifications for the curve C_2^*.

5.4 Using Degenerate Divisors and Denominator Elimination

For a hyperelliptic curve of genus $g > 1$, using a degenerate divisor as the image divisor is more efficient than using a general divisor when evaluating line functions. Frey and Lange [13] discussed in detail when it is permissible to choose a degenerate divisor as the second argument of Miller's algorithm. They also note that, when the embedding degree k is even, one might choose the second pairing argument from a set $S = \{(x, y) \in C(\mathbb{F}_{q^k}) \mid x \in \mathbb{F}_{q^{k/2}}, y \in \mathbb{F}_{q^k} \backslash \mathbb{F}_{q^{k/2}}\}$. Note that the point in the set S is on the quadratic twist of $C/\mathbb{F}_{q^{k/2}}$. When considering C_2^* as a curve defined over $\mathbb{F}_{p^2}$, we can define a quadratic twist of C_2^* over $\mathbb{F}_{p^2}$, denoted by $C_{2,t}^*$, as follows

$$C_{2,t}^* : y^2 = x^5 + 9c^4 x,$$

where $c \in \mathbb{F}_{p^2}$ is a quadratic non-residue over $\mathbb{F}_{p^2}$. It is known that $C_{2,t}^*(\mathbb{F}_{p^4}) \cong C_2^*(\mathbb{F}_{p^4})$ and the isomorphism of $C_{2,t}^*(\mathbb{F}_{p^4})$ and $C_2^*(\mathbb{F}_{p^4})$ also induces an isomorphism ϕ of $\mathcal{J}_{C_{2,t}^*}(\mathbb{F}_{p^4})$ and $\mathcal{J}_{C_2^*}(\mathbb{F}_{p^4})$ [26]. As in [11] we first generate a degenerate divisor class $\overline{D}_t = [x - x_t, y_t] \in \mathcal{J}_{C_{2,t}^*}(\mathbb{F}_{p^2})$ on the twisted curve $C_{2,t}^*/\mathbb{F}_{p^2}$. Then the isomorphism ϕ will map $\overline{D}_t$ to a degenerate divisor class $\overline{D}_2 = \phi(\overline{D}_t) = [x - c^{-1}x_t, c^{-5/2}y_t] \in \mathcal{J}_{C_2^*}(\mathbb{F}_{p^4})$ on the curve C_2^* over $\mathbb{F}_{p^4}$. Note that the denominator elimination technique applies in this case since $x - c^{-1}x_t$ is defined over $\mathbb{F}_{p^2}$. Furthermore, we do not need to compute $f_5 = u_1(D_2) \in \mathbb{F}_{p^2}$ in Algorithm 2 either, for the same reason.

5.5 Evaluating Line Functions at a Degenerate Divisor

Here we consider the evaluation of line functions at a degenerate divisor $D_2 = [x - x_2, y_2] \in \mathcal{J}_{C_2^*}(\mathbb{F}_{p^4})$ generated by the method in Section 5.4, where $x_2 = c^{-1}x_t \in \mathbb{F}_{p^2}$ and $y_2 = c^{-5/2}y_t \in \mathbb{F}_{p^4} \backslash \mathbb{F}_{p^2}$. Moreover, we further assume that in this work $c = \sqrt{-3}$ is taken as a quadratic non-residue over $\mathbb{F}_{p^2}$. Therefore, y_2 has only two non-zero coefficients instead of four in a polynomial basis representation of $\mathbb{F}_{p^4}$. Furthermore, since the denominator elimination technique applies in this case, we only need to evaluate the numerators of the rational functions at D_2. From [11] we know that in new coordinates we can respectively work with the numerators $c_1(x, y) = (Z_{31}Z_{32})y - ((s_1 z_{11})x^3 + l_2 x^2 + l_1 x + l_0)$ for group doubling and $c_2(x, y) = (\tilde{r} z_{21})y - ((s_1' z_{21})x^3 + l_2 x^2 + l_1 x + l_0)$ for group addition, where $Z_{31}, Z_{32}, \tilde{r}, z_{11}, z_{21}, s_1, s_1', l_2, l_1$ and l_0 are from Table 4 and Table 5 in the appendix. Note that in Algorithm 2 we need to evaluate the function $c_i(x, y), i = 1$

or 2 at four image divisors $D_2 = [x - x_2, y_2]$, $\widehat{\psi}_2(D_2) = [x - \xi_8^{-2}x_2, \xi_8^{-1}y_2]$, $\widehat{\psi}_2^2(D_2) = [x - \xi_8^4 x_2, \xi_8^{-2}y_2] = [x + x_2, \xi_8^{-2}y_2]$ and $\widehat{\psi}_2^3(D_2) = [x - \xi_8^2 x_2, \xi_8^{-3}y_2]$ for each iteration of the loop. Hence we have the following relations

$$\begin{aligned}
c_i(D_2) &= (\tilde{r}z_{11})y_2 - [((s_1' z_{11})x_2^3 + l_1 x_2) + (l_2 x_2^2 + l_0)],\\
c_i(\widehat{\psi}_2(D_2)) &= ((\tilde{r}z_{11})y_2)\xi_8^{-1} - [((s_1' z_{11})x_2^3 - l_1 x_2)\xi_8^2 - (l_2 x_2^2 - l_0)],\\
c_i(\widehat{\psi}_2^2(D_2)) &= ((\tilde{r}z_{11})y_2)\xi_8^{-2} + [((s_1' z_{11})x_2^3 + l_1 x_2) - (l_2 x_2^2 + l_0)],\\
c_i(\widehat{\psi}_2^3(D_2)) &= ((\tilde{r}z_{11})y_2)\xi_8^{-3} + [((s_1' z_{11})x_2^3 - l_1 x_2)\xi_8^2 + (l_2 x_2^2 - l_0)].
\end{aligned}$$

We assume that x_2^2, x_2^3, ξ_8^{-1} and ξ_8^2 are precomputed with $7M + 2S$ over $\mathbb{F}_p$. Since x_2, x_2^2 and x_2^3 are in $\mathbb{F}_{p^2}$ and y_2 has only two non-zero coefficients in the polynomial basis representation of $\mathbb{F}_{p^4}$, $c_i(D_2)$ can be computed with $10M$ over $\mathbb{F}_p$. By reusing the intermediate computation results, we can compute $c_i(\widehat{\psi}_2(D_2))$, $c_i(\widehat{\psi}_2^2(D_2))$ and $c_i(\widehat{\psi}_2^3(D_2))$ with $4M$, $2M$ and $2M$ over $\mathbb{F}_p$, respectively. Therefore, the total cost of evaluating the function $c_i(x, y)$ at the degenerate divisor D_2 is $18M$ over $\mathbb{F}_p$ per iteration, with a precomputation of $7M + 2S$. For the case of evaluating the rational functions at a general divisor, the reader is referred to [11].

5.6 Final Exponentiation

For a genus 2 curve with an embedding degree of $k = 4$, the output of Miller's algorithm needs to be raised to the power of $(p^4 - 1)/n$. Calculating this exponentiation can follow two steps as shown in [21]. Letting $f \in \mathbb{F}_{p^4}$ be the output of Miller's algorithm, the first step is to compute f^{p^2-1} which can be obtained with a conjugation with respect to $\mathbb{F}_{p^2}$ and $1I + 1M$ in $\mathbb{F}_{p^4}$. The remaining exponentiation to $(p^2 + 1)/n$ is an expensive operation which can be efficiently computed with the Lucas laddering algorithm [35] for the curve C_2^* in question.

5.7 Performance Analysis and Comparison

In this section, we analyze the computational complexity of the Algorithm 2 for calculating the Tate pairing on non-supersingular genus 2 hyperelliptic curves C_2' and compare the performance of pairing computations on supersingular and non-supersingular genus 2 curves over prime fields with the same embedding degree of $k = 4$.

We first analyze the algebraic complexity of the pairing computation on curves C_2' with our new algorithm (see Section 4.2). Recall that n is the subgroup order and λ_2 is a root of the equation $\lambda^4 + 1 \equiv 0 \bmod n$. We assume that the embedding degree k is even and the line functions in Algorithm 2 are evaluated at a degenerate divisor D_2 instead of a general divisor for efficiency reasons. We also assume that λ_2 has a random Hamming weight, meaning that about $(\frac{1}{2}\log_2 \lambda_2)$ additions take place in Algorithm 2 on average. Then the algebraic cost for computing the Tate pairing is given by (without including the cost of the final exponentiation)

$$T_1 + (\log_2 \lambda_2)(T_D + T_G + 4T_{sk} + 8T_{mk}) + \left(\frac{1}{2}\log_2 \lambda_2\right)(T_A + T_G + 8T_{mk}) + T_{10},$$

where

1. T_2 – the cost of precomputing f_5 in Step 1 of Algorithm 2.
2. T_D – the cost of doubling a general divisor.
3. T_A – the cost of adding two general divisors.
4. T_G – the cost of evaluating rational function $G(x, y)$ at the image divisors $D_2, \widehat{\psi_2}(D_2), \widehat{\psi_2^2}(D_2)$ and $\widehat{\psi_2^3}(D_2)$.
5. T_{sk} – the cost of squaring in $\mathbb{F}_{p^k}$.
6. T_{mk} – the cost of multiplication in $\mathbb{F}_{p^k}$.
7. T_{10} – the cost of computing f in Step 10 of Algorithm 2.

When applying various optimization techniques detailed in previous subsections to the particular curve C_2^*, we can further reduce the above cost of computing the Tate pairing to

$$43 \cdot (T_D + T_G + 4T_{sk} + 4T_{mk}) + (T_A + T_G + 4T_{mk}) + T_{10},$$

where $T_D = 32M + 6S, T_A = 36M + 5S, T_G = 18M, T_{sk} = 6M, T_{mk} = 9M$ and $T_{10} = 828M$. Furthermore, we also need $7M + 2S$ for precomputations (see Section 5.5). Note that all multiplications and squarings here are over $\mathbb{F}_p$. Therefore, the total cost of computing the Tate pairing with our optimizations is given by $5655M + 265S$ in $\mathbb{F}_p$.

In [6,11,21], the authors examined the implementation of the Tate pairing on a family of supersingular genus 2 hyperelliptic curves C_1 (see Section 3.1) over prime fields with embedding degree 4 in affine and projective coordinates, respectively. We compare the performance of pairing computations on the supersingular curve C_1 and the non-supersingular curve C_2^* in the following Table 1. Note that both curves have the same embedding degree of $k = 4$.

From Table 1, we note that for the same security level the computation of the Tate pairing on the non-supersingular genus 2 curve C_2^* is algebraically about 55.8% faster than on the supersingular genus 2 curve C_1, under the assumption that field squarings have cost $S = 0.8M$. Therefore, our algorithm improves previous work for pairing computations on genus 2 hyperelliptic curves over prime fields by a considerable margin.

Table 1. Performance Comparison of Pairing Computation on Curves C_1 and C_2^*

Reference	Curve Equation	Coordinate Type	Cost
Choie and Lee [6]	$C_1 : y^2 = x^5 + a,$	Affine	$240I, 17688M, 2163S$
Ó hÉigeartaigh & Scott [21]	$a \in \mathbb{F}_p^*, p \equiv 2, 3 \bmod 5$	Affine	$162I, 10375M, 645S$
Fan, Gong and Jao [11]		Projective	$13129M, 967S$
		New	$12487M, 971S$
This work	$C_2^* : y^2 = x^5 + 9x,$ $p \equiv 1 \bmod 8$	New	$\mathbf{5655M, 265S}$

5.8 Experimental Results

For verifying our theoretical analysis in Section 5.7, we report implementation results of computing the Tate pairing on the supersingular genus 2 curve C_1 and non-supersingular genus 2 curve C_2^* in this section. Both curves are defined over $\mathbb{F}_p$ and have an embedding degree of $k = 4$. The code was written in C and *Microsoft Developer Studio 6* was used for compilation and debugging on a Core 2 Duo^{TM}@2.67 GHz processor. For the curve C_1 and the (160/1024) bit security level we use the curve parameters that are generated in [21], where the subgroup order $n = 2^{159} + 2^{17} + 1$ is a Solinas prime [37] and the characteristic p of the finite field $\mathbb{F}_p$ is a 256-bit prime. Recall that the curve C_2^* is defined over a prime field of size 329 bits (see Section 5.1). The operations in the above two prime fields are implemented with various efficient algorithms in [20]. Table 2 shows the timings of our finite field library and the corresponding IM-ratio. From Table 2 we note that the IM-ratios are sufficiently large for the two prime fields in our implementation that using new coordinates and encapsulated group operations [11] can provide better performance than using affine coordinates in this case.

Table 3 summarizes previous work and our experimental results for the implementation of the Tate pairing on the curve C_1 and C_2^* for the (160/1024) bit security level. All of the timings are given in milliseconds.

From Table 3, we note that in our implementation the pairing computation on the curve C_2^* is about 10% faster than that on the curve C_1, in contrast to the algebraic complexity analysis in Section 5.7. The reason is that the sizes of the fields over which both curves are defined are different. Observe that the curve C_2^* is defined over a larger prime field than C_1, which significantly decreases the speed of computing the final exponentiation of the Tate pairing when the curve C_2^* is used. This explains why our new algorithm only obtains a 10% performance improvement in the implementation. Unfortunately, under current

Table 2. Timings of Prime Field $\mathbb{F}_p$ Library

Curve	# of bits of p	Multiplication (M)	Squaring (S)	Inversion (I)	IM-ratio
C_1	256	$0.84\mu s$	$0.78\mu s$	$41.9\mu s$	53.7
C_2^*	329	$1.40\mu s$	$1.30\mu s$	$64.6\mu s$	46.1

Table 3. Experimental Results – (160/1024) Security Level

Reference	Curve	Coordinate Type	Subgroup Order	Running Time (ms)
Choie and Lee [6]	C_1	Affine	Random	515
Ó hÉigeartaigh and Scott [21]	C_1	Affine	$n = 2^{159} + 2^{17} + 1$	16
This work	C_1	New	$n = 2^{159} + 2^{17} + 1$	14.6
	C_2^*	New	$\lambda_2 = 2^{43} + 2^{10}$	13.1

techniques for generating pairing-friendly non-supersingular genus 2 hyperelliptic curves, we cannot find such a curve of the form $y^2 = x^5 + ax$ defined over a 256-bit prime field with an embedding degree of $k = 4$. Nevertheless, despite the unequal field size, our implementation on the curve C_2^* is still slightly faster, even though strictly speaking a direct comparison between fields of different size is complicated as many factors could affect the comparison one way or another.

6 Conclusion

In this paper, we have proposed new variants of Miller's algorithm for computing the Tate pairing on two families of non-supersingular genus 2 hyperelliptic curves over prime fields with efficiently computable automorphisms. We describe how to use the automorphisms to unroll the main loop of Miller's algorithm. As a case study, we combine our new algorithm with various optimization techniques in the literature to efficiently implement the Tate pairing on a non-supersingular genus 2 curve $y^2 = x^5 + 9x$ over $\mathbb{F}_p$ with an embedding degree of $k = 4$. We also analyze the performance for the new algorithm in detail. When compared with pairing computations on supersingular genus 2 curves at the same security level, our new algorithm can obtain 55.8% performance improvements algebraically. Furthermore, favorable experimental results have been obtained for the implementation of the Tate pairing on both a supersingular and a non-supersingular genus 2 curve with embedding degree 4.

References

1. Avanzi, R.M., Cohen, H., Doche, C., Frey, G., Lange, T., Nguyen, K., Vercauteren, F.: Handbook of Elliptic and Hyperelliptic Curve Cryptography. Chapman & Hall/CRC, Boca Raton (2006)
2. Barreto, P.L.S.M., Galbraith, S., Ó'hÉigeartaigh, C., Scott, M.: Efficient Pairing Computation on Supersingular Abelian Varieties. Design, Codes and Cryptography 42, 239–271 (2007)
3. Barreto, P.L.S.M., Kim, H.Y., Lynn, B., Scott, M.: Efficient Algorithm for Pairing-Based Cryptosystems. In: Yung, M. (ed.) CRYPTO 2002. LNCS, vol. 2442, pp. 354–368. Springer, Heidelberg (2002)
4. Barreto, P.L.S.M., Lynn, B., Scott, M.: On the Selection of Pairing-Friendly Groups. In: Matsui, M., Zuccherato, R.J. (eds.) SAC 2003. LNCS, vol. 3006, pp. 17–25. Springer, Heidelberg (2004)
5. Boneh, D., Franklin, M.: Identity-Based Encryption from the Weil Pairing. SIAM Journal of Computing 32(3), 586–615 (2003)
6. Choie, Y., Lee, E.: Implementation of Tate Pairing on Hyperelliptic Curve of Genus 2. In: Lim, J.-I., Lee, D.-H. (eds.) ICISC 2003. LNCS, vol. 2971, pp. 97–111. Springer, Heidelberg (2004)
7. Choie, Y., Jeong, E., Lee, E.: Supersingular Hyperelliptic Curves of Genus 2 over Finite Fields. Journal of Applied Mathematics and Computation 163(2), 565–576 (2005)

8. Cocks, C., Pinch, R.G.E.: Identity-based Cryptosystems Based on the Weil Pairing (unpublished manuscript, 2001)
9. Duursma, I., Gaudry, P., Morain, F.: Speeding up the Discrete Log Computation on Curves with Automorphisms. In: Lam, K.-Y., Okamoto, E., Xing, C. (eds.) ASIACRYPT 1999. LNCS, vol. 1716, pp. 103–121. Springer, Heidelberg (1999)
10. Duursma, I., Lee, H.S.: Tate Pairing Implementation for Hyperelliptic Curves $y^2 = x^p - x + d$. In: Laih, C.-S. (ed.) ASIACRYPT 2003. LNCS, vol. 2894, pp. 111–123. Springer, Heidelberg (2003)
11. Fan, X., Gong, G., Jao, D.: Efficient Pairing Computation on Genus 2 Curves in Projective Coordinates. Centre for Applied Cryptographic Research (CACR) Technical Reports, CACR 2008-03, `http://www.cacr.math.uwaterloo.ca/techreports/2008/cacr2008-03.pdf`
12. Freeman, D.: Constructing Pairing-Friendly Genus 2 Curves over Prime Fields with Ordinary Jacobians. In: Takagi, T., Okamoto, T., Okamoto, E., Okamoto, T. (eds.) Pairing 2007. LNCS, vol. 4575, pp. 152–176. Springer, Heidelberg (2007)
13. Frey, G., Lange, T.: Fast Bilinear Maps from The Tate-Lichtenbaum Pairing on Hyperelliptic Curves. In: Hess, F., Pauli, S., Pohst, M. (eds.) ANTS 2006. LNCS, vol. 4076, pp. 466–479. Springer, Heidelberg (2006)
14. Frey, G., Rück, H.-G.: A Remark Concerning m-Divisibility and the Discrete Logarithm Problem in the Divisor Class Group of Curves. Mathematics of Computation 62(206), 865–874 (1994)
15. Furukawa, E., Kawazoe, M., Takahashi, T.: Counting Points for Hyperelliptic Curves of Type $y^2 = x^5 + ax$ over Finite Prime Fields. In: Matsui, M., Zuccherato, R.J. (eds.) SAC 2003. LNCS, vol. 3006, pp. 26–41. Springer, Heidelberg (2004)
16. Galbraith, S.D., McKee, J.F., Valença, P.C.: Ordinary Abelian Varieties Having Small Embedding Degree. Finite Fields and Their Applications 13(4), 800–814 (2007)
17. Gaudry, P.: An Algorithm for Solving the Discrete Log Problem on Hyperelliptic Curves. In: Preneel, B. (ed.) EUROCRYPT 2000. LNCS, vol. 1807, pp. 19–34. Springer, Heidelberg (2000)
18. Granger, R., Hess, F., Oyono, R., Thériault, N., Vercauteren, F.: Ate Pairing on Hyperelliptic Curves. In: Naor, M. (ed.) EUROCRYPT 2007. LNCS, vol. 4515, pp. 430–447. Springer, Heidelberg (2007)
19. Haneda, M., Kawazoe, M., Takahashi, T.: Suitable Curves for Genus-4 HEC over Prime Fields: Point Counting Formulae for Hyperelliptic Curves of Type $y^2 = x^{2k+1} + ax$. In: Caires, L., Italiano, G.F., Monteiro, L., Palamidessi, C., Yung, M. (eds.) ICALP 2005. LNCS, vol. 3580, pp. 539–550. Springer, Heidelberg (2005)
20. Hankerson, D., Menezes, A., Vanstone, S.: Guide to Elliptic Curve Cryptography. Springer, New York (2004)
21. Ó'hÉigeartaigh, C., Scott, M.: Pairing Calculation on Supersingular Genus 2 Curves. In: Biham, E., Youssef, A.M. (eds.) SAC 2006. LNCS, vol. 4356, pp. 302–316. Springer, Heidelberg (2007)
22. Hess, F., Smart, N.P., Vercauteren, F.: The Eta Pairing Revisited. IEEE Transactions on Information Theory 52(10), 4595–4602 (2006)
23. Hitt, L.: Families of Genus 2 Curves with Small Embedding Degree, Cryptology ePrint Archive, Report 2007/001 (2007), `http://eprint.iacr.org/2007/001`
24. Joux, A.: A One-Round Protocol for Tripartite Diffie-Hellman. In: Bosma, W. (ed.) ANTS 2000. LNCS, vol. 1838, pp. 385–394. Springer, Heidelberg (2000)

25. Kawazoe, M., Takahashi, T.: Pairing-friendly Hyperelliptic Curves of Type $y^2 = x^5 + ax$, Cryptology ePrint Archive, Report 2008/026 (2008) http://eprint.iacr.org/2008/026
26. Kozaki, S., Matsuo, K., Shimbara, Y.: Skew-Frobenius Maps on Hyperelliptic Curves. In: The 2007 Symposium on Cryptography and Information Security - SCIS 2007, IEICE Japan, pp. 1D2–4 (January 2007)
27. Lee, E., Lee, H.-S., Lee, Y.: Eta Pairing Computation on General Divisors over Hyperelliptic Curves $y^2 = x^7 - x \pm 1$. In: Takagi, T., Okamoto, T., Okamoto, E., Okamoto, T. (eds.) Pairing 2007. LNCS, vol. 4575, pp. 349–366. Springer, Heidelberg (2007)
28. Menezes, A., Okamoto, T., Vanstone, S.A.: Reducing Elliptic Curve Logarithms to a Finite Field. IEEE Transactions on Information Theory 39(5), 1639–1646 (1993)
29. Miller, V.S.: Short Programs for Functions on Curves (unpublished manuscript, 1986), http://crypto.stanford.edu/miller/miller.pdf
30. Miller, V.S.: The Weil Pairing and Its Efficient Calculation. Journal of Cryptology 17(4), 235–261 (2004)
31. Park, Y.-H., Jeong, S., Lim, J.: Speeding Up Point Multiplication on Hyperelliptic Curves with Efficiently-Computable Endomorphisms. In: Knudsen, L.R. (ed.) EUROCRYPT 2002. LNCS, vol. 2332, pp. 197–208. Springer, Heidelberg (2002)
32. Rubin, K., Silverberg, A.: Supersingular Abelian Varieties in Cryptography. In: Yung, M. (ed.) CRYPTO 2002. LNCS, vol. 2442, pp. 336–353. Springer, Heidelberg (2002)
33. Sakai, R., Ohgishi, K., Kasahara, M.: Cryptosystems Based on Pairings. In: Proceedings of the 2000 Symposium on Cryptography and Information Security - SCIS 2002, Okinawa, Japan, pp. 26–28 (2000)
34. Scott, M.: Faster Pairings Using an Elliptic Curve with an Efficient Endomorphism. In: Maitra, S., Veni Madhavan, C.E., Venkatesan, R. (eds.) INDOCRYPT 2005. LNCS, vol. 3797, pp. 258–269. Springer, Heidelberg (2005)
35. Scott, M., Barreto, P.L.S.M.: Compressed Pairings. In: Franklin, M. (ed.) CRYPTO 2004. LNCS, vol. 3152, pp. 140–156. Springer, Heidelberg (2004)
36. Silverman, J.H.: The Arithmetic of Elliptic Curves. Graduate Texts in Mathematics 106. Springer, Heidelberg (1986)
37. Solinas, J.: Generalized Mersenne Primes, Centre for Applied Cryptographic Research (CACR) Technical Reports, CORR 99-39, http://www.cacr.math.uwaterloo.ca/techreprots/1999/corr99-39.pdf
38. Takashima, K.: Scaling Security of Elliptic Curves with Fast Pairing Using Efficient Endomorphism. IEICE Transactions on Fundamentals of Electronics, Communications and Computer Science E90-A(1), 152–159 (2007)
39. Zhao, C., Zhang, F., Huang, J.: Speeding Up the Bilinear Pairings Computation on Curves with Automorphisms, Cryptology ePrint Archive, Report 2006/474 (2006), http://eprint.iacr.org/2006/474

Appendix: Explicit Formulae for Genus 2 Curves over $\mathbb{F}_p$

In this appendix, we give efficient explicit formulae for group operations on genus 2 curves over $\mathbb{F}_p$ in new coordinates in the context of pairing computations. Table 4 and Table 5 address the cases of new coordinates. Given two divisor classes $\overline{E}_1$ and $\overline{E}_2$, Table 4 computes the divisor class $\overline{E}_3 = [u_3(x), v_3(x)]$ and the rational function $l(x)$ such that $E_1 + E_2 = E_3 + \text{div}\left(\frac{y-l(x)}{u_3(x)}\right)$ in the new coordinate system, where $l(x) = \frac{s_1'}{rz_{23}}x^3 + \frac{l_2}{rz_{24}}x^2 + \frac{l_1}{rz_{24}}x + \frac{l_0}{rz_{24}}$. For doubling a reduced divisor class E_1, Table 5 calculates the divisor class $\overline{E}_3 = [u_3(x), v_3(x)]$ and the rational function $l(x)$ such that $2E_1 = E_3 + \text{div}\left(\frac{y-l(x)}{u_3(x)}\right)$ in projective coordinates, where $l(x) = \frac{s_1}{s_1'Z_{32}}x^3 + \frac{l_2}{Z_{31}Z_{32}}x^2 + \frac{l_1}{Z_{31}Z_{32}}x + \frac{l_0}{Z_{31}Z_{32}}$.

Table 4. Mixed-Addition Formula on a Genus 2 Curve over $\mathbb{F}_p$ (New Coordinates) [11]

Input	Genus 2 HEC $C: y^2 = x^5 + ax$ $\overline{E}_1 = [U_{11}, U_{10}, V_{11}, V_{10}, 1, 1, 1, 1]$ and $\overline{E}_2 = [U_{21}, U_{20}, V_{21}, V_{20}, Z_{21}, Z_{22}, z_{21}, z_{22}]$	
Output	$\overline{E}_3 = [U_{31}, U_{30}, V_{31}, V_{30}, Z_{31}, Z_{32}, z_{31}, z_{32}] = \overline{E}_1 \oplus \overline{E}_2$ $l(x)$ such that $E_1 + E_2 = E_3 + \text{div}\left(\frac{y-l(x)}{u_3(x)}\right)$	
Step	Expression	Cost
1	**Compute resultant and precomputations:** $z_{23} = Z_{21}Z_{22}, z_{24} = z_{21}z_{23}, \tilde{U}_{11} = U_{11}z_{21}$ $\tilde{U}_{10} = U_{10}z_{21}, y_1 = \tilde{U}_{11} - U_{21}, y_2 = U_{20} - \tilde{U}_{10}$ $y_3 = U_{11}y_1, y_4 = y_2 + y_3, r = y_2y_4 + y_1^2U_{10}$	$7M, 1S$
2	**Compute almost inverse of u_2 mod u_1:** $inv_1 = y_1, inv_0 = y_4$	–
3	**Compute s':** $w_0 = V_{10}z_{24} - V_{20}, w_1 = V_{11}z_{24} - V_{21}, w_2 = inv_0w_0$ $w_3 = inv_1w_1, s_1' = y_1w_0 + y_2w_1, s_0' = w_2 - U_{10}w_3$	$7M$
4	**Precomputations:** $\tilde{r} = rz_{23}, R = \tilde{r}^2, Z_{31} = s_1'Z_{21}, Z_{32} = \tilde{r}Z_{21}$ $z_{31} = Z_{31}^2, z_{32} = Z_{32}^2, \tilde{s}_0' = s_0'z_{21}$	$4M, 3S$
5	**Compute l:** $l_2 = s_1'U_{21} + \tilde{s}_0', l_0 = s_0'U_{20} + rV_{20}$ $l_1 = (s_1' + s_0')(U_{21} + U_{20}) - s_1'U_{21} - s_0'U_{20} + rV_{21}$	$5M$
6	**Compute U_3:** $w_1 = \tilde{U}_{11} + U_{21}, U_{31} = s_1'(2\tilde{s}_0' - s_1'y_1) - z_{32}, l_1' = l_1s_1'$ $U_{30} = \tilde{s}_0'(s_0' - 2s_1'U_{11}) + s_1'^2(y_3 - \tilde{U}_{10} - U_{20}) + 2l_1' + Rw_1$	$7M, 1S$
7	**Compute V_3:** $w_1 = l_2s_1' - U_{31}, V_{30} = U_{30}w_1 - z_{31}(l_0s_1')$ $V_{31} = U_{31}w_1 + z_{31}(U_{30} - l_1')$	$6M$
Sum		$36M, 5S$

Table 5. Doubling Formula on a Genus 2 Curve over $\mathbb{F}_p$ (New Coordinates) [11]

Input	Genus 2 HEC $C : y^2 = x^5 + ax$ $\overline{E}_1 = [U_{11}, U_{10}, V_{11}, V_{10}, Z_{11}, Z_{12}, z_{11}, z_{12}]$	
Output	$\overline{E}_3 = [U_{31}, U_{30}, V_{31}, V_{30}, Z_{31}, Z_{32}, z_{31}, z_{32}] = [2]\overline{E}_1$ $l(x)$ such that $2E_1 = E_3 + \text{div}\left(\frac{y-l(x)}{u_3(x)}\right)$	
Step	**Expression**	**Cost**
1	**Compute resultant:** $w_0 = V_{11}^2, w_1 = U_{11}^2, w_2 = V_{10}z_{11}, w_3 = w_2 - U_{11}V_{11}$ $r = U_{10}w_0 + V_{10}w_3$	$4M, 2S$
2	**Compute almost inverse:** $inv_1' = -V_{11}, inv_0' = w_3$	–
3	**Compute k':** $\tilde{U}_{10} = U_{10}z_{11}, k_1' = z_{12}(2(w_1 - \tilde{U}_{10}) + w_1)$ $k_0' = (z_{12}U_{11})(4\tilde{U}_{10} - w_3) - w_0$	$4M$
4	**Compute s':** $w_0 = k_0' inv_0', w_1 = k_1' inv_1'$ $s_1' = w_2k_1' - V_{11}k_0', s_0' = w_0 - \tilde{U}_{10}w_1$	$5M$
5	**Precomputations:** $Z_{31} = s_1'z_{11}, z_{31} = Z_{31}^2, w_0 = rz_{11}, w_1 = w_0Z_{12}$ $Z_{32} = 2w_1Z_{11}, z_{32} = Z_{32}^2, w_2 = w_1^2, R = rZ_{31}$ $S_0 = s_0'^2, S = s_0'Z_{31}, s_0 = s_0's_1', s_1 = s_1'Z_{31}$	$8M, 4S$
6	**Compute l:** $l_2 = s_1U_{11} + s_0z_{11}, V_{10}' = RV_{10}$ $l_0 = s_0U_{10} + 2V_{10}', V_{11}' = RV_{11}$ $l_1 = (s_1 + s_0)(U_{11} + U_{10}) - s_1U_{11} - s_0U_{10} + 2V_{11}'$	$6M$
7	**Compute U_3:** $U_{30} = S_0 + 4(V_{11}' + 2w_2U_{11}), U_{31} = 2S - z_{32}$	$1M$
8	**Compute V_3:** $w_0 = l_2 - U_{31}, w_1 = w_0U_{30}, w_2 = w_0U_{31}$ $V_{31} = w_2 + z_{31}(U_{30} - l_1), V_{30} = w_1 - z_{31}l_0$	$4M$
Sum		$32M, 6S$

Pairings on Hyperelliptic Curves with a Real Model

Steven D. Galbraith[1], Xibin Lin[1,2], and David J. Mireles Morales[1]

[1]Mathematics Department
Royal Holloway, University of London
United Kingdom
{steven.galbraith,d.mireles-morales}@rhul.ac.uk
[2]School of Mathematics and Computational Science
Sun Yat-Sen University
P.R.China
linxibin@mail2.sysu.edu.cn

Abstract. We analyse the efficiency of pairing computations on hyperelliptic curves given by a real model using a balanced divisor at infinity. Several optimisations are proposed and analysed. Genus two curves given by a real model arise when considering pairing friendly groups of order dividing $p^2 - p + 1$. We compare the performance of pairings on such groups in both elliptic and hyperelliptic versions. We conclude that pairings can be efficiently computable in real models of hyperelliptic curves.

1 Introduction

The study of efficient pairing computation on hyperelliptic curves has focused exclusively on the analysis of hyperelliptic curves given by an imaginary model. With the development of new divisor addition algorithms on hyperelliptic curves given by a real model [6], it is natural to ask if pairings can be implemented on these curves competitively.

The authors of [7] construct a genus 2 curve C, defined over $\mathbf{F}_p$ for p a prime $p \equiv 5 \mod 6$. The Jacobian $\mathrm{Jac}(C)$ of this curve has $p^2 - p + 1$ points, and embedding degree 6 with respect to any subgroup with prime order $r > 3$. The curve C is given by a real model (see [6]), which in particular means that it has 2 points at infinity.

In [18], Verheul presents the construction of an elliptic curve with embedding degree 3. This curve is defined over a field $\mathbf{F}_{p^2}$ for p a prime $p \equiv 5 \mod 6$, and has $p^2 - p + 1$ $\mathbf{F}_{p^2}$-rational points. Pairings on these elliptic curves have been studied by Hu et.al. in [11].

The similarities between these curves make them natural candidates for a comparison between elliptic and hyperelliptic curve pairing implementations. In this article we explore several optimisation techniques on these curves, implement pairings and compare their performance. Among the optimisations used in the implementation is the recent R-ate pairing presented by Lee, Lee and

S.D. Galbraith and K.G. Paterson (Eds.): Pairing 2008, LNCS 5209, pp. 265–281, 2008.

Park in [12], and the well-known *denominator elimination* technique, which is combined with the R-ate pairing thanks to Theorem 2.

A crucial step towards a competitive implementation of pairings on hyperelliptic curves given by a real model is having efficient divisor addition algorithms that result in simple Miller functions. The addition algorithms presented in [6] allow for a fast implementation not only because the operation count in the addition and doubling algorithms is smaller than that in previous proposals [15], but also because the Miller function, whose evaluation is the bottleneck in pairing computations on high genus curves, is simpler using the algorithms of [6]. We give a theoretical and practical comparison of the efficiency of our pairings in the elliptic and hyperelliptic cases. We conclude that pairings can be efficiently implemented on hyperelliptic curves given by a real model, however it seems that elliptic curves still offer better performance.

The article is organized as follows: Section 2 describes the representation of divisors (and hence the addition algorithms) that we will use for genus 2 curves given by a real model. In this section we also present the embedding degree 6 construction of Galbraith, Pujolas, Ritzenthaler and Smith [7]. Section 3 presents a brief overview of pairing computation techniques, including the recently presented R-ate pairing. Section 4 describes our parameter generation algorithms and the optimisations used in the implementation. In Section 5 we report our implementation results and compare them with pairing computation results obtained for similar elliptic or hyperelliptic curves. Some conclusions are discussed in Section 6.

2 Curves

Given an algebraic curve C and two divisors D_0 and D_1 on C, we say that D_0 and D_1 are *linearly equivalent*, denoted $D_0 \sim D_1$, if there is a function f such that

$$\mathrm{div}(f) = D_1 - D_0$$

where $\mathrm{div}(f)$ is the divisor of f.

Definition 1. *The* divisor class group *of C is the group of divisor classes modulo linear equivalence. We will denote it as $\mathrm{Cl}(C)$. The class of a divisor D in $\mathrm{Cl}(C)$ will be denoted by $[D]$. We define $\mathrm{Cl}^0(C)$ to be the degree zero subgroup of $\mathrm{Cl}(C)$.*

Notice that the degree of the divisor $\mathrm{div}(f)$ associated to a function f is always zero, and thus it makes sense to talk of the degree of a divisor class $[D]$ in $\mathrm{Cl}(C)$. In this article we will work exclusively with curves C which are elliptic or hyperelliptic curves of genus 2.

2.1 Arithmetic on Hyperelliptic Curves

Let K be a field such that $\mathrm{char}(K) \neq 2, 3$. Let C be a genus 2 hyperelliptic curve over K given by

$$C : y^2 = F(x),$$

where $F(x) \in K[x]$ is a square-free degree 6 polynomial. We say that this is a *real model* for C. If $\deg(F(x)) = 5$ then we say it is an *imaginary model.* If $P = (x, y)$ is a point on C then we write $\bar{P} = (x, -y)$ for the hyperelliptic conjugate. The desingularization of C has 2 different points at infinity, which we will denote ∞^+ and ∞^-. Let $D_\infty = \infty^+ + \infty^-$, note that this divisor is K-rational even if the points ∞^+ and ∞^- are not independently so.

Proposition 1 (Proposition 1 in [6]). *Let $D_\infty = \infty^+ + \infty^-$, and let $D \in \mathrm{Div}_0(C)$ be a K-rational divisor on the curve C. Then $[D]$ has a unique representative in $\mathrm{Cl}^0(C)$ of the form $[D_0 - D_\infty]$, where $D_0 = P_1 + P_2$ is an effective K-rational divisor of degree 2 such that $P_1 \neq \bar{P}_2$.*

A generic divisor class has a representative $D_0 - D_\infty$ where $D_0 = P_1 + P_2$ with $P_1, P_2 \notin \{\infty^+, \infty^-\}$. Hence, for the remainder of the paper we discuss arithmetic only for generic divisors. This is not a serious restriction for the pairing applications: there will exist divisor classes of the required prime order which are of the generic form. Full details of how to handle the special cases are given in [6].

We will use Mumford's representation to represent divisors of the form

$$D = P_1 + P_2 - D_\infty, \quad P_1, P_2 \notin \{\infty^+, \infty^-\}.$$

Let $P_i = (x_i, y_i)$ for $i \in \{1, 2\}$. Mumford's representation is a pair of polynomials $(u(x), v(x))$, where $u(x) = (x - x_1)(x - x_2)$ and where $v(x)$ satisfies $v(x)^2 - F(x) \equiv 0 \mod u(x)$. This last condition implies that $y_i = v(x_i)$. The polynomial v is only determined modulo u; if a canonical representative is needed, the unique representative with $\deg v < \deg u$ can be used.

We will denote the divisor $D = P_1 + P_2 - D_\infty$ associated to the pair of polynomials (u, v) as $D = \mathrm{div}(u, v)$. Traditionally this notation has been used to denote the affine divisor $P_1 + P_2$ but we will extend it since there is no risk of confusion.

Let $D_1 = P_1 + P_2 - D_\infty$ and $D_2 = P_3 + P_4 - D_\infty$ be two divisors. An explicit interpretation of the results of [6] in the case of a genus 2 curve implies that if $p(x)$ denotes the unique polynomial of degree at most 3 such that $y - p(x)$ passes through P_1, P_2, P_3 and P_4, and we let P_5, P_6 be the remaining intersection points of $y - p(x)$ with C, then

$$\mathrm{div}(y - p(x)) = \sum_{i=1}^{6} P_i - 3D_\infty. \tag{1}$$

If we write $D_3 = \bar{P}_5 + \bar{P}_6 - D_\infty$, Equation (1) can be rewritten as

$$[D_1] + [D_2] = [D_3].$$

If u_3 is the first polynomial in the Mumford representation of D_3, the function

$$g_{D_1,D_2} = \frac{y - p(x)}{u_3} \tag{2}$$

has associated divisor $D_1 + D_2 - D_3$. This will be used later to compute pairings.

In our pairing implementation we will use the addition formulae presented in [4], which we include in an appendix for completeness. The polynomial $p(x)$ in Equation (1) can be easily computed from the intermediate results in the addition formulae from [4] and presented in the Appendix.

When the divisor at infinity used is the traditional $D_\infty = 2\infty^+$, the function g_{D_1,D_2} with divisor $D_1 + D_2 - D_3$ has the form $g_{D_1,D_2} = (y - p_1(x))(y - p_2(x))/(u_3(x)u_4(x))$, where again $p_1(x)$ and $p_2(x)$ are cubic polynomials and $u_3(x), u_4(x)$ are quadratic polynomials. Since the bottleneck of pairing calculations is precisely the evaluation of this function, the speed-up obtained from using the representation of $\mathrm{Cl}^0(C)$ described in [6] goes beyond the operations saved in the addition algorithm.

2.2 Hyperelliptic Curves with Embedding Degree 6

In this section we will substitute the notation $\mathrm{Cl}^0(C)$ we had been using for the more geometric (and equivalent) $\mathrm{Jac}(C)$, better suited when dealing with endomorphism rings.

In [7, Section 7], the authors present a family of genus 2 curves with embedding degree 6 and generators of a subring R of the endomorphism ring of $\mathrm{Jac}(C)$, such that R contains a distortion map for any non-trivial pair (D_1, D_2) of divisors.

The curves in this family will have 2 points at infinity and our addition algorithm is well-suited to perform efficient arithmetic on them. We now briefly describe the construction of the curves given in [7, Section 7].

Let $p \neq 2$ a prime such that $p \equiv 2 \pmod 3$. Denote by ζ_6 a root of $x^2 - x + 1$ and by $\zeta_3 = \zeta_6^2$, let $\gamma \in \mathbf{F}_{p^6}$ be such that $\gamma^{p^2-1} = \zeta_3$. The curve C is defined to be

$$C : y^2 = (ax + b)^6 + (cx + d)^6$$

where $a = \gamma^p, b = \zeta_3^2\gamma^p, c = \gamma$ and $d = \zeta_3\gamma$. The following Lemma shows that this is not an imaginary model of a curve (i.e., it does have two points at infinity).

Lemma 1. *The model of the curve C defined above has 2 points at infinity.*

Proof. The curve C is given by $y^2 = F(x)$ where the leading coefficient of F is

$$F_6 = a^6 + c^6 = \gamma^{6p} + \gamma^6.$$

To prove the lemma we only need to prove that $F_6 \neq 0$. Since $p^2 - 1$ is a multiple of 3 and $\gamma^{p^2-1} = \zeta_3$ the multiplicative order of γ is a multiple of 9. So $F_6 = \gamma^6(\gamma^{6p-6} + 1)$ cannot be zero as this would imply that $\gamma^{12p-12} = 1$, but $12p - 12$ is not a multiple of 9 as $p \equiv 2 \pmod 3$. □

Note that if $a^6 + c^6$ is not a square, we can take two rational points on C and move them to the line at infinity, and get a curve isomorphic to C given by a monic polynomial. This will let us use the addition formulae presented in [4], which only work on curves given by an equation of the form $y^2 = x^6 + f_4x^4 + f_3x^3 + f_2x^2 + f_1x + f_0$.

The characteristic polynomial of Frobenius on C is $T^4 - pT^2 + p^2$, so $\mathrm{Jac}(C)$ has $p^2 - p + 1$ elements. This implies that if $r > 3$ is a prime that divides $p^2 - p + 1$, then the embedding degree of C with respect to r is 6. Note that if C' is the curve $C' : y^2 = x^6 + 1$, then C is a twist of C' by the automorphism $u : (x, y) \mapsto (\frac{\zeta_3}{x}, \frac{y}{x^3})$. Furthermore, there is an isomorphism $\phi : C \longrightarrow C'$ given by

$$\phi(x, y) = \left(\frac{ax + b}{cx + d}, \frac{y}{(cx + d)^3} \right).$$

The authors of [7] then define the following endomorphisms of C':

$$\pi(x, y) = (x^p, y^p)$$
$$\chi(x, y) = \left(\frac{1}{x}, \frac{y}{x^3} \right)$$
$$\zeta_6(x, y) = (\zeta_6 x, y).$$

We will abuse notation and extend these endomorphisms to $\mathrm{Jac}(C')$. These endomorphisms are enough to find a distortion map on $\mathrm{Jac}(C)$ (see Definition 3), as the following result shows.

Theorem 1 (Theorem 7.2 in [7]). *Let r be a prime different from 2, 3 and p. Then for all pairs of divisors D_1 and D_2 on C of order r, there exists a distortion map in the ring $\phi^{-1}\mathbb{Z}[\pi, \chi, \zeta_6]\phi$.*

It is well known that if the first coordinate of the Mumford representation of a divisor lies in a proper subfield of $\mathbf{F}_{p^6}$, then the function g_{D_1,D_2} in Equation (2) can be substituted by $y - p(x)$ (p as in Equation (2)) in the Miller loop of the pairing computation. The following Lemma shows that the automorphisms χ and ζ_6 can be used to this end.

Lemma 2. *Let $P \in C$ be a point with a $\mathbf{F}_p$-rational x-coordinate. Then:*

- *The x-coordinate of $(\phi^{-1} \circ \zeta_6 \circ \phi)(P)$ is $\mathbf{F}_p$-rational.*
- *The x-coordinate of $(\phi^{-1} \circ \chi \circ \phi)(P)$ is $\mathbf{F}_{p^3}$-rational.*
- *The x-coordinate of $(\phi^{-1} \circ \chi \circ \zeta_6 \circ \phi)(P)$ is $\mathbf{F}_{p^3}$-rational.*

Proof. Let $P = (x, y)$ be the coordinates of P. A tedious but simple calculation shows that the x-coordinate of $(\phi^{-1} \circ \zeta_6 \circ \phi)(x, y)$ is given by

$$\frac{-x - 1}{x - 2},$$

which is $\mathbf{F}_p$-rational whenever x is an element of $\mathbf{F}_p$.

The x-coordinate of $(\phi^{-1} \circ \chi \circ \phi)(x, y)$ is given by

$$x_\chi = \frac{(\zeta_3\gamma^2 - \zeta_3^2\gamma^{2p})x + (\zeta_3^2\gamma^2 - \zeta_3\gamma^{2p})}{(\gamma^{2p} - \gamma^2)x + (\zeta_3^2\gamma^{2p} - \zeta_3\gamma^2)},$$

and again, it is straightforward to prove that $x_\chi^{p^3} = x_\chi$. The third claim follows from the first two. □

The previous Lemma shows that using χ and ζ_6 as distortion maps (see Definition 3) makes it possible to use *denominator elimination*. We will now prove that the image of $\mathbf{F}_p$-rational divisors under the distortion map $(\phi^{-1} \circ \chi \circ \zeta_6 \circ \phi)$ lies in the p-eigenspace of Frobenius, thus allowing us to directly use loop-shortening techniques.

Theorem 2. *Let $\psi = \phi^{-1} \circ \chi \circ \zeta_6 \circ \phi$. Let $D_1 \in \mathrm{Cl}^0(C)[r]$ be a $\mathbf{F}_p$-rational divisor. Then $D_2 = \psi(D_1)$ lies in the p-eigenspace of the p-power Frobenius on $\mathrm{Cl}^0(C)[r]$.*

Proof. The r-torsion subgroup $\mathrm{Cl}^0(C)[r]$ can be decomposed as the direct sum of four 1-dimensional eigenspaces with respect to the p-power Frobenius π_p, with eigenvalues $1, -1, p$ and $-p$. The polynomial $T^2 - T + 1$ is divisible by $T - p$ mod r, hence the endomorphism $(\pi_p^2 - \pi_p + 1)$ annihilates the p-eigenspace, and is invertible when restricted to the other eigenspaces. It follows that D_2 lies in the p-eigenspace if and only if $(\pi_p^2 - \pi_p + 1)(D_2) = 0$.

To prove that this is the case, it suffices to show that the unique cubic polynomial passing through the four points in the affine support of D_2 and $\pi_p^2(D_2)$ also passes through the points in the affine support of $\pi_p(D_2)$. This can be proven symbolically simply by defining formal variables γ and γ^p over $\mathbb{Q}(\zeta_6)$, and formally defining the action of Frobenius as $\pi_p(\gamma) = \gamma^p$, $\pi_p(\gamma^p) = \zeta_6^2\gamma$ and $\pi_p(\zeta^6) = \zeta_6^5$. The verification of our claim boils down to a trivial, albeit tedious calculation, which we performed using Magma [2]. □

2.3 Elliptic Curves with Embedding Degree 3

In this subsection we describe the construction of elliptic curves with embedding degree $k = 3$ given in [18]. We will report our pairing implementation results on these curves in later sections.

Let p be a prime, $p \equiv 5 \bmod 6$, let E be an elliptic curve defined over $\mathbf{F}_{p^2}$ by $y^2 = x^3 + \rho^2$, where $\rho \in \mathbf{F}_{p^2}$ is an element such that ρ^2 is not a cube in $\mathbf{F}_{p^2}$. The number of $\mathbf{F}_{p^2}$ rational points of E is $p^2 - p + 1$ (see Lemma 7 of [8] for a proof). Let r be the largest prime dividing $p^2 - p + 1$, then E has embedding degree $k = 3$ with respect to r. Define the following map:

$$\begin{aligned} \phi_E : E(\mathbf{F}_{p^2}) &\to E(\mathbf{F}_{p^6}) \\ (x, y) &\to (a\beta x^p, by^p) \end{aligned}$$

where $a = \rho^{-(2p-1)/3}$, $b = \rho^{-(p-1)}$, and β is a cubic root of ρ in $\mathbf{F}_{p^6}$. If we let $(x', y') = \phi_E(x, y)$, it is not hard to see that $x' \in \mathbf{F}_{p^6}$ and $y' \in \mathbf{F}_{p^2}$. By Lemma 8 of [8] the endomorphism ϕ_E maps the 1-eigenspace of Frobenius to the p^2-eigenspace of Frobenius in $E(\mathbf{F}_{p^2})$, so this map will be used as a distortion map in our pairing implementation.

3 Pairings

3.1 Background on the Tate Pairing

We will briefly recall the definition of the Tate pairing (see [5] for a more detailed description) and describe the applications of the results in [6] to the computation of pairings on hyperelliptic curves given by a real model. Let $\mathbf{F}_q$ be a finite field with $q = p^n$ elements and let C be a smooth, irreducible curve over $\mathbf{F}_q$. Denote the degree zero divisor class group of C over $\mathbf{F}_q$ by $\mathrm{Cl}^0_{\mathbf{F}_q}(C)$. Let r be an integer such that $r \mid \#\mathrm{Cl}^0_{\mathbf{F}_q}(C)$ and denote by $\mathrm{Cl}^0_{\mathbf{F}_q}(C)[r]$ the group of divisor classes of order dividing r. Let k be the smallest integer such that $r \mid (q^k - 1)$. We say that k is the embedding degree of C.

Let $D_1 \in \mathrm{Cl}^0_{\mathbf{F}_q}(C)[r]$ and $D_2 \in \mathrm{Cl}^0_{\mathbf{F}_{q^k}}(C)$ be two divisors with disjoint support. Since rD_1 is principal, there is a function f_{r,D_1} defined over $\mathbf{F}_q$ such that $\mathrm{div}(f_{r,D_1}) = rD_1$. The Tate pairing is defined as

$$\langle D_1, D_2\rangle_r = f_{r,D_1}(D_2),$$

and one can prove that it is a non-degenerate, bilinear pairing:

$$\mathrm{Cl}^0_{\mathbf{F}_{q^k}}(C)[r] \times \mathrm{Cl}^0_{\mathbf{F}_{q^k}}(C)/r\,\mathrm{Cl}^0_{\mathbf{F}_{q^k}}(C) \longrightarrow \mathbf{F}^*_{q^k}/(\mathbf{F}^*_{q^k})^r.$$

The result is only defined up to an r-th power, hence to obtain a unique representative, one defines the reduced Tate pairing as

$$e(D_1, D_2) = \langle D_1, D_2\rangle_r^{(q^k-1)/r} = f_{r,D_1}(D_2)^{(q^k-1)/r}.$$

In practice to compute the Tate pairing one uses Miller's algorithm, which we now describe.

Definition 2. *Let C be a curve for which there exists a way to select a canonical representative for every element of $\mathrm{Cl}^0_{\mathbf{F}_{q^k}}(C)$. Given a degree 0 divisor D on C and an integer n, let D_n be the canonical representative of the class $[nD]$. We will denote the unique function (up to scalar multiples) with associated divisor $nD - D_n$ as $f_{n,D}$.*

The function $f_{n,D}$ is usually chosen to be normalised at infinity (i.e., the leading coefficient with respect to a fixed uniformizer at infinity is 1; see [10]). For real models there are two points at infinity and so two leading coefficients to consider. For our application it is enough to insist that the functions $f_{n,D}$ are such that the product of the two leading coefficients at infinity lies in the subfield $\mathbf{F}_{q^{k/2}}$. One can then avoid evaluating $f_{n,D}$ at D_∞ and relax the requirement that D_1 and D_2 have disjoint support.

By definition, given two degree 0 divisors D_1, D_2 on C, if D_3 is the canonical representative of $[D_1 + D_2]$, there is a function whose associated divisor is $D_1 + D_2 - D_3$. Denote this function as g_{D_1,D_2}. Miller's fundamental observation is that

$$f_{n_1+n_2,D} = f_{n_1,D} \cdot f_{n_2,D} \cdot g_{n_1D,n_2D}, \tag{3}$$

which allows us to compute $f_{r,D}$ (and hence the Tate-pairing) using a square and multiply calculation with $O(\log r)$ steps. Note that in the case of a genus 2 hyperelliptic curve, g_{n_1D,n_2D} is given by Equation (2). We will refer to the process of calculating $f_{n_1+n_2,D}$ from $f_{n_1,D}$ and $f_{n_2,D}$ as a Miller step.

The ate pairings in the coming subsections are defined on the product $\mathbf{G}_1 \times \mathbf{G}_2$ where $\mathbf{G}_1$ is the 1-eigenspace of the q-power Frobenius element and $\mathbf{G}_2$ is the q-eigenspace. We make the following definition.

Definition 3. *Let $\mathbf{G}_1$ and $\mathbf{G}_2$ be groups. A surjective morphism $\psi : \mathbf{G}_1 \longrightarrow \mathbf{G}_2$ is called a* distortion map.

We have already seen distortion maps which are suitable for the elliptic and hyperelliptic curves studied in this paper.

3.2 Elliptic Ate Pairing

Let E be the supersingular elliptic curve defined over $\mathbf{F}_{p^2}$ from Section 2.3. Then $\#E(\mathbf{F}_{p^2}) = p^2 - p + 1$. Let $r \mid (p^2 - p + 1)$ be prime and write $T = t - 1 = p - 1$.

Let π_{p^2} be the p^2-power Frobenius on E and define

$$\mathbf{G}_1 = E[r] \cap \operatorname{Ker}(\pi_{p^2} - \mathrm{id}), \text{ and } \mathbf{G}_2 = E[r] \cap \operatorname{Ker}(\pi_{p^2} - p^2).$$

Since E is supersingular, by Section 4.2 of [10] it follows that

$$e(P, Q) = f_{T,P}(Q)^{(p^6-1)/r} = f_{p-1,P}(Q)^{(p^6-1)/r} \quad (4)$$

is a non-degenerate bilinear pairing on $\mathbf{G}_1 \times \mathbf{G}_2$.

Since $k = 3$, the denominator elimination method of [1] does not apply. We now describe a way to replace the denominator with a few multiplications.

When executing Miller's algorithm to compute pairings on an elliptic curve, the denominator of the function $g_{n_1,n_2,D}$ in Equation (3) has the form $(x_R - x_Q)$, where R and Q are points on the elliptic curve. Note that $x_R \in \mathbf{F}_{p^2}$ and $x_Q \in \mathbf{F}_{p^6}$. We replace

$$\frac{1}{x_R - x_Q} = \frac{x_R(x_R + x_Q) + x_Q^2}{y_R^2 - y_Q^2},$$

and since $y_R^2 - y_Q^2$ lies in the proper subfield $\mathbf{F}_{p^2}$ of $\mathbf{F}_{p^6}$, we can discard its value as it will become 1 after the final exponentiation.

So the function $g_{n_1,n_2,D}$ in Equation (3) can be substituted by

$$l_{R,P}(Q) \cdot (x_R(x_R + x_Q) + x_Q^2), \quad (5)$$

where $l_{R,P}$ denotes the line passing through the points P and R. If x_Q^2 is pre-computed then the saving compared with the standard method (i.e., writing the Miller variable f as a numerator and a denominator) is to replace a squaring in $\mathbf{F}_{p^6}$ by a multiplication of an element in $\mathbf{F}_{p^2}$ with an element in $\mathbf{F}_{p^6}$.

3.3 Hyperelliptic Ate Pairings

We have seen that in some cases it is possible to compute pairings using a function $f_{n,D}$ where n is much smaller than required for the Tate-pairing. We will revisit some of these techniques in the case of hyperelliptic curves.

Let C be a hyperelliptic curve defined over a finite field $\mathbf{F}_q$. Denote the Frobenius endomorphism of C as π_q, and extend this notation to $\mathrm{Cl}^0_{\mathbf{F}_{q^k}}(C)$. Let

$$\mathbf{G}_1 = \mathrm{Cl}^0_{\mathbf{F}_{q^k}}(C)[r] \cap \mathrm{Ker}(\pi_q - \mathrm{id}) \text{ and } \mathbf{G}_2 = \mathrm{Cl}^0_{\mathbf{F}_{q^k}}(C)[r] \cap \mathrm{Ker}(\pi_q - q)$$

denote the 1- and q-eigenspaces of π_q in the r-torsion subgroup of $\mathrm{Cl}^0_{\mathbf{F}_{q^k}}(C)$. If $D_1 \in \mathbf{G}_1$ and $D_2 \in \mathbf{G}_2$ are divisors on C, the authors of [9] proved:

Theorem 3. *The function $e_q : \mathbf{G}_1 \times \mathbf{G}_2 \longrightarrow \mu_r$, given by*

$$e_q(D_1, D_2) = f_{q,D_1}(D_2)^{(q^k-1)/r},$$

defines a non-degenerate bilinear pairing on $\mathbf{G}_1 \times \mathbf{G}_2$.

3.4 *R*-Ate Pairings

Let $\mathbf{G}_1$ and $\mathbf{G}_2$ be subgroups of the class group of a curve C. If $D_1 \in \mathbf{G}_1$ and $D_2 \in \mathbf{G}_2$, Lee, Lee and Park prove in [12] the following:

Theorem 4. *[Theorem 3.2 in [12]] Let A, B, a, b be integers such that $A = aB + b$, where the functions $f_{A,D_1}(D_2)$ and $f_{B,D_1}(D_2)$ define bilinear maps in $\mathbf{G}_1 \times \mathbf{G}_2$. Then the function*

$$f_{a,BD_1}(D_2) \cdot f_{b,D_1}(D_2) \cdot g_{aBD_1,bD_1}(D_2)$$

defines a bilinear map in $\mathbf{G}_1 \times \mathbf{G}_2$.

Corollary 1. *Let C be an elliptic curve over $\mathbf{F}_{p^2}$ or a genus 2 curve over $\mathbf{F}_p$ whose divisor class group has $p^2 - p + 1$ points. Let $r \mid (p^2 - p + 1)$ and write $b = (p-1) \pmod r$ in the elliptic case and $b = p \pmod r$ in the genus 2 case. Then the function*

$$f_{b,D_1}(D_2)^{(p^6-1)/r}$$

defines a non-degenerate bilinear pairing on $\mathbf{G}_1 \times \mathbf{G}_2$.

Proof. We already know that $f_{r,D_1}(D_2)^{(p^6-1)/r}$ and (for $T = p - 1$ or p respectively) $f_{T,D_1}(D_2)^{(p^6-1)/r}$ are non-degenerate and bilinear. Write $B = r$ and $A = p - 1$ or p and apply Theorem 4. One gets the pairing

$$f_{a,rD_1}(D_2) \cdot f_{b,D_1}(D_2) \cdot g_{arD_1,bD_1}(D_2).$$

Since $rD_1 = 0$ it follows that one can choose $f_{a,rD_1}(D_2) = 1$ and $g_{arD_1,bD_1}(D_2) = 1$. The result follows. □

Remark 1. Choosing (p, r) so that b is small will speed up the pairing computations. We show how to do this in the following section.

4 Pairing Implementation and Efficiency Analysis

For the rest of the paper we restrict to the two curves from Sections 2.2 and 2.3. Let $\mathbf{G}_1$ be the 1-eigenspace of Frobenius. We consider pairings

$$e : \mathbf{G}_1 \times \mathbf{G}_1 \rightarrow \mu_r \subset \mathbf{F}_{p^6}^*$$

obtained from Corollary 1 together with the distortion maps given earlier.

In this section, we describe some optimisations for pairing implementation on these curves given above, including generation of parameters to shorten the Miller loop, denominator elimination, and we give a suitable representation for the required finite field.

4.1 Generation of Parameters

In this subsection, we describe a method to generate parameters for the curves constructed in Subsections 2.2 and 2.3, which will allow the pairings to be computed quickly by exploiting Corollary 1. As can be seen from the algorithm, b can be chosen to have very low hamming weight and half the bit-length of r.

Algorithm 1. Parameter Generation

INPUT: Integers n, l_{max}.
OUTPUT: Integers b, r, l_1, l_0 and a prime p such that $r \mid p^2 - p + 1, p \equiv b \mod r$, and $p^2 - p + 1 = r(l_1 p + l_0)$.

1: **repeat**
2: Choose b of size n bits and low hamming weight.
3: Let $r = b^2 - b + 1$.
4: **until** r is prime or nearly prime.
5: **for** l from 1 to l_{max}. **do**
6: let $p = l \cdot r + b$.
7: **if** p is a prime and $p \equiv 11 \mod 12$. **then**
8: Break.
9: **end if**
10: **end for**
11: **if** $l = l_{max}$, goto step 1.
12: let $l_1 = l$ and $l_0 = l(b-1) + 1$. (as $p^2 - p + 1 = r(lp + l(b-1) + 1)$)
13: **return** p, r, b, l_1 and l_0.

The following is a set of parameters generated by Algorithm 1, using $n = 80$. These are the parameters used in our implementation, which will be described in the following section.

Example 1. **A set of parameters for 80-bit security**

- p =B000000000000000112600000000000000006AEFB
- r =1000000000000000018F0000000000000009B79
- b =10000000000000000000C8

Remark 2. The parameters in Example 1 are given for the sake of comparison with elliptic curves of embedding degree 3. Note that Algorithm 1 can efficiently generate parameters for higher security levels. The relative performance of the pairing computation in the elliptic and hyperelliptic cases at higher security level should be very similar to that obtained in this article.

Remark 3. Algorithm 1 can be generalized to find parameters for many other types of curves. For example, a similar algorithm can be used to generate parameters for supersingular genus 2 curves given by an equation of the form $y^2 = x^5 + a$, where $a \in \mathbf{F}_p^*$, $p \equiv 2, 3 \mod 5$. Ó hÉigeartaigh and Scott efficiently implemented pairings on these curves in [14], achieving some of the fastest pairing computations on genus 2 curves. Using a parameter selection algorithm similar to Algorithm 1 could further improve their results.

4.2 Finite Field Construction and Arithmetic

The following field construction was presented by Hu et al. in [11].

We restrict to $p \equiv 3 \mod 4$ so that -1 is not a quadratic residue modulo p. In other words, we require $p \equiv 11 \mod 12$. The finite fields are represented as follows:

$$\begin{aligned}\mathbf{F}_{p^2} &\cong \mathbf{F}_p[\alpha]/(\alpha^2+1) = \{w_1\alpha + w_2 | w_1, w_2 \in \mathbf{F}_p\} = \{a_1 + a_2\beta^3 | a_1, a_2 \in \mathbf{F}_p\}.\\ \mathbf{F}_{p^6} &\cong \mathbf{F}_{p^2}[\beta]/(\beta^3-\rho) = \{b_0 + b_1\beta + b_2\beta^2 + b_3\beta^3 + b_4\beta^4 + b_5\beta^5 | b_i \in \mathbf{F}_p\}\\ &= \{c_0 + c_1\beta + c_2\beta^2 | c_i \in \mathbf{F}_{p^2}\},\end{aligned}$$

where $\rho = \alpha + w_0$ and w_0 is a small integer such that $x^3 - \rho$ is irreducible over $\mathbf{F}_{p^2}$.

Let M_i, S_i, and I_i denote the cost of multiplication, squaring, and inversion in $\mathbf{F}_{p^i}$ for $i = 1, 2, 6$ using the above representation. It is standard (see Section 7 of [10]) that $M_2 = 3M_1$ and $I_2 = 2M_1 + 2S_1 + 1I_1$. For purposes of comparison we follow [10] and assume that $M_1 = S_1$, and $1I_1 = 10M_1$. Finally, as explained in [3] one has $S_2 = 2M_1$, $M_6 = 15M_1$ and $S_6 = 11M_1$.

Let $e_{ij} \in \mathbf{F}_p$ be defined by $\beta^{ip} = e_{i0} + e_{i1}\beta + \cdots + e_{i5}\beta^5$ for $1 \le i \le 5$. We have that $\beta^{ip} = \beta^{2i}\rho^{i(p-2)/3}$. Since $\beta^3 = \rho$ and $\rho \in \mathbf{F}_{p^2}$, there are at most two non-zero terms in the coefficient vector $(e_{i0}, e_{i1}, \cdots e_{i5})$. Specifically, we have $(e_{30}, e_{31}, \cdots e_{35}) = (2w_0, 0, 0, -1, 0, 0)$. Hence, raising a random element to the pth power is given by

$$(b_0 + b_1\beta + b_2\beta^2 + b_3\beta^3 + b_4\beta^4 + b_5\beta^5)^p = b_0 + \sum_{i=1}^{5} b_i(e_{i0} + e_{i1}\beta + \cdots + e_{i5}\beta^5).$$

This computation costs only 8 $\mathbf{F}_p$−multiplications (remember w_0 is a small integer).

The final exponentiation is often computed using a base p expansion. In the cases $k = 6$, the final exponentiation can be represented as

$$\frac{p^6-1}{r} = (p^3-1)(p+1)\frac{p^2-p+1}{r} = (p^3-1)(p+1)(l_1 p + l_0)$$

where l_1 is small. Thus, the construction above allows for very fast exponentiation.

4.3 Optimised Pairing Computation

The cost of Miller's algorithm to compute pairings is determined by the length of the Miller loop, the cost of the calculations inside the loop, and the final exponentiation. To compute pairings on hyperelliptic genus 2 curves given by a real model, we used the techniques described above to speed up the computation, that is:

- Algorithm 1 generates suitable parameters to get a short, low Hamming weight Miller loop.
- Use $D_\infty = \infty^+ + \infty^-$ to represent elements of $\mathrm{Cl}^0(C)$ to get fast addition and a simple Miller funciton.
- The distortion map $(\phi^{-1} \circ \chi \circ \zeta_6 \circ \phi)$ described in Theorem 2 allows for denominator elimination while using the R-ate pairing [12] technique.
- The field construction in Subsection 4.2 provides the arithmetic for a very efficient final exponentiation.

5 Efficiency Analysis and Implementation Results

The optimisation techniques described above make the computation of pairings on hyperelliptic genus 2 curves practical and efficient. In this section we analyse the efficiency, and compare it with pairing implementations on elliptic curves with similar characteristics.

5.1 Comparison with Elliptic Curves with $k = 3$

As mentioned in the introduction, the curves constructed in Subsections 2.2 and 2.3 have very similar characteristics, so implementation results on the embedding degree 3 elliptic curve provide a useful benchmark to analyse our pairing implementation on hyperelliptic curves given by a real model.

As mentioned before, (the class groups of) both curves have the same number of $\mathbf{F}_p$-rational points, and the *embedding field* for both curves is the same, as is the bandwith requirement. A point $P = (x, y) \in E(\mathbf{F}_{p^2})$ is represented by 4 elements of $\mathbf{F}_p$, which is the same number of coefficients required to represent a divisor $D = (x^2+u_1x+u_0, x^3+v_1x+v_0)$ over $\mathbf{F}_p$. Here we use the representation different from that of Section 2.1, please see the appendix for details. Since the target field is the same, both pairing values can be compressed at the same rate by using traces (as with the XTR public key cryptosystem [13]) or tori [16,17].

In the notation of Theorem 4, we need to calculate f_{b,D_1}. Since b is an integer generated by Algorithm 1, it will have very low Hamming weight and we will only analyse the cost of the doubling steps in the Miller loop.

We are pairing two divisors D_1 and D_2 defined over $\mathbf{F}_p$. We assume that these divisors are generic (as mentioned earlier). We further assume that the Mumford representation of D_2 factors over $\mathbf{F}_p$ and that we know the factorisation; this is a restriction to roughly half the divisor classes. Before computing the pairing we apply the distortion map ψ from Theorem 2 to map D_2 to a divisor over $\mathbf{F}_{p^6}$. In other words, we know $x_1, x_2 \in \mathbf{F}_{p^3}$ and $y_1, y_2 \in \mathbf{F}_{p^6}$ such that $\psi(D_2) = (x_1, y_1)+(x_2, y_2)-D_\infty$. The Miller functions are evaluated at (x_1, y_1) and (x_2, y_2) rather than performing a resultant computation. Since D_∞ and $p(x)$ are defined over $\mathbf{F}_p$ we can omit the evaluation $(y - p(x))(D_\infty)$.

To compare the efficiency of our pairing implementations on elliptic and hyperelliptic curves, we first estimate the cost of each doubling step. We will let f denote the intermediate value in the Miller loop. The update of f is similar to that used in other standard implementations of Miller's algorithm, such as Algorithm 1 in Section 2 of [9], except that the denominator of $g_{n_1 D, n_2 D}$ in Equation (3) can be removed as described by Equation (5) in the elliptic curve case, and by Lemma 2 in the hyperelliptic curve case.

elliptic : $f \leftarrow f^2 \cdot l_{R,P}(Q) \cdot (x_R(x_R + x_Q) + x_Q^2)$ and $R \leftarrow 2R$
hyperelliptic: $f \leftarrow f^2 \cdot (y_1 - p(x_1)) \cdot (y_2 - p(x_2))$ and $D_1 \leftarrow 2D_1$.

Here $l_{R,P}$ is the line through R and P, and $y - p(x)$ is as in Equation (1). Note that $p(x)$ is a cubic polynomial with coefficients in $\mathbf{F}_p$.

We use affine coordinates for our implementation. In the elliptic case, doubling a point costs $1I_2 + 2M_2 + 2S_2$, which makes each doubling step in the Miller loop cost about $77M_1$ (see the formulae for M_2, S_2 etc in Section 4.2). In the hyperelliptic case, doubling a divisor costs $1I_1 + 32M_1 \approx 42M_1$ [6], which makes each doubling step in the Miller loop cost about $101M_1$. There are a total of 84 doubling steps using the parameters given in Example 1. So the costs of the Miller loops are roughly $6468M_1$ and $8484M_1$ respectively (counting only doubling steps since b has Hamming weight 4).

The final exponentiation step is identical in both cases, and costs about $1621M_1$.

This shows that pairings on real hyperelliptic genus 2 curves with $k = 6$ are competitive to parings on elliptic curves with $k = 3$, though slower.

5.2 Theoretical Comparison with Imaginary Hyperelliptic Curves with $k = 4$

To complement our efficiency analysis, we will also make an abstract comparison of our implementation results with those reported in [14], using genus 2 hyperelliptic curves with embedding degree $k = 4$. The implementation results in [14] are amongst the best reported in the literature.

In curves with embedding degree $k = 4$, the underlying prime field needs to be 86 bits larger than our implementation to achieve an 80-bit security level. The representation of each divisor will then need 344 more bits.

The computation of pairings on genus 2 curves with embedding degree $k = 4$ given by an imaginary model is studied in [14], where timings for the pairing computation of degenerate divisors (i.e., divisors of the form $(P)-(\infty)$) are given. As mentioned in Remark 3, the use of an algorithm similar to Algorithm 1 to find curve parameters could improve the results of [14]. Since the computation requires multiplication modulo a 256-bit prime rather than modulo a 170-bit prime (as in the $k = 6$ case), for general divisors we expect pairings on curves with embedding degree $k = 4$ to be slower than our results, although they will probably be faster for degenerate divisors (there is no analogue of degenerate divisors for hyperelliptic curves given by a real model).

We can see that pairings on hyperelliptic curves given by a real model can be competitive with pairings on curves given by an imaginary model, in terms of bandwidth and computation requirements.

5.3 Implementation Results

This section reports some implementation results. The implementation uses the parameters given in Example 1. The timings are obtained using the Magma Online Platform [2].

The following table summarizes the results. The first row shows our implementation result for hyperelliptic curves, and the second row shows our implementation result for elliptic curves.

Table 1. Efficiency Comparison with an AES 80 Security Level

Curve	size of p	Operation Count	time(ms)
$C(\mathbf{F}_p)$ $k = 6$	160	$10105M_1$	21.6
$E(\mathbf{F}_{p^2})$ $k = 3$	160	$8089M_1$	15.3

6 Conclusion

In this article we presented several techniques to speed-up the calculation of pairings on hyperelliptic curves given by a real model. We showed that computing pairings on real genus 2 curves is practical. We compared the efficiency of two similar elliptic and hyperelliptic curves, and conclude that pairings on elliptic curves with $k = 3$ require 20% less field multiplications than pairings on real hyperelliptic genus 2 curves with $k = 6$. The timing difference in our implementation was that elliptic curves are 28% faster than genus 2 curves.

Acknowledgments

S. Galbraith was supported by EPSRC grant EP/D069904/1. X. Lin thanks the Chinese Scholarship Council. D. Mireles thanks CONACyT for its financial support.

References

1. Barreto, P.S.L.M., Kim, H.Y., Lynn, B., Scott, M.: Efficient algorithms for pairing-based cryptosystems. In: Yung, M. (ed.) CRYPTO 2002. LNCS, vol. 2442, pp. 354–368. Springer, Heidelberg (2002)
2. Bosma, W., Cannon, J., Playoust, C.: The Magma algebra system. I. The user language. J. Symbolic Comput. 24(3-4), 235–265 (1997)
3. Devegili, A.J., Ó'hÉigeartaigh, C., Scott, M., Dahab, R.: Multiplication and squaring on pairing-friendly fields (2006); Cryptology ePrint Archive: Report 2006/471
4. Erickson, S., Jacobson, M.J., Shang, N., Shen, S., Stein, A.: Explicit formulas for real hyperelliptic curves of genus 2 in affine representation. In: Carlet, C., Sunar, B. (eds.) WAIFI 2007. LNCS, vol. 4547, pp. 202–218. Springer, Heidelberg (2007)
5. Frey, G., Rück, H.-G.: A remark concerning m-divisibility and the discrete logarithm in the divisor class group of curves. Math. Comp. 62(206), 865–874 (1994)
6. Galbraith, S.D., Harrison, M., Mireles Morales, D.J.: Efficient hyperelliptic arithmetic using balanced representation for divisors. In: van der Poorten, A.J., Stein, A. (eds.) ANTS-VIII 2008. LNCS, vol. 5011, pp. 342–356. Springer, Heidelberg (2008)
7. Galbraith, S.D., Pujolas, J., Ritzenthaler, C., Smith, B.: Distortion maps for genus two curves (2006); Arxiv/math.NT/0611471
8. Galbraith, S.D., Verheul, E.R.: An analysis of the vector decomposition problem. In: Cramer, R. (ed.) PKC 2008. LNCS, vol. 4939, pp. 308–327. Springer, Heidelberg (2008)
9. Granger, R., Hess, F., Oyono, R., Thériault, N., Vercauteren, F.: Ate pairing on hyperelliptic curves. In: Naor, M. (ed.) EUROCRYPT 2007. LNCS, vol. 4515, pp. 430–447. Springer, Heidelberg (2007)
10. Hess, F., Smart, N.P., Vercauteren, F.: The eta pairing revisited. IEEE Transactions on Information Theory 52(10), 4595–4602 (2006)
11. Hu, L., Dong, J.-W., Pei, D.: Implementation of cryptosystems based on tate pairing. J. Comput. Sci. Technol. 20(2), 264–269 (2005)
12. Lee, E., Lee, H.-S., and Park, C.-M. Efficient and generalized pairing computation on abelian varieties. Cryptology ePrint Archive, Report 2008/040 (2008), `http://eprint.iacr.org/`
13. Lenstra, A.K., Verheul, E.R.: The XTR public key system. In: Bellare, M. (ed.) CRYPTO 2000. LNCS, vol. 1880, pp. 1–19. Springer, Heidelberg (2000)
14. Ó'hÉigeartaigh, C., Scott, M.: Pairing calculation on supersingular genus 2 curves. In: Biham, E., Youssef, A.M. (eds.) SAC 2006. LNCS, vol. 4356, pp. 302–316. Springer, Heidelberg (2007)
15. Paulus, S., Rück, H.-G.: Real and imaginary quadratic representations of hyperelliptic function fields. Math. Comp. 68(227), 1233–1241 (1999)
16. Granger, R., Page, D., Stam, M.: On small characteristic algebraic tori in pairing-based cryptography. LMS Journal of Computation and Mathematics 9, 64–85 (2006)

17. Rubin, K., Silverberg, A.: Torus-based cryptography. In: Boneh, D. (ed.) CRYPTO 2003. LNCS, vol. 2729, pp. 349–365. Springer, Heidelberg (2003)
18. Verheul, E.R.: Evidence that XTR is more secure than supersingular elliptic curve cryptosystems. In: Pfitzmann, B. (ed.) EUROCRYPT 2001. LNCS, vol. 2045, pp. 195–210. Springer, Heidelberg (2001)

A Appendix: Addition Formulae

We now present the formulae from [4], which are explicit formulae for the sub-algorithms used in [6] to build an efficient algorithm for divisor arithmetic on hyperelliptic curves with two points at infinity. These formulae require that the curve have model of the form

$$y^2 = x^6 + f_4x^4 + f_3x^3 + f_2x^2 + f_1x + f_0.$$

To make the polynomial monic one takes a random pair of $\mathbf{F}_p$-rational points $(x, \pm y)$ on the curve, moves them to infinity, and absorbs the square root of the leading coefficient into y. Since we are working in large characteristic there is no problem setting $f_5 = 0$.

To be compatible with the divisor representation used in [4] the second polynomial in the Mumford representation is the unique polynomial $v' \equiv v \mod u$ of the form $v' = x^3 + v_1x + v_0$. Notice that v' can be represented only by 2 coefficients even though it has degree 3.

Algorithm 2. Addition Formulae

INPUT: Divisors $D_1 = \text{div}(u_1, v_1)$ and $D_2 = \text{div}(u_2, v_2)$.

$z_0 = u_{10} - u_{20}, z_1 = u_{11} - u_{21}$.
$z_2 = u_{11} \cdot z_1 - z_0, z_3 = u_{10} \cdot z_1$.
$r = z_1 \cdot z_3 - z_0 \cdot z_2$.
$w_0 = v_{10} - v_{20}, w_1 = v_{11} - v_{21}$.
$s_1' = w_0 \cdot z_1 - w_1 \cdot z_0, s_0' = w_0 \cdot z_2 - w_1 \cdot z_3$.
$k_2 = f_4 - 2v_{21}$.
$r_2 = r^2, \hat{w}_0 = r_2 - (s_1' + r)^2, \hat{w}_1 = (r \cdot \hat{w}^{-1})$.
$\hat{w}_2 = \hat{w}_0 \cdot \hat{w}_1, \hat{w}_3 = r \cdot r_2 \cdot w_1$.
$s_1 = s_1' \cdot \hat{w}_2, s_0 = s_0'\hat{w}_2$.
$\tilde{w}_0 s_0 \cdot u_{20}, \tilde{w}_1 = s_1 \cdot u_{21}, l_2 = s_0 + \tilde{w}_1$.
$l_1 = (s_0 + s_1)(u_{21} + u_{20}) - \tilde{w}_1 - \tilde{w}_0, l_0 = \tilde{w}_0$.
$m_3' = \hat{w}_3 \cdot (-s_1 \cdot (s_0 + l_2) - 2s_0)$.
$m_2' = \hat{w}_3 \cdot (k_2 - s_1 \cdot (l_1 + 2v_{21}) - s_0 l_2)$.
$u_1' = m_3' - u_{11}, u_0' = m_2' - u_{10} - u_{11} \cdot u_1'$.
$\underline{w}_1 = u_1' \cdot (s_1 + 2), \underline{w}_0 = u_0' \cdot (l_2 - \underline{w}_1)$.
$v_1' = (u_0' + u_1') \cdot (s_1 + -\underline{w}_1 + l_2) - v_{21} - l_1 - \underline{w}_0 - \underline{w}_1$.
$v_0' = \underline{w}_0 - v_{20} - l_0$.

When adding divisors D_1 and D_2, the cubic polynomial $p(x)$ given by Equation (1) can be calculated as $p(x) = v_2(x) + u_2(x)s(x)$, where $s(x) = s_1x + s_0$ in Algorithm 2.

Algorithm 3. Doubling Formulae

INPUT: $D = \mathrm{div}(u, v)$.

$w_1 = u_1^2, \tilde{v}_1 = 2(v_1 + w_1 - u_0), \tilde{v}_0 = 2(v_0 + u_0 \cdot u_1)$.

$w_2 = u_0 \cdot \tilde{v}_1, w_3 = u_1 \cdot \tilde{v}_1$.

$\mathrm{inv}_1 = \tilde{v}_1, \mathrm{inv}_0 = w_3 - \tilde{v}_0$.

$r = \tilde{v}_0 \cdot \mathrm{inv}_0 - w_2 \cdot \tilde{v}_1$.

$k_2' = f_4 - 2v_1$

$k_1' = f_3 - 2v_0 - 2k_2' \cdot u_1$.

$k_0' = f_2 - v_1^2 - k_1' \cdot u_1 - k_2'(w_1 + 2u_0)$.

$s_1' = \mathrm{inv}_1 \cdot k_0' - \tilde{v}_0 \cdot k_1', s_0' = \mathrm{inv}_0 \cdot k_0' - w_2 \cdot k_1'$.

$r_2 = r^2, \hat{w}_0 = (s_1' + r)^2 - r_2, \hat{w}_1 = (r \cdot \hat{w}_0)^{-1}$.

$\hat{w}_2 = \hat{w}_0 \cdot \hat{w}_1, \hat{w}_3 = r \cdot r_2 \cdot \hat{w}_1$.

$s_1 = \hat{w}_2 \cdot s_1', s_0 = \hat{w}_2 \cdot s_0'$.

$u_1' = 2\hat{w}_3 \cdot ((s_0 - u_1) \cdot s_1 + s_0)$.

$u_0' = \hat{w}_3 \cdot ((s_0 2u_1) \cdot s_0 + \tilde{v}_1 \cdot s_1 - k_2')$.

$z_0 = u_0' - u_0, z_1 = u_1' - u_1$.

$\underline{w}_0 = z_0 \cdot s_0, \underline{w}_1 = z_1 \cdot s_1$.

$v_1' = 2u_0' - v_1 + (s_0 + s_1) \cdot (z_0 + z_1) - \underline{w}_0 - \underline{w}_1 - u_1' \cdot (2u_1' + \underline{w}_1)$.

$v_0' = \underline{w}_0 - v_0 - u_0' \cdot (2u_1' + \underline{w}_1)$.

The cubic polynomial from Equation (1) used in Miller's algorithm when doubling a divisor D is given by $p(x) = v(x) + u(x)s(x)$, where $s(x) = s_1 x + s_0$ in Algorithm 3.

Faster Implementation of η_T Pairing over $GF(3^m)$ Using Minimum Number of Logical Instructions for GF(3)-Addition

Yuto Kawahara[1], Kazumaro Aoki[2], and Tsuyoshi Takagi[1]

[1] Future University-Hakodate, Japan
[2] NTT Information Sharing Platform Laboratories, NTT Corporation, Japan

Abstract. The η_T pairing in characteristic three is implemented by arithmetic in $GF(3) = \{0, 1, 2\}$. Harrison et al. reported an efficient implementation of the GF(3)-addition by using seven logical instructions (consisting of AND, OR, and XOR) with the two-bit encoding $\{(0,0) \mapsto 0, (0,1) \mapsto 1, (1,0) \mapsto 2\}$. It has not yet been proven whether seven is the minimum number of logical instructions for the GF(3)-addition. In this paper, we show many implementations of the GF(3)-addition using only six logical instructions with different encodings such as $\{(1,1) \mapsto 0, (0,1) \mapsto 1, (1,0) \mapsto 2\}$ or $\{(0,0) \mapsto 0, (0,1) \mapsto 1, (1,1) \mapsto 2\}$. We then prove that there is no implementation of the GF(3)-addition using five logical instructions with any encoding of GF(3) by two bits. Moreover, we apply the new GF(3)-additions to an efficient software implementation of the η_T pairing. The running time of the η_T pairing over $GF(3^{509})$, that is considered to be realized as 128-bit security, using the new GF(3)-addition with the encoding $\{(0,0) \mapsto 0, (0,1) \mapsto 1, (1,1) \mapsto 2\}$ is 16.3 milliseconds on an AMD Opteron 2.2-GHz processor. This is approximately 7% faster than the implementation using the previous GF(3)-addition with seven logical instructions.

Keywords: η_T pairing, GF(3)-addition, logical instruction.

1 Introduction

Bilinear pairings on elliptic curves over finite fields have attracted much attention in cryptography, since pairing-based cryptosystems can provide many novel applications, such as ID-based cryptosystems [9,27], keyword-searchable encryption [8], and efficient broadcast encryption [10]. The first algorithm for computing the Tate pairing was proposed by Miller [24]. Duursma and Lee [12] proposed an efficient algorithm for computing the Tate pairing on supersingular elliptic curves over the binary field $GF(2^m)$ or the ternary field $GF(3^m)$. Barreto et al. [3] proposed an η_T pairing, which is a different version of the Duursma-Lee algorithm. This pairing is about twice as fast as the Duursma-Lee algorithm. Currently, one of the fastest pairings is the η_T pairing over $GF(3^m)$.

The arithmetic in $GF(3^m)$ requires the GF(3)-addition, which cannot be directly computed by virtually any typical CPU, such as one based on the x86-architecture [20]. In an efficient implementation in GF(3), each element in GF(3)

S.D. Galbraith and K.G. Paterson (Eds.): Pairing 2008, LNCS 5209, pp. 282–296, 2008.

is represented by two bits, and the GF(3)-addition is constructed using logical instructions. Galbraith et al. [13] demonstrated a GF(3)-addition using 12 logical instructions, consisting of AND, OR, XOR, and NOT. Then, Harrison et al. [19] improved this to seven logical instructions, consisting of OR and XOR. Until now, this has been the minimum number of logical instructions for computing the GF(3)-addition.

In this paper, we describe our search for implementations of the GF(3)-addition that use seven or fewer logical instructions. Every GF(3) element is assigned to two bits in GF$(2)^2$, and the GF(3)-addition is considered as a map GF$(2)^2 \times$ GF$(2)^2 \rightarrow$ GF$(2)^2$. An exhaustive search was performed for sequences of logical instructions that can construct the GF(3)-addition for the map. Indeed, although we found many implementations of the GF(3)-addition with only six logical instructions, the representation of GF(3) elements is not the natural assignment $\{(0,0) \mapsto 0, (0,1) \mapsto 1, (1,0) \mapsto 2\}$, or the special logical instructions such as ANDN are used. Moreover, we found no instruction sequence that can compute the GF(3)-addition with five logical instructions. In other words, we have proven that the minimum number of logical instructions required for the GF(3)-addition is six for any assignment of elements in GF(3) using two bits.

To demonstrate the cryptographic implications of the new GF(3)-addition implementations using six logical instructions, we implemented arithmetic in GF(3^m) and the η_T pairing over GF(3^m), on an AMD Opteron processor model 275 (2.2 GHz). For comparison, we chose the extension degree $m = 509$, because NIST indicates that the key size should have more than 80-bit through 2011 [2], and the η_T pairing over GF(3^{509}) has 128-bit security [1,23]. With the assignment $\{(0,0) \mapsto 0, (0,1) \mapsto 1, (1,1) \mapsto 2\}$, the addition in GF($3^{509}$) was about 12% ($\simeq 1/7$) faster than that with the natural assignment. Similarly, the multiplication in GF(3^{509}), computed by the left-to-right comb method with window size $w = 4$, was about 8% faster than that with the natural assignment. As a result, the running time of the η_T pairing over GF(3^{509}) for the new GF(3)-addition with six logical instructions was 16.3 milliseconds, which is about 7% faster than the running time for the previous GF(3)-addition with seven logical instructions.

The rest of this paper is organized as follows. In Section 2, we describe the addition in GF(3) and discuss previous results on this topic. Section 3 gives the details of our search for instruction sequences to compute the GF(3)-addition, along with the search results. In Section 4, we describe our implementations and running time results for arithmetic in GF(3^m) and the η_T pairing. Finally, Section 5 concludes the paper.

2 Addition in GF(3)

Let GF$(3) = \{0, 1, 2\}$ be a finite field of three elements. Here we can represent $e \in$ GF(3) by using two bits:

$$e = (e_h, e_l),$$

where $e_h, e_l \in \mathrm{GF}(2)$. The assignment of GF(3) to $\mathrm{GF}(2)^2$ is the following:

$$\{(0,0) \mapsto 0,\ (0,1) \mapsto 1,\ (1,0) \mapsto 2\}. \tag{1}$$

We call this *natural assignment*.

For given a and b in GF(3), we can compute addition $c \leftarrow a + b$ in GF(3) as $c = a + b \bmod 3$. In the assignment of equation (1), a negative element, $-e$, of $e = (e_h, e_l)$ in GF(3) is obtained as

$$\begin{cases} (-e)_h \leftarrow e_l \\ (-e)_l \leftarrow e_h. \end{cases}$$

Let a and b be elements in GF(3). The subtraction $a - b$ can be performed as $a + (-b)$, with the same cost as the GF(3)-addition.

In this paper, we construct the GF(3)-addition by using the logical instructions below.

$|$: bitwise OR operation
$\&$: bitwise AND operation
$\wedge$: bitwise XOR operation
$\bar{x}$: bitwise complement (NOT)

Implementation of an efficient GF(3)-addition using these logical instructions is a non-trivial problem. For efficient implementation, it is preferable to keep the number of logical instructions as small as possible. Galbraith et al. [13] presented the following GF(3)-addition using 12 logical instructions.

$$c_h \leftarrow ((a_h \wedge b_h)\&\overline{(a_l \mid b_l)}) \mid (a_l\&b_l),$$
$$c_l \leftarrow ((a_l \wedge b_l)\&\overline{(a_h \mid b_h)}) \mid (a_h\&b_h).$$

Harrison et al. [19] improved the implementation to seven logical instructions by using an auxiliary variable t, as indicated below.

$$\begin{aligned} t &\leftarrow (a_h \mid b_l) \wedge (b_h \mid a_l), \\ c_h &\leftarrow t \wedge (a_l \mid b_l), \\ c_l &\leftarrow t \wedge (a_h \mid b_h). \end{aligned} \tag{2}$$

There are many other implementations of the addition in GF(3); all, however, are constructed using exactly the same logical instructions given in equation (2) [1,5,16,25]. It is not yet proved whether seven is the minimum number of logical instructions for the GF(3)-addition.

3 Search for Minimum Number of Logical Instructions for GF(3)-Addition

This section describes the process and results of our attempt to search for the minimum number of logical instructions for computing the GF(3)-addition.

3.1 Choice of Assignment in GF(3)

The assignment R of GF(3) to GF$(2)^2$ is denoted by

$$R = \{(e_h, e_l) \mapsto e \mid e_h, e_l \in \mathrm{GF}(2), e \in \mathrm{GF}(3)\}.$$

The addition $c \leftarrow a + b$ in GF(3) is a map GF(3) $\leftarrow$ GF(3) $\times$ GF(3). To specify each assignment of GF(3) in the map, we denote by R_a, R_b, and R_c the assignment for the right part of the input, the left part of the input, and the output, respectively, of the addition $c \leftarrow a + b$ in GF(3). The pair of R_a, R_b, and R_c, called the assignment set, determines the implementation of the GF(3)-addition $c \leftarrow a + b$. The natural assignment in the previous GF(3)-addition [19] uses the pair $R_a = R_b = R_c = \{(0,0) \mapsto 0,\ (0,1) \mapsto 1,\ (1,0) \mapsto 2\}$. Note that there are many possible assignment sets, such as these:

$$\begin{aligned} R_a &= \{(0,0) \mapsto 0,\ (0,1) \mapsto 1,\ (1,0) \mapsto 2\}, \\ R_b &= \{(1,0) \mapsto 0,\ (0,1) \mapsto 1,\ (1,1) \mapsto 2,\ (0,0) \mapsto 2\}, \\ R_c &= \{(1,1) \mapsto 0,\ (0,1) \mapsto 1,\ (1,0) \mapsto 2\}. \end{aligned} \tag{3}$$

In this paper, we consider the following assignments.

- Redundant representation. One element e_0 in GF(3) has two different representations $e_0 = (e_{0h}, e_{0l}) = (e'_{0h}, e'_{0l})$ in GF$(2)^2$, but the other two elements e_1 and e_2 in GF(3) are uniquely assigned as $e_1 = (e_{1h}, e_{1l})$ and $e_2 = (e_{2h}, e_{2l})$ in GF$(2)^2$, where $(e_{0h}, e_{0l}), (e'_{0h}, e'_{0l}), (e_{1h}, e_{1l})$, and (e_{2h}, e_{2l}) are pairwise different in GF$(2)^2$.
- Independent assignments. All assignments R_a, R_b, and R_c for the addition $c \leftarrow a + b$ in GF(3) are not necessarily the same.

The cardinality of the elements in R_i is denoted by $\#R_i$, and we can define the inclusion relation $R_i \subseteq R_j$, where $i, j \in \{a, b, c\}$. Then, all patterns of the assignment set R_a, R_b, and R_c can be categorized according to the cardinality and inclusion relation, as follows.

- $\#R_a = \#R_b = \#R_c = 3$
  - 1–i Natural assignment set $R_a = R_b = R_c = \{(0,0) \mapsto 0,\ (0,1) \mapsto 1,\ (1,0) \mapsto 2\}$
  - 1–ii Assignment set of common $R_a = R_b = R_c$ (This assignment set includes the natural assignment set.)
  - 1–iii Assignment set of independent R_a, R_b, R_c
- $\#R_a = \#R_b = 3,\ \#R_c = 4$
  - 2–i $R_a = R_b \subseteq R_c$
  - 2–ii Assignment set of independent R_a, R_b, R_c
- $\#R_a = \#R_c = 3,\ \#R_b = 4$
  - 3–i $R_a = R_c \subseteq R_b$
  - 3–ii Assignment set of independent R_a, R_b, R_c
- $\#R_a = 3,\ \#R_b = \#R_c = 4$

4–i $R_a \subseteq R_b = R_c$
4–ii Assignment set of independent R_a, R_b, R_c
- $\#R_a = \#R_b = 4$, $\#R_c = 3$
5–i $R_a = R_b \supseteq R_c$
5–ii Assignment set of independent R_a, R_b, R_c
- $\#R_a = \#R_b = \#R_c = 4$
6–i Assignment set of common $R_a = R_b = R_c$ using the redundant representation
6–ii Assignment set of independent R_a, R_b, R_c

In this paper, we use this categorization of the assignments R_a, R_b, and R_c in the search algorithm for efficient implementation of the GF(3)-addition.

3.2 Logical Instruction Sets

Many CPUs are equipped with AND, OR, and XOR as logical instructions. In some architectures, however, other logical instructions are also available. Given x, y in GF(2), ANDN (x ANDN $y = x \& \bar{y}$) can be used in MMX and SSE implementations. The SPARC and Alpha architectures provide ORN (x ORN $y = x \mid \bar{y}$) and XORN (x XORN $y = x \wedge \bar{y}$) [21,11]. Combining the last two with the NOT operation, we further obtain NAND ($\overline{x \& y}$) and NOR ($\overline{x \mid y}$).

Our search algorithm deals with the following two logical instruction sets.

LISet 3 = {AND, OR, XOR}
LISet 8 = {AND, OR, XOR, ANDN, ORN, XORN, NAND, NOR}

LISet 8 contains all binary operations except for trivial operation such as $x * y = x$ for the binary operation $*$. Note that LISet 8 includes non-commutative instructions such as ANDN (x ANDN $y \neq y$ ANDN x).

3.3 Search Algorithm

In this section, we describe the search method for computing the addition $c \leftarrow a+b$ in GF(3) by using the assignments $a = (a_h, a_l)$, $b = (b_h, b_l)$, and $c = (c_h, c_l)$. We prepare bit strings a_h, a_l, b_h, b_l, c_h, and c_l consisting of all calculated patterns for GF(3)-addition using the assignment set R_a, R_b, and R_c.

Recall that in the redundant representation, one element in GF(3) has two different representations in $GF(2)^2$. When both R_a and R_b do not use the redundant representation, the bit strings a_h, a_l, b_h, b_l, c_h, and c_l are constructed using nine bits. These bit lengths depend, however, on whether assignments R_a and R_b use the redundant representation. If either R_a or R_b uses the redundant representation, then the bit lengths are 12 bits. Similarly, if both R_a and R_b use the redundant representation, then the bit lengths are 16 bits. Furthermore, we need to consider the case in which assignment R_c uses the redundant representation. In this case, one GF(3) element is assigned two non-unique representations. Therefore, the result for one element (c_h, c_l) in GF(3) must be checked twice.

Table 1. Truth Table for Assignment Set R_a, R_b, and R_c in Equation (3)

a_h	0	0	0	0	0	0	0	0	1	1	1	1
a_l	0	0	0	0	1	1	1	1	0	0	0	0
b_h	1	0	1	0	1	0	1	0	1	0	1	0
b_l	0	1	1	0	0	1	1	0	0	1	1	0
c_h	1	0	1	1	0	1	1	1	1	1	0	0
c_l	1	1	0	0	1	0	1	1	0	1	1	1

Table 1 gives the truth table for the bit strings created by the assignment set R_a, R_b, and R_c in equation (3).

Using the above bit strings a_h, a_l, b_h, b_l, c_h, and c_l, we use depth-first search to search for instruction sequences to compute the GF(3)-addition. The search algorithm proceeds as follows.

1. Using Table 1, initialize the search set S as $S = \{a_h, a_l, b_h, b_l\}$.
2. Choose arbitrary two bit strings x and y in S.
3. Choose an arbitrary logical instruction $*$, and compute $z \leftarrow x * y$.
4. Add the resultant bit string z to S.
5. Iterate Steps 2–4 for a limited number of logical instructions.
6. Check whether both c_h and c_l are included in S.

The search algorithm is not efficient if all instruction sequences are computed, since most instruction sequences cannot be computed on the GF(3)-addition. Let N be the limited number of logical instructions. Then we stop the search deeper than the present iteration when it satisfies the following conditions.

- The computed result z is already contained in S.
- Neither c_h nor c_l is contained in S, when the limited number of instructions is $N - 1$.

3.4 Search Algorithm Cost

This section discusses the computational cost of our search algorithm. The number of assignment patterns for $\#R_i = 3$ is 24 ($= 4 \times 3 \times 2$). Similarly, the number of assignment patterns for $\#R_i = 4$ is 72 ($= 24 \times 3$), since one element in GF(3) is assigned to two representations. Let r be the number of assignments of $\#R_i = 4$ in $\#R_a, \#R_b$, and $\#R_c$, where $0 \leq r \leq 3$. Then, there are $24^{3-r} \times 72^r$ patterns using each assignment R_a, R_b, and R_c. For example, with $\#R_a = \#R_b = \#R_c = 3$, the number of assignment set patterns is 13,824 ($= 24^3$).

Here, we estimate the search algorithm cost for $\#R_a = \#R_b = \#R_c = 3$, $N = 5$, and LISet 8, including the details of the computational cost. All possible instruction sequences give

$$\prod_{j=1}^{5} ((j+3) \cdot (j+2) \cdot 10) = 1693440000000 \approx 2^{40.6}$$

Table 2. Search Results for Computing the GF(3)-Addition

$\#R_a$	$\#R_b$	$\#R_c$	Assignment set	LISet	Number of instructions	Existence
3	3	3	1-iii	8	Less than 6	No
3	3	3	1-i	3	Less than 7	No
3	3	3	1-ii	3	6	Yes
3	3	4	2-ii	8	Less than 6	No
3	3	4	2-i	3	6	Yes
3	4	3	3-ii	8	Less than 6	No
3	4	3	3-ii	3	Less than 7	No
3	4	3	3-i	3	7	Yes
3	4	4	4-ii	8	Less than 6	No
3	4	4	4-i	3	6	Yes
4	4	3	5-ii	8	Less than 6	No
4	4	3	5-ii	3	Less than 7	No
4	4	3	5-i	3	7	Yes
4	4	4	6-ii	8	Less than 6	No
4	4	4	6-i	3	6	Yes

LISet. LISet 3 = {AND, OR, XOR}.
LISet 8 = {AND, OR, XOR, ANDN, ORN, XORN, NAND, NOR}.
Assignment Set. The assignment set was R_a, R_b, and R_c as indicated in Section 3.1.

patterns. Most of the instruction sequences do not contain the results c_h and c_l for computing the GF(3)-addition. Thus, we can reduce the computational cost by using the conditions described in Section 3.3.

We performed an experiment using 24 Pentium 4, 96 Pentium D and 6 Xeon processors in parallel. The search took about half a day for $N = 5$ and LISet 3. It required about a day for $\#R_a = \#R_b = \#R_c = 3$, $N = 5$, and LISet 8. We completed all experiments in Table 2 about one month.

3.5 Search Results and Some Examples

Table 2 lists the specific results of our search, leading to the following general results.

1. There is no implementation of the addition $c \leftarrow a + b$ in GF(3) that uses less than six logical instructions from the set of {AND, OR, XOR, ANDN, ORN, XORN, NAND, NOR} for any assignment set R_a, R_b, and R_c.
2. There are many implementations of the addition $c \leftarrow a+b$ in GF(3) that use six logical instructions from the set of {AND, OR, XOR} in the assignment set R_a, R_b, and R_c with $\#R_a = \#R_b = \#R_c = 3$.

Hence, we have proved that the minimum number of logical instructions for computing the GF(3)-addition is six.

In the following, we show some examples of computing the GF(3)-addition $c \leftarrow a + b$ with six logical instructions. Let $a = (a_h, a_l)$, $b = (b_h, b_l)$, and $c = (c_h, c_l)$ be elements in GF(3).

In the case of $R_a = R_b = R_c = \{(1,1) \mapsto 0, (0,1) \mapsto 1, (1,0) \mapsto 2\}$, the GF(3)-addition can be computed by

$$\begin{cases} c_h \leftarrow (a_l \wedge b_l) \mid ((a_h \wedge b_h) \wedge a_l) \\ c_l \leftarrow (a_h \wedge b_h) \mid ((a_l \wedge b_l) \wedge a_h). \end{cases} \tag{4}$$

In this assignment, the negative element, $-a$, can be converted to exchange a_h and a_l.

In the case of $R_a = R_b = R_c = \{(0,0) \mapsto 0, (0,1) \mapsto 1, (1,1) \mapsto 2\}$, the GF(3)-addition can be computed by

$$\begin{cases} c_h \leftarrow (a_l \wedge b_h) \& (a_h \wedge b_l) \\ c_l \leftarrow (a_l \wedge b_l) \mid ((a_h \wedge b_l) \wedge b_h). \end{cases} \tag{5}$$

In this assignment, the negative element, $-a$, can be computed simply by $a_h \leftarrow a_h \wedge a_l$. In this case, although additional cost is required to compute the negative element, the subtraction $a - b$ can be computed using six logical instructions:

$$\begin{cases} c_h \leftarrow (a_h \wedge b_l) \& ((a_l \wedge b_l) \wedge b_h) \\ c_l \leftarrow (a_h \wedge b_h) \mid (a_l \wedge b_l). \end{cases} \tag{6}$$

Moreover, we have found that the GF(3)-addition can be computed using six logical instructions from the set of {AND, OR, XOR, ANDN} even for the natural assignment set $R_a = R_b = R_c = \{(0,0) \mapsto 0, (0,1) \mapsto 1, (1,0) \mapsto 2\}$:

$$\begin{cases} c_h \leftarrow \overline{(a_l \wedge b_l)} \& ((a_h \wedge b_h) \wedge a_l) \\ c_l \leftarrow \overline{(a_h \wedge b_h)} \& ((a_l \wedge b_l) \wedge a_h). \end{cases}$$

This instruction sequence can efficiently compute the GF(3)-addition by using four registers, requiring no additional register. Therefore, it is suitable for an implementation, such as SSE, in which the computed result overwrites one of the operands. It can be performed using registers r_i $(i = 0, 1, 2, 3)$ as indicated below.

1. $r_0 \leftarrow a_h,\ r_1 \leftarrow a_l,\ r_2 \leftarrow b_h,\ r_3 \leftarrow b_l,$
2. $r_2 \leftarrow r_0 \wedge r_2,\ r_3 \leftarrow r_1 \wedge r_3,$
3. $r_1 \leftarrow r_2 \wedge r_1,\ r_0 \leftarrow r_3 \wedge r_0,$
4. $r_3 \leftarrow r_1$ ANDN $r_3,\ r_2 \leftarrow r_0$ ANDN $r_2,$
5. $c_h \leftarrow r_3,\ c_l \leftarrow r_2.$

4 Application to η_T Pairing over GF(3^m)

In this section, we explain how to implement operations in GF(3^m) and the η_T pairing by using the new addition implementation proposed in the previous section. Then, we give the operations' running times on an AMD Opteron processor model 275 (2.2 GHz), using GCC 4.1.2 with the `-O3` option under Linux/x86_64.

4.1 Bit Representation of GF(3^m)

Let the ternary field GF(3^m) be given by GF(3)[x]/($f(x)$) where GF(3)[x] is the set of all polynomials over GF(3) and $f(x)$ is an irreducible polynomial over GF(3). An element A in GF(3^m) can be represented as the polynomial $\sum_{i=0}^{m-1} a_i x^i$, where each coefficient a_i is an element in GF(3). Recall that, for the purposes of this paper, each coefficient is converted to two bits, such as $a_i = ((a_i)_h, (a_i)_l) \in \mathrm{GF}(2)^2$. By using bit-sliced representation [25], element A in GF(3^m) can be represented as two bit strings (A_h, A_l) of GF(3) elements as indicated below.

$$\begin{aligned} A_h &= ((a_{m-1})_h, (a_{m-2})_h, \cdots, (a_0)_h), \\ A_l &= ((a_{m-1})_l, (a_{m-2})_l, \cdots, (a_0)_l). \end{aligned} \tag{7}$$

In our implementation, the bit strings of the GF(3^m) element are stored in two arrays of size $\lceil m/W \rceil$, where W is the word size of the target processor. For the purpose of this paper, the word size $W = 64$.

Here, we use the following three assignments to represent the coefficients of a_i in GF(3). The first assignment is the natural assignment shown by Harrison et al., which can compute the GF(3)-addition with seven logical instructions. The second and third ones are assignments in which the GF(3)-addition can be computed using six logical instructions, as indicated in Section 3.5. These three assignments are written as follows:

$$\begin{aligned} &\textit{Natural}\text{ assignment} : \{(0,0) \mapsto 0,\ (0,1) \mapsto 1,\ (1,0) \mapsto 2\}, \\ &\textit{Type 1}\text{ assignment} : \{(1,1) \mapsto 0,\ (0,1) \mapsto 1,\ (1,0) \mapsto 2\}, \\ &\textit{Type 2}\text{ assignment} : \{(0,0) \mapsto 0,\ (0,1) \mapsto 1,\ (1,1) \mapsto 2\}. \end{aligned}$$

In the following subsections, we compare the running times for addition, subtraction, and multiplication in GF(3^m) and the η_T pairing with each of these assignments.

4.2 Addition and Subtraction in GF(3^m)

When the elements $A, B, C \in \mathrm{GF}(3^m)$ are represented according to equation (7), the addition $C \leftarrow A + B$ in GF(3^m) can be computed by the GF(3)-addition $c_i \leftarrow a_i + b_i$ for each coefficient of elements A and B. In the *Natural* and *Types 1* and *2* assignments, the GF(3)-addition can be implemented according to equations (2), (4) and (5), respectively. A logical instruction can compute W coefficients at one time. Therefore, the GF(3^m)-addition is implemented using $7\lceil m/W \rceil$ logical instructions in the *Natural* assignment, but in the *Types 1* and *2* assignments, it is implemented using $6\lceil m/W \rceil$ logical instructions.

Similarly, the subtraction $C \leftarrow A - B$ in GF(3^m) can be constructed using logical instructions. In the *Natural* and *Type 1* assignments, the subtraction can be performed as $A + (-B)$, and its cost is the same as that of the GF(3^m)-addition. In the *Type 2* assignment, the subtraction can be implemented using $6\lceil m/W \rceil$ logical instructions according to equation (6).

To compare the running times, we used the finite field GF(3^{509}), since the discrete logarithm problem for the extension field GF($3^{6\cdot 509}$) achieves the 128-bit security level [1,23]. Table 3 lists the running time results for addition and

Table 3. Running Time Comparison for Addition and Subtraction in GF(3^{509}) (μsec)

Assignment	*Natural*	*Type 1*	*Type 2*
Addition	0.0197	0.0166	0.0175
Subtraction	0.0199	0.0166	0.0176

subtraction in GF(3^{509}). Each listed running time is the average of 100,000,000 executions. For fair comparison, we used the same program code for implementing the logical instructions, and we did not use extensions such as SSE.

As a theoretical estimation, the running times of addition and subtraction in GF(3^m) for the *Types 1* and *2* assignments should be about 6/7 of that for the *Natural* assignment, since the number of logical instructions for computing the GF(3)-addition is reduced from seven to six. In our experiment, both addition and subtraction in GF(3^m) using $6\lceil m/W \rceil$ logical instructions (*Types 1* and *2*) were about 12 $\sim$ 16% ($\simeq 1/7$) faster than using $7\lceil m/W \rceil$ logical instructions (*Natural*). These operations are very simple and require hardly any computational cost except for computing $6\lceil m/W \rceil$ or $7\lceil m/W \rceil$ logical instructions. Therefore, the running times of these operations became faster, as predicted from the theoretical estimation, by reducing the number of logical instructions.

4.3 Multiplication in GF(3^m)

The multiplication $A \cdot B$ in GF(3^m) consists of the polynomial multiplication $C' \leftarrow A \cdot B$ in GF(3)[x] and the reduction $C \leftarrow C' \bmod f(x)$, where $f(x)$ is the irreducible polynomial defining GF(3^m). If $f(x)$ is an irreducible trinomial $f(x) = x^m + x^k + 2$, then the computational cost of the reduction is negligibly small compared to that of the polynomial multiplication [18]. Therefore, in the following, we only discuss the polynomial multiplication algorithms.

The polynomial multiplication is constructed using the addition in GF(3^m) and the shift Ax^i computed by $\sum_{j=0}^{m-1} a_j x^{j+i}$ from x^i and $A = \sum_{j=0}^{m-1} a_j x^j$. A simple algorithm is the shift-and-add method [18, Alg. 2.33], performed as $C' \leftarrow C' + Ab_i x^i$ for i from 0 to $m-1$. The comb method [18, Alg. 2.35] is performed as $C' \leftarrow C' + Ab_{jW+i} x^{jW+i}$ $(0 < j < \lceil m/W \rceil)$ after $C' \leftarrow C' + Ab_i x^i$ $(0 \le i < W)$ for i from 0 to $W-1$. The word size shift Ax^{jW+i} computation is virtually free, so it is faster than the shift-and-add method.

In conjunction with the shift-and-add and comb methods, we deployed the window method [18, Alg. 2.36], an algorithm that reduces the number of GF(3^m)-additions by using an online precomputed table. Let w be the window size. The precomputation requires $(3^w - 2w - 1)/2$ additions in GF(3^m), $(3^w - 1)/2$ negative elements, and $w - 1$ shifts. The number of the GF(3^m)-additions required for evaluation is about $1/w$ times smaller than with a non-window method.

In our experiment, we implemented the shift-and-add method and the comb method with window size $w = 4$. Table 4 lists the running time results for multiplication in GF(3^{509}). We used the trinomial $f(x) = x^{509} + x^{358} + 2$ for

Table 4. Running Time Comparison for Multiplication in $GF(3^{509})$ (μsec)

Assignment	*Natural*	*Type 1*	*Type 2*
Shift-and-add method with $w = 4$	7.0910	6.9111	6.6184
Comb method with $w = 4$	4.9810	4.7391	4.5933

the polynomial basis of $GF(3^{509})$. Each listed running time is the average of 1,000,000 executions.

The running time with the *Type 2* assignment was about a few percent faster than that with the *Type 1* assignment. For the *Type 1* assignment, the shift requires some additional cost. When the shift Ax^i is computed, the bits less than x^i are padded by zeroes. Since $0 \in GF(3)$ is assigned as $(1, 1)$ in this assignment, it is necessary to change these bits from zero to one. On the other hand, the negative element is computed by $A_h \leftarrow A_h \wedge A_l$ for the *Type 2* assignment, so that it requires $\lceil m/W \rceil$ logical instructions.

The running time of the comb method with $w = 4$ for the *Type 2* assignment is about 8% faster than that for the *Natural* assignment. On the other hand, the shift-and-add method requires more shifts than does the comb method, and thus, the running time of the shift-and-add method with $w = 4$ for the *Type 2* assignment was only about 7% faster than that for the *Natural* assignment.

As a result, we found that the comb method with $w = 4$ for the *Type 2* assignment was the fastest multiplication implementation under these conditions.

4.4 η_T Pairing over $GF(3^m)$ and Its Efficient Implementation

This section describes how we applied the new GF(3)-addition to the η_T pairing over $GF(3^m)$ proposed, by Barreto et al. [3].

Many efficient algorithms for implementing the η_T pairing have been proposed [6,14,28,29,30]. We implemented the pairing with the following algorithms: the η_T pairing algorithm without cube root, proposed by Shirase et al. [28]; the loop unrolling technique proposed by Beuchat et al. [6]; an efficient multiplication in $GF(3^{6m})$ proposed by Gorla et al. [14]; reuse of the precomputed table for the window multiplication in $GF(3^m)$, proposed by Takahashi et al. [30]; and final exponentiation using the torus T_2 over $GF(3^m)$, proposed by Shirase et al. [29].

Cubing was performed with an 11-bit lookup table that computes $\sum_{i=t}^{t+10} a_i x^{3i}$ for positive integer t, and the reduction by the irreducible polynomial. Inversion was implemented by the extended Euclidean algorithm [18, Alg. 2.48] for a ternary polynomial.

Table 5 lists the running time results for the η_T pairing over $GF(3^{509})$. Each listed running time is the average of 100,000 executions. We used the comb method with $w = 4$ to compute the multiplication in $GF(3^m)$. As a result, we obtained a running time of about 16.3 milliseconds for computing the η_T pairing over $GF(3^{509})$.

Table 5. Running Time Comparison for η_T Pairing over GF(3^{509}) (μsec)

Assignment	*Natural*	*Type 1*	*Type 2*
η_T pairing	17524.99	17238.31	16295.33

Table 6. Running Times for Arithmetic in GF(3^m) and the η_T Pairing with the *Type 2* Assignment (μsec)

Degree(m)	97	167	193	239	353	509
Addition	0.0057	0.0075	0.0098	0.0098	0.0137	0.0175
Subtraction	0.0057	0.0075	0.0098	0.0098	0.0135	0.0176
Multiplication [a]	0.7706	1.3740	1.8223	1.8237	3.3079	4.5933
Cubing	0.0734	0.0955	0.1222	0.1297	0.1821	0.2820
Inversion	6.8932	13.6856	19.1599	23.8463	49.4362	94.3061
η_T Pairing [b]	614.96	1687.62	2611.45	3157.36	8299.48	16295.33

[a] Comb method with the window size $w = 4$ [18, Alg. 2.36].
[b] Loop unrolling [6], fast multiplication in GF(3^{6m}) [14], without cube root [28], final exponentiation using torus T_2 [29], reuse of precomputed table [30].

The efficiency improvement by using the *Type 2* assignment instead of the *Natural* assignment had almost the same ratio as that for the multiplication in GF(3^{509}) using the comb method with $w = 4$, since the dominant computational cost of the η_T pairing is the multiplication in GF(3^m). The η_T pairing for the *Type 1* assignment, however, was only 2% faster than that for the *Natural* assignment, because the shift requires additional cost in the cubing and inversion for the *Type 1* assignment.

Finally, we present the running times for arithmetic in GF(3^m) and the η_T pairing, using the *Type 2* assignment with several degrees. We used irreducible trinomials $f(x) = x^m + x^k + 2$ with a given $(m, k) = \{ (97, 12), (167, 96), (193, 12), (239, 24), (353, 142), (509, 358)\}$ [1]. Table 6 lists the running time results for computing arithmetic in GF(3^m) and the η_T pairing.

For comparison, we give some recently reported efficient implementations of arithmetic in GF(3^{509}). Ahmadi et al. [1] reported the following timings. The addition in GF(3^{509}) requires 0.09μsec, and the running time of the multiplication in GF(3^{509}), that is implemented by the comb method with $w = 3$, is 15.5μsec. They depolyed an irreducible pentanomial $f(x) = x^{509} - x^{477} + x^{445} + x^{32} - 1$ for the polynomial basis of GF(3^{509}). These running times were obtained on a Pentium 4 processor (2.4 GHz), under Linux/x86 with GCC 3.3. Very recently, Hankerson et al. [17] presented the following implementation. The timing of the multiplication and the η_T pairing over GF(3^{509}) is 10.6×10^3 cycles (about 3.79μsec@2.8GHz) and 46×10^6 cycles (about 16.43msec@2.8GHz), respectively.

[1] Page et al. [26] estimated that the discrete logarithm problems in GF(3^{6m}) with $m = 97, 163, 193, 239, 353$ have the security level of factoring $845, 912, 1080, 1338, 1976$ bits, respectively.

$f(x) = x^{509} - x^{318} - x^{191} + x^{127} + 1$ was used for the polynomial basis of GF(3^{509}). These timings were obtained on 64-bit Opteron processor (2.8 GHz) using GCC 4.1 with neither SSE nor MMX.

5 Conclusion

In this paper, we have described our search for logical instruction sequences implementing the GF(3)-addition in software. We considered a redundant representation of GF(3) in GF$(2)^2$ and an independent assignment of each GF(3) via the addition map GF(3) × GF(3) → GF(3). We also deployed several logical instructions provided in many CPUs, namely {AND, OR, XOR}, as well as {ANDN, ORN, XORN, NAND, NOR}, available in the SPARC and Alpha architectures, MMX, SSE, and other implementations.

As a result, we found many implementations of the GF(3)-addition with only six logical instructions. Their representations in GF$(2)^2$, however, do not use the natural assignment $\{(0,0) \mapsto 0, (0,1) \mapsto 1, (1,0) \mapsto 2\}$, or their logical instructions include not only {AND, OR, XOR} but also other ones. Moreover, we proved that the minimum number of logical instructions for constructing the GF(3)-addition is six in any representation of GF(3) in GF$(2)^2$ by using the logical instructions {AND, OR, XOR, ANDN, ORN, XORN, NAND, NOR}. Finally, we implemented the arithmetic in GF(3^m) and the η_T pairing by applying the new GF(3)-addition with six logical instructions. The resulting running time of the η_T pairing over GF(3^{509}) was approximately 7% faster than that using the GF(3)-addition with seven logical instructions, on an AMD Opteron processor model 275 (2.2 GHz). Therefore, for an efficient implementation of the η_T pairing, using the minimum number of logical instructions for computing the GF(3)-addition is effective.

References

1. Ahmadi, O., Hankerson, D., Menezes, A.: Software implementation of arithmetic in $\mathbb{F}_{3^m}$. In: Carlet, C., Sunar, B. (eds.) WAIFI 2007. LNCS, vol. 4547, pp. 85–102. Springer, Heidelberg (2007)
2. Barker, E., Barker, W., Burr, W., Polk, W., Smid, M.: Recommendation for Key Management – Part 1: General (Revised). NIST Special Publication 800–57 (2007)
3. Barreto, P.S.L.M., Galbraith, S., Ó'hÉigeartaigh, C., Scott, M.: Efficient pairing computation on supersingular abelian varieties. Designs, Codes and Cryptography 42(3), 239–271 (2007)
4. Barreto, P.S.L.M., Kim, H.Y., Lynn, B., Scott, M.: Efficient algorithms for pairing-based cryptosystems. In: Yung, M. (ed.) CRYPTO 2002. LNCS, vol. 2442, pp. 354–368. Springer, Heidelberg (2002)
5. Bertoni, G., Guajardo, J., Kumar, S., Orland, G., Paar, C., Wollinger, T.: Efficient GF(p^m) arithmetic architectures for cryptographic applications. In: Joye, M. (ed.) CT-RSA 2003. LNCS, vol. 2612, pp. 158–175. Springer, Heidelberg (2003)
6. Beuchat, J., Brisebarre, N., Detrey, J., Okamoto, E., Shirase, M., Takagi, T.: Algorithms and arithmetic operators for computing the η_T pairing in characteristic three. Cryptology ePrint Archive, Report 2007/417 (2007)

7. Beuchat, J., Shirase, M., Takagi, T., Okamoto, E.: An algorithm for the η_T pairing calculation in characteristic three and its hardware implementation. In: 18th IEEE International Symposium on Computer Arithmetic ARITH-18, pp. 97–104 (2007)
8. Boneh, D., Crescenzo, D., Ostrovsky, R., Persiano, G.: Public key encryption with keyword search. In: Cachin, C., Camenisch, J.L. (eds.) EUROCRYPT 2004. LNCS, vol. 3027, pp. 506–522. Springer, Heidelberg (2004)
9. Boneh, D., Franklin, M.: Identity based encryption from the Weil pairing. SIAM Journal on Computing 32(3), 586–615 (2003)
10. Boneh, D., Gentry, C., Waters, B.: Collusion resistant broadcast encryption with short ciphertexts and private keys. In: Shoup, V. (ed.) CRYPTO 2005. LNCS, vol. 3621, pp. 258–275. Springer, Heidelberg (2005)
11. Compaq Computer Corporation: Alpha Architecture Handbook (Version 4) (1998)
12. Duursma, I., Lee, H.: Tate pairing implementation for hyperelliptic curves $y^2 = x^p - x + d$. In: Laih, C.-S. (ed.) ASIACRYPT 2003. LNCS, vol. 2894, pp. 111–123. Springer, Heidelberg (2003)
13. Galbraith, S., Harrison, K., Soldera, D.: Implementing the Tate pairing. In: Fieker, C., Kohel, D.R. (eds.) ANTS 2002. LNCS, vol. 2369, pp. 324–337. Springer, Heidelberg (2002)
14. Gorla, E., Puttmann, C., Shokrollahi, J.: Explicit formulas for efficient multiplication in $\mathbb{F}_{3^{6m}}$. In: Adams, C., Miri, A., Wiener, M. (eds.) SAC 2007. LNCS, vol. 4876, pp. 173–183. Springer, Heidelberg (2007)
15. Granger, R., Page, D., Smart, N.P.: High security pairing-based cryptography revisited. In: Hess, F., Pauli, S., Pohst, M. (eds.) ANTS 2006. LNCS, vol. 4076, pp. 480–494. Springer, Heidelberg (2006)
16. Granger, R., Page, D., Stam, M.: Hardware and software normal basis arithmetic for pairing-based cryptography in characteristic three. IEEE Transactions on Computers 54(7), 852–860 (2005)
17. Hankerson, D., Menezes, A., Scott, M.: Software implementation of pairings. Centre for Applied Cryptographic Research (CACR) Technical Reports, CACR 2008-08 (2008), `http://www.cacr.math.uwaterloo.ca/techreports/2008/cacr2008-08.pdf`
18. Hankerson, D., Menezes, A., Vanstone, S.: Guide to Elliptic Curve Cryptography. Springer, Heidelberg (2004)
19. Harrison, K., Page, D., Smart, N.P.: Software implementation of finite fields of characteristic three, for use in pairing-based cryptosystems. LMS Journal of Computation and Mathematics 5, 181–193 (2002)
20. Intel Corporation: Intel Architecture Software Developer's Manual, vol. 2, Instruction Set Reference (1999)
21. SPARC International, Inc.: The SPARC Architecture Manual, Version 9 (2000)
22. Kerins, T., Marnane, W., Popovici, E., Barreto, P.S.L.M.: Efficient hardware for the Tate pairing calculation in characteristic three. In: Rao, J.R., Sunar, B. (eds.) CHES 2005. LNCS, vol. 3659, pp. 412–426. Springer, Heidelberg (2005)
23. Koblitz, N., Menezes, A.: Pairing-based cryptography at high security levels. In: Smart, N. (ed.) Cryptography and Coding 2005. LNCS, vol. 3796, pp. 13–36. Springer, Heidelberg (2005)
24. Miller, V.S.: Short program for functions on curves (unpublished manuscript, 1986)
25. Page, D., Smart, N.P.: Hardware implementation of finite fields of characteristic three. In: Kaliski Jr., B.S., Koç, Ç.K., Paar, C. (eds.) CHES 2002. LNCS, vol. 2523, pp. 529–539. Springer, Heidelberg (2003)

26. Page, D., Smart, N., Vercauteren, F.: A comparison of MNT curves and supersingular curves. Applicable Algebra in Engineering. Communication and Computing 17(5), 379–392 (2006)
27. Sakai, R., Kasahara, M.: ID based cryptosystems with pairing on elliptic curve. Cryptology ePrint Archive, Report 2003/054 (2003)
28. Shirase, M., Kawahara, Y., Takagi, T., Okamoto, E.: Universal η_T pairing algorithm over arbitrary extension degree. In: Kim, S., Yung, M., Lee, H.-W. (eds.) WISA 2007. LNCS, vol. 4867, pp. 1–15. Springer, Heidelberg (2008)
29. Shirase, M., Takagi, T., Okamoto, E.: Some efficient algorithms for the final exponentiation of η_T pairing. In: Dawson, E., Wong, D.S. (eds.) ISPEC 2007. LNCS, vol. 4464, pp. 254–268. Springer, Heidelberg (2007)
30. Takahashi, G., Hoshino, F., Kobayashi, T.: Efficient GF(3^m) multiplication algorithm for ηT pairing. Cryptology ePrint Archive, Report 2007/463 (2007)

A Comparison between Hardware Accelerators for the Modified Tate Pairing over $\mathbb{F}_{2^m}$ and $\mathbb{F}_{3^m}$

Jean-Luc Beuchat[1], Nicolas Brisebarre[2], Jérémie Detrey[3], Eiji Okamoto[1], and Francisco Rodríguez-Henríquez[4]

[1] Graduate School of Systems and Information Engineering, University of Tsukuba, 1-1-1 Tennodai, Tsukuba, Ibaraki, 305-8573, Japan
[2] LIP/Arénaire (CNRS – ENS Lyon – INRIA – UCBL), ENS Lyon, 46, allée d'Italie, F-69364 Lyon Cedex 07, France
[3] Cosec group, B-IT, Dahlmannstraße 2, D-53113 Bonn, Germany
[4] Computer Science Section, Electrical Engineering Department, Centro de Investigación y de Estudios Avanzados del IPN, Av. Instituto Politécnico Nacional No. 2508, 07300 México City, México

Abstract. In this article we propose a study of the modified Tate pairing in characteristics two and three. Starting from the η_T pairing introduced by Barreto *et al.* [1], we detail various algorithmic improvements in the case of characteristic two. As far as characteristic three is concerned, we refer to the survey by Beuchat *et al.* [5]. We then show how to get back to the modified Tate pairing at almost no extra cost. Finally, we explore the trade-offs involved in the hardware implementation of this pairing for both characteristics two and three. From our experiments, characteristic three appears to have a slight advantage over characteristic two.

Keywords: Modified Tate pairing, reduced η_T pairing, finite field arithmetic, elliptic curve, hardware accelerator, FPGA.

1 Introduction

Over the past few years, bilinear pairings over elliptic and hyperelliptic curves have been the focus of an ever increasing attention in cryptology. Since their introduction to this domain by Menezes, Okamoto & Vanstone [23] and Frey & Rück [9], and the first discovery of their constructive properties by Mitsunari, Sakai & Kasahara [26], Sakai, Oghishi & Kasahara [31], and Joux [17], a large number of pairing-based cryptographic protocols have already been published. For those reasons, efficient computation of pairings is crucial and, according to the recommendations of [12, 21], the Tate pairing, rather than the Weil pairing, appears to be the most appropriate choice.

Miller [24, 25] proposed in 1986 the first algorithm for iteratively computing the Weil and Tate pairings. In the case of the Tate pairing, a further final exponentiation of the Miller's algorithm result is required to obtain a uniquely defined value. Various improvements were published in [2, 7, 10, 22] and we will

S.D. Galbraith and K.G. Paterson (Eds.): Pairing 2008, LNCS 5209, pp. 297–315, 2008.

consider in this paper the modified Tate pairing as defined in [2]. Generalizing some results by Duursma & Lee [7], Barreto *et al.* then introduced the η_T pairing [1], in which the number of iterations in Miller's algorithm is halved. This nondegenerate bilinear pairing can also be used as a tool for computing the modified Tate pairing, at the expense of an additional exponentiation.

General purpose microprocessors are intrinsically not suited for computations on finite fields of small characteristic, hence software implementations are bound to be quite slow and the need for special purpose hardware coprocessors is strong [5, 6, 11, 16, 18, 19, 20, 28, 29, 30, 33]. In this context, we extend here to the characteristic two the results by Beuchat *et al.* [5] in the case of the hardware implementation of the reduced η_T pairing in characteristic three.

In Section 2, we detail the algorithms required to compute the reduced η_T pairing in characteristic two. Some algorithmic improvements in both the pairing computation and the tower-field arithmetic are also presented, and an accurate cost analysis in terms of operations over the base field $\mathbb{F}_{2^m}$ is given. We then study in Section 3 the relation between the η_T and Tate pairings, and show that the modified Tate pairing can be computed from the reduced η_T pairing at almost no extra cost in characteristics two and three. Section 4 gives hardware implementation results of the modified Tate pairing in both characteristics and for various field extension degrees. Comparisons between $\mathbb{F}_{2^m}$ and $\mathbb{F}_{3^m}$ are presented at equivalent levels of security and they show a slight advantage in favor of characteristic three. Finally, some comparisons with already published solutions are also given to attest the meaningfulness of our results. Supplementary material is available in a research report version of this paper [4].

2 Computation of the Reduced η_T Pairing in Characteristic Two

2.1 Preliminary Definitions

We consider the supersingular curve E over $\mathbb{F}_{2^m}$ defined by the equation

$$y^2 + y = x^3 + x + b, \tag{1}$$

where $b \in \{0, 1\}$ and m is an odd integer. We define $\delta = b$ when $m \equiv 1, 7 \pmod 8$; in all other cases, $\delta = 1 - b$. The number of rational points of E over $\mathbb{F}_{2^m}$ is given by $N = \#E(\mathbb{F}_{2^m}) = 2^m + 1 + \nu 2^{(m+1)/2}$, where $\nu = (-1)^\delta$ [2]. The embedding degree of this curve, which is the least positive integer k such that N divides $2^{km} - 1$, is 4.

Choosing $T = 2^m - N$ and a prime ℓ dividing N, Barreto *et al.* [1] defined the η_T pairing of two points P and $Q \in E(\mathbb{F}_{2^m})[\ell]$ as:

$$\eta_T(P, Q) = f_{T',P'}(\psi(Q)),$$

where $T' = -\nu T$, $P' = [-\nu]P$, and $E(\mathbb{F}_{2^m})[\ell]$ denotes the ℓ-torsion subgroup of $E(\mathbb{F}_{2^m})$. The distortion map ψ is defined from $E(\mathbb{F}_{2^m})[\ell]$ to $E(\mathbb{F}_{2^{4m}})[\ell]$ as

$$\psi(x, y) = (x + s^2, y + sx + t),$$

for all $(x, y) \in E(\mathbb{F}_{2^m})[\ell]$ [1]. The elements s and t of $\mathbb{F}_{2^{4m}}$ satisfy $s^2 = s + 1$ and $t^2 = t + s$. This allows for representing $\mathbb{F}_{2^{4m}}$ as an extension of $\mathbb{F}_{2^m}$ using the basis $(1, s, t, st)$: $\mathbb{F}_{2^{4m}} = \mathbb{F}_{2^m}[s, t] \cong \mathbb{F}_{2^m}[X, Y]/(X^2 + X + 1, Y^2 + Y + X)$. Finally, $f_{T',P'}$ is an element of $\mathbb{F}_{2^m}(E)$, where $\mathbb{F}_{2^m}(E)$ denotes the function field of the curve, and is given by

$$\begin{aligned} f_{T',P'} : E(\mathbb{F}_{2^{4m}})[\ell] &\longrightarrow \mathbb{F}^*_{2^{4m}} \\ \psi(Q) &\longmapsto \left(\prod_{i=0}^{\frac{m-1}{2}} g_{[2^i]P'}(\psi(Q))^{2^{\frac{m-1}{2}-i}} \right) l_{P'}(\psi(Q)), \end{aligned} \tag{2}$$

where:

- The point doubling formula is given by

  $$\left[2^i\right] P' = \left(x_{P'}^{2^{2i}} + i, y_{P'}^{2^{2i}} + i x_{P'}^{2^{2i}} + \tau(i) \right),$$

  with

  $$\tau(i) = \begin{cases} 0 & \text{if } i \equiv 0, 1 \pmod 4, \\ 1 & \text{otherwise.} \end{cases}$$

- g_V, for all $V = (x_V, y_V) \in E(\mathbb{F}_{2^m})[\ell]$, is the rational function defined over $E(\mathbb{F}_{2^{4m}})[\ell]$ corresponding to the doubling of V. For all $(x, y) \in E(\mathbb{F}_{2^{4m}})[\ell]$, we have $g_V(x, y) = (x_V^2 + 1)(x_V + x) + y_V + y$ [1]. According to the equation of the elliptic curve (1), $x_V^3 + x_V + y_V$ is equal to $y_V^2 + b$ and we obtain [33]:

  $$g_V(x, y) = x(x_V^2 + 1) + y_V^2 + y + b. \tag{3}$$

  We considered both forms of $g_V(x, y)$ when studying η_T pairing algorithms over $\mathbb{F}_{2^m}$ and discovered that the second one always leads to the fastest algorithms.
- l_V, for all $V = (x_V, y_V) \in E(\mathbb{F}_{2^m})[\ell]$, is the equation of the line corresponding to the addition of $\left[2^{\frac{m+1}{2}}\right] V$ with $[\nu]V$, and defined for all $(x, y) \in E(\mathbb{F}_{2^{4m}})[\ell]$ as follows:

  $$\begin{aligned} l_V(x, y) = x_V^2 + (x_V + \alpha)(x + \alpha) + x + y_V + y + \delta + 1 + \\ (x_V + x + 1 - \alpha)s + t, \end{aligned} \tag{4}$$

  where

  $$\alpha = \begin{cases} 0 & \text{if } m \equiv 3 \pmod 4, \\ 1 & \text{if } m \equiv 1 \pmod 4. \end{cases}$$

2.2 Computation of the η_T Pairing in Characteristic Two

Barreto *et al.* suggested reversing the loop to compute the η_T pairing [1]. They introduced the new index $j = \frac{m-1}{2} - i$ and obtained

$$f_{T',P'}(\psi(Q)) = l_{P'}(\psi(Q)) \prod_{j=0}^{\frac{m-1}{2}} \left(g_{\left[2^{\frac{m-1}{2}-j}\right]P'}(\psi(Q)) \right)^{2^j}.$$

A tedious case-by-case analysis allows one to prove that:

$$\left(g_{\left[2^{\frac{m-1}{2}-j}\right]P'}(\psi(Q))\right)^{2^j} = (x_{P'}^{2^{-j}} + \alpha)\cdot(x_Q^{2^j} + \alpha) + y_{P'}^{2^{-j}} + y_Q^{2^j} + \beta + (x_{P'}^{2^{-j}} + x_Q^{2^j} + \alpha)s + t,$$

where

$$\beta = \begin{cases} b & \text{if } m \equiv 1,3 \pmod 8, \\ 1-b & \text{if } m \equiv 5,7 \pmod 8). \end{cases}$$

This equation differs from the one given by Barreto *et al.* [1]: taking advantage of the second form of g_V (3), we obtain a slight reduction in the number of additions over $\mathbb{F}_{2^m}$.

We suggest a second improvement to save a multiplication over $\mathbb{F}_{2^m}$. At first glance multiplying $l_{P'}(\psi(Q))$ by $g_{\left[2^{\frac{m-1}{2}}\right]P'}(\psi(Q))$ involves three multiplications over $\mathbb{F}_{2^m}$. However, when $j = 0$, we have:

$$g_{\left[2^{\frac{m-1}{2}}\right]P'}(\psi(Q)) = (x_{P'} + \alpha)(x_Q + \alpha) + y_{P'} + y_Q + \beta + (x_{P'} + x_Q + \alpha)s + t.$$

Seeing that $\alpha + \beta = \delta + 1$, we rewrite $l_{P'}(\psi(Q))$ as follows:

$$l_{P'}(\psi(Q)) = g_{\left[2^{\frac{m-1}{2}}\right]P'}(\psi(Q)) + x_{P'}^2 + x_Q + \alpha + s.$$

Defining $g_0 = (x_{P'} + \alpha)(x_Q + \alpha) + y_{P'} + y_Q + \beta$, $g_1 = x_{P'} + x_Q + \alpha$, and $g_2 = x_{P'}^2 + x_Q + \alpha$, we eventually obtain:

$$g_{\left[2^{\frac{m-1}{2}}\right]P'}(\psi(Q)) = g_0 + g_1 s + t \quad \text{and} \quad l_{P'}(\psi(Q)) = (g_0 + g_2) + (g_1 + 1)s + t.$$

The product $l_{P'}(\psi(Q)) \cdot g_{\left[2^{\frac{m-1}{2}}\right]P'}(\psi(Q))$ can be computed by means of two multiplications over $\mathbb{F}_{2^m}$ (see [4, Appendix D.2]). Algorithm 1 describes the computation of the η_T pairing according to this construction. Addition over $\mathbb{F}_{2^m}$ involves m bitwise exclusive-OR operations that can be implemented in parallel. We refer to this operation as addition (A) when we give the cost of an algorithm. However, the addition of an element of $\mathbb{F}_2$ requires a single exclusive-OR operation, denoted by XOR. Additionally, M denotes multiplications, S squarings and R square roots. We also introduce $\bar{\delta} = 1 - \delta$.

The first step consists in computing $P' = [-\nu]P$ (line 1). Multiplication over $\mathbb{F}_{2^{4m}}$ usually requires nine multiplications and twenty additions over $\mathbb{F}_{2^m}$. However, the sparsity of G (as given line 13) allows one to compute the product $F \cdot G$ (line 14) by means of only six multiplications and fourteen additions over $\mathbb{F}_{2^m}$ (see [4, Appendix D.2] for further details). Contrary to what was suggested by Ronan *et al.* [29], the loop unrolling technique introduced by Granger *et al.* [13]

in the context of the Tate pairing in characteristic three turns out to be useless in our case. Let G_j and G_{j+1} denote the values of G at iterations j and $j+1$, respectively. Algorithm 1 computes $(F \cdot G_j) \cdot G_{j+1}$ by means of twelve multiplications and some additions over $\mathbb{F}_{2^m}$. The loop unrolling trick consists in taking advantage of the sparsity of G_j and G_{j+1}: only three multiplications over $\mathbb{F}_{2^m}$ are required to compute the product $G_j \cdot G_{j+1}$. Unfortunately, the result is not a sparse polynomial, and the multiplication by F involves nine multiplications over $\mathbb{F}_{2^m}$. Thus, computing $(G_j \cdot G_{j+1}) \cdot F$ instead of $(F \cdot G_j) \cdot G_{j+1}$ does not decrease the number of multiplications over the underlying field.

Algorithm 1. Computation of the η_T pairing in characteristic two: reversed-loop approach with square roots.

Input: $P, Q \in \mathbb{F}_{2^m}[\ell]$.
Output: $\eta_T(P,Q) \in \mathbb{F}^*_{2^{4m}}$.

$y_P \leftarrow y_P + \bar{\delta}$;	($\bar{\delta}$ XOR)
$u \leftarrow x_P + \alpha$; $v \leftarrow x_Q + \alpha$	(2α XOR)
$g_0 \leftarrow u \cdot v + y_P + y_Q + \beta$;	(1 M, 2 A, β XOR)
$g_1 \leftarrow u + x_Q$; $g_2 \leftarrow v + x_P^2$;	(1 S, 2 A)
$G \leftarrow g_0 + g_1 s + t$;	
$L \leftarrow (g_0 + g_2) + (g_1 + 1)s + t$;	(1 A, 1 XOR)
$F \leftarrow L \cdot G$;	(2 M, 1 S, 5 A, 2 XOR)
for $j = 1$ to $\frac{m-1}{2}$ **do**	
$x_P \leftarrow \sqrt{x_P}$; $y_P \leftarrow \sqrt{y_P}$; $x_Q \leftarrow x_Q^2$; $y_Q \leftarrow y_Q^2$;	(2 R, 2 S)
$u \leftarrow x_P + \alpha$; $v \leftarrow x_Q + \alpha$	(2α XOR)
$g_0 \leftarrow u \cdot v + y_P + y_Q + \beta$;	(1 M, 2 A, β XOR)
$g_1 \leftarrow u + x_Q$;	(1 A)
$G \leftarrow g_0 + g_1 s + t$;	
$F \leftarrow F \cdot G$;	(6 M, 14 A)
end for	
return F^M;	

The square roots in Algorithm 1 could be computed according to the technique described by Fong *et al.* [8]. However, this approach would require dedicated hardware and could potentially slow down a pairing coprocessor. Thus, it is attractive to study square-root-free algorithms which allow one to design simpler arithmetic and logic units. Another argument preventing the usage of square roots is that the complexity of their computation heavily depends on the particular irreducible polynomial selected for representing the field $\mathbb{F}_{2^m}$. On the other hand, the complexity of squarings is somehow more independent of the irreducible polynomial [27, 32]. To get rid of the square roots, we remark that

$$\eta_T(P,Q) = \eta_T\left(\left[2^{-\frac{m-1}{2}}\right] P, Q\right)^{2^{\frac{m-1}{2}}}.$$

Let $[2^j]Q = (x_{[2^j]Q}, y_{[2^j]Q})$. Since

$$g_{\left[2^{\frac{m-1}{2}-j}\right]\left(\left[2^{-\frac{m-1}{2}}\right]P'\right)}(\psi(Q)) = g_{[2^{-j}]P'}(\psi(Q)),$$

the η_T pairing is equal to

$$f_{T',P'}(\psi(Q)) = l_{\left[2^{-\frac{m-1}{2}}\right]P'}(\psi(Q))^{2^{\frac{m-1}{2}}} \prod_{j=0}^{\frac{m-1}{2}} \left(\left(g_{[2^{-j}]P'}(\psi(Q))\right)^{2^{2j}}\right)^{2^{\frac{m-1}{2}-j}},$$

where

$$\begin{aligned} l_{\left[2^{-\frac{m-1}{2}}\right]P'}(\psi(Q)) &= x_{P'}^2(x_{P'}^2 + x_Q + \alpha) + (\alpha+1)x_{P'}^2 + y_{P'}^2 + y_Q + \\ &\quad \gamma + \delta + (x_{P'}^2 + x_Q)s + t, \\ \left(g_{[2^{-j}]P'}(\psi(Q))\right)^{2^{2j}} &= (x_{P'}^2 + 1)\cdot\left(x_{[2^j]Q} + 1\right) + \\ &\quad y_{P'}^2 + y_{[2^j]Q} + b + \left(x_P^2 + x_{[2^j]Q} + 1\right)s + t, \end{aligned}$$

and

$$\gamma = \begin{cases} 0 & \text{if } m \equiv 1, 7 \pmod 8, \\ 1 & \text{if } m \equiv 3, 5 \pmod 8. \end{cases}$$

Again, one can simplify the computation of the product $l_{\left[2^{-\frac{m-1}{2}}\right]P'}(\psi(Q)) \cdot g_{P'}(\psi(Q))$. Noting that $\gamma + \delta = b$ and defining $g_0' = x_{P'}^2 x_Q + x_{P'}^2 + x_Q + y_{P'}^2 + y_q + b + 1$, $g_1' = x_{P'}^2 + x_Q + 1$, and $g_2' = x_{P'}^4 + x_Q + 1$, we obtain

$$l_{\left[2^{-\frac{m-1}{2}}\right]P'}(\psi(Q)) \cdot g_{P'}(\psi(Q)) = ((g_0' + g_2') + (g_1' + 1)s + t) \cdot (g_0' + g_1's + t).$$

An implementation of the η_T pairing following this construction is given in Algorithm 2.

We also studied direct approaches based on Equation (2). However, they turned out to be slower and we will not consider such algorithms in this paper (see [4, Appendix A] for details).

2.3 Final Exponentiation

The η_T pairing has to be reduced in order to be uniquely defined. We have to raise $\eta_T(P, Q)$ to the Mth power, where

$$M = \frac{2^{4m} - 1}{N} = (2^{2m} - 1)(2^m + 1 - \nu 2^{\frac{m+1}{2}}).$$

Two algorithms have been proposed in the open literature for $\nu = 1$ and $\nu = -1$, respectively:

Algorithm 2. Computation of the η_T pairing in characteristic two: reversed-loop approach without square roots.

Input: $P, Q \in \mathbb{F}_{2^m}[\ell]$.
Output: $\eta_T(P,Q) \in \mathbb{F}^*_{2^{4m}}$.

$y_P \leftarrow y_P + \bar{\delta}$; ($\bar{\delta}$ XOR)
$x_P \leftarrow x_P^2$; $y_P \leftarrow y_P^2$; (2 S)
$y_P \leftarrow y_P + b$; $u \leftarrow x_P + 1$; ($b + 1$ XOR)
$g_1 \leftarrow u + x_Q$; (1 A)
$g_0 \leftarrow x_P \cdot x_Q + y_P + y_Q + g_1$; (1 M, 3 A)
$x_Q \leftarrow x_Q + 1$; (1 XOR)
$g_2 \leftarrow x_P^2 + x_Q$; (1 S, 1 A)
$G \leftarrow g_0 + g_1 s + t$;
$L \leftarrow (g_0 + g_2) + (g_1 + 1)s + t$; (1 A, 1 XOR)
$F \leftarrow L \cdot G$; (2 M, 1 S, 5 A, 2 XOR)
for $j \leftarrow 1$ to $\frac{m-1}{2}$ **do**
  $F \leftarrow F^2$; (4 S, 4 A)
  $x_Q \leftarrow x_Q^4$; $y_Q \leftarrow y_Q^4$; (4 S)
  $x_Q \leftarrow x_Q + 1$; $y_Q \leftarrow y_Q + x_Q$; (1 A, 1 XOR)
  $g_0 \leftarrow u \cdot x_Q + y_P + y_Q$; (1 M, 2 A)
  $g_1 \leftarrow x_P + x_Q$; (1 A)
  $G \leftarrow g_0 + g_1 s + t$;
  $F \leftarrow F \cdot G$; (6 M, 14 A)
end for
Return F^M;

- Ronan *et al.* [29] assumed that $\nu = 1$, unrolled the different powering, and grouped the inversions together. Thus, their final exponentiation algorithm involves a single inversion over $\mathbb{F}_{2^{4m}}$.
- Shu *et al.* [33] noted that raising the η_T pairing to the power of $2^{2m} - 1$ requires only one inversion over $\mathbb{F}_{2^{2m}}$. When $\nu = -1$, the second part of the final exponentiation consists in raising this intermediate result to the power of $2^m + 1 + 2^{\frac{m+1}{2}}$.

In the following, we show that the final exponentiation of the η_T pairing in characteristic two always involves a single inversion over $\mathbb{F}_{2^{2m}}$. Since $M = (2^{2m} - 1)(2^m + 1) + \nu(1 - 2^{2m})2^{\frac{m+1}{2}}$, we compute

$$\eta_T(P,Q)^M = \left(\eta_T(P,Q)^{2^{2m}-1}\right)^{2^m+1} \cdot \left(\eta_T(P,Q)^{\nu(1-2^{2m})}\right)^{2^{\frac{m+1}{2}}},$$

and we remark that the final exponentiation requires a single inversion over $\mathbb{F}^{2^{2m}}$. Let $U = \eta_T(P,Q) \in \mathbb{F}^*_{2^{4m}}$. Writing $U = U_0 + U_1 t$, where U_0 and $U_1 \in \mathbb{F}_{2^{2m}}$ and noting that $t^{2^{2m}} = t + 1$, we obtain $U^{2^{2m}} = U_0 + U_1 + U_1 t$. Therefore, we have:

$$\begin{aligned} U^{2^{2m}-1} &= \frac{U_0 + U_1 + U_1 t}{U_0 + U_1 t} = \frac{(U_0 + U_1 + U_1 t)^2}{(U_0 + U_1 t) \cdot (U_0 + U_1 + U_1 t)} \\ &= \frac{U_0^2 + U_1^2 + U_1^2 s + U_1^2 t}{U_0^2 + U_0 U_1 + U_1^2 s}, \text{ and} \\ U^{1-2^{2m}} &= \frac{U_0 + U_1 t}{U_0 + U_1 + U_1 t} = \frac{U_0^2 + U_1^2 s + U_1^2 t}{U_0^2 + U_0 U_1 + U_1^2 s}, \end{aligned}$$

where $U_0^2 + U_0 U_1 + U_1^2 s \in \mathbb{F}_{2^{2m}}$. Algorithm 3 summarizes the computation of the $\eta_T(P, Q)^M$:

- According to our notation, we have $U = U_0 + U_1 t$, where $U_0 = u_0 + u_1 s$ and $U_1 = u_2 + u_3 s$. Since $s^2 = s + 1$, we remark that:

$$\begin{aligned} U_0^2 &= (u_0^2 + u_1^2) + u_1^2 s, \\ U_1^2 &= (u_2^2 + u_3^2) + u_3^2 s, \quad U_1^2 s = u_3^2 + u_2^2 s. \end{aligned}$$

  Therefore, 4 squarings and 2 additions over $\mathbb{F}_{2^m}$ allow us to get $T_0 = U_0^2$, $T_1 = U_1^2$, and $T_2 = U_1^2 s$.
- Multiplication over $\mathbb{F}_{2^{2m}}$ on line 3 is performed according to the Karatsuba-Ofman's scheme and involves three multiplications and four additions over $\mathbb{F}_{2^m}$:

$$T_3 = U_0 U_1 = u_0 u_2 + u_1 u_3 + ((u_0 + u_1)(u_2 + u_3) + u_0 u_2) s.$$

- Thanks to the tower field, inversion of $D = U_0^2 + U_0 U_1 + U_1^2 s \in \mathbb{F}_{2^{2m}}$ is replaced by an inversion (denoted by I), a squaring, three multiplications, and two additions over $\mathbb{F}_{2^m}$ (see [4, Appendix C] for details).
- The next step consists in computing $V = V_0 + V_1 t = U^{2^{2m}-1}$ and $W = W_0 + W_1 t = U^{\nu(1-2^{2m})}$, where V_0, V_1, W_0, and $W_1 \in \mathbb{F}_{2^{2m}}$. Defining $T_5 = \frac{U_0^2 + U_1^2 s}{U_0^2 + U_0 U_1 + U_1^2 s}$ and $T_6 = \frac{U_1^2}{U_0^2 + U_0 U_1 + U_1^2 s}$ (line 6), we easily check that $U^{2^{2m}-1} = (T_5 + T_6) + T_6 t$ and $U^{1-2^{2m}} = T_5 + T_6 t$. Thus,

$$V_0 = T_5 + T_6, \qquad W_0 = \begin{cases} T_5 + T_6 & \text{if } \nu = -1, \\ T_6 & \text{if } \nu = 1, \end{cases} \qquad V_1 = W_1 = T_6.$$

- Raising $V = V_0 + V_1 t \in \mathbb{F}_{2^{4m}}^*$ to the $(2^m + 1)$th power over $\mathbb{F}_{2^{4m}}$ (line 15) consists in multiplying V^{2^m} by V. This operation turns out to be less expensive than the usual multiplication over $\mathbb{F}_{2^{4m}}$ (see [4, Appendix D.3] for details).

2.4 Overall Cost Evaluations

Table 1 summarizes the costs of the algorithms studied in this section in terms of arithmetic operations over $\mathbb{F}_{2^m}$. Software implementations benefit from the

Algorithm 3. Final exponentiation of the reduced η_T pairing.

Input: $U = u_0 + u_1 s + u_2 t + u_3 st \in \mathbb{F}^*_{2^{4m}}$.
The intermediate variables m_i belong to $\mathbb{F}_{2^m}$. The T_i's, V_i's, W_i's, and D belong to $\mathbb{F}_{2^{2m}}$. V and $W \in \mathbb{F}_{2^{4m}}$.

Output: $V = U^M \in \mathbb{F}^*_{2^{4m}}$, with $M = (2^{2m}+1)(2^m - \nu 2^{\frac{m+1}{2}} + 1)$.

$m_0 \leftarrow u_0^2$; $m_1 \leftarrow u_1^2$; $m_2 \leftarrow u_2^2$; $m_3 \leftarrow u_3^2$; (4 S)
$T_0 \leftarrow (m_0 + m_1) + m_1 s$; $T_1 \leftarrow (m_2 + m_3) + m_3 s$; (2 A)
$T_2 \leftarrow m_3 + m_2 s$; $T_3 \leftarrow (u_0 + u_1 s) \cdot (u_2 + u_3 s)$; (3 M, 4 A)
$T_4 \leftarrow T_0 + T_2$; $D \leftarrow T_3 + T_4$; (4 A)
$D \leftarrow D^{-1}$; (1 I, 3 M, 1 S, 2 A)
$T_5 \leftarrow T_1 \cdot D$; $T_6 \leftarrow T_4 \cdot D$; (6 M, 8 A)
$V_0 \leftarrow T_5 + T_6$; (2 A)
$V_1, W_1 \leftarrow T_5$;
if $\nu = -1$ **then**
  $W_0 \leftarrow V_0$;
else
  $W_0 \leftarrow T_6$;
end if
$V \leftarrow V_0 + V_1 t$; $W \leftarrow W_0 + W_1 t$;
$V \leftarrow V^{2^m+1}$ (5 M, 2 S, 9 A)
for $i \leftarrow 1$ to $\frac{m+1}{2}$ **do**
  $W \leftarrow W^2$; (4 S, 4 A)
end for
Return $V \cdot W$; (9 M, 20 A)

Extended Euclidean Algorithm (EEA) to perform the inversion over $\mathbb{F}_{2^m}$. However, supplementing a pairing coprocessor with dedicated hardware for the EEA is not the most appropriate solution. Computing the inverse of $a \in \mathbb{F}_{2^m}$ by means of multiplications and squarings over $\mathbb{F}_{2^m}$ according to Fermat's little theorem and Itoh and Tsujii's work [15] allows one to keep the circuit area as small as possible without impacting too severely on the performances [3]. Since $a^{-1} = \left(a^{2^{m-1}-1}\right)^2$, we first raise a to the power of $2^{m-1}-1$ using a Brauer-type addition chain for $m-1$. Then, a squaring over $\mathbb{F}_{2^m}$ suffices to obtain a^{-1}. We reported the cost of this inversion scheme for typical values of m in Table 2.

3 Computation of the Modified Tate Pairing

Several researchers designed hardware accelerators over $\mathbb{F}_{2^m}$ and $\mathbb{F}_{3^m}$ for the modified Tate pairing. According to Barreto *et al.* [1], a second exponentiation allows one to compute the modified Tate pairing from the reduced η_T pairing. Thus, the modified Tate pairing is believed to be slower and a comparison between architectures for the modified Tate and η_T pairings would be unfair. Here, we take advantage of the bilinearity of the reduced η_T pairing and show how to get the modified Tate pairing almost for free.

Table 1. Cost of the presented algorithms for computing the reduced η_T pairing in characteristic two in terms of operations over the underlying field $\mathbb{F}_{2^m}$

	η_T pairing with square roots (Algorithm 1)	η_T pairing without square root (Algorithm 2)	Final Exponentiation (Algorithm 3)
Additions	$10 + 17 \cdot \frac{m-1}{2}$	$11m$	$2m + 53$
XORs	$3 + \bar{\delta} + (2\alpha + \beta) \cdot \frac{m+1}{2}$	$5 + \bar{\delta} + b + \frac{m-1}{2}$	–
Multiplications	$3 + 7 \cdot \frac{m-1}{2}$	$3 + 7 \cdot \frac{m-1}{2}$	26
Squarings	$m + 1$	$4m$	$2m + 9$
Square roots	$m - 1$	–	–
Inversions	–	–	1

Table 2. Cost of inversion over $\mathbb{F}_{2^m}$ according to Itoh and Tsujii's algorithm in terms of multiplications and squarings

Field	$\mathbb{F}_{2^{239}}$	$\mathbb{F}_{2^{251}}$	$\mathbb{F}_{2^{283}}$	$\mathbb{F}_{2^{313}}$
Cost	10 M, 238 S	10 M, 250 S	11 M, 282 S	10 M, 312 S

3.1 Modified Tate Pairing in Characteristic Two

The modified Tate pairing in characteristic two is given by $\hat{e}(P,Q)^M = \eta_T(P,Q)^{MT}$, where $M = \frac{2^{4m}-1}{N}$ and $T = 2^m - N$ [1]. Let $V = \eta_T(P,Q)^M$. We have $V^N = \eta_T(P,Q)^{2^{4m}-1} = 1$. Since $\eta_T(P,Q)^M$ is a bilinear pairing, we obtain:

$$\hat{e}(P,Q)^M = V^T = V^{2^m-N} = V^{2^m} = \eta_T(P,Q)^{M\cdot 2^m} = \eta_T([2^m]P,Q)^M,$$

where $[2^m]P = (x_P+1, x_P+y_P+\alpha+1)$. Thus, it suffices to provide a hardware accelerator for the reduced η_T pairing with $[2^m]P$ and Q to get the modified Tate pairing. Since this preprocessing step involves an XOR operation and an addition over $\mathbb{F}_{2^m}$, it can be computed in software. Conversely, a processor for the modified Tate pairing computes the η_T pairing if its inputs are $[2^{-m}]P$ and Q:

$$\eta_T(P,Q)^M = \hat{e}([2^{-m}]P,Q)^M,$$

where $[2^{-m}]P = (x_P+1, x_P+y_P+\alpha)$.

3.2 Modified Tate Pairing in Characteristic Three

The same approach allows one to compute the modified Tate pairing in characteristic three. Let m be a positive integer coprime to 6 and E be the supersingular elliptic curve defined by $E : y^2 = x^3 - x + b$, where $b \in \{-1, 1\}$. The number of rational points of E over $\mathbb{F}_{3^m}$ is given by $N = \#E(\mathbb{F}_{3^m}) = 3^m + 1 + \mu b 3^{\frac{m+1}{2}}$ [2], with

$$\mu = \begin{cases} 1 & \text{if } m \equiv 1, 11 \pmod{12}, \\ -1 & \text{if } m \equiv 5, 7 \pmod{12}. \end{cases}$$

In characteristic three, we have the following relation between the reduced η_T and modified Tate pairings [1]:

$$\left(\eta_T(P,Q)^M\right)^{3T^2} = \left(\hat{e}(P,Q)^M\right)^L,$$

with $M = \frac{3^{6m}-1}{N}$, $T = 3^m - N$, and $L = -\mu b 3^{\frac{m+3}{2}}$. Defining $V = \eta_T(P,Q)^M \in \mathbb{F}^*_{3^{6m}}$ and seeing that $V^N = 1$, we obtain

$$V^{3T^2} = V^{3^{2m+1} - 2\cdot 3^{m+1}\cdot N + 3N^2} = V^{3^{2m+1}}.$$

Dividing by L at the exponent level, we finally get the following relation between the reduced η_T and modified Tate pairings:

$$\begin{aligned} \hat{e}(P,Q)^M &= V^{\frac{3^{2m+1}}{L}} \\ &= V^{-\mu b 3^{\frac{3m-1}{2}}} = \eta_T\left(\left[-\mu b 3^{\frac{3m-1}{2}}\right] P, Q\right)^M, \end{aligned}$$

where $\left[-\mu b 3^{\frac{3m-1}{2}}\right] P = (\sqrt[3]{x_P} - b, -\mu b \lambda \sqrt[3]{y_P})$ and

$$\lambda = (-1)^{\frac{m+1}{2}} = \begin{cases} 1 & \text{if } m \equiv 7, 11 \pmod{12}, \\ -1 & \text{if } m \equiv 1, 5 \pmod{12}. \end{cases}$$

Again, the overhead introduced is negligible compared to the calculation time of the reduced η_T pairing. Consider now the cube-root-free reversed-loop algorithm proposed by Beuchat *et al.* (Algorithm 4 in [5]). In this case, we suggest to compute $\eta_T\left([-\mu b]P, \left[3^{\frac{3m-1}{2}}\right] Q\right)^M$. Surprisingly, the modified Tate pairing in characteristic three turns out to be slightly less expensive than the η_T pairing: we save two cubings and one addition over $\mathbb{F}_{3^m}$ (see [4, Appendix B] for details). Conversely, a processor for the modified Tate pairing provided with $[-\mu b]P$ and $\left[3^{\frac{-3m+1}{2}}\right] Q$ will return the reduced η_T pairing.

4 Implementation Results and Comparisons

4.1 A Unified Operator for the Arithmetic over $\mathbb{F}_{2^m}$ and $\mathbb{F}_{3^m}$

In [3], Beuchat *et al.* presented an FPGA-based accelerator for the computation of the η_T pairing in characteristic three. The coprocessor was based on a unified operator capable of handling all the necessary arithmetic operations over the base field $\mathbb{F}_{3^m}$. This streamlined design led to smaller circuits while retaining competitive performances with respect to the other published architectures.

For these reasons, we chose to use such a unified operator for our own implementations in characteristic three. We also adapted the operator for supporting finite-field arithmetic in characteristic two.

The core of this unified operator is an array multiplier [34] for computing the product of two elements of $\mathbb{F}_{p^m}$ (where $p = 2$ or 3), represented in a polynomial basis using a degree-m polynomial $f(x)$ irreducible over $\mathbb{F}_p$: $\mathbb{F}_{p^m} \cong \mathbb{F}_p[x]/(f(x))$. D coefficients of the multiplicand are processed at each clock cycle. The D corresponding partial products are then shifted and reduced modulo $f(x)$ according to their respective weight, and finally summed into a register thanks to a tree of adders over $\mathbb{F}_{p^m}$. A feedback loop allows the accumulation of the previous partial products. A product over $\mathbb{F}_{p^m}$ is therefore computed in $\lceil m/D \rceil$ clock cycles.

With only slight modifications, it is possible for this multiplier to also support the other operations required by the computation of the modified Tate pairing. For instance, bypassing the shift/modulo-$f(x)$ reduction stage allows for additions, subtractions and accumulations. Similarly, the Frobenius endomorphism (*i.e.* squaring in characteristic two or cubing in characteristic three) only amounts to a linear combination of the coefficients of the polynomial. This linear combination can be computed at design time and then directly hard-wired as an alternative datapath during the shift/modulo stage.

4.2 Characteristic Two Versus Characteristic Three

It is common knowledge that arithmetic over $\mathbb{F}_{2^m}$ is more compact and efficient than over $\mathbb{F}_{3^m}$. However, due to the different embedding degrees enjoyed by the elliptic curves of interest, competitive levels of security for pairing implementations in characteristic two are only achieved at the price of working over extension degrees much larger than what their counterparts in characteristic three require.

For a better understanding of this trade-off, we present here FPGA implementation results of a coprocessor for the modified Tate pairing in both characteristics two and three. The coprocessor is based on the previously described unified operator and implements the square- and cube-root-free reversed-loop algorithms (Algorithm 2, and Algorithm 4 in [5]) along with the corresponding final exponentiation. We also experimented with several values for D, aiming at a more exhaustive study of the trade-off between cost and performances.

Tables 3 and 4 present the post-place-and-route results for characteristic two and three respectively. These results were obtained for a Xilinx Virtex-II Pro 20 FPGA with average speedgrade, using the Xilinx ISE 9.2i tool suite. The two tables are also summarized in Figure 1.

The given results show a slight advantage of characteristic three over characteristic two, for all the studied levels of security. This goes against the performances obtained by Barreto *et al.* in the case of software implementation [1], but also against the hardware results published by Shu *et al.* in [33]. Of course, this observation remains closely related to our unified architecture. However, as detailed in the following, our coprocessors perform better than the previously published solutions in terms of area-time product, which leads us to believe this observation to be accurate.

Table 3. Implementation results of the modified Tate pairing in characteristic two using our unified operator (on a Xilinx Virtex-II Pro xc2vp20, speedgrade -6)

Field	Security [bits]	D	Area [slices]	Frequency [MHz]	#cycles	Estimated calc. time [μs]
$\mathbb{F}_{2^{239}}$	956	7	2366	199	39075	196
		15	2736	165	20830	127
		31	4557	123	13147	107
$\mathbb{F}_{2^{251}}$	1004	7	2270	185	41969	227
		15	3140	145	22846	157
		31	4861	126	14794	117
$\mathbb{F}_{2^{283}}$	1132	7	2517	169	52820	313
		15	3481	140	27942	200
		31	5350	127	17765	140
$\mathbb{F}_{2^{313}}$	1252	7	2661	182	63167	347
		15	3731	156	33283	213
		31	6310	111	20831	186
$\mathbb{F}_{2^{459}}$	1836	7	3809	168	129780	771
		15	5297	135	66589	492
		31	8153	115	37601	327

Table 4. Implementation results of the modified Tate pairing in characteristic three using our unified operator (on a Xilinx Virtex-II Pro xc2vp20, speedgrade -6)

Field	Security [bits]	D	Area [slices]	Frequency [MHz]	#cycles	Estimated calc. time [μs]
$\mathbb{F}_{3^{97}}$	922	3	1896	156	27800	178
		7	2711	128	14954	117
		15	4455	105	9657	92
$\mathbb{F}_{3^{103}}$	980	3	2003	151	32649	217
		7	2841	126	16633	132
		15	4695	103	10227	99
$\mathbb{F}_{3^{119}}$	1132	3	2223	140	41788	299
		7	3225	125	20814	166
		15	5293	99	12607	127
$\mathbb{F}_{3^{127}}$	1208	3	2320	149	47234	317
		7	3379	129	24028	186
		15	5596	99	14349	145
$\mathbb{F}_{3^{193}}$	1835	3	3266	147	100668	682
		7	4905	111	48205	433
		15	8266	90	26937	298

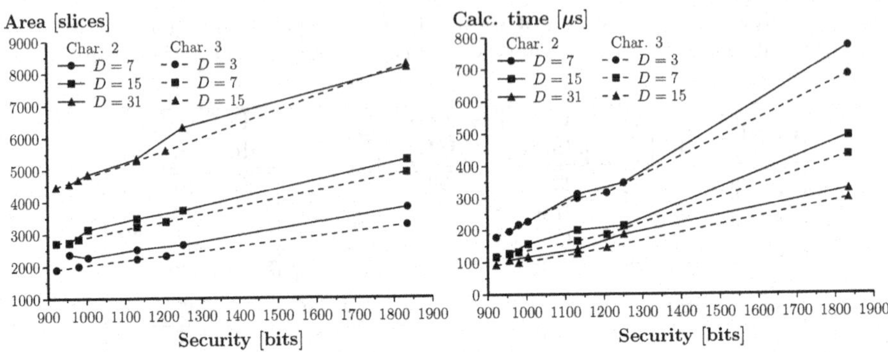

Fig. 1. Area (left) and calculation time (right) for the modified Tate pairing on our unified operator, in both characteristics two and three, for various extension degrees and different values for the parameter D

Moreover, the optimal number D of coefficients processed per clock cycle for the array multiplier appears to be 15 in characteristic two and 7 in characteristic three. However, modifying the value of this parameter changes only marginally the overall area-time product. According to the requirements of each application in terms of area and speed, one can then select the most appropriate value for D.

4.3 Comparisons

Tables 5, 6 and 7 present the cost and performances of other coprocessors for the computation of the modified Tate and reduced η_T pairings in characteristics two and three as published in the open literature. The results are summarized in Figure 2 as a comparison of these solutions against our proposed architecture in terms of their area-time product.

Despite its inherent lack of parallelism between operations, our unified operator greatly benefits from its compact design in order to reach higher frequencies. Combined with the algorithmic improvements described in this paper and in [5], this leads to competitive calculation times. Additionally, the streamlined design allows for reaching higher extension degrees and levels of security without risking to exhaust the FPGA resources: the slow increase of the area-time product with the security level of the system hints at the high scalability of the coprocessor.

Finally, the good performances of our solution against the previously published works vouches for a strong confidence in the outcome of our comparison between characteristics two and three for the hardware implementation of the modified Tate pairing.

5 Conclusion

We discussed several algorithms to compute the η_T pairing and its final exponentiation in characteristic two. We then showed how to get back to the modified

Table 5. FPGA-based accelerators for the modified Tate pairing over $\mathbb{F}_{2^m}$ in the literature. The parameter D refers to the number of coefficients processed at each clock cycle by a multiplier. The architectures by Shu *et al.* [33] include four kinds of multipliers.

	Curve	FPGA	#mult.	D	Area [slices]	Freq. [MHz]	Calculation time [μs]
Shu *et al.* [33]	$E(\mathbb{F}_{2^{239}})$	xc2vp100	6	16	25287	84	41
			1	4			
			1	1			
			1	2			
Keller *et al.* [18]	$E(\mathbb{F}_{2^{251}})$	xc2v6000	13	1	16621	50	6440
				6	21955	43	2580
				10	27725	40	2370
Keller *et al.* [19]	$E(\mathbb{F}_{2^{251}})$	xc2v6000	1	6	3788	40	4900
			3	6	6181	40	3200
			9	6	13387	40	2600
Keller *et al.* [18]	$E(\mathbb{F}_{2^{283}})$	xc2v6000	13	1	18599	50	7980
				4	22636	49	3230
				6	24655	47	2810
Keller *et al.* [19]	$E(\mathbb{F}_{2^{283}})$	xc2v6000	1	6	4273	40	6000
			3	6	6981	40	3800
			9	6	15065	40	3000
Shu *et al.* [33]	$E(\mathbb{F}_{2^{283}})$	xc2vp100	6	32	37803	72	61
			1	4			
			1	1			
			1	2			
Ronan *et al.* [29]	$E(\mathbb{F}_{2^{313}})$	xc2vp100	14	4	34675	55	203
				8	41078	50	124
				12	44060	33	146
Ronan *et al.* [30]	$C(\mathbb{F}_{2^{103}})$	xc2vp100	20	4	21021	51	206
				8	24290	46	152
				16	30464	41	132

Table 6. FPGA-based accelerators for the modified Tate pairing over $\mathbb{F}_{3^{97}}$ in the literature. The parameter D refers to the number of coefficients processed at each clock cycle by a multiplier.

	FPGA	#mult.	D	Area [slices]	Freq. [MHz]	Calculation time [μs]
Grabher and Page [11]	xc2vp4	1	4	4481	150	432.3
Kerins *et al.* [20]	xc2vp125	18	4	55616	15	850

Table 7. FPGA-based accelerators for reduced η_T pairing over $\mathbb{F}_{3^{97}}$ in the literature. The parameter D refers to the number of coefficients processed at each clock cycle by a multiplier.

	FPGA	#mult.	D	Area [slices]	Freq. [MHz]	Calculation time [μs]
Ronan *et al.* [28]	xc2vp100	5	4	10540	84.8	187
Jiang [16]	xc4vlx200	Not specified	7	74105	77.7	20.9
Beuchat *et al.* [5]	xc2vp4	1	3	1833	145	192
Beuchat *et al.* [6]	xc2vp30	9	3	10897	147	33.0
	xc4vlx25	9	3	11318	200	24.2

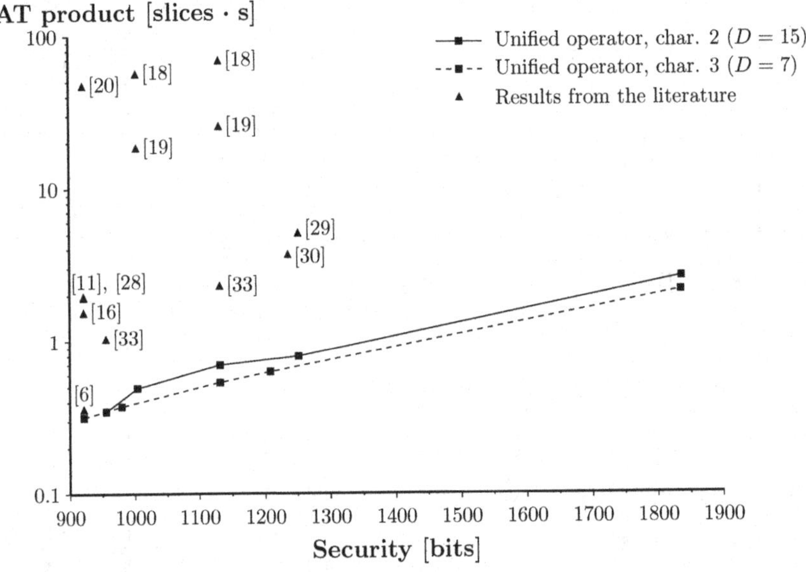

Fig. 2. Area-time product of the proposed coprocessor for the modified Tate pairing in characteristics two and three against the other solutions published in the literature

Tate pairing at almost no extra cost. Finally, we explored the trade-offs involved in the hardware implementation of the modified Tate pairing for both characteristic two and three. Our architectures are based on the unified arithmetic operator introduced in [3], and achieve a better area-time trade-off compared to previously published solutions [11, 16, 18, 19, 20, 28, 29, 30, 33].

Our modified Tate pairing coprocessors embed a single multiplier. A challenge consists in designing parallel architectures with the same (or even a smaller) area-time product. Future work should also include a study of the η_T pairing over genus-2 curves. The Ate pairing [14] would also be of interest, for it generalizes to ordinary curves the improvements introduced by the η_T pairing in the case of supersingular curves.

Acknowledgment

The authors would like to thank Guillaume Hanrot and the anonymous referees for their valuable comments. This work was supported by the New Energy and Industrial Technology Development Organization (NEDO), Japan.

The authors would also like to express their deepest gratitude to the Carthusian Monks of the Grande Chartreuse in the French Alps for their succulent herbal products which fueled our efforts in writing this article.

References

1. Barreto, P.S.L.M., Galbraith, S.D., ÓhÉigeartaigh, C., Scott, M.: Efficient pairing computation on supersingular Abelian varieties. Designs, Codes and Cryptography 42, 239–271 (2007)
2. Barreto, P.S.L.M., Kim, H.Y., Lynn, B., Scott, M.: Efficient algorithms for pairing-based cryptosystems. In: Yung, M. (ed.) CRYPTO 2002. LNCS, vol. 2442, pp. 354–368. Springer, Heidelberg (2002)
3. Beuchat, J.-L., Brisebarre, N., Detrey, J., Okamoto, E.: Arithmetic operators for pairing-based cryptography. In: Paillier, P., Verbauwhede, I. (eds.) CHES 2007. LNCS, vol. 4727, pp. 239–255. Springer, Heidelberg (2007)
4. Beuchat, J.-L., Brisebarre, N., Detrey, J., Okamoto, E., Rodríguez-Henríquez, F.: A comparison between hardware accelerators for the modified Tate pairing over $\mathbb{F}_{2^m}$ and $\mathbb{F}_{3^m}$. Cryptology ePrint Archive, Report 2008/115 (2008)
5. Beuchat, J.-L., Brisebarre, N., Detrey, J., Okamoto, E., Shirase, M., Takagi, T.: Algorithms and arithmetic operators for computing the η_T pairing in characteristic three. IEEE Transactions on Computers 57(11) (November 2008) (to appear) An extended version is available as Report 2007/417 of the Cryptology ePrint Archive
6. Beuchat, J.-L., Shirase, M., Takagi, T., Okamoto, E.: An algorithm for the η_T pairing calculation in characteristic three and its hardware implementation. In: Kornerup, P., Muller, J.-M. (eds.) Proceedings of the 18th IEEE Symposium on Computer Arithmetic, pp. 97–104. IEEE Computer Society, Los Alamitos (2007)
7. Duursma, I., Lee, H.S.: Tate pairing implementation for hyperelliptic curves $y^2 = x^p - x + d$. In: Laih, C.S. (ed.) ASIACRYPT 2003. LNCS, vol. 2894, pp. 111–123. Springer, Heidelberg (2003)
8. Fong, K., Hankerson, D., López, J., Menezes, A.: Field inversion and point halving revisited. IEEE Transactions on Computers 53(8), 1047–1059 (2004)
9. Frey, G., Rück, H.-G.: A remark concerning m-divisibility and the discrete logarithm in the divisor class group of curves. Mathematics of Computation 62(206), 865–874 (1994)
10. Galbraith, S.D., Harrison, K., Soldera, D.: Implementing the Tate pairing. In: Fieker, C., Kohel, D.R. (eds.) ANTS 2002. LNCS, vol. 2369, pp. 324–337. Springer, Heidelberg (2002)
11. Grabher, P., Page, D.: Hardware acceleration of the Tate pairing in characteristic three. In: Rao, J.R., Sunar, B. (eds.) CHES 2005. LNCS, vol. 3659, pp. 398–411. Springer, Heidelberg (2005)
12. Granger, R., Page, D., Smart, N.P.: High security pairing-based cryptography revisited. In: Hess, F., Pauli, S., Pohst, M. (eds.) ANTS 2006. LNCS, vol. 4076, pp. 480–494. Springer, Heidelberg (2006)

13. Granger, R., Page, D., Stam, M.: On small characteristic algebraic tori in pairing-based cryptography. LMS Journal of Computation and Mathematics 9, 64–85 (2006)
14. Hess, F., Smart, N., Vercauteren, F.: The Eta pairing revisited. IEEE Transactions on Information Theory 52(10), 4595–4602 (2006)
15. Itoh, T., Tsujii, S.: A fast algorithm for computing multiplicative inverses in $GF(2^m)$ using normal bases. Information and Computation 78, 171–177 (1988)
16. Jiang, J.: Bilinear pairing (Eta_T Pairing) IP core. Technical report, City University of Hong Kong – Department of Computer Science (May 2007)
17. Joux, A.: A one round protocol for tripartite Diffie-Hellman. In: Bosma, W. (ed.) ANTS 2000. LNCS, vol. 1838, pp. 385–394. Springer, Heidelberg (2000)
18. Keller, M., Kerins, T., Crowe, F., Marnane, W.P.: FPGA implementation of a $GF(2^m)$ Tate pairing architecture. In: Bertels, K., Cardoso, J.M.P., Vassiliadis, S. (eds.) ARC 2006. LNCS, vol. 3985, pp. 358–369. Springer, Heidelberg (2006)
19. Keller, M., Ronan, R., Marnane, W.P., Murphy, C.: Hardware architectures for the Tate pairing over $GF(2^m)$. Computers and Electrical Engineering 33(5–6), 392–406 (2007)
20. Kerins, T., Marnane, W.P., Popovici, E.M., Barreto, P.S.L.M.: Efficient hardware for the Tate pairing calculation in characteristic three. In: Rao, J.R., Sunar, B. (eds.) CHES 2005. LNCS, vol. 3659, pp. 412–426. Springer, Heidelberg (2005)
21. Koblitz, N., Menezes, A.: Pairing-based cryptography at high security levels. In: Smart, N.P. (ed.) Cryptography and Coding 2005. LNCS, vol. 3796, pp. 13–36. Springer, Heidelberg (2005)
22. Kwon, S.: Efficient Tate pairing computation for elliptic curves over binary fields. In: Boyd, C., González Nieto, J.M. (eds.) ACISP 2005. LNCS, vol. 3574, pp. 134–145. Springer, Heidelberg (2005)
23. Menezes, A., Okamoto, T., Vanstone, S.A.: Reducing elliptic curves logarithms to logarithms in a finite field. IEEE Transactions on Information Theory 39(5), 1639–1646 (1993)
24. Miller, V.S.: Short programs for functions on curves (1986), `http://crypto.stanford.edu/miller`
25. Miller, V.S.: The Weil pairing, and its efficient calculation. Journal of Cryptology 17(4), 235–261 (2004)
26. Mitsunari, S., Sakai, R., Kasahara, M.: A new traitor tracing. IEICE Trans. Fundamentals E85-A(2), 481–484 (2002)
27. Rodríguez-Henríquez, F., Morales-Luna, G., López, J.: Low-complexity bit-parallel square root computation over $GF(2^m)$ for all trinomials. IEEE Transactions on Computers 57(4), 472–480 (2008)
28. Ronan, R., Murphy, C., Kerins, T., ÓhÉigeartaigh, C., Barreto, P.S.L.M.: A flexible processor for the characteristic 3 η_T pairing. Int. J. High Performance Systems Architecture 1(2), 79–88 (2007)
29. Ronan, R., ÓhÉigeartaigh, C., Murphy, C., Scott, M., Kerins, T.: FPGA acceleration of the Tate pairing in characteristic 2. In: Proceedings of the IEEE International Conference on Field Programmable Technology – FPT 2006, pp. 213–220. IEEE, Los Alamitos (2006)
30. Ronan, R., ÓhÉigeartaigh, C., Murphy, C., Scott, M., Kerins, T.: Hardware acceleration of the Tate pairing on a genus 2 hyperelliptic curve. Journal of Systems Architecture 53, 85–98 (2007)
31. Sakai, R., Ohgishi, K., Kasahara, M.: Cryptosystems based on pairing. In: 2000 Symposium on Cryptography and Information Security (SCIS 2000), Okinawa, Japan, pp. 26–28 (January 2000)

32. Scott, M.: Optimal irreducible polynomials for GF(2^m) arithmetic. Cryptology ePrint Archive, Report 2007/192 (2007)
33. Shu, C., Kwon, S., Gaj, K.: FPGA accelerated Tate pairing based cryptosystem over binary fields. In: Proceedings of the IEEE International Conference on Field Programmable Technology – FPT 2006, pp. 173–180. IEEE, Los Alamitos (2006)
34. Song, L., Parhi, K.K.: Low energy digit-serial/parallel finite field multipliers. Journal of VLSI Signal Processing 19(2), 149–166 (1998)

One-Round ID-Based Blind Signature Scheme without ROS Assumption*

Wei Gao[1,2], Guilin Wang[3], Xueli Wang[4], and Fei Li[5]

[1] College of Mathematics and Information, Ludong University, Yantai 264025, China
[2] Guangdong Key Laboratory of Information Security Technology, Guangzhou 510275, China
sdgaowei@yahoo.com.cn
[3] School of Computer Science, University of Birmingham Birmingham B15 2TT, UK
G.Wang@cs.bham.ac.uk
[4] School of Mathematics Science, South China Normal University, Guangzhou 510631, China
wangxuyuyan@yahoo.com.cn
[5] College of Mathematics and Information, Ludong University, Yantai 264025, China
miss_lifei@163.com

Abstract. In this paper, we propose the first one-round identity-based blind signature (IDBS) scheme without ROS assumption, which supposes that it is infeasible to find an overdetermined, solvable system of linear equations modulo q with random inhomogenities [25]. Our construction has the following features. First, it achieves the optimal bound of round complexity for blind signatures, i.e., each signature can be generated with one round (or two moves) of message exchanges between the signer and signature requesting user. Second, the proposed IDBS scheme is provably secure against generic parallel attack without relying on the ROS assumption. This means our scheme can guarantee the same security level with smaller security parameter, in contrast to some IDBS schemes with ROS assumptions, such as the IDBS deduced from the blind Schnorr signature. Third, our construction is based on bilinear pairings from scratch (i.e. without using existing identity-based signature schemes, and without using existing computational assumptions). Finally, the security of our IDBS is based on a new formalized assumption, called one-more bilinear Diffie-Hellman inversion (1m-BDHI) assumption.

1 Introduction

Background. In 1984, Shamir [26] introduced the concept of identity-based (ID-based for short) public key cryptosystems to simplify key management procedures in certificate-based public key setting. ID-based cryptosystems have a

* This work is partially supported by National Natural Science Foundation of China CNF10771078 and Open Fund of Guangdong Key Laboratory of Information Security Technology.

S.D. Galbraith and K.G. Paterson (Eds.): Pairing 2008, LNCS 5209, pp. 316–331, 2008.

property that a user's public key can be easily derived from his identity by a publicly available function, while his private key can be calculated for him by a trusted authority, called Private Key Generator (PKG). They enable any pair of users to communicate securely without exchanging public key certificates, without keeping a public key directory, and without using online service of a third party, as long as the trusted PKG issues a private key to each user when he first joins the network. So they can be a good alternative for certificate-based public key infrastructure (PKI), especially when efficient key management and moderate security are required.

Bilinear pairings are the main tools to construct new ID-based cryptographic primitives. In 2000, Joux [20] used the Weil pairing to construct a one-round tripartite Diffie-Hellman key agreement protocol. After Joux's breakthrough, many ID-based cryptographic schemes have been proposed using bilinear pairings [15]. In Crypto 2001, Boneh and Franklin [8] presented an ID-based encryption scheme based on bilinear pairings which is the first fully functioning, efficient and provably secure ID-based encryption scheme. In Asiacrypt 2001, Boneh, Lynn and Shacham [9] proposed a basic signature scheme using pairings which has the shortest length among signature schemes in classical cryptography.

Blind signature, introduced by Chaum [12] in Crypto'82, is a variant of digital signature, which allows the user to get a signature without giving the signer any information about the actual message or the resulting signature. Formally, blindness means that the signer's view and the resulting signature are statistically independent, where the signer's view is the set of all values that can be gotten by the signer during the execution of the signature issuing protocol. This blindness property plays a central role in applications such as electronic voting and electronic cash systems.

MOTIVATION. Before the generic result of Galindo et al. [17], three ID-based blind signature (IDBS) schemes [19, 28, 29] based on bilinear pairings have been proposed. However, for all these schemes, the security against one more signature forgery under the generic parallel attack [22] requires that the following ROS problem is intractable [19, 25, 28, 29]: find an overdetermined, solvable system of linear equations modulo q with random inhomogenities (right sides). Unfortunately, in Crypto 2002, Wagner [27] claimed that there is a subexponential time algorithm to break the ROS-problem. To resist this attack, the size of q (security parameter) may need to be at least 1,600 bits long. In contrast, for common cryptographic primitives based on bilinear pairings such as [8, 9], the size of q is only about 160 bits. Since even the slightly larger security parameter will result in the dramatically larger amount of computation, all these existing schemes can not be efficiently implemented, and hence be of little interest in practice. In fact, until Galindo et al.'s generic result [17], it remained *an open problem* to construct an ID-based blind signature scheme whose security does not depend on the ROS assumption.

On the other hand, all of the aforementioned IDBS schemes require two rounds (more specifically, three moves) of message exchange between the signer and user, who requests a blind signature from the signer, since these protocols have the signer go first which typically is a server. Of course, round complexity is the most important efficiency factor for an ID-based blind signature scheme, especially for some applications like E-voting and E-cash. And one-round is the optimal bound of round complexity. In fact, there are only four PKI-based blind signature schemes [7, 12, 16, 21] with an optimal two-move signature generation protocol. However, to the best of our knowledge, there exists no ID-based signature scheme with two-move signature generation protocol. This paper is motivated to construct the first one-round IDBS scheme without relying on the ROS assumption. To this end, we need to tackle three challenges here: (a) Since almost all ID-based signature schemes are constructed by using the proof of knowledge paradigm [5], it seems difficult to extend them into ID-based blind signature schemes with optimal round complexity [19, 24, 28, 29]; (b) The ID-based blind signature schemes constructed by Galindo et al.[17] need at least 4 moves (see Section 6 for more details); and (c) How to realize a concrete IDBS scheme whose security does not rely on ROS assumption (so essentially improves the results in [19, 28, 29]).

Our Contribution. In this paper, we propose the first one-round ID-based blind signature scheme without ROS assumption. This construction is based on bilinear pairings from scratch, i.e., it relies on new computational assumptions, new basic ID-based signature scheme and/or new blind signature scheme. In more details, our contribution can be summarized as follows. (1) The round complexity of our IDBS is optimal. Namely, each interactive signature generation requires the requesting user and the signer to transmit only one message to the other. (2) The provable security against generic parallel attack doesn't depend on the difficulty of ROS problem (see Definition 4 in Section 2). (3) To prove its security, we introduce a new plausible computational assumption, called *one-more bilinear Diffie-Hellman inversion assumption* (**1m-BDHI**, for short). This new assumption may be of independent interest, since other recently proposed computation assumptions in one-more flavor, such as one-more-RSA-inversion [3], one-more CDH [7], one-more discrete logarithm [4], have found many applications in provable security for blind signatures [3, 7], transitive signatures [4], identification protocols [2] and so on. (4) The underlying ID-based signature scheme may be of independent interest, since it avoids using the proof of knowledge paradigm and has a loose algebraic structure which already allows the efficient extension to blind signatures.

2 Preliminaries

In this section, we present the definitions of bilinear pairings and some relative assumptions.

Definition 1. *Let $\mathbb{G}_1$ and $\mathbb{G}_2$ be groups of prime order q and let P be a generator of $\mathbb{G}_1$. The map $e : \mathbb{G}_1 \times \mathbb{G}_1 \to \mathbb{G}_2$ is said to be an bilinear pairing if the following three conditions hold: (i) e is bilinear, i.e. $e(aP, bP) = e(P, P)^{ab}$ for all $a, b \in \mathbb{Z}_q$; (ii) e is non-degenerate, i.e. $e(P, P) \neq 1$; (iii) e is efficiently computable. Such a group $\mathbb{G}_1$ is called a bilinear group.*

Note that throughout this paper, without special descriptions, the groups $\mathbb{G}_1, \mathbb{G}_2$, the prime order q, the generator P of $\mathbb{G}_1$ and the bilinear pairing e are as defined in the above definition. Next, we review the following problems with respect to $(\mathbb{G}_1, \mathbb{G}_2, e, P, q)$:

- **Computational Diffie-Hellman (CDH) Problem**: Given random $P, aP, bP \in \mathbb{G}_1$, output $abP \in \mathbb{G}_1$, where $a, b \in_R \mathbb{Z}_q$.
- **Bilinear Diffie-Hellman (BDH) Problem** [8]: Given random $P, aP, bP, cP \in \mathbb{G}_1$, output $e(P,P)^{abc}$, where $a, b, c \in_R \mathbb{Z}_q$.
- **Generalized Tate Inversion (GTI) Problem** [20]: Given $h \in \mathbb{G}_2$, find a pair $(S,T) \in \mathbb{G}_1$ such that $e(S,T) = h$, where $e : \mathbb{G}_1 \times \mathbb{G}_1 \to \mathbb{G}_2$ denotes the Tate pairing.
- **Modified Generalized Bilinear Inversion (MGBI)**[1]: Given $h \in \mathbb{G}_2$ and the generator $P \in \mathbb{G}_1$, find a point $S \in \mathbb{G}_1$ such that $e(P,S) = h$, where e denotes the bilinear pairing.

Based on the above problems, we propose a new computational problem:

Definition 2 (Bilinear Diffie-Hellman Inversion (BDHI) Problem). *Given three random elements $aP, bP, cP \in \mathbb{G}_1$, compute two elements $S, T \in \mathbb{G}_1$ such that $e(S,T) = e(P,P)^{abc}$. Accordingly, the Bilinear Diffie-Hellman Inversion (BDHI) assumption states that: there is no PPT algorithm that can solve the BDHI problem with non-negligible probability.*

It is obvious that the BDH problem can be solved if the BDHI problem can be solved. And it is also obvious that the BDHI problem can solved if the CDH problem can be solved. So BDHI assumption is somewhere between CDH assumption and BDH assumption. That is, BDHI assumption is weaker than BDH assumption, but stronger than CDH assumption.

Furthermore, we propose another new computational assumption called one-more bilinear Diffie-Hellman Inversion (1m-BDHI) assumption. In fact, there exist many computational assumptions in the one-more flavor, such as One-more-RSA-inversion [3], one-more CDH [7], one more discrete logarithm [4]. These one-more assumptions can be used to prove security of many cryptographic schemes, such as the GQ identification scheme [2], blind signature schemes [4, 7], transitive signatures [3].

Definition 3 (1m-BDHI Assumption). *Let $e : \mathbb{G}_1 \times \mathbb{G}_1 \to \mathbb{G}_2$ be a bilinear pairing, where $\mathbb{G}_1$ and $\mathbb{G}_2$ be groups of prime order q and P be a generator of $\mathbb{G}_1$. Let x, y be random elements in $\mathbb{Z}_q$ and let $X = xP, Y = yP$. The adversary $\mathcal{A}$ is given $(e, \mathbb{G}_1, \mathbb{G}_2, q, P, X, Y)$ and has access to two oracles.*

- *The first one is a target oracle $\mathcal{TO}$ that, each time it is invoked (it takes no inputs), returns a random point from $\mathbb{G}_1$.*
- *The second one is the helper oracle $\mathcal{HO}$ which given $Z \in \mathbb{G}_1$, returns $S, T \in \mathbb{G}_1$ such that $e(S,T) = e(Y,Z)^x$. Additionally, this help oracle $\mathcal{HO}$ returns an auxiliary information piece R which can be used to check whether the equation $e(S,T) = e(Y,Z)^x$ holds. An example of the form of (R, S, T) used in this paper is given in the following remark.*

We say that $\mathcal{A}$ wins if its output is a sequence of points $S_1, T_1, \ldots, S_n, T_n \in \mathbb{G}_1$ satisfying $e(S_1, T_1) = e(Y, Z_1)^x, \ldots, e(S_n, T_n) = e(Y, Z_n)^x$, where all different $Z_1, \ldots, Z_n$ are random points returned by $\mathcal{TO}$ and the number of queries made by $\mathcal{A}$ to its helper oracle $\mathcal{HO}$, is strictly less than n. The 1m-BDHI advantage of $\mathcal{A}$, denoted $Adv_{\mathcal{A}}^{1m-BDHI}(k)$, is the probability that $\mathcal{A}$ wins, taken over the coins used in the generation of $(e, \mathbb{G}_1, \mathbb{G}_T, q, P, X, Y)$, the coins of $\mathcal{A}$, and the coins used by the target oracle across its invocations. We say that the one-more BDHI problem is hard if the function $Adv_{\mathcal{A}}^{1m-BDHI}(k)$ is negligible for all polynomial-time adversaries $\mathcal{A}$.

Remark 1: In this paper, a valid answer (R, S, T) of the helper oracle $\mathcal{HO}$ should satisfy:

$$e(R, S) = e(xP, yP), e(R, Z) = e(P, T).$$

Indeed, suppose that $R = rP$. Then the above two equations imply the following equations respectively:

$$S = r^{-1}xyP, T = rZ.$$

So we have $e(S, T) = e(yP, Z)^x$.

Finally, we review the ROS problem.

Definition 4 (ROS Problem [25]). *Given an oracle random function $F : \mathbb{Z}_q^l \to \mathbb{Z}_q$, find coefficients $a_{k,i} \in \mathbb{Z}_q$ and a solvable system of $l+1$ distinct equations (1) in the unknowns $c_1, c_2, \ldots, c_l$ over $\mathbb{Z}_q$:*

$$a_{k,1}c_1 + \ldots + a_{k,l}c_l = F(a_{k,1}, \ldots, a_{k,l}), \text{for } k = 1, 2, \ldots, t. \quad (1)$$

Accordingly, the ROS assumption states that: there is no PPT algorithm that can solve the ROS problem with non-negligible probability.

As Schnorr states, the intractability of the ROS problem is "a plausible but novel complexity assumption". At Crypto 2002, D. Wagner [27] claimed that the ROS problem can be broken in subexponential time. As argued in [28], to be resistant against this new attack, q may need to be at least 1600 bits long.

3 Frameworks of ID-Based Blind Signatures

Definition 5. *An identity-based blind signature scheme $\mathcal{IDBS}$ can be described as a collection of the following four algorithms (or protocols):*

- Setup. *This algorithm is run by the trusted party called PKG on input a security parameter, and generates the public parameters params of the scheme and a master secret. PKG publishes params and keeps the master secret to itself.*
- Extract. *Given an identity ID, the master secret and params, this algorithm generates the private key D_{ID} of ID.*

- Issue. *The signer blindly issues a signature for the user by this protocol, which is often divided into three sub-protocols or algorithms* (Blind, BSign, Unblind)*:*
  - Blind. *Given the message m and a random string r, it outputs the blinded message m' and sends it the signer. In this process, the user sometimes needs the interactive help from the signer.*
  - BSign. *Given the blinded message m' and the signer's private signing key D_{ID} as the input, it outputs a blind signature σ' and sends it to the user. This procedure may be an interactive sub-protocol between the user and the signer.*
  - Unblind. *Given a signature σ' and the previous used random string r, it outputs the unblinded signature σ.*
- Verify. *Given a signature σ, a message m, an identity ID and params, this algorithm outputs 1 if σ is a valid signature on m for identity ID, or 0 otherwise.*

The security of an ID-based blind signature scheme consists of two requirements: the blindness property and the unforgeability of additional signatures. We say a blind signature scheme is secure if it satisfies these two requirements.

Definition 6 (Blindness). *Let $\mathcal{A}$ be a probabilistic polynomial-time adversary which plays the role of the signer, $\mathcal{U}_0$ and $\mathcal{U}_1$ be two honest users. $\mathcal{U}_0$ and $\mathcal{U}_1$ engage in the blind signature issuing protocol with $\mathcal{A}$ on messages m_b and m_{1-b}, and output signatures σ_b and σ_{1-b}, respectively, where $b \in \{0,1\}$ is a random bit chosen uniformly. $(m_0, m_1, \sigma_b, \sigma_{1-b})$ are sent to $\mathcal{A}$ and then $\mathcal{A}$ outputs $b' \in \{0,1\}$. For all such $\mathcal{A}$, $\mathcal{U}_0$ and $\mathcal{U}_1$, for any constant c, and for sufficiently large n,*

$$|Pr[b = b'] - 1/2| < n^{-c}.$$

To define unforgeability, let us introduce the following game among the adversary $\mathcal{A}$ which plays the role of the user, and the challenger $\mathcal{C}$ which plays the role of the honest signer.

- Setup. The challenger $\mathcal{C}$ takes a security parameter 1^k and runs the algorithm Setup to generate common public parameters *params* and also the master secret key s. $\mathcal{C}$ sends *params* to $\mathcal{A}$.
- Queries. The adversary $\mathcal{A}$ can perform a polynomially bounded number of queries in a concurrent and interleaving way as follows.
  - Hash function query. If the security is analyzed in the random oracle model [6], $\mathcal{C}$ computes the values of the hash functions for the requested input and sends the values to $\mathcal{A}$.
  - Extract query. $\mathcal{A}$ chooses an identity ID and sends it to $\mathcal{C}$. $\mathcal{C}$ computes Extract$(ID) = D_{ID}$ and sends the result to $\mathcal{A}$.
  - Issue query. $\mathcal{A}$ chooses an identity ID, a plaintext m. To blindly obtain a signature on m with respect to ID, $\mathcal{A}$ engages in the blind signature issuing protocol with $\mathcal{C}$ in a concurrent and interleaving way.
- Forgery. $\mathcal{A}$ wins the game if $\mathcal{A}$ outputs n valid signatures (m_1, σ_1), $\ldots, (m_n, \sigma_n)$ with respect to the identity ID^* such that

- $m_i \neq m_j$ for any pair (i, j), where $i \neq j$ $i, j \in \{1, \ldots, n\}$.
- n is strictly larger than the number of the executions (with respect to the identity ID^*) of the protocol Issue between $\mathcal{C}$ and $\mathcal{A}$.
- $\mathcal{A}$ has not made an extract query on the identity ID^*.

The advantage $Adv_{IDBS}^{unforge}$ of $\mathcal{A}$ is defined as the probability that it wins the above game, taken over the coin tosses made by $\mathcal{C}, \mathcal{A}$, Setup. In the above attack model, $\mathcal{A}$ is called *one-more forger under parallel chosen message and ID attacks.*

Definition 7 (Unforgeability). *An adversary $\mathcal{A}$ (t, q_E, q_S, ϵ)-breaks an ID-based blind signature scheme, if (1) $\mathcal{A}$ runs in time at most t, (2) $\mathcal{A}$ queries private keys for at most q_E identities and execute at most q_S times the blind signature issuing protocol, (3) $Adv_{IDBS}^{unforge}$ is at least ϵ. We say an ID-based blind signature scheme is (t, q_E, q_S, ϵ)-secure against one-more forgery under parallel chosen message and ID attacks if no adversary $\mathcal{A}$ (t, q_E, q_S, ϵ)-breaks the scheme.*

Remark 2: In the forgery step of the above attack game, if $(m_i, \sigma_i) \neq (m_j, \sigma_j)$ instead of $m_i \neq m_j$ holds for message-signature pairs output by the adversary, then we get the definition of the strong unforgeability of blind signature schemes. As mentioned in [10], for the main application of blind signatures, i.e., electronic cash, unforgeability (rather than strong unforgeability) suffices.

In fact, the above forger $\mathcal{A}$ against ID-based blind signatures is the natural analogy of the one-more forger under parallel attack which is the most powerful attack for blind signatures. Unfortunately, before our schemes, there is no ID-based blind signature scheme based on bilinear pairings which can be proved secure in this model.

4 Construction

Our proposed scheme is described as follows:

- Setup. The Private Key Generator (PKG) generates parameters and master keys as follows:
  - generates groups $\mathbb{G}_1$ and $\mathbb{G}_2$ of prime order q with bilinear pairing $e : \mathbb{G}_1 \times \mathbb{G}_1 \rightarrow \mathbb{G}_2$;
  - chooses an arbitrary generator $P \in \mathbb{G}_1$;
  - picks a random $s \in \mathbb{Z}_q$ and sets $P_{pub} = sP$;
  - chooses cryptographic hash functions $H_1, H_2 : \{0,1\}^* \rightarrow \mathbb{G}_1$. The PKG's public parameter is $params = (\mathbb{G}_1, \mathbb{G}_2, e, q, P, P_{pub}, H_1, H_2)$; its master secret is $s \in \mathbb{Z}_q$.
- Extract. The signer with identity ID receives the value $D_{ID} = sQ_{ID}$ from the PKG as its private key, where $Q_{ID} = H_1(ID) \in \mathbb{G}_1$.
- Issue.
  - Blind. The user randomly chooses a number $r_1 \in \mathbb{Z}_q$ as the blinding factor, computes $P'_m = r_1 H_2(m)$ and sends it to the signer.
  - BSign. The signer sends back (A', B', C'), where $A' = x_{ID} P'_m, B' = x_{ID}^{-1} D_{ID}, C' = x_{ID} P, x_{ID} \xleftarrow{R} \mathbb{Z}_q$.

- Unblind. First, the user verifies the blind signature (A', B', C') by checking whehter
$$e(A', P) = e(P'_m, C'), e(Q_{ID}, P_{pub}) = e(B', C').$$
Next, the user selects a random number $r_2 \in \mathbb{Z}_q$ and computes the signature as (A, B, C), where $A = r_2 r_1^{-1} A', B = r_2^{-1} B', C = r_2 C'$.

– Verify. Let (A, B, C) be the signature on the message m and $P_m = H_2(m)$. The verifier checks that:

$$e(A, P) = e(P_m, C), e(Q_{ID}, P_{pub}) = e(B, C).$$

Correctness. If an entity with identity ID blindly issues a signature $\sigma = (A, B, C)$ on a message m to a user as described in the Issue protocol above, it is easy to see that σ will be accepted by a verifier:

$$\begin{aligned} e(A, P) &= e(r_2 r_1^{-1} A', P) = (r_2 r_1^{-1} x_{ID} P'_m, P) \\ &= e(r_2 r_1^{-1} x_{ID} r_1 P_m, P) \\ &= e(r_2 x_{ID} P_m, P) = e(P_m, r_2 x_{ID} P) \\ &= e(P_m, r_2 C') \\ &= e(P_m, C), \\ e(B, C) &= e(r_2^{-1} B', r_2 C') \\ &= e(B', C') \\ &= e(x_{ID}^{-1} D_{ID}, x_{ID} P) \\ &= e(D_{ID}, P) = e(Q_{ID}, sP) \\ &= e(Q_{ID}, P_{pub}). \end{aligned}$$

Similarly, we can see that the blind signature generated by the honest signer in Bsign must be accepted by the user in the step Unblind.

Remark 3: After we submitted this work to ePrint (archive 2007/007), Sherman S.M. Chow informed us that our above IDBS scheme shares some similar ideas with the ID-based signature scheme [14]. As we stated in the first section, our proposal is motivated by solving some open problems related to ID-based blind signature schemes, while the similar result in [14] is proposed as one of applications of the so-called *verifiable pairing*.

5 Security

First, we claim that our scheme has the *blindness property*. This is obvious since the signer receives only random elements in $\mathbb{G}_1$ which are independent of the outputs of the user. In fact, as we stated in the first section, the idea of our proposal is motivated by solving some open problems relative to ID-based blind signature schemes.

Theorem 1. *The proposed ID-based blind signature scheme is blind.*

Proof. The blindness property will be proved according to Definition 6. We assume that when the signature $\sigma_b = (A_b, B_b, C_b)$ on the message m_b (resp. $\sigma_{1-b} = (A_{1-b}, B_{1-b}, C_{1-b})$ on m_{1-b}) is generated, the user $\mathcal{U}_0$ (resp. $\mathcal{U}_1$) sends P'_{m_b} (resp. $P'_{m_{1-b}}$) to the adversary $\mathcal{A}$ which then returns the blinded signature $\sigma'_b = (A'_b, B'_b, C'_b)$ (resp. $\sigma'_{1-b} = (A'_{1-b}, B'_{1-b}, C'_{1-b})$).

For σ_b, if we can prove that there exist two integers $r'_1, r'_2 \in \mathbb{Z}_q$ such that

$$P'_{m_{1-b}} = r'_1 H_2(m_b), A_b = r'_2 {r'}_1^{-1} A'_{1-b}, B_b = {r'}_2^{-1} B'_{1-b}, C_b = r'_2 C'_{1-b},$$

then it is obtained that for the adversary, σ_b may be linked to the process relative to the messages $(P'_{m_{1-b}}, A'_{1-b}, B'_{1-b}, C'_{1-b})$ and the user $\mathcal{U}_1$. In other words, the adversary $\mathcal{A}$ can not determine which of the two user generated the signature σ_b.

In fact, since (A_b, B_b, C_b) and $(A'_{1-b}, B'_{1-b}, C'_{1-b})$ are valid, we have

$$e(A_b, P) = e(P_{m_b}, C_b), e(Q_{ID}, P_{pub}) = e(B_b, C_b);$$

$$e(A'_{1-b}, P) = e(P'_{m_{1-b}}, C'_{1-b}), e(Q_{ID}, P_{pub}) = e(B'_{1-b}, C'_{1-b}).$$

Let $c_b, c'_{1-b} \in \mathbb{Z}_q$ be integers satisfying $C_b = c_b P$, $C'_{1-b} = c'_{1-b} P$ respectively. By the bilinear property of the pairing, then we have

$$A_b = c_b P_{m_b}, B_b = c_b^{-1} s Q_{ID};$$

$$A'_{1-b} = c'_{1-b} P'_{m_{1-b}}, B'_{1-b} = {c'}_{1-b}^{-1} s Q_{ID}.$$

Let r'_1, r'_2 be integers satisfying $C_b = r'_2 C'_{1-b}$ (i.e. $r'_2 = c_b {c'}_{1-b}^{-1} \mod q$) and $P'_{m_{1-b}} = r'_1 P_{m_b} (= r'_1 H_2(m_b))$ respectively , then they also satisfy

$$A_b = r'_2 {r'}_1^{-1} A'_{1-b}, B_b = {r'}_2^{-1} B'_{1-b}.$$ □

Next, we analyze the unforgeability of our scheme as follows. Here note that it is obvious that our blind signature scheme is not strongly unforgeable (see Remark 2 in Section 3). Instead, we will prove that its security satisfies the standard definition given in Section 3. As in [11], the proof is divided into two steps.

Consider the following variant of the attacking game for unforgeability in Section 3. First we fix an identity ID^*. In Setup Step, $\mathcal{C}$ gives to $\mathcal{A}$ system parameters together with ID^*, and in Step Forgery, $\mathcal{A}$ must output the given ID^* (together with n pairs (m_i, σ_i)) as its final result. If no polynomial time algorithm $\mathcal{A}$ has non-negligible advantage in this game, we say that the blind signature scheme is secure against *one-more forgery under parallel chosen message and given ID attacks.* The first step of our proof is to reduce the problem to this case.

Lemma 1. *For our scheme, if there is a one-more forger $\mathcal{A}_0$ under a parallel chosen message and ID attack with running time t_0 and advantage ϵ_0, then there is a one-more forger $\mathcal{A}_1$ under a parallel chosen message and given ID attack, which has running time $t_1 \le t_0$ and advantage $\epsilon_1 \ge \epsilon_0(1-\frac{1}{q})/q_{H_1}$, where q_{H_1} is the maximum number of queries to H_1 asked by $\mathcal{A}_0$. In addition, the numbers of queries to hash functions,* Extract, *and* Issue *asked by $\mathcal{A}_1$ are the same as those of $\mathcal{A}_0$.*

Proof. Without any loss of generality, we can assume that for any ID, $\mathcal{A}_0$ queries $H_1(ID)$ and $\mathsf{Extract}(ID)$ at most once. Let the fixed identity for $\mathcal{A}_1$ be ID^*. Our algorithm $\mathcal{A}_1$ is as follows:

- Choose $r \in \{1, \ldots, q_{H_1}\}$ randomly. Denote by ID_i the input of the i-th query to H_1 asked by $\mathcal{A}_0$. Let ID_i' be ID^* if $i = r$, and ID_i otherwise. Define $H_1'(ID_i), \mathsf{Extract}'(ID_i), \mathsf{Issue}'(ID_i, m)$ to be $H_1(ID_i')$, $\mathsf{Extract}(ID_i')$, $\mathsf{Issue}(ID_i', m)$, respectively.
- Run $\mathcal{A}_0$ with the given system parameters. $\mathcal{A}_1$ responds to $\mathcal{A}_0$'s queries to H_1, H_2, $\mathsf{Extract}$, and Issue by evaluating H_1', H_2, $\mathsf{Extract}'$, and Issue', respectively. Let the output of $\mathcal{A}_0$ be n valid signatures $(m_1, \sigma_1), \ldots, (m_n, \sigma_n)$ with respect to ID_{out}, where n is strictly larger than the number of executions of the Issue' protocol.
- If $ID_{out} = ID^*$, then output n valid signatures $(m_1, \sigma_1), \ldots, (m_n, \sigma_n)$ together with the corresponding identity ID^*. Otherwise output *fail*.

Since the distributions produced by $H_1', \mathsf{Extract}'$, and Issue' are indistinguishable from those produced by $H_1, \mathsf{Extract}$, and Issue of our scheme, $\mathcal{A}_0$ learns nothing from query results, and hence

$$Pr[\mathcal{A}_0 \ succeeds] \geq \epsilon_0.$$

Since H_1 is a random oracle, if $\mathcal{A}_0$ has not made the the query $H_1'(ID_{out})$, the probability that the $\mathcal{A}_0$'s output is valid is negligible. Explicitly,

$$Pr[ID_{out} = ID_i \ for \ some \ i | \mathcal{A}_0 \ succeeds] \geq 1 - \tfrac{1}{q}.$$

Since r is independently and randomly chosen, we have

$$Pr[ID_{out} = ID_r = ID^* | ID_{out} = ID_i \ for \ some \ i] \geq \tfrac{1}{q_{H_1}}$$

Combining these,

$$Pr[\mathcal{A}_1 \ succeeds] \geq \epsilon_0 (1 - \tfrac{1}{q}) \tfrac{1}{q_{H_1}}$$

as desired. □

Lemma 2. *For our scheme, if there is a one-more forger $\mathcal{A}$ under a parallel chosen message and given ID attack with running time t_1 and advantage ϵ_1, then there is an adversary $\mathcal{B}$ attacking the one-more BDHI problem, which has running time $t_2 \leq t_1 + 4c_{\mathbb{G}_1}(q_{H_1} + q_{H_2} + q_S + q_E)$ and advantage $\epsilon_2 \geq \epsilon_1$, where $c_{\mathbb{G}_1}$ is a constant that depends on $\mathbb{G}_1$, and $q_{H_1}, q_{H_2}, q_E, q_S$ are the numbers of queries to the hash functions H_1, H_2,* $\mathsf{Extract}$*, and* Issue *asked by $\mathcal{A}_1$ respectively.*

Proof. Suppose that $\mathcal{A}$ is a one-more forger against our scheme under a parallel chosen message and given ID attack. We describe the algorithm $\mathcal{B}$ which will simulate the challenger for $\mathcal{A}$ in order to solve the one-more BDHI problem. The adversary $\mathcal{B}$ is given $(e, \mathbb{G}_1, \mathbb{G}_2, q, P, X, Y)$, the target oracle and the helper oracle. $\mathcal{B}$ simulates the challenger and interacts with forger $\mathcal{A}$ as follows.

- Setup. $\mathcal{B}$ first provides $\mathcal{A}$ with the public parameter $(e, \mathbb{G}_1, \mathbb{G}_2, q, P, P_{pub})$ and the fixed identity ID^*, where $P_{pub} = X$.
- H_1-queries. To respond to these queries, $\mathcal{B}$ maintains a list of tuples $(ID_i, H_1(ID_i), r_i)$ as explained below. We refer to this list as H_1-list. The list is initially empty. When $\mathcal{A}$ queries the oracle H_1 at an identity ID_i, $\mathcal{B}$ responds as follows.
  - If the query ID_i appears on the H_1-list in a tuple $(ID_i, H_1(ID_i), r_i)$ (or $(ID_i, H_1(ID_i), *)$), then $\mathcal{B}$ responds with $H_1(ID_i)$.
  - If $ID_i = ID^*$, $\mathcal{B}$ sets $H_1(ID_i) = Y$ and sends it to $\mathcal{A}$. Additionally, $\mathcal{B}$ appends the tuple $(ID_i, H_1(ID_i), *)$ to the H_1-list.
  - If $ID_i \neq ID^*$, $\mathcal{B}$ randomly selects $r_i \in \mathbb{Z}_q$ and sends $H_1(ID_i) = r_iP$ to $\mathcal{A}$. Additionally, $\mathcal{B}$ appends the tuple $(ID_i, H_1(ID_i), r_i)$ to the H_1-list.

  Since H_1 is a random oracle, $\mathcal{A}$ obtains no information on $H_1(ID)$ before he queries the H_1-oracle on ID. So, without loss of generality, we assume that $\mathcal{A}$ has already queried the H_1 oracle on an identity ID before he makes the issue query or extract query with respect to the ID.
- H_2-queries. When given the new query m_j, that is distinct from the previous hash queries, $\mathcal{B}$ obtains a point $Z_j \in \mathbb{G}$ as the hash value $H_2(m_j)$ from its target oracle $\mathcal{TO}$ and sends it to $\mathcal{A}$.
- Extract queries. Suppose that $\mathcal{A}$ makes an extract query on the identity $ID_i \neq ID^*$. Let $(ID_i, H_1(ID_i), r_i)$ be the tuple on the H_1-list containing ID_i. $\mathcal{B}$ answers this query by sends to $\mathcal{A}$ $D_{ID_i} = r_iX$. By assuming $X = xP$ for some unknown x, it is obvious that $D_{ID_i} = xH_1(ID_i) = r_iX$, since $H_1(ID_i) = r_iP$.
- Issue queries. Assume that $\mathcal{A}$ chooses the identity ID_i and the plaintext m_i and wants to blindly obtain the signature on m_i with respect to the identity ID_i. Note that the signer has only one move in the Issue protocol. Let P'_{m_i} be the blinded message that $\mathcal{A}$ sends to $\mathcal{B}$. $\mathcal{B}$ answer this query as follows.
  - If $ID_i \neq ID^*$, $\mathcal{B}$ computes the private key $D_{ID_i} = r_iX$, where $(ID_i, H_1(ID_i), r_i)$ is the corresponding tuple on the H_1-list. Then $\mathcal{B}$ uses the private key D_{ID_i} to compute the corresponding blinded signature as in BSign.
  - If $ID_i = ID^*$, $\mathcal{B}$ sends P'_{m_i} to its helper oracle $\mathcal{HO}$. Let (R_i, S_i, T_i) be the corresponding answer. $\mathcal{B}$ sets the blinded signature as (A'_i, B'_i, C'_i), where $A' = T_i, B'_i = S_i, C'_i = R_i$. It is obvious that this simulated signature is valid (see remark 1 in Section 2 and the algorithm Verifiy in Section 4).
- Outputs. At last, $\mathcal{A}$ outputs a list of message-signature pairs $((m_1, (A_1, B_1, C_1)), \ldots, (m_n, (A_n, B_n, C_n))$ with respect to the identity ID^*, where n is strictly larger than the number of executions of the protocol Issue with respect to the identity ID^*, and hence strictly larger than the number of queries made by $\mathcal{B}$ to its helper oracle $\mathcal{HO}$. $\mathcal{B}$ outputs $A_1, B_1, A_2, B_2, \ldots, A_n, B_n$. Here note that a valid signature (A_i, B_i, C_i) satisfies $e(A_i, B_i) = e(H_1(ID^*), H_2(m_i))^x = (Y, H_2(m_i))^x$ (see Remark 1 in Section 2), and $H_2(m_i)$ is obtained from the target oracle. So the one-more BDHI problem is solved by $\mathcal{B}$.

It is easy to see that the view of $\mathcal{A}$ in the simulated experiment is indistinguishable from its view in the real experiment, and that $\mathcal{B}$ is successful only if $\mathcal{A}$ is successful. Thus, the probability ϵ_2 that $\mathcal{B}$ succeeds is at least the probability ϵ_1 that $\mathcal{A}$ succeeds. Algorithm $\mathcal{B}$'s running time is the same as $\mathcal{A}$'s running time plus the time it takes to respond to q_{H_1} H_1-hash queries, q_{H_2} H_2-hash queries, q_E extract queries and q_S signature issue queries. Each query requires at most four exponentiations (corresponding to issue queries for $ID_i \neq ID^*$) in $\mathbb{G}_1$ which we assume takes time $c_{\mathbb{G}_1}$. Hence, the total running time t_2 is at most $t_1 + 4c_{\mathbb{G}_1}(q_{H_1} + q_{H_2} + q_S + q_E)$ as required. This completes the proof of Theorem 1. □

Combing the above lemmas, we obtain the following theorem:

Theorem 2. *If the one-more BDHI assumption is true in the group $\mathbb{G}_1$, then the proposed ID-based blind signature scheme is secure against one-more forgery under parallel chosen message and ID attacks in the random oracle model.*

6 A Comparison of ID-Based Blind Signatures

In this Section, we give a brief comparison of ID-based blind signatures (IDBS) (see Table 1 below). The purpose is to show the advantages of our scheme compared with existing solutions. Namely, as we claimed before, the proposed scheme is first one-round ID-base blind signature scheme, which is secure against generic parallel attack without relying on the intractability of ROS problem.

Table 1. A Comparison of ID-based Blind Signatures

Schemes	Signer	User	Verifier	Move	Security Model
Ours	3M	4M+4e	4e	2	ROM+1m-BDHI
ZK02 [28]	3M	3M+3e	1E+2e	3	ROM+CDH+ROS(?)
ZK03 [29]	2M	4M+2e	1M+2e	3	ROM+CDH+ROS(?)
HCW05 [19]	2M+1e	1M+3E+3e	1M+2e	3	ROM+CDH+ROS(?)
Schnorr [22, 25]	1E	3E	2E	4	ROM+DLP+ROS
Chaum [12, 17]	1E	2E	2E	4	ROM+1m-RSA
Boldyreva [7, 17]	1M	2M+4e	4e	4	ROM+1m-CDH
CKW04 [10, 17]	25E	38E	2E	10	SM+CRS+SRSA+Seqn.
KZ05 [17, 21]	5M+10E+6e	7M+15E+18e	1M+6e	6	SM+CRS+DLDH+LRSW
Okamoto [17, 23]	6M+3E	10M+5E+4e	3M+4e	6	SM+CRS+DCR+2SDH
Fischlin [16, 17]	1E	NIZK	NIZK	4	SM+CRS+GC

First of all, we remark that the first four schemes (including our construction) in Table 1 are explicit IDBS schemes, while all other schemes are deduced from the underlying blind signatures by using the certificate-based generic transformation [17], which extends the result given in [5]. More specifically, we get these ID-based blind signature schemes from the corresponding blind signatures [7, 10, 12, 16, 22, 23, 25]. Galindo et al.'s generic approach transforms a standard blind signature scheme into an ID-based blind signature scheme as follows. The

PKG selects a key pair (sk_i, pk_i) for a signer ID_i, issues a certificate $Cert_i$ to certify the string $ID_i||pk_i$ by using the PKG's PKI-based secret key, and then forwards $(sk_i, Cert_i)$ to the signer ID_i. To get an ID-based blind signature, a user first enquires the signer ID_i for its $Cert_i$ and checks the validity of $Cert_i$. If this procedure is successful, the user and the signer can engage in the stand blind signature issuing protocol to output a signature σ for a message m under public key pk_i. The final ID-based signature is a pair $(\sigma, Cert_i)$, which is valid if $Cert_i$ is a certificate for ID_i together with some public key pk_i issued by the PKG, and σ is a valid signature for message m with respect to pk_i.

As the main computational overheads, we only consider modular exponentiations (denote by **E**), scalar multiplications (denote by **M**), and bilinear mappings (denote by **e**). Since simultaneous exponentiations can be efficiently carried out by means of an exponent array, for simplicity, we treat the cost for $a_1^{x_1}a_2^{x_2}$ or $a_1^{x_1}a_2^{x_2}a_3^{x_3}$ as just one single exponentiation. To count the computational costs of the signer, user and verifier in the above deduced IDBS schemes, we assume the PKG use a similar underlying signature to issue certificates for signers. That is, the PKG uses Schnorr signature in the ID-based blind Schnorr signature [25], the RSA signature with a full domain hash in the ID-based Chaum [12] and CKW04 [10] blind signature schemes, and the BLS short signature [9] in the ID-based Boldyreva [7], KZ [21], and Okamoto [23] blind signature schemes. For the generic scheme proposed by Fishlin [16], there are no concrete values since his scheme relies on general NIZK to prove the correctness of a ciphertext. Due to the usage of certificates in Galindo et al.'s approach, the round complexity, the communication complexity and the signature size are also increased in all deduced IDBS schemes. For example, though the standard blind signature schemes in [7, 12, 16] are round-optimal (i.e., they are one-round or 2-moves solutions), the correspond ID-based blind signatures become 4-move schemes.

About the security model, we mainly consider the following aspects: (1) whether a scheme is secure in the random oracle model (ROM) or standard model (SM); (2) whether a scheme needs common reference string (CRS); (3) whether a scheme relies on the ROS assumption; and (4) what are the computational assumptions required. According to Table 1, we can see that the last four schemes are all provably secure in the standard model but need common reference strings. At the same time, these schemes are not very efficient, since in the blind signature issuing protocols some kinds of ZK proofs are involved. In addition, note that the CKW04 scheme is only claimed to be secure in the scenario of sequential attacks (weaker than generic parallel attacks). The directly constructed schemes in [19, 28, 29] are computationally efficient, but their security against one-more forgery is not formally proved even under the ROS assumption. Due to this reason, we make a question mark "?" to these three schemes under the column of "Security Model". Based on the result in [17, 22, 25], the ID-based Schnorr blind signature scheme is secure against one-more forgery, but needs the ROS assumption, which results the loss of practical efficiency since to guarantee the security one has to select q as large as 1600 bits. Compared with efficient ID-based blind signatures deduced from [7, 12], our scheme is round-optimal (i.e. two moves rather than 4 moves) and has shorter signatures (without using a certificate to binding a random public key with each signer).

According to the above discussion, we can conclude that the proposed scheme is the first one-round ID-based blind signature, which is provably secure against generic parallel attack without relying on the ROS problem and any set-up assumptions, in the the random oracle model. Compared with ID-based blind scheme deduced from Galindo et al.'s generic transformation, which can be secure in the stand model, our solution is much more efficient in all aspects of round complexity, computational complexity, and signature size.

Remark 4: We are especially grateful to one reviewer of Pairing 2008 who pointed out that one IDBS scheme from the generic construction [17], with some improvements, could also be round-optimal and more efficient than our scheme. However, we remark that such a scheme with these potential improvements is not mentioned in [17], though this scheme mentioned by the referee is very interesting and deserves further study. In addition, as stated in Section 1, both the motivation and method of our work are different from that in [17].

7 Other Considerations

First, the new formalized 1m-BDHI assumption may be of independent interest, since other recently proposed computation assumptions in one-more flavor, such as one-more-RSA-inversion [3], one-more CDH [7], one-more discrete logarithm [4], have found many applications in provable security for blind signatures [3, 7], transitive signatures [4], identification protocols [2] and so on.

Second, the underlying ID-based signature scheme may be of independent interest, since it avoids to use the proof of knowledge paradigm and has a loose algebraic structure which already allows the efficient extension to blind signatures. In fact, the underlying ID-based signature scheme is not strongly unforgeable, but satisfy the well-known standard definition of unforgeability. However, a non-strongly unforgeable signature may have other advantages over the strongly unforgeable one. For example, the authors of [18] constructed the first constant-length ID-based aggregate signature scheme based on an non-strongly unforgeable ID-based signature scheme.

8 Conclusion

In this paper, we proposed a new identity-based blind signature scheme based on bilinear pairings, which contributes the first one-round identity-based blind signature without the ROS assumption. This means that the proposed construction is not only optimal in the sense of round complexity, but also practically efficient in contrast to the existing solutions [19, 28, 29], which are actually inefficient and rely on the difficulty of ROS problem, and other potential schemes that can be deduced from a generic result [17]. In addition, we showed that the proposed scheme is provably secure in the random oracle model, under the one-more bilinear Diffie-Hellman inversion (1m-BDHI) assumption, a new computational assumption introduced in this work.

Acknowledgements. The authors thank the anonymous referees of Pairing 2008 for their very helpful comments, and Sherman S.M. Chow for informing us his work in [14].

References

1. Baek, J., Zheng, Y.: Identity-based Threshold Signature Scheme From the Bilinear Pairings. In: Proc. of IAS 2004 track of ITCC 2004, pp. 124–128. IEEE Computer Society, Los Alamitos (2004)
2. Bellare, M., Palacio, A.: GQ and Schnorr Identification Schemes: Proofs of Security against Impersonation under Active and Concurrent Attack. In: Yung, M. (ed.) CRYPTO 2002. LNCS, vol. 2442, pp. 162–177. Springer, Heidelberg (2002)
3. Bellare, M., Namprempre, C., Pointcheval, D., Semanko, M.: The Power of RSA Inversion Oracles and the Security of Chaum's RSA-Based Blind Signature Scheme. In: Syverson, P.F. (ed.) FC 2001. LNCS, vol. 2339, pp. 319–338. Springer, Heidelberg (2002)
4. Bellare, M., Neven, G.: Transitive Signatures Based on Factoring and RSA. In: Zheng, Y. (ed.) ASIACRYPT 2002. LNCS, vol. 2501, pp. 397–414. Springer, Heidelberg (2002)
5. Bellare, M., Namprempre, C., Neven, G.: Security Proofs for Identity-Based Identification and Signature Schemes. In: Cachin, C., Camenisch, J.L. (eds.) EUROCRYPT 2004. LNCS, vol. 3027, pp. 268–286. Springer, Heidelberg (2004)
6. Bellare, M., Rogaway, P.: Random Oracles Are Practical: a Paradigm for Designing Efficient Protocols. In: Proc. of the 1st CCS, pp. 62–73. ACM Press, New York (1993)
7. Boldyreva, A.: Efficient Threshold Signatures, Multisignatures and Blind Signatures Based on the Gap-Diffie-Hellman-group Signature Scheme. In: Desmedt, Y.G. (ed.) PKC 2003. LNCS, vol. 2567, pp. 31–46. Springer, Heidelberg (2002)
8. Boneh, D., Franklin, M.: Identity-based Encryption from the Weil Pairing. In: Kilian, J. (ed.) CRYPTO 2001. LNCS, vol. 2139, pp. 213–229. Springer, Heidelberg (2001)
9. Boneh, D., Lynn, B., Shacham, H.: Short Signatures from the Weil Pairing. In: Boyd, C. (ed.) ASIACRYPT 2001. LNCS, vol. 2248, pp. 514–532. Springer, Heidelberg (2001)
10. Camenisch, J., Koprowski, M., Warinschi, B.: Efficient Blind Signatures Without Random Oracles. In: Blundo, C., Cimato, S. (eds.) SCN 2004. LNCS, vol. 3352, pp. 134–148. Springer, Heidelberg (2005)
11. Cha, J.C., Cheon, J.H.: An Identity-based Signature from Gap Diffie-Hellman Groups. In: Desmedt, Y.G. (ed.) PKC 2003. LNCS, vol. 2567, pp. 18–30. Springer, Heidelberg (2002)
12. Chaum, D.: Blind Signatures for Untraceable Payments. In: Proc. of Advances in Cryptology - CRYPTO 1982, pp. 199–203. Plenum Press, New York (1983)
13. Cheon, J.H.: Security Analysis of the Strong Diffie-Hellman Problem. In: Vaudenay, S. (ed.) EUROCRYPT 2006. LNCS, vol. 4004, pp. 1–11. Springer, Heidelberg (2006)

14. Chow, S.S.M.: Verifiable Pairing and its Applications. In: Lim, C.H., Yung, M. (eds.) WISA 2004. LNCS, vol. 3325, pp. 170–187. Springer, Heidelberg (2005)
15. Dutta, R., Barua, R., Sarkar, P.: Pairing-based Cryptography: a Survey. IACR preprint sever, submission 2004/064 (2004)
16. Fischlin, M.: Round-Optimal Composable Blind Signatures in the Common Reference String Model. In: Dwork, C. (ed.) CRYPTO 2006. LNCS, vol. 4117, pp. 60–77. Springer, Heidelberg (2006)
17. Galindo, D., Herranz, J., Kiltz, E.: On the Generic Construction of Identity-Based Signatures with Additional Properties. In: Lai, X., Chen, K. (eds.) ASIACRYPT 2006. LNCS, vol. 4284, pp. 178–193. Springer, Heidelberg (2006)
18. Gentry, C., Ramzan, Z.: Identity-Based Aggregate Signatures. In: Yung, M., Dodis, Y., Kiayias, A., Malkin, T. (eds.) PKC 2006. LNCS, vol. 3958, pp. 257–273. Springer, Heidelberg (2006)
19. Huang, Z., Chen, K., Wang, Y.: Efficient Identity-Based Signatures and Blind Signatures. In: Desmedt, Y.G., Wang, H., Mu, Y., Li, Y. (eds.) CANS 2005. LNCS, vol. 3810, pp. 120–133. Springer, Heidelberg (2005)
20. Joux, A.: The Weil and Tate Pairings as Building Blocks for Public Key Cryptosystems. In: Fieker, C., Kohel, D.R. (eds.) ANTS 2002. LNCS, vol. 2369, pp. 20–32. Springer, Heidelberg (2002)
21. Kiayias, A., Zhou, H.: Two-Round Concurrent Blind Signatures without Random Oracles. In: De Prisco, R., Yung, M. (eds.) SCN 2006. LNCS, vol. 4116, pp. 49–62. Springer, Heidelberg (2006)
22. Pointcheval, D., Stern, J.: Security Arguments for Digital Signatures and Blind Signatures. Journal of Cryptology 13(3), 361–396 (2000)
23. Okamoto, T.: Efficient Blind and Partially Blind Signatures Without Random Oracles. In: Halevi, S., Rabin, T. (eds.) TCC 2006. LNCS, vol. 3876, pp. 80–99. Springer, Heidelberg (2006)
24. Qiu, W.: Converting Normal DLP-based Signatures into Blind. Applied Mathematics and Computation 170(1), 657–665 (2005)
25. Schnorr, C.P.: Security of Blind Discrete Log Signatures against Interactive Attacks. In: Qing, S., Okamoto, T., Zhou, J. (eds.) ICICS 2001. LNCS, vol. 2229, pp. 1–12. Springer, Heidelberg (2001)
26. Shamir, A.: Identity-based Cryptosystems and Signature Schemes. In: Blakely, G.R., Chaum, D. (eds.) CRYPTO 1984. LNCS, vol. 196, pp. 47–53. Springer, Heidelberg (1985)
27. Wagner, D.: A Generalized Birthday Problem. In: Yung, M. (ed.) CRYPTO 2002. LNCS, vol. 2442, pp. 288–303. Springer, Heidelberg (2002)
28. Zhang, F., Kim, K.: ID-based Blind Signature and Ring Signature from Pairings. In: Zheng, Y. (ed.) ASIACRYPT 2002. LNCS, vol. 2501, pp. 533–547. Springer, Heidelberg (2002)
29. Zhang, F., Kim, K.: Efficient ID-based Blind Signature and Proxy Signature from Bilinear Pairings. In: Safavi-Naini, R., Seberry, J. (eds.) ACISP 2003. LNCS, vol. 2727, pp. 312–323. Springer, Heidelberg (2003)

Tracing Malicious Proxies in Proxy Re-encryption

Benoît Libert[1] and Damien Vergnaud[2]

[1] Université Catholique de Louvain, Crypto Group
Place du Levant, 3 – 1348 Louvain-la-Neuve – Belgium
[2] École normale supérieure – C.N.R.S. – I.N.R.I.A.
45, Rue d'Ulm – 75230 Paris CEDEX 05 – France

Abstract. In 1998, Blaze, Bleumer and Strauss put forth a cryptographic primitive, termed *proxy re-encryption*, where a semi-trusted proxy is given some piece of information that enables the re-encryption of ciphertexts from one key to another. Unidirectional schemes only allow translating from the delegator to the delegatee and not in the opposite direction. In all constructions described so far, although colluding proxies and delegatees cannot expose the delegator's long term secret, they can derive and disclose sub-keys that suffice to open all translatable ciphertexts sent to the delegator. They can also generate new re-encryption keys for receivers that are not trusted by the delegator. In this paper, we propose *traceable proxy re-encryption* systems, where proxies that leak their re-encryption key can be identified by the delegator. The primitive does not preclude illegal transfers of delegation but rather strives to deter them. We give security definitions for this new primitive and a construction meeting the formalized requirements. This construction is fairly efficient, with ciphertexts that have logarithmic size in the number of delegations, but uses a non-black-box tracing algorithm. We discuss how to provide the scheme with a black box tracing mechanism at the expense of longer ciphertexts.

Keywords: unidirectional proxy re-encryption, transferability issues, collusion detection and traceability.

1 Introduction

Ten years ago, Blaze, Bleumer and Strauss proposed a cryptographic primitive called *proxy re-encryption* (PRE), in which a proxy transforms – without being able to infer any information on the corresponding plaintext – a ciphertext computed under Alice's public key into one that can be opened using Bob's secret key. In all known constructions, if Bob and a malicious proxy cooperate, they can derive new re-encryption keys without Alice's consent. The purpose of this paper is to coin a new notion, that we call *traceable proxy re-encryption* (TPRE) in which such misbehaving proxies can be identified by the delegator. We formalize security notions for this new primitive and give a reasonably efficient construction fitting this model under appropriate complexity assumptions.

S.D. Galbraith and K.G. Paterson (Eds.): Pairing 2008, LNCS 5209, pp. 332–353, 2008.

Related work. Blaze *et al.* [8] proposed the first PRE scheme, where plaintexts and secret keys remain hidden from the proxy. Unfortunately, their scheme has inherent limitations: the proxy key also allows translating ciphertexts from Bob to Alice, which may be undesirable, and the proxy and the delegatee can collude to expose the delegator's private key.

In 2005, Ateniese, Fu, Green and Hohenberger [4,5] showed how to construct *unidirectional* schemes using bilinear maps and simultaneously prevent proxies from colluding with delegatees in order to expose the delegator's long term secret. Their schemes involve two distinct encryption algorithms: *first-level* encryptions are not translatable whilst *second-level* encryptions can be re-encrypted by proxies into ciphertexts that are openable by delegatees. Let $(\mathbb{G}_1, \mathbb{G}_2, \mathbb{G}_T, e, \psi)$ be a cryptographic bilinear structure (denoted multiplicatively) of prime order p and let g be a generator of $\mathbb{G}_1$ (see § 2.2 for a definition). Alice and Bob publish the public keys $y_A = g^a$ and $y_B = g^b$ (respectively) and keep secret their discrete logarithms a and b. To encrypt a message $m \in \mathbb{G}_T$ to Alice at the second level, a sender picks a random $r \in \mathbb{Z}_p^*$ and transmits the pair (c_1, c_2) where $c_1 = y_A^r$ and $c_2 = m \cdot e(g, h)^r$ where $h = \psi(g)$. The proxy is given the re-encryption key $h^{b/a}$ and can translate ciphertexts from Alice to Bob by computing $(e(c_1, h^{b/a}), c_2) = (e(g, h)^{br}, m \cdot e(g, h)^r)$. The decryption operations are somewhat similar to those of the Elgamal [18] cryptosystem. This strategy does not completely withstand collusions since, if Bob and the proxy cooperates, they obtain the element $h^{1/a}$ which suffices to decrypt any second-level ciphertext intended to Alice. Even if the last few years saw a renewed interest in proxy re-encryption [4,5,25,16,19,24], all known constructions entail to trust proxies not to collude with certain participants. Otherwise, sub-keys such as $h^{1/a}$ or new re-encryption keys can be derived and disclosed over the Internet.

Transferability issues in proxy re-encryption. Following [21], a PRE scheme is said *non-transferable* if the proxy and a set of colluding delegatees cannot re-delegate their decryption rights. The first question that comes to mind is whether transferability is really preventable since the delegatee can always decrypt and forward the plaintext. However, the difficulty in retransmitting data restricts this behavior. The security goal is therefore to prevent the delegatee and the proxy to provide another party with a secret value that can be used *offline* to decrypt the delegator's ciphertexts. Obviously, the delegatee can always send its secret key to this party, but in doing so, it assumes a security risk that is potentially injurious to itself. In the simple aforementioned unidirectional system, colluders can unfortunately disclose $h^{1/a}$ which is clearly harmless to the cheating delegatee and allows for the offline opening of second level ciphertexts encrypted for the delegator. All other existing unidirectional [5] schemes are actually vulnerable to this kind of attack.

A desirable security goal is therefore to prevent a malicious proxy (or a collusion of several rogue proxies) interacting with users to take such actions. To the best of our knowledge, this non-transferability property has been elusive in the literature. This is not surprising since, given that proxies and delegatees can always decrypt level 2 ciphertexts by combining their secrets, they must be able to jointly compute data that allows decrypting and, once revealed to a malicious third party, ends up with a transfer of delegation. Therefore, discouraging such behaviors seems much easier than preventing them.

Our contributions. We introduce a new notion, that we call *traceable proxy re-encryption* (TPRE), where proxies that reveal their re-encryption key to third parties can be identified by the delegator. The primitive does not preclude illegal transfers of delegation but provides a disincentive to them. Unlike prior unidirectional PRE systems, when delegators come across an illegally formed re-encryption key, they can determine its source among potentially malicious proxies. It also allows tracing delegatees and proxies that pool their secrets to disclose a pirate decryption sub-key which suffices to decipher ciphertexts originally intended for the delegator. Identifying dishonest delegatees is useful in applications such as PRE-based file storage systems [4,5] where there is a single proxy (i.e. the access control server) and many delegatees (i.e. end users). When a pirate decryption sub-key is disclosed in such a situation, we can find out which client broke into the access control server to generate it.

Deterring potentially harmful actions from parties that are *a priori* trustworthy may seem overburden: no one would elect a delegatee without having high confidence in his honesty. In these regards, the present work is somehow related to ideas from Goyal [20] that aim at avoiding to place too much trust in entities (i.e. trusted authorities in identity-based encryption schemes) that must be trusted anyway. Arguably, users are less reluctant to grant their trust when abuses of delegated power are detectable and discouraged.

We formalize security notions for TPRE and give efficient implementations meeting these requirements under different pairing-related assumptions. Our constructions borrow techniques from *traitor tracing* schemes [17]. We also make use of a special kind of *identity-based encryption* (IBE) system (where arbitrary strings such as email addresses [27,11] can act as a public keys so as to avoid costly digital certificates), introduced in 2006 by Abdalla *et al.* and called *wildcard identity-based encryption* (WIBE) [1].

Our main scheme is fairly efficient, with ciphertexts of logarithmic size in the number of delegations, but the tracing system is non-black-box. Its security relies on (formerly used) mild pairing-related assumptions and the security analysis takes place in the standard model (without the random oracle heuristic [7]).

We also discuss how the scheme can be equipped with a black-box tracing mechanism at the expense of longer ciphertexts. The design principle is to associate re-encryption keys with codewords from a collusion-secure code [14]. This scheme is inspired from a WIBE-based identity-based traitor tracing scheme [2] and inherits its disadvantages: its computational overhead and the size of ciphertexts are linear in the length of the underlying code.

Roadmap. In the upcoming sections, we first define the concept of TPRE scheme and its security model. Then, we describe the intractability assumption that our scheme relies on. In section 3, we detail our scheme and first provide some intuition of the underlying idea. We finally give security results. Section 4 briefly explains how to obtain a black-box tracing mechanism.

2 Preliminaries

2.1 Model and Security Notions

Definition 1. *A (single hop) unidirectional PRE scheme is a tuple of algorithms* $(\mathsf{Global\text{-}setup}, \mathsf{Keygen}, \mathsf{ReKeygen}, \mathsf{CheckKey}, \mathsf{Enc}_1, \mathsf{Enc}_2, \mathsf{ReEnc}, \mathsf{Dec}_1, \mathsf{Dec}_2)$*:*

- $\mathsf{Global\text{-}setup}(\lambda) \rightarrow \mathsf{par}$*: on input of a security parameter* λ*, this algorithm produces public parameters* par *to be used by all parties.*
- $\mathsf{Keygen}(\lambda, \mathsf{par}) \rightarrow (sk, pk)$*: on input of common public parameters* par *and a security parameter* λ*, all parties use this randomized algorithm to generate a private/public key pair* (sk, pk)*.*
- $\mathsf{ReKeygen}(\mathsf{par}, sk_i, pk_j) \rightarrow R_{ij}$*: given public parameters* par*, user* i*'s private key* sk_i *and user* j*'s public key* pk_j*, this (possibly randomized) algorithm outputs a key* R_{ij} *that allows re-encrypting second level ciphertexts intended to* i *into first level ciphertexts encrypted for* j*.*
- $\mathsf{CheckKey}(\mathsf{par}, sk_i, pk_j, R_{ij}) \rightarrow b \in \{0,1\}$*: is a deterministic algorithm checking the well-formedness of* R_{ij} *as a proxy key for re-encrypting messages from user* i *to user* j*.*
- $\mathsf{Enc}_1(\mathsf{par}, pk, m) \rightarrow C$*: on input of public parameters* par*, a receiver's public key* pk *and a plaintext* m*, this probabilistic algorithm outputs a first level ciphertext that cannot be re-encrypted for another party.*
- $\mathsf{Enc}_2(\mathsf{par}, pk, m) \rightarrow C$*: given public parameters* par*, a receiver's public key* pk *and a plaintext* m*, this randomized algorithm outputs a second level ciphertext that can be re-encrypted into a first level ciphertext (intended to a possibly different receiver) using the appropriate re-encryption key.*
- $\mathsf{ReEnc}(\mathsf{par}, R_{ij}, C) \rightarrow C'$*: this (possibly randomized) algorithm takes as input public parameters* par*, a re-encryption key* R_{ij} *and a second level ciphertext* C *encrypted for user* i*. The output is a first level ciphertext* C' *re-encrypted for user* j*. In single hop schemes,* C' *cannot be re-encrypted any further.*
- $\mathsf{Dec}_1(\mathsf{par}, sk, C) \rightarrow m$*: given a private key* sk*, a first level ciphertext* C *and system-wide parameters* par*, this algorithm outputs a plaintext* $m \in \{0,1\}^*$*.*
- $\mathsf{Dec}_2(\mathsf{par}, sk, C) \rightarrow m$*: given a private key* sk*, a second level ciphertext* C *and public parameters* par*, this algorithm returns a plaintext* $m \in \{0,1\}^*$*.*

For any common public parameters par*, any message* $m \in \{0,1\}^*$ *and any couple of private/public key pair* (sk_i, pk_i)*,* (sk_j, pk_j) *these algorithms should satisfy the following correctness conditions:*

$$\mathsf{Dec}_1(\mathsf{par}, sk_i, \mathsf{Enc}_1(\mathsf{par}, pk_i, m)) = m;$$
$$\mathsf{Dec}_2(\mathsf{par}, sk_i, \mathsf{Enc}_2(\mathsf{par}, pk_i, m)) = m;$$
$$\mathsf{Dec}_1(\mathsf{par}, sk_j, \mathsf{ReEnc}(\mathsf{par}, \mathsf{ReKeygen}(\mathsf{par}, sk_i, pk_j), \mathsf{Enc}_2(\mathsf{par}, pk_i, m))) = m;$$
$$\mathsf{CheckKey}(\mathsf{par}, sk_i, pk_j, \mathsf{ReEnc}(\mathsf{par}, \mathsf{ReKeygen}(\mathsf{par}, sk_i, pk_j))) = 1.$$

In a traceable PRE scheme, there is an additional procedure Trace which, given user i's private key sk_i as well as a pirate proxy key R_{ij}^{bad} allowing for illegal translations from i to another user j, outputs the identity of at least one of the malicious proxies that made up R_{ij}^{bad}. Algorithm Trace can also take as input a pirate decryption key $R_{i\star}^{\mathsf{bad}}$ that, instead of re-encrypting second level ciphertexts intended for user i, simply directly recovers the underlying plaintext. In this case, the tracing algorithm should also determine which malicious delegatee has colluded with the incriminated proxy to generate of $R_{i\star}^{\mathsf{bad}}$.

Semantic security. As in [4,5,16], we require that users publicize public keys only if they hold the corresponding private keys. This amounts to adopt a trusted key generation model or a model where all parties have to prove knowledge of their secret keys when registering their public keys upon certification.

Like [4,5,16], we also assume a static model where adversaries do not choose whom to corrupt depending on the information gathered so far.

Definition 2. *A (single-hop) unidirectional PRE scheme is semantically secure at level* 2 *if the probability*

$$\begin{aligned}\Pr[(pk^\star, sk^\star) \leftarrow \mathsf{Keygen}(\lambda), \{(pk_x, sk_x) \leftarrow \mathsf{Keygen}(\lambda)\}, \{(pk_h, sk_h) \leftarrow \mathsf{Keygen}(\lambda)\},\\ \{R_{x\star} \leftarrow \mathsf{ReKeygen}(sk_x, pk^\star)\},\\ \{R_{\star h} \leftarrow \mathsf{ReKeygen}(sk^\star, pk_h)\}, \{R_{h\star} \leftarrow \mathsf{ReKeygen}(sk_h, pk^\star)\},\\ \{R_{hx} \leftarrow \mathsf{ReKeygen}(sk_h, pk_x)\}, \{R_{xh} \leftarrow \mathsf{ReKeygen}(sk_x, pk_h)\},\\ \{R_{hh'} \leftarrow \mathsf{ReKeygen}(sk_h, pk_{h'})\}, \{R_{xx'} \leftarrow \mathsf{ReKeygen}(sk_x, pk_{x'})\},\\ (m_0, m_1, St) \leftarrow \mathcal{A}(pk^\star, \{(pk_x, sk_x)\}, \{pk_h\}, \{R_{x\star}\}, \{R_{h\star}\},\\ \{R_{\star h}\}, \{R_{xh}\}, \{R_{hx}\}, \{R_{hh'}\}, \{R_{xx'}\}),\\ d^\star \xleftarrow{R} \{0,1\}, C^\star = \mathsf{Enc}_2(m_{d^\star}, pk^\star), d' \leftarrow \mathcal{A}(C^\star, St) :\\ d' = d^\star]\end{aligned}$$

is negligibly (as a function of the security parameter λ) close to $1/2$ for any PPT adversary $\mathcal{A}$. In our notation, St is a state information maintained by $\mathcal{A}$ while $(pk^\star, sk^\star)$ is the target user's key pair generated by the challenger that also chooses other keys for corrupt and honest parties. For other honest parties, keys are subscripted by h or h' and we subscript corrupt keys by x or x'. The adversary is granted access to all re-encryption keys but those for re-encrypting from the target user to a corrupt one. $\mathcal{A}$ is said to have advantage ε if this probability, taken over all coin tosses, is at least $1/2 + \varepsilon$.

Security of first level ciphertexts. Definition 2 provides adversaries with a second level challenge ciphertext. An orthogonal definition captures $\mathcal{A}$'s inability to distinguish first level ciphertexts as well. For *single-hop* schemes, the adversary is allowed to see *all* re-encryption keys in this definition. As first level ciphertexts cannot be re-encrypted any further, there is no reason to hold specific honest-to-corrupt re-encryption keys back from the adversary. A unidirectional scheme fitting this definition is said semantically secure at the first level.

Digital-identity security in PRE. In [4], Ateniese *et al.* define an important security requirement for unidirectional PRE schemes. This notion, termed *master secret security* or *digital-identity security*, demands that no coalition of dishonest delegatees and proxies be able to pool their keys in order to expose the private key of their delegator. More formally, the following probability should be negligible as a function of the security parameter λ. In our notations, we superscript pk and sk with $\star$ to denote the keys of the target honest user whereas adversarial users' keys are subscripted by x.

$$\begin{aligned}\Pr[\ &(pk^\star, sk^\star) \leftarrow \mathsf{Keygen}(\lambda), \{(pk_x, sk_x) \leftarrow \mathsf{Keygen}(\lambda)\},\\ &\{R_{\star x} \leftarrow \mathsf{ReKeygen}(sk^\star, pk_x)\}, \{R_{x\star} \leftarrow \mathsf{ReKeygen}(sk_x, pk^\star)\},\\ &\gamma \leftarrow A(pk^\star, \{(pk_x, sk_x)\}, \{R_{\star x}\}, \{R_{x\star}\}) : \gamma = sk^\star\ ]\end{aligned}$$

While reasonable in many applications, this definition does not consider colluding delegatees and proxies who attempt to produce a new re-encryption key $R_{\star x'}$ that was not originally given and allows re-encrypting from the target user to another malicious party x'. As already stressed, all known unidirectional PRE schemes fail to resist such attacks. Although colluders are unable to expose the delegator's long term secret $sk^\star$, they can still compute a sub-key sk^{bad} that allows decrypting ciphertexts at level 2. To address this issue, our model asks that the cheated delegator be able to determine – at least partially and with high probability – where the illegal transfer of delegation stems from or who crafted the pirate sub-key sk^{bad}. In our scheme, this unfortunately comes at the expense of sacrificing the key and ciphertext optimality properties met in [4,5].

Traceability. Consider a set of proxies $P_1, P_2, \ldots, P_N$ that receive re-encryption keys allowing for the translation of ciphertexts from user A to his delegatees $B_1, B_2, \ldots, B_N$. We say that a PRE scheme is *traceable* if any subset of these proxies colluding with delegatees $B_1, B_2, \ldots, B_N$ is unable to generate a new re-encryption key that cannot be traced back to one of them.

Definition 3. *A unidirectional PRE scheme is* **traceable** *if no PPT adversary $\mathcal{A}$ has non-negligible probability of success in the following game:*

1. *The challenger provides $\mathcal{A}$ with the target user's public key pk_0, public keys pk_i for other honest parties and key pairs (sk_i, pk_i) for corrupt users.*
2. *On multiple occasions, $\mathcal{A}$ may invoke a re-encryption key generation oracle $\mathcal{O}_{rkey}$. When queried on input of public keys (pk_i, pk_j) that were both obtained from the challenger, this oracle returns the re-encryption key $R_{ij} =$ $\mathsf{ReKeygen}(sk_i, pk_j)$. Let T be the set of proxy keys obtained by $\mathcal{A}$.*
3. *$\mathcal{A}$ outputs a pirate re-encryption key $R^\star_{0t}$ together with a public key pk_t that belongs to the public key space of the scheme (i.e. for which an associated private key exists) and differs from public keys of the target user's delegatees. The adversary is declared successful if the following two conditions hold:*

a. $\mathsf{CheckKey}(sk_0, pk_t, R^\star_{0t}) = 1$ *(i.e. $R^\star_{0t}$ is a valid re-encryption key).*
b. The tracing procedure (run by the challenger on $R^\star_{0t}$ using the target user's secret sk_0) fails to identify a correct proxy key $R^{\mathsf{bad}}_{0j} \in T$. That is, if $R^{\mathsf{bad}}_{0j} = \mathsf{Trace}(sk_0, R^\star_{0t}, pk_t)$, we have either $R^{\mathsf{bad}}_{0j} = \perp$ or if $R^{\mathsf{bad}}_{0j} \notin T$.

The pirate key $R^\star_{0t}$ should re-encrypt from user 0 to a user having public key pk_t. For simplicity, we assume that the latter is supplied by $\mathcal{A}$ at the end of the game. When the target user finds a suspicious re-encryption key $R^\star$ in practice, he does not *a priori* know to whom ciphertexts can be re-encrypted using $R^\star$. However, he can determine it by simply testing whether $\mathsf{CheckKey}(sk_0, \tilde{pk}_j, R^\star) = 1$, for $j = 1, \ldots, \eta$, given a set of suspicious public keys $\{\tilde{pk}_1, \ldots, \tilde{pk}_\eta\}$.

We insist that pk_t may differ from public keys that are generated by the challenger at step 1 of the game. Besides, the definition does not force $\mathcal{A}$ to reveal the matching private key sk_t to the challenger: the only requirement is that such a private key exists.

At first glance, one may wonder why $\mathcal{A}$ should be allowed to come up with an arbitrary pk_t of her choosing whilst delegation queries to $\mathcal{O}_{rkey}$ are only permitted for delegatees' public keys that were chosen by the challenger.

We actually find it natural to assume that honest users only delegate to parties whose public key has been properly certified and for which knowledge of the underlying secret key has been demonstrated to the CA at key registration. In contrast, pk_t is not meant to have a legal use and simply provides a way to covertly translate the target user's communications. Hence, there is no reason to assume that the challenger learns sk_t whatsoever. Finally, when the proxy is compromised but the delegatee j remains honest, the adversary obtains R_{0j} such that $\mathsf{CheckKey}(sk_0, pk_j, R_{0j}) = 1$. Then, she might be able to compute $R^\star_{0t}$ and pk_t (as a function of pk_j) such that $\mathsf{CheckKey}(sk_0, pk_t, R^\star_{0t}) = 1$. In this case, the adversary clearly does not know sk_t. The property that we require is that $R^\star_{0t}$ can be traced back to the proxy involved in its creation. Then, if pk_t happens to be a registered public key (for which a proof of knowledge of the underlying private key was provided), the delegator figures out that the original delegatee was also part of the collusion, as well as the user holding sk_t.

Bounded Traceability. Similarly to common situations in traitor tracing schemes, it may happen that traceability is guaranteed only if the adversary makes at most k re-encryption key queries involving the secret sk_0 of the target user acting as a delegator (regardless of whether the delegatee is honest). On the other hand, she is granted as many re-encryption key queries involving other honest delegators as she likes. Schemes that are secure in this scenario are said k-traceable.

Black Box Traceability. A new analogy with traitor tracing primitives suggests to strengthen the definition by assuming that the adversary only outputs a re-encryption device $\mathbb{P}$ that translates ciphertexts with non-negligible probability but cannot be reverse-engineered so as to extract the built-in key. Indeed, it has been reported [22] that proxy re-encryption systems can be safely obfuscated. It would thus be desirable to have a black-box tracing procedure to recover the identity of colluding parties using $\mathbb{P}$ as a re-encryption oracle. A variant of our scheme can be equipped with a limited black-box tracing mechanism. Due to the use of collusion-secure codes [14], this variant unfortunately features unreasonably large ciphertexts and cannot be considered as being practical. Moreover, it only tolerates a bounded number of traitors k. Lastly, it does not allow to determine who the dishonest delegatees are when running a pirate decryption device $\mathbb{D}$ in tracing mode: only colluding proxies can be traced.

2.2 Bilinear Maps and Complexity Assumptions

We consider a configuration of *bilinear map groups* $(\mathbb{G}_1, \mathbb{G}_2, \mathbb{G}_T)$ of prime order p with a mapping $e : \mathbb{G}_1 \times \mathbb{G}_2 \to \mathbb{G}_T$ and an isomorphism $\psi : \mathbb{G}_2 \to \mathbb{G}_1$ satisfying the following properties:

1. bilinearity: $e(g^a, h^b) = e(g,h)^{ab}$ for any $(g,h) \in \mathbb{G}_1 \times \mathbb{G}_2$ and $a, b \in \mathbb{Z}$;
2. efficient computability for any input pair;
3. non-degeneracy: $e(g,h) \neq 1_{\mathbb{G}_T}$ whenever $g \neq 1_{\mathbb{G}_1}$ and $h \neq 1_{\mathbb{G}_2}$.

We will need an extension of the Decision Bilinear Diffie-Hellman (DBDH) assumption which is the intractability of distinguishing $e(g,h)^{abc}$ given (h^a, h^b, h^c).

Definition 4. *In bilinear map groups* $(\mathbb{G}_1, \mathbb{G}_2, \mathbb{G}_T)$*, The* **Augmented Decision Bilinear Diffie-Hellman Problem** *(ADBDH) is to distinguish* $e(g,h)^{abc}$ *from random elements of* $\mathbb{G}_T$ *given* $(g, h, h^a, h^b, h^c, h^{a^2 b}) \in \mathbb{G}_1 \times \mathbb{G}_2^5$*. A distinguisher* $\mathcal{D}$ (τ, ε)*-breaks the assumption if it has running time* τ *and*

$$\begin{aligned} Adv(\mathcal{D}) = |\Pr[\mathcal{D}(h^a, h^b, h^c, h^{a^2 b}, e(g,h)^{abc}) = 1 | a, b, c \xleftarrow{R} \mathbb{Z}_p^*] \\ - \Pr[\mathcal{D}(h^a, h^b, h^c, h^{a^2 b}, e(g,h)^{z}) = 1 | a, b, c, z \xleftarrow{R} \mathbb{Z}_p^*]| \geq \varepsilon \end{aligned}$$

This problem is not easier than breaking the ℓ-Bilinear Diffie-Hellman Exponent (ℓ-BDHE) assumption of [12] that implies the infeasibility of recognizing $e(\psi(h'), h)^{(a^{\ell+1})}$ given $(h', h, h^a, h^{(a^2)}, \ldots, h^{(a^\ell)}, h^{(a^{\ell+2})}) \in \mathbb{G}_2^{\ell+4}$. When $a = b$, ADBDH boils down to a special case[1] of ℓ-BDHE with $\ell = 1$. The generic hardness ADBDH is thus implied by that of ℓ-BDHE, which was shown in [10].

Our proof of traceability relies on a problem named *2-out-of-3 Diffie-Hellman* in [23], where its generic intractability was shown in prime order groups. A not harder version of this problem was previously considered in [3].

Definition 5. *The* **2-out-of-3 Diffie-Hellman** *problem (2-3-CDH) is, given* $(h, h^a, h^b) \in \mathbb{G}^3$*, to find a pair* $(C, C^{ab}) \in \mathbb{G} \times \mathbb{G}$ *with* $C \neq 1_{\mathbb{G}}$*.*

3 A Scheme with Logarithmic Complexity

This section presents our main scheme providing non-black-box traceability. It borrows ideas from the identity-based traitor tracing described in [2].

[1] It is actually the hardness of deciding if $T \stackrel{?}{=} e(g,h)^{a^2 c}$ given $(h' = h^c, h^a, h^{(a^3)})$.

3.1 Intuition

To provide a better intuition of the scheme, we need the recall the Waters IBE [29] and the notion of wildcard IBE [1]. The former involves a trusted party that publishes a master public key $mpk = (Z = e(g,h)^z, V_0, V_1, \ldots, V_n) \in \mathbb{G}_T \times \mathbb{G}_2^{n+1}$ where $z \stackrel{R}{\leftarrow} \mathbb{Z}_p^*$ and n is the length of identity strings. The trusted authority keeps a master secret $msk = h^z$ to itself. This secret is used to derive private keys from user's identities $id = i_1 \ldots i_n \in \{0,1\}^n$ by computing

$$d_{id} = (d_1, d_2) = \big(msk \cdot (V_0 \cdot \prod_{\ell=1}^{n} V_\ell^{i_\ell})^r, h^r\big)$$

for a randomly chosen exponent $r \stackrel{R}{\leftarrow} \mathbb{Z}_p^*$. Such a private key always satisfies

$$e(g, d_1) = Z \cdot e(U_0 \cdot \prod_{\ell=1}^{n} U_\ell, d_2) \tag{1}$$

where $U_\ell = \psi(V_\ell)$ for $\ell = 0, \ldots, n$. Therefore, a ciphertext encrypted as

$$C_0 = m \cdot Z^s \qquad C_1 = g^s \qquad C_2 = \big(U_0 \cdot \prod_{\ell=1}^{n} U_\ell^{i_\ell}\big)^s,$$

for a random $s \stackrel{R}{\leftarrow} \mathbb{Z}_p^*$, can be deciphered by computing $m = C_0 \cdot e(C_2, d_2)/e(C_1, d_1)$ (this is easily observed by raising both members of (1) to the power s).

Wildcard IBE schemes [1] (or WIBE for short) are hierarchical IBE systems where certain levels of the hierarchy can be left unspecified by a sender willing to allow decryption by any hierarchy member whose identity fits a certain pattern. These WIBE systems were notably used to construct multi-receiver encryption systems. In the case of Waters' IBE, the unique level of the hierarchy can be left unspecified by replacing the ciphertext component C_2 with a vector $(U_0^s, U_1^s, \ldots U_n^s)$ so that $C_2 = \big(U_0 \cdot \prod_{\ell=1}^{n} U_\ell^{i_\ell}\big)^s$ can be reconstructed at decryption for any identity $id \in \{0,1\}^n$. Placing such a "wildcard" at the unique level of the hierarchy permits decryption by anyone holding a decryption key for some identity. The same underlying idea was used in [2] to devise an identity-based traitor tracing scheme from a 2-level WIBE built on [29].

At high level, our scheme can be seen as using a multi-receiver encryption scheme derived from the single level $\mathcal{W}a$-$\mathcal{WIBE}$ of [1]. Instead of assigning a unique identifier to decryption keys as in [2], we embed it in re-encryption keys.

These re-encryption keys are generated by binding decryption keys of the multi-receiver scheme to delegatees' public keys. Identity-based private keys are associated with serial numbers (seen as identities) and tied up to the public keys of entities to whom messages must be re-encrypted. More precisely, we let each party j generate an additional public key component $Y_j = h^{y_j}$ and a delegation from user i to user j is made effective by the re-encryption key

$$R_{ij} = (id, A_{ij}, B_{ij}) = \big(id, Y_j^{z_i} \cdot (V_{i,0} \cdot \prod_{\ell=1}^{n} V_{i,\ell}^{i_\ell})^r, h^r\big)$$

where $pk_i = (Z_i = e(g,h)^{z_i}, Y_i = h^{y_i}, U_{i,0}, \ldots, U_{i,n})$ is user i's public key and $U_{i,\ell} = \psi(V_{i,\ell})$ for $\ell = 0, \ldots, n$. The re-encryption algorithm can actually be thought of as translating WIBE ciphertexts into regular public key encryptions under the delegatee's public key.

The tracing system is non-black-box. It takes as input a pirate re-encryption key and merely extracts the built-in serial number from it. With a non-black-box tracing algorithm, we do not need collusion-secure codes [14]. The proof of traceability takes advantage of the collusion-resistance of the underlying WIBE and we have logarithmic-size ciphertexts in the number of delegations.

3.2 The Scheme

For simplicity, we assume that all users have at most N delegatees. Public keys and second level ciphertexts consist of $O(n) = O(\log N)$ group elements.

Global-setup(λ): on input of a security parameter λ, choose bilinear map groups $(\mathbb{G}_1, \mathbb{G}_2, \mathbb{G}_T, e, \psi)$ of prime order $p > 2^\lambda$ with generators $h \xleftarrow{R} \mathbb{G}_2$, $g = \psi(h)$.

Keygen(λ): user i sets his public key as

$$pk_i = \big(Z_i = e(g,h)^{z_i}, Y_i = h^{y_i}, U_{i,0} = g^{u_{i,0}}, U_{i,1} = g^{u_{i,1}}, \ldots, U_{i,n} = g^{u_{i,n}}\big)$$

for random values $(z_i, y_i, u_{i,0}, u_{i,1}, \ldots, u_{i,n}) \xleftarrow{R} (\mathbb{Z}_p^*)^{n+3}$. For $\ell = 0, \ldots, n$, group elements $V_{i,\ell} = h^{u_{i,\ell}} \in \mathbb{G}_2$ (that satisfy $U_{i,\ell} = \psi(V_{i,\ell})$) are also computed and included in the private key $sk_i = (z_i, y_i, V_{i,0}, \ldots, V_{i,n})$. Let $w_{i,j} \in \{0,1\}^n$ be a unique identifier to be assigned by user i to the re-encryption key R_{ij} translating to user j. Elements $U_{i,\ell}, V_{i,\ell}$ define functions

$$F_{V_i} : \{0,1\}^n \to \mathbb{G}_2 : F_{V_i}(w_{i,j}) = V_{i,0} \cdot \prod_{\ell=1}^{n} V_{i,\ell}^{w_{i,j,\ell}}$$

and $F_{U_i} : \{0,1\}^n \to \mathbb{G}_1 : F_{U_i}(w_{i,j}) = \psi\big(F_{V_i}(w_{i,j})\big)$.

ReKeygen(sk_i, pk_j): given user i's private key $sk_i = (z_i, y_i, V_{i,0}, \ldots, V_{i,n})$ and user j's public key $pk_j = (Z_j, Y_j, U_{j,0}, U_{j,1}, \ldots, U_{j,n})$, choose[2] a previously unemployed string $w_{i,j} = w_{i,j,1} \ldots w_{i,j,n} \in \{0,1\}^n$ and a random exponent $r \xleftarrow{R} \mathbb{Z}_p^*$ to generate the unidirectional key

$$R_{ij} = (w_{i,j}, A_{ij}, B_{ij}) = (w_{i,j}, Y_j^{z_i} \cdot F_{V_i}(w_{i,j})^r, h^r).$$

CheckKey(sk_i, pk_j, R_{ij}): given $sk_i = (z_i, y_i, V_{i,0}, \ldots, V_{i,n})$, parse user j's public key pk_j as $(Z_j, Y_j, U_{j,0}, U_{j,1}, \ldots, U_{j,n})$ and R_{ij} as $(w_{i,j}, A_{ij}, B_{ij})$. Return 1 if

$$e(g, A_{ij}) = e(g, Y_j)^{z_i} \cdot e(F_{U_i}(w_{i,j}), B_{ij}) \tag{2}$$

and 0 otherwise.

---

[2] In order to avoid to store w_{ij} and r, the delegator can compute them as a pseudo-random function of a short secret key and the public key pk_j.

Enc$_1(m, pk_i, \mathsf{par})$: to encrypt a message $m \in \mathbb{G}_T$ under the public key $pk_i = (Z_i, Y_i, U_{i,0}, U_{i,1}, \ldots, U_{i,n})$ at the first level, choose $s \stackrel{R}{\leftarrow} \mathbb{Z}_p^*$ and output

$$\mathbf{C} = (C_0, C_1) = \big(m \cdot e(g,h)^s, e(g, Y_i)^s\big)$$

Enc$_2(m, pk_i, \mathsf{par})$: to encrypt a message $m \in \mathbb{G}_T$ under the public key pk_i at level 2, the sender picks a random exponent $s \stackrel{R}{\leftarrow} \mathbb{Z}_p^*$ and computes

$$\mathbf{C} = (C_0, C_1, C_{2,0}, C_{2,1}, \ldots, C_{2,n}) = \big(m \cdot Z_i^s, g^s, U_{i,0}^s, U_{i,1}^s, \ldots, U_{i,n}^s\big)$$

ReEnc$(R_{ij}, \mathbf{C}_i)$: given the translation key $R_{ij} = (w_{i,j}, A_{ij}, B_{ij}) \in \{0,1\}^n \times \mathbb{G}_2^2$ and a ciphertext $\mathbf{C}_i = (C_0, C_1, C_{2,0}, \ldots, C_{2,n}) \in \mathbb{G}_T \times \mathbb{G}_1^{n+2}$, compute

$$F_{U_i}(w_{i,j})^s = C_{2,0} \cdot \prod_{\ell=1}^{n} C_{2,\ell}^{w_{i,j,\ell}} = \big(U_{i,0} \cdot \prod_{\ell=1}^{n} U_{i,\ell}^{w_{i,j,\ell}}\big)^s$$

and output

$$\mathbf{C}'_j = (C'_0, C'_1) = \Big(C_0, \frac{e(C_1, A_{ij})}{e(F_{U_i}(w_{i,j})^s, B_{ij})}\Big) \tag{3}$$

$$= \Big(m \cdot e(g,h)^{z_i s}, e(g, Y_j)^{z_i s}\Big) = \Big(m \cdot e(g,h)^{\tilde{s}}, e(g, Y_j)^{\tilde{s}}\Big) \tag{4}$$

with $\tilde{s} = s z_i$.

Dec$_1(\mathbf{C}_j, sk_j)$: given $sk_j = (z_j, y_j, V_{j,0}, \ldots, V_{j,n})$, parse the ciphertext $\mathbf{C}_j$ as $(C_0, C_1) \in \mathbb{G}_T^2$. Return $m = C_0 / C_1^{1/y_j}$.

Dec$_2(\mathbf{C}_i, sk_i)$: parse $\mathbf{C}_i$ as $C = (C_0, C_1, C_{2,0}, \ldots, C_{2,n}) \in \mathbb{G}_T \times \mathbb{G}_1^{n+2}$ and sk_i as $(z_i, y_i, V_{i,0}, \ldots, V_{i,n})$. Return $m = C_0 / e(C_1, h)^{z_i}$.

Trace(sk_i, R_{it}, pk_t): on input of a public key $pk_t = (Z_t, Y_t, U_{t,0}, \ldots, U_{t,n})$ and a re-encryption key $R_{it} = (w, A_{it}, B_{it}) \in \{0,1\}^n \times \mathbb{G}_2 \times \mathbb{G}_2$ such that $\mathsf{CheckKey}(sk_i, pk_t, R_{it}) = 1$, this algorithm incriminates the proxy that has been provided with a re-encryption key including w as identifier.

The correctness of the re-encryption algorithm is easily checked by observing that re-encryption keys $R_{ij} = (w_{i,j}, A_{ij}, B_{ij})$ always satisfy relation (2). Raising both members of the latter to the power $s \in \mathbb{Z}_p^*$ gives

$$e(g^s, A_{ij}) = e(g, Y_j)^{z_i s} \cdot e(F_{U_i}(w_{i,j})^s, B_{ij})$$

which explains the transition between relations (3) and (4).

As in prior unidirectional schemes, the proxy and the delegator can collude to compute and disclose a quantity that allows opening all second level ciphertexts: given $R_{ij} = (w_{i,j}, A_{ij}, B_{ij})$ and y_j s.t. $Y_j = h^{y_j}$, they can obtain

$$R_{i\star}^{\mathsf{bad}} = (w_{i,j}, A_{ij}^{1/y_j}, B_{ij}^{1/y_j}) = \big(w_{i,j}, h^{z_i} \cdot F_{V_i}(w_{i,j})^{r'}, h^{r'}\big),$$

with $r' = r/y_j$, that allows for the off-line decryption of level 2 ciphertexts. However, when presented with $R_{i\star}^{\mathsf{bad}} = (w_{i,j}, A'_{ij}, B'_{ij})$, the tracing algorithm runs the validity check $e(g, A'_{ij}) \stackrel{?}{=} Z_i \cdot e(F_{U_i}(w_{i,j}), B'_{ij})$. If the latter test is successful, the the proxy identified by $w_{i,j}$ and its associated delegatee are *both* found guilty for having conspired to produce $R^{\mathsf{bad}}_{i\star}$. The serial number $w_{i,j}$ makes the source of the collusion evident and provides a deterrent for abuses of trust.

When the tracing system takes as input a pair $(R_{it} = (w, A_{it}, B_{it}), pk_t)$, the original delegatee j associated the serial number $w = w_{ij}$ cannot be incriminated as the corrupt proxy may have maliciously chosen pk_t as a function of pk_j (possibly in an attempt to trick user i into believing that j is not trustworthy).

3.3 Security

Theorem 1. *The scheme is semantically secure at the second level under the Augmented DBDH assumption.*

Proof. Let $(A = h^a, B = h^b, C = h^c, D = h^{a^2 b}, T) \in \mathbb{G}_2^4 \times \mathbb{G}_T$ be an Augmented DBDH instance. We construct an algorithm $\mathcal{B}$ that decides if $T = e(g, h)^{abc}$ using its interaction with a chosen-plaintext adversary $\mathcal{A}$.

All public keys that $\mathcal{A}$ gets to see are indexed by an integer $i \in \{0, \ldots, N_{max}\}$, where $N_{max} + 1$ denotes the maximal number of users in the system. Let us call $HU \subset \{0, \ldots, N_{max}\}$ the set of honest players, including the target receiver whose public key has index 0. Let also $CU \subset \{1, \ldots, N_{max}\}$ denote the set of corrupt receivers. The attack environment is emulated as follows.

- *Key generation*:
  - The public key $pk_0 = (Z_0, Y_0, U_{0,0}, U_{0,1}, \ldots, U_{0,n})$ of the target user is chosen as $Z_0 = e(\psi(A), B) = e(g, h)^{ab}$ and $Y_0 = h^{y_0}$, $U_{0,\ell} = g^{u_{0,\ell}}$ with $y_0, u_{0,\ell} \stackrel{R}{\leftarrow} \mathbb{Z}_p^*$ for $\ell = 0, \ldots, n$.
  - For users $i \in HU \backslash \{0\}$, public keys are defined by randomly choosing $z_i, y_i, u_{i,0}, \ldots, u_{i,n} \stackrel{R}{\leftarrow} \mathbb{Z}_p^*$ and setting $Z_i = e(g, h)^{z_i}$, $Y_i = A^{y_i} = h^{a y_i}$ and $U_{i,\ell} = g^{u_{i,\ell}}$ for $\ell = 0, \ldots, n$.
  - For corrupt users $i \in CU$, $\mathcal{B}$ generates pk_i according to the specification of the scheme and discloses private elements $z_i, y_i, u_{i,0}, \ldots, u_{i,n} \in \mathbb{Z}_p^*$.
- *Re-encryption key generation*: to generate re-encryption keys R_{ij} from player i to player j, $\mathcal{B}$ has to distinguish several situations.
  - If $i \in CU$ or $i \in HU \backslash \{0\}$, $\mathcal{B}$ knows user i's private key component z_i such that $Z_i = e(g, h)^{z_i}$ and generates a re-encryption key as specified by the re-encryption algorithm.
  - If $i = 0$ and $j \in HU \backslash \{0\}$, $\mathcal{B}$ picks a new string $w_{0,j} \in \{0,1\}^n$ and a random exponent $r \stackrel{R}{\leftarrow} \mathbb{Z}_p^*$ to return

    $$R_{0j} = \Big(w_{0,j}, D^{y_j} \cdot F_{V_0}(w_{0,j})^r, h^r\Big),$$

    for a random $r \stackrel{R}{\leftarrow} \mathbb{Z}_p^*$. Observe that R_{0j} has the correct shape since $Z_0 = e(g, h)^{ab}$, $Y_j = A^{y_j} = h^{a y_j}$ and $D^{y_j} = (h^{a^2 b y_j}) = (h^{a y_j})^{ab}$.

- *Challenge*: when $\mathcal{A}$ comes up with messages $m_0, m_1 \in \mathbb{G}_T$, $\mathcal{B}$ flips a fair coin $d^\star \stackrel{R}{\leftarrow} \{0,1\}$ and sets the challenge ciphertext as

$$C_0 = m_{d^\star} \cdot T \qquad C_1 = \psi(C) \qquad C_{2,\ell} = \psi(C)^{u_{0,\ell}} \quad \text{for } \ell = 0, \ldots, n.$$

Since $C = h^c$ and $Z_0 = e(g,h)^{ab}$, $\mathbf{C} = (C_0, C_1, C_{2,0}, \ldots, C_{2,n})$ is a valid encryption of $m_{d^\star}$ under pk_0 with the encryption exponent $s = c$ whenever $T = e(g,h)^{abc}$. When T is random in $\mathbb{G}_T$, $\mathbf{C}$ leaks no information on $d^\star$ and $\mathcal{A}$ can only guess it with probability 1/2. Therefore, $\mathcal{B}$ outputs 1 (meaning that $T = e(g,h)^{abc}$) if $\mathcal{A}$ successfully guesses $d^\star$ and 0 otherwise. □

Theorem 2. *The scheme is semantically secure at the first level under the DBDH assumption.*

Proof. Given in appendix A. □

Theorem 3. *The scheme is traceable under the 2-3-CDH assumption in $\mathbb{G}_2$.*

Proof. For the sake of contradiction, assume that an adversary $\mathcal{A}$ defeats the non-black-box tracing algorithm (in the sense of definition 3) with probability ε. We build an algorithm $\mathcal{B}''$ solving a 2-3-CDH instance $(A = h^a, B = h^b)$ with probability $O(\varepsilon/q_{rk})$, where q_{rk} is the number of re-encryption key queries.

- *Key generation*: a set of public keys is prepared by $\mathcal{B}''$. For the target user 0, it first defines $Z_0 = e(\psi(A), B) = e(g,h)^{ab}$ and $Y_0 = h^{y_0}$ for a random $y_0 \stackrel{R}{\leftarrow} \mathbb{Z}_p^*$. The vector $(V_{0,0}, V_{0,1}, \ldots, V_{0,n})$ is defined as $V_{0,0} = A^{\alpha_0 - \kappa\tau} \cdot h^{\beta_0}$, $V_{0,\ell} = A^{\alpha_\ell} \cdot h^{\beta_\ell}$ for $\ell \in \{1, \ldots, n\}$ using random vectors $(\alpha_0, \alpha_1, \ldots, \alpha_n) \stackrel{R}{\leftarrow} \mathbb{Z}_\tau^{n+1}$, $(\beta_0, \beta_1, \ldots, \beta_n) \stackrel{R}{\leftarrow} \mathbb{Z}_p^{n+1}$, where $\kappa \stackrel{R}{\leftarrow} \{0, \ldots, n\}$ is chosen at random and $\tau = 2q_{rk}$. For any string $w_{0,j} = w_{0,j,1} \ldots w_{0,j,n} \in \{0,1\}^n$, we have

$$F_{V_0}(w_{0,j}) = V' \cdot \prod_{\ell=1}^{n} V_{0,\ell}^{w_{0,j,\ell}} = A^{J(w_{0,j})} h^{K(w_{0,j})}$$

for functions $J : \{0,1\}^n \to \mathbb{Z}$, $K : \{0,1\}^n \to \mathbb{Z}_p$ respectively defined as $J(w_{0,j}) = \alpha_0 + \sum_{\ell=1}^{n} \alpha_\ell w_{0,j,\ell} - \kappa\tau$ and $K(w_{0,j}) = \beta_0 + \sum_{\ell=1}^{n} \beta_\ell w_{0,j,\ell}$. For $\ell = 0, \ldots, n$, $\mathcal{B}''$ also sets $U_{0,\ell} = \psi(V_{0,\ell})$. As in [29], the simulator will be successful if $J(w_{0,j}) \neq 0$ for all strings $w_{0,j} \neq w^\star$ involved in delegation queries whereas $J(w^\star) = 0$ for the identifier $w^\star$ of the re-encryption key produced by $\mathcal{A}$ at the tracing stage. Since $|J(.)| \leq \tau(n+1) \ll p$, we have $J(w^\star) = 0$ with non-negligible probability $O(1/\tau(n+1))$. For all other (honest or corrupt) users $i \in \{1, \ldots, N_{max}\}$, public keys are honestly generated by $\mathcal{B}''$ that chooses the private keys $(z_i, y_i, u_{i,0}, \ldots, u_{i,n}) \in \mathbb{Z}_p^{n+3}$. The latter secrets are given to $\mathcal{A}$ for indices $i \in CU \subset \{1, \ldots, N_{max}\}$ of corrupt users.
- *Re-encryption key queries*: at any time, $\mathcal{A}$ may ask for re-encryption keys R_{ij} of her choosing. When $i \neq 0$, $\mathcal{B}''$ knows user i's private key and can normally handle the delegation query. Otherwise, following the technique of

[9,29], it constructs a re-encryption key by sampling a fresh random string $w_{0,j} \stackrel{R}{\leftarrow} \{0,1\}^n$ and a random exponent $r \stackrel{R}{\leftarrow} \mathbb{Z}_p$ to compute

$$R_{0j} = (w_{0,j}, A_{0j}, B_{0j}) = \left(w_{0,j}, B^{-y_j \frac{K(w_{0,j})}{J(w_{0,j})}} \cdot F_{V_0}(w_{0,j})^r, B^{-\frac{y_j}{J(w_{0,j})}} \cdot h^r \right),$$

where $y_j \in \mathbb{Z}_p^*$ is part of user j's private key, which is returned to $\mathcal{A}$. If we define $\tilde{r} = r - (by_j)/J(w_{0,j})$, R_{0j} has the correct distribution since

$$\begin{aligned} A_{0j} &= B^{-y_j \frac{K(w_{0,j})}{J(w_{0,j})}} \cdot F(w_{0,j})^r \\ &= B^{-y_j \frac{K(w_{0,j})}{J(w_{0,j})}} \cdot F(w_{0,j})^{\tilde{r}} \cdot (A^{J(w_{0,j})} \cdot h^{K(w_{0,j})})^{\frac{by_j}{J(w_{0,j})}} = (h^{y_j})^{ab} \cdot F(w_{0,j})^{\tilde{r}} \end{aligned}$$

and $B_{0j} = h^{\tilde{r}}$. If $J(w_{0,j}) = 0$, $\mathcal{B}''$ aborts as it cannot answer the query.

- *Tracing stage*: a successful attacker must output a pair $(R^\star_{0t}, pk_t)$ such that $\mathsf{CheckKey}(sk_0, R^\star_{0t}, pk_t) = 1$ and $R^\star_{0t} = (w^\star, A^\star_{0t}, B^\star_{0t})$ cannot be traced to a member of the coalition T. This implies that $w^\star$ must differ from all the serial numbers w_{0j} that were associated with user 0's delegatees. At this point, $\mathcal{B}''$ declares failure if $J(w^\star) \neq 0$. With probability at least $1/4q_{rk}(n+1)$ (see [29] for a detailed analysis of this probability) such a failure state is avoided. In this case, $\mathcal{B}''$ parses pk_t as $(Z_t, Y_t, U_{t,0}, \ldots, U_{t,n})$ and outputs

$$\left(Y_t, A^\star_{0t}/{B^\star_{0t}}^{K(w^\star)}\right) = (Y_t, Y_t^{ab})$$

which solves the 2-3-CDH problem in $\mathbb{G}_2$. □

4 A Variant with Black Box k-Traceability

The scheme can be endowed with a black-box tracing mechanism which is similar to the one described in [2]. The idea is to associate identity-based private keys with the codewords (seen as identities) of a collusion-secure code [14] instead of serial numbers. These keys are bound to delegatees' public keys to form fingerprinted re-encryption keys. Assuming the hardness of the Decision Diffie-Hellman problem in $\mathbb{G}_1$ for configurations where $\mathbb{G}_1 \neq \mathbb{G}_2$ (and no isomorphism from $\mathbb{G}_1$ to $\mathbb{G}_2$ is computable), well-formed ciphertexts are not publicly recognizable. Then, pirate re-encryption devices $\mathbb{P}$ can be probed with invalid ciphertexts so as to determine the codeword of one of the pirate re-encryption keys.

As in [2], this comes at the expense of prohibitively large ciphertexts, the size of which becomes proportional to the length of the collusion-secure code. We need a binary (k, N, ε)-collusion-secure code (as defined in appendix B), where N is the the maximal number of delegatees per user, k is the maximal number of colluding proxies against a delegator and ε is the maximal probability that a colluder avoids being traced. Such a code can be obtained with codewords of length $n = O\big(k^2(\log N + \log(\varepsilon^{-1}))\big)$ [28], which is also the number of group elements in a ciphertext. If users have at most $N = 100$ delegatees, in the case $k \approx 10$, we end up with ciphertexts made of about 700 group elements (which amounts to

13 Kb using curves [6] where elements of $\mathbb{G}_1$ have a 161-bit representation). We leave open the problem of constructing an efficient black-box traceable scheme.

The tracing system, borrowed from [2], probes re-encryption devices with second level ciphertexts wherein certain components have been altered and eventually retrieves bits at all positions where words in the feasible set of the coalition (see appendices B and C for details) are identical. More precisely, the tracing algorithm checks whether the pirate device successfully re-encrypts ciphertexts where components $C_{2,\ell}$ (for all $\ell \in \{1, \ldots, n\}$), have been tampered with. If it does, the tracer deduces that $C_{2,\ell}$ was not used by the pirate device, which means that the associated bit is 0 in all codewords that were assigned to re-encryption keys available to the coalition. Once a n-bit word in the feasible set of the coalition has been found, the tracing procedure of the collusion-secure code allows recovering the fingerprint of one of the involved re-encryption keys, which identifies a misbehaving proxy.

5 Conclusion

In all PRE schemes proposed so far, proxies and delegatees can derive new re-encryption keys for receivers that are not trusted by the delegator. In this paper, we proposed traceable proxy re-encryption systems, in which proxies that leak their re-encryption key can be identified by the delegator and we presented an efficient realization of this concept. An interesting open issue is to design a more efficient TPRE scheme with black-box traceability.

Acknowledgements. We thank Duong Hieu Phan for his comments. The first author is supported by the Belgian National Fund for Scientific Research (F.R.S.-F.N.R.S.). The second author is supported by the European Commission through the IST Program under Contract IST-2002-507932 ECRYPT and by the French *Agence Nationale de la Recherche* through the PACE project.

References

1. Abdalla, M., Catalano, D., Dent, A., Malone-Lee, J., Neven, G., Smart, N.: Identity-Based Encryption Gone Wild. In: Bugliesi, M., Preneel, B., Sassone, V., Wegener, I. (eds.) ICALP 2006. LNCS, vol. 4052, pp. 300–311. Springer, Heidelberg (2006)
2. Abdalla, M., Dent, A., Malone-Lee, J., Neven, G., Phan, D., Smart, N.: Identity-Based Traitor Tracing. In: Okamoto, T., Wang, X. (eds.) PKC 2007. LNCS, vol. 4450, pp. 361–376. Springer, Heidelberg (2007)
3. Al-Riyami, S., Paterson, K.: Certificateless Public Key Cryptography. In: Laih, C.-S. (ed.) ASIACRYPT 2003. LNCS, vol. 2894, pp. 452–473. Springer, Heidelberg (2003)
4. Ateniese, G., Fu, K., Green, M., Hohenberger, S.: Improved Proxy Re-Encryption Schemes with Applications to Secure Distributed Storage. In: NDSS (2005)
5. Ateniese, G., Fu, K., Green, M., Hohenberger, S.: Improved Proxy Re-Encryption Schemes with Applications to Secure Distributed Storage. ACM TISSEC 9(1), 1–30 (2006)

6. Barreto, P.S.L.M., Naehrig, M.: Pairing-Friendly Elliptic Curves of Prime Order. In: Preneel, B., Tavares, S. (eds.) SAC 2005. LNCS, vol. 3897, pp. 319–331. Springer, Heidelberg (2006)
7. Bellare, M., Rogaway, P.: Random oracles are practical: A paradigm for designing efficient protocols. In: 1st ACM Conference on Computer and Communications Security, pp. 62–73. ACM Press, New York (1993)
8. Blaze, M., Bleumer, G., Strauss, M.: Divertible Protocols and Atomic Proxy Cryptography. In: Nyberg, K. (ed.) EUROCRYPT 1998. LNCS, vol. 1403, pp. 127–144. Springer, Heidelberg (1998)
9. Boneh, D., Boyen, X.: Efficient selective-ID secure identity based encryption without random oracles. In: Cachin, C., Camenisch, J.L. (eds.) EUROCRYPT 2004. LNCS, vol. 3027, pp. 223–238. Springer, Heidelberg (2004)
10. Boneh, D., Boyen, X., Goh, E.-J.: Hierarchical Identity Based Encryption with Constant Size Ciphertext. In: Cramer, R. (ed.) EUROCRYPT 2005. LNCS, vol. 3494, pp. 440–456. Springer, Heidelberg (2005)
11. Boneh, D., Franklin, M.: Identity-based encryption from the Weil pairing. In: Kilian, J. (ed.) CRYPTO 2001. LNCS, vol. 2139, pp. 213–229. Springer, Heidelberg (2001)
12. Boneh, D., Gentry, C., Waters, B.: Collusion Resistant Broadcast Encryption with Short Ciphertexts and Private Keys. In: Shoup, V. (ed.) CRYPTO 2005. LNCS, vol. 3621, pp. 258–275. Springer, Heidelberg (2005)
13. Boneh, D., Sahai, A., Waters, B.: Fully Collusion Resistant Traitor Tracing with Short Ciphertexts and Private Keys. In: Vaudenay, S. (ed.) EUROCRYPT 2006. LNCS, vol. 4004, pp. 573–592. Springer, Heidelberg (2006)
14. Boneh, D., Shaw, J.: Collusion-Secure Fingerprinting for Digital Data. In: Coppersmith, D. (ed.) CRYPTO 1995. LNCS, vol. 963, pp. 452–465. Springer, Heidelberg (1995)
15. Camenisch, J., Hohenberger, S., Lysyanskaya, A.: Compact E-Cash. In: Cramer, R. (ed.) EUROCRYPT 2005. LNCS, vol. 3494, pp. 302–321. Springer, Heidelberg (2005)
16. Canetti, R., Hohenberger, S.: Chosen-Ciphertext Secure Proxy Re-Encryption. In: ACM CCS 2007, pp. 185–194. ACM Press, New York (2007)
17. Chor, B., Fiat, A., Naor, M.: Tracing Traitors. In: Desmedt, Y.G. (ed.) CRYPTO 1994. LNCS, vol. 839, pp. 257–270. Springer, Heidelberg (1994)
18. ElGamal, T.: A public key cryptosystem and a signature scheme based on discrete logarithms. In: Blakely, G.R., Chaum, D. (eds.) CRYPTO 1984. LNCS, vol. 196, pp. 10–18. Springer, Heidelberg (1985)
19. Green, M., Ateniese, G.: Identity-Based Proxy Re-encryption. In: Katz, J., Yung, M. (eds.) ACNS 2007. LNCS, vol. 4521, pp. 288–306. Springer, Heidelberg (2007)
20. Goyal, V.: Reducing Trust in the PKG in Identity Based Cryptosystems. In: Menezes, A. (ed.) CRYPTO 2007. LNCS, vol. 4622, pp. 430–447. Springer, Heidelberg (2007)
21. Hohenberger, S.: Advances in Signatures, Encryption, and E-Cash from Bilinear Groups. Ph.D. Thesis, MIT (May 2006)
22. Hohenberger, S., Rothblum, G.N., Shelat, A., Vaikuntanathan, V.: Securely Obfuscating Re-encryption. In: Vadhan, S.P. (ed.) TCC 2007. LNCS, vol. 4392, pp. 233–252. Springer, Heidelberg (2007)
23. Kunz-Jacques, S., Pointcheval, D.: About the Security of MTI/C0 and MQV. In: De Prisco, R., Yung, M. (eds.) SCN 2006. LNCS, vol. 4116, pp. 156–172. Springer, Heidelberg (2006)

24. Libert, B., Vergnaud, D.: Unidirectional Chosen-Ciphertext Secure Proxy Re-Encryption. In: Cramer, R. (ed.) PKC 2008. LNCS, vol. 4939, pp. 360–379. Springer, Heidelberg (2008)
25. Matsuo, T.: Proxy Re-encryption Systems for Identity-based Encryption. In: Takagi, T., Okamoto, T., Okamoto, E., Okamoto, T. (eds.) Pairing 2007. LNCS, vol. 4575, pp. 247–267. Springer, Heidelberg (2007)
26. Scott, M.: Authenticated ID-based Key Exchange and remote log-in with simple token and PIN number. Cryptology ePrint Archive: Report 2002/164 (2002)
27. Shamir, A.: Identity based cryptosystems and signature schemes. In: Blakely, G.R., Chaum, D. (eds.) CRYPTO 1984. LNCS, vol. 196, pp. 47–53. Springer, Heidelberg (1985)
28. Tardos, G.: Optimal probabilistic fingerprint codes. In: STOC 2003, pp. 116–125. ACM Press, New York (2003)
29. Waters, B.: Efficient Identity-Based Encryption Without Random Oracles. In: Cramer, R. (ed.) EUROCRYPT 2005. LNCS, vol. 3494, pp. 114–127. Springer, Heidelberg (2005)

A Proof of Theorem 2

Let $(A = h^a, B = h^b, C = h^c, T \stackrel{?}{=} e(g,h)^{abc})$ be a DBDH instance. We show a simple distinguisher $\mathcal{B}'$ built from an adversary $\mathcal{A}$ against first level challenge ciphertexts. For the target user, the public key pk_0 is made of $Z_0 = e(g,h)^{z_0}$, $Y_0 = C = h^c$, $U_{0,\ell} = g^{u_{0,\ell}}$ for $\ell = 0,\ldots,n$ with $z_0, u_{0,0},\ldots,u_{0,n} \stackrel{R}{\leftarrow} \mathbb{Z}_p^*$. All other users' public keys are honestly generated and $\mathcal{B}'$ knows the corresponding secret key $sk_i = (z_i, y_i, u_{i,0},\ldots,u_{i,n})$. Recall that all re-encryption keys must be given to the adversary. Since $\mathcal{B}'$ knows $z_i \in \mathbb{Z}_p$ such that $Z_i = e(g,h)^{z_i}$ for all users (including user 0), it can handle all delegation queries on behalf of all parties acting as delegators.

At the challenge step, $\mathcal{A}$ outputs messages $m_0, m_1 \in \mathbb{G}_T$ and expects to receive a challenge ciphertext encrypted for user 0. To generate it, $\mathcal{B}'$ flips a fair coin $d^\star \stackrel{R}{\leftarrow} \{0,1\}$ and sets

$$C_0 = m_{d^\star} \cdot e(\psi(A), B) \qquad C_1 = T.$$

Since $Y_0 = h^c$, it can be readily observed that $\mathbf{C} = (C_0, C_1)$ is a proper encryption of $m_{d^\star}$ with the encryption exponent $s = ab$ if $T = e(g,h)^{abc}$. If T is random, the bit $d^\star$ is perfectly hidden from $\mathcal{A}$. As usual, $\mathcal{B}'$ decides that $T = e(g,h)^{abc}$ if and only if $\mathcal{A}$'s guess is correct. □

B Binary Collusion-Secure (Fingerprinting) Codes

In order to make the description of the scheme with black-box traceability self-contained, we review in this appendix the definition of *collusion-secure (fingerprinting) codes* from [14]. We only consider *binary codes* (*i.e.* codes defined over $\{0,1\}$) and for more details on collusion-secure codes, we refer the reader to [14,28] and references therein.

We begin by defining some notation:

- $x \in \{0,1\}^n$ is called a *binary word of length n*. For such a word, we write $x = x_1 \ldots x_n$ where $x_i \in \{0,1\}$ is the ith bit of x (for $i \in \{1, \ldots, n\}$).
- Let $I = \{1 \leq i_1 < \ldots < i_j \leq n\}$ be a set of indices. For a word $x \in \{0,1\}^n$, $x_{|I}$ denotes the subword $x_{i_1} \ldots x_{i_j} \in \{0,1\}^n$ made of bits at positions in I.
- Let $W = \{w_1, \ldots, w_j \in \{0,1\}^n\}$ be a set of words, and let I be the set of all positions where all strings in W are equal, *i.e.* I is the maximal set such that $w_{1|I} = \cdots = w_{k|I}$. The *feasible set* $FS(W)$ of W is defined as the set of all strings that are equal to $w_1, \ldots, w_k$ at positions in I, *i.e.*

$$FS(W) = \{x \in \{0,1\}^n : x_{|I} = w_{1|I} = \cdots = w_{k|I}\}.$$

The formal definition of collusion-secure codes proposed by Boneh and Shaw in [14] is the following:

Definition 6. *Let* $0 < k \leq N$ *be positive integers and* $\varepsilon \in (0,1]$. *A* binary (k, N, ε) collusion-secure code of length n *consists of a tracing algorithm* $\mathcal{T}$, *a set* $\mathbb{C}$ *called the* codebook, *of indexed* codeswords w_i *for* $1 \leq i \leq N$ *and a trapdoor* τ. *These are such that for all collusions* $C \subset \{1, \ldots, N\}$ *of size at most* k, $W = \{w_i : i \in C\}$, *and for all (unbounded) algorithms* $\mathcal{A}$ *it holds that*

$$\Pr\left[\mathcal{T}(x, \tau) \in C | x \in FS(W); x \leftarrow \mathcal{A}(W)\right] > 1 - \varepsilon,$$

where the probability is taken over the random coins of $\mathcal{T}$ *and* $\mathcal{A}$.

C Details of the Scheme with Black-Box Tracing

The variant with black-box traceability is very close to the scheme of section 3 and we just outline the simple modifications that are required.

As in [2], we assume that pirate devices do not retain state information from prior re-encryptions when run in tracing mode.

Unlike what occurs in the scheme of section 3, the black-box tracing algorithm does not allow to incriminate delegatees when we run it on input of a pirate sub-key that decrypts at level 2. The reason is that the reconstructed word eventually lies in the feasible set of codewords assigned to *all* re-encryption keys (i.e. those assigned to dishonest delegatees as well as those corresponding to honest ones) that were made available to the coalition.

Global-setup(λ)**:** is the same as in section 3.2.

Keygen(λ)**:** is as in section 3.2 with the difference that user i also selects a set $\mathbb{C}_i$ of N binary words $w_{i,1}, \ldots, w_{i,N}$ of length n that form a (k, N, ε) collusion-secure code. The latter is generated with an underlying trapdoor τ_i to be used by its tracing procedure and that is also part of user i's private key. For codewords, elements $U_{i,\ell}, V_{i,\ell}$ define functions $F_{V_i} : \{0,1\}^n \rightarrow \mathbb{G}_2$ and $F_{U_i} : \{0,1\}^n \rightarrow \mathbb{G}_1$ as in section 3.2.

ReKeygen, Enc_2, Enc_1, **ReEnc**, Dec_2 and Dec_1 also remain unchanged.

Trace$(sk_i, \mathbb{P})$: given oracle access to a pirate proxy $\mathbb{P}$ that correctly re-encrypts with probability δ, the tracing algorithm conducts the following steps.

Let $pk_t = (Z_t, Y_t, U_{t,0}, \ldots, U_{t,n})$ be the public key under which $\mathbb{P}$ re-encrypts ciphertexts. For $\ell = 1, \ldots, n$, initialize a counter $ctr_\ell \leftarrow 0$ and run the following test $L = 16\lambda/\delta$ times:

1. Choose a random message $m \in \mathbb{G}_T$ and encrypt it using a random exponent $s \xleftarrow{R} \mathbb{Z}_p^*$ to get a ciphertext $\mathbf{C} = (C_0, C_1, C_{2,0}, C_{2,1}, \ldots, C_{2,n})$.
2. Replace element $C_{2,\ell}$ with a random element from $\mathbb{G}_1$.
3. Query the pirate proxy $\mathbb{P}$ on the altered ciphertext.
4. If $\mathbb{P}$ actually re-encrypts $\mathbf{C}$ as a first level ciphertext $\mathbf{C}' = (C_0, C_1')$ with $C_1' = e(g^s, Y_t)^{z_i}$, increase ctr_ℓ.

After these L iterations, set $w_\ell^{\mathbb{P}} \leftarrow 1$ if $ctr_\ell < 4\lambda$ and $w_\ell^{\mathbb{P}} \leftarrow 0$ otherwise.

The decoded n-bit word $w^{\mathbb{P}}$ is finally taken as input by the tracing procedure of the collusion-secure code that uses the trapdoor τ_i to uncover the identity of a rogue proxy with probability ε.

If I denotes the set of positions where all codewords of the coalition are identical, bits of $w^{\mathbb{P}}$ outside I can be arbitrarily chosen by the pirate device (that can notice the ill-formedness of the ciphertext when its altered component is $C_{2,\ell}$ for $\ell \notin I$). But it does not matter since, as in [2], the tracing system of the code only needs a word $w^{\mathbb{P}} \in \{0,1\}^n$ *inside* the feasible set.

It is essentially routine to prove the black-box traceability property using ideas from [2] but a slightly different assumption is needed. As in [2], we first have to count on the difficulty of DDH in $\mathbb{G}_1$ within asymmetric pairing configuration. This assumption obviously requires the infeasibility of inverting $\psi : \mathbb{G}_2 \rightarrow \mathbb{G}_1$ and found several applications (see [26,15,2] for instance).

Definition 7. *The* ***eXternal Diffie-Hellman assumption*** *(XDH) in asymmetric bilinear groups $(\mathbb{G}_1, \mathbb{G}_2)$ posits the hardness of the Decisional Diffie-Hellman problem in $\mathbb{G}_1$: given $(g^a, g^b) \in {\mathbb{G}_1}^2$, distinguishing g^{ab} from random should be hard. A distinguisher's advantage can be defined as in definition 4.*

The second assumption that we make is a generalization – introduced in [3] – of the computational BDH assumption (CBDH).

Definition 8. *The* ***Generalized Bilinear Diffie-Hellman Problem*** *(GBDH) is, given $(h^a, h^b, h^c) \in \mathbb{G}_2^3$, to come up with a pair $(g', e(g', h)^{abc}) \in \mathbb{G}_1 \times \mathbb{G}_T$.*

The GBDH assumption is non-standard but it is worth mentioning that any algorithm breaking it would also be able to solve the Decision Tripartite Diffie-Hellman problem in $\mathbb{G}_2$ which is to distinguish h^{abc} from random given (h^a, h^b, h^c) and that has been more widely used (see [13] for instance).

Theorem 4. *The modified scheme is black-box k-traceable assuming that the code is a (k, N, ε)-collusion-secure code of length n, that the XDH assumption* *holds in $\mathbb{G}_1$ and that the GBDH problem is hard. More concretely, the advantage of any PPT adversary $\mathcal{A}$ in constructing an untraceable re-encryption device that translates ciphertexts with probability δ after having obtained k re-encryption keys is at most*

$$Adv(\mathcal{A})^{\mathsf{TPRE}} \leq \varepsilon + n \cdot (Adv^{\mathsf{GDBH}}(\mathcal{B}'') + \exp(-\lambda))$$

if $\delta > 2 \cdot Adv^{(}\mathcal{B}')^{\mathsf{XDH}}$ where $\mathcal{B}'$, $\mathcal{B}''$ are PPT algorithm that are built on $\mathcal{A}$.

Proof. Given an adversary $\mathcal{A}$ that outputs a pirate device $\mathbb{P}$ translating ciphertexts with probability δ, we construct an attacker $\mathcal{A}'$ against the collusion-secure code. The latter adversary takes a set of k codewords and outputs a new one w'. As in [2], we show that, with all but negligible probability, $\mathcal{A}'$ avoids being traced whenever $\mathcal{A}$ does. Algorithm $\mathcal{A}'$ takes as input a set of random codewords $W = \{w_1, \ldots, w_k\}$ and generates public keys on behalf of all honest and corrupt users $i \in HU \cup CU$. Codewords of W are used to define the target user's codebook while $\mathcal{A}'$ generates itself the codebooks that are part of other users' private keys. At the jth re-encryption key of the shape (pk_0, pk_j) (i.e. involving user 0 as a delegator and pk_j as a delegatee's public key), $\mathcal{A}'$ fetches a fresh codeword from W and assigns it to the re-encryption key R_{0j} which is returned to $\mathcal{A}$.

Eventually, $\mathcal{A}$ outputs a pirate translation device $\mathbb{P}$ which is run in tracing mode so as to finally reconstruct a n-bit word w'. As in [2], it can be shown that w' falls outside $FS(W)$ with probability smaller than

$$n \cdot (Adv^{\mathsf{GDBH}}(\mathcal{B}'') + \exp(-\lambda)). \tag{5}$$

Let I be the set of positions that are identical in all words of W. For indices $\ell^\star \in I$ such that $w_{\ell^\star} = 0$, lemma 1 first shows that $\mathbb{P}$ re-encrypts ciphertexts where $C_{2,\ell^\star}$ is random with probability negligibly close to δ unless the XDH assumption is false. For indices $\ell^\star \in I$ where $w_{\ell^\star} = 1$, lemma 2 gives an upper bound on $\mathbb{P}$'s chance to succeed in translating ciphertexts where $C_{2,\ell^\star}$ is perturbed. The claimed bound (5) is obtained through a similar analysis to [2]. □

Lemma 1. *For any $\ell^\star \in \{1, \ldots, n\}$, if $w_{0,j,\ell^\star} = 0$ in all codewords $w_{0,j}$ associated with re-encryption keys available to the coalition, $\mathbb{P}$ has probability at least $p_0 \geq \delta - Adv^{XDH}(\lambda)$ to re-encrypt ciphertexts where $C_{2,\ell^\star}$ was tampered with.*

Proof. Towards a contradiction, assume that an adversary $\mathcal{A}$ comes up with a pirate device $\mathbb{P}$, where $w_{0,j,\ell^\star} = 0$ in all underlying codewords $w_{0,j}$, that re-encrypts ciphertexts with probability $p_0 \leq \delta - \gamma$ for some $\gamma > 0$. Then, there exists an algorithm $\mathcal{B}'$ breaking the XDH assumption with advantage γ.

On input of an XDH instance $(A = g^a, B = g^b, \eta \stackrel{?}{=} g^{ab})$, this algorithm $\mathcal{B}'$ first prepares a set of public keys by defining the target user's public key $pk_0 = (Z_0, Y_0, U_{0,0}, U_{0,1}, \ldots, U_{0,n})$ as $Z_0 = e(g,h)^{z_0}$, $Y_0 = h^{y_0}$, with $z_0, y_0 \stackrel{R}{\leftarrow} \mathbb{Z}_p^*$, $U_{0,\ell^\star} = A = g^a$ and $U_{0,\ell} = g^{u_{0,\ell}}$ with $u_{0,\ell} \stackrel{R}{\leftarrow} \mathbb{Z}_p^*$ for $\ell \in \{0, \ldots, n\} \backslash \{\ell^\star\}$. Note that pre-images $V_{0,\ell} = h^{u_{0,\ell}}$ so that $\psi(V_{0,\ell}) = U_{0,\ell}$ are also available for all

$\ell \in \{0, \ldots, n\} \backslash \{\ell^\star\}$. For other public keys pk_i with $i \in \{1, \ldots, N\}$, $\mathcal{B}'$ simply runs the key generation algorithm according to its specification.

As $w_{0,j,\ell^\star} = 0$ for all codewords $w_{0,j}$ assigned to re-encryption keys R_{0j} queried by $\mathcal{A}$, $\mathcal{B}'$ is able to compute such keys $R_{0j} = (w_{0,j}, Y_j^{z_0} \cdot F_{V_0}(w_{0,j})^r, h^r)$ by running ReKeygen (although it does not know $V_{0,\ell^\star} = \psi^{-1}(g^a)$). When $\mathcal{A}$ outputs a pirate ciphertext translator $\mathbb{P}$, $\mathcal{B}'$ feeds it with a ciphertext

$$C_0 = m \cdot e(B,h)^{z_0} \quad C_1 = B \quad C_{2,\ell^\star} = \eta \quad C_{2,\ell} = B^{u_{0,\ell}} \text{ for } \ell \in \{0, \ldots, n\} \backslash \{\ell^\star\}$$

for a random message $m \stackrel{R}{\leftarrow} \mathbb{G}_T$. The device $\mathbb{P}$ then generates a re-encryption $\mathbf{C}' = (C_0, C_1')$. Given the public key $pk_t = (Z_t, Y_t, U_{t,0}, \ldots, U_{t,n})$ of the user receiving re-encryptions from $\mathbb{P}$, $\mathcal{B}'$ can check whether $\mathbf{C}'$ was successfully translated by testing if $C_1' = e(B, Y_t)^{z_0}$. If yes, $\mathcal{B}'$ outputs 1 (meaning that $\eta = g^{ab}$). Otherwise, it returns 0 and bets that η is random. □

Lemma 2. *For any $\ell^\star \in \{1, \ldots, n\}$, if $w_{0,j,\ell^\star} = 1$ in all codewords $w_{0,j}$ embedded in re-encryption keys of colluding proxies, then $\mathbb{P}$ has probability at most $p_1 \leq Adv^{GBDH}(\lambda)$ to re-encrypt ciphertexts where $C_{2,\ell^\star}$ was tampered with.*

Proof. Assume that $\mathcal{A}$ is an adversary producing a re-encryption box $\mathbb{P}$ that has non-negligible probability p_1 of re-encrypting ciphertexts where $C_{2,\ell^\star}$ has been replaced by a random element of $\mathbb{G}_1$. We construct a distinguisher $\mathcal{B}''$ solving a computational GBDH instance $(A = h^a, B = h^b, C = h^c)$.

$\mathcal{B}''$ first generates a set of public keys. The target user's public key is set as $pk_0 = (Z_0, Y_0, U_{0,0}, U_{0,1}, \ldots, U_{0,n})$ where $Z_0 = e(\psi(A), B) = e(g,h)^{ab}$, $Y_0 = h^{y_0}$ and $U_{0,\ell} = g^{u_{0,\ell}}$ with $y_0, u_{0,\ell} \stackrel{R}{\leftarrow} \mathbb{Z}_p^*$ for $\ell \in \{0, \ldots, n\} \backslash \{\ell^\star\}$. The remaining public key component is chosen as $U_{0,\ell^\star} = g^{\alpha_{0,\ell^\star}} \cdot \psi(A)^{\beta_{0,\ell^\star}}$ for random integers $\alpha_{0,\ell^\star}, \beta_{0,\ell^\star} \stackrel{R}{\leftarrow} \mathbb{Z}_p^*$. Note that $V_{0,\ell^\star} = h^{\alpha_{0,\ell^\star}} \cdot A^{\beta_{0,\ell^\star}}$ is also computable as well as $V_{0,\ell} = h^{u_{0,\ell}}$ for $\ell \neq \ell^\star$. For other users $i \in \{1, \ldots, n\}$, public keys $pk_i = (Z_i, Y_i, U_{i,0}, U_{i,1}, \ldots, U_{i,n})$ are calculated as specified by the key generation algorithm and private elements $(z_i, y_i, u_{i,0}, \ldots, u_{i,n})$ are known to $\mathcal{B}''$.

Given that $w_{0,j,\ell^\star} = 1$ for all of the k codewords $w_{0,j}$ contained in re-encryption keys R_{0j} that $\mathcal{A}$ must be provided with, these keys can be generated by choosing $r \stackrel{R}{\leftarrow} \mathbb{Z}_p^*$ and setting

$$A_{0j} = V_{0,\ell^\star}^r \cdot B^{-\frac{y_j \alpha_{0,\ell^\star}}{\beta_{0,\ell^\star}}} \cdot \prod_{\ell=0,\ell\neq\ell^\star}^{n} \left(V_{0,\ell}^r \cdot B^{-\frac{y_j u_{0,\ell}}{\beta_{0,\ell^\star}}}\right)^{w_{0,j,\ell}}, \qquad B_{0j} = h^r \cdot B^{-\frac{y_j}{\beta_{0,\ell^\star}}}$$

which provides a valid re-encryption key $R_{0j} = (w_{0,j}, A_{0j}, B_{0j})$ since $X_j = h^{x_j}$ and, if we define $\tilde{r} = r - by_j/\beta_{0,\ell^\star}$, we have $B_{0j} = h^{\tilde{r}}$ and

$$\begin{aligned} A_{0j} &= V_{0,\ell^\star}^{\tilde{r}} \cdot (h^{\alpha_{0,\ell^\star}} \cdot A^{\beta_{0,\ell^\star}})^{\frac{by_j}{\beta_{0,\ell^\star}}} \cdot B^{\frac{y_j \alpha_{0,\ell^\star}}{\beta_{0,\ell^\star}}} \cdot \prod_{\ell=0,\ell\neq\ell^\star}^{n} \left(V_{0,\ell}^{\tilde{r}} \cdot h^{u_{0,\ell}\frac{by_j}{\beta_{0,\ell^\star}}} \cdot B^{-\frac{y_j u_{0,\ell}}{\beta_{0,\ell^\star}}}\right)^{w_{0,j,\ell}} \\ &= (h^{y_j})^{ab} \cdot V_{0,\ell^\star}^{\tilde{r}} \cdot \prod_{\ell=0,\ell\neq\ell^\star}^{n} \left(V_{0,\ell}^{\tilde{r}}\right)^{w_{0,j,\ell}}. \end{aligned}$$

When $\mathcal{B}''$ obtains a pirate device $\mathbb{P}$ from $\mathcal{A}$, it probes it with a ciphertext

$$C_0 \stackrel{R}{\leftarrow} \mathbb{G}_T \quad C_1 = \psi(C) \quad C_{2,\ell^\star} \stackrel{R}{\leftarrow} \mathbb{G}_1 \quad C_{2,\ell} = \psi(C)^{u_{0,\ell}} \text{ for } \ell \in \{0,\ldots,n\}\backslash\{\ell^\star\}$$

which is a valid ciphertext (with the encryption exponent $s = c$) where $C_{2,\ell^\star}$ has been replaced by a random element. By assumption, $\mathbb{P}$ is assumed to re-encrypt it under some public key $pk_t = (X_t, Y_t, U_{t,0}, U_{t,1}, \ldots, U_{t,n})$ that was not involved in a re-encryption key query with user 0 acting as a delegator. When obtaining a re-encryption $\mathbf{C}'_t = (C_0, C'_1) = \big(C_0, e(g, Y_t)^{abc}\big) = \big(C_0, e(\psi(Y_t), h)^{abc}\big)$, $\mathcal{B}''$ outputs a pair $(\psi(Y_t), C'_1)$ which violates the GBDH assumption. □

Security and Anonymity of Identity-Based Encryption with Multiple Trusted Authorities

Kenneth G. Paterson and Sriramkrishnan Srinivasan

Information Security Group,
Royal Holloway, University of London,
Egham, Surrey TW20 0EX, U.K.
{kenny.paterson,s.srinivasan}@rhul.ac.uk

Abstract. We consider the security of Identity-Based Encryption (IBE) in the setting of multiple Trusted Authorities (TAs). In this multi-TA setting, we envisage multiple TAs sharing some common parameters, but each TA generating its own master secrets and master public keys. We provide security notions and security models for the multi-TA setting which can be seen as natural extensions of existing notions and models for the single-TA setting. In addition, we study the concept of TA anonymity, which formally models the inability of an adversary to distinguish two ciphertexts corresponding to the same message and identity but generated using different TA master public keys. We argue that this anonymity property is a natural one of importance in enhancing privacy and limiting traffic analysis in multi-TA environments. We study a modified version of a Fujisaki-Okamoto conversion in the multi-TA setting, proving that our modification lifts security and anonymity properties from the CPA to the CCA setting. Finally, we apply these results to study the security of the Boneh-Franklin and Sakai-Kasahara IBE schemes in the multi-TA setting.

Keywords: identity-based encryption, multi-TA IBE, anonymity, multiple trusted authorities.

1 Introduction

The concept of Identity-Based Encryption (IBE) was first introduced by Shamir in [23]. In identity-based cryptography(IBC), arbitrary identifying strings such as e-mail addresses or IP addresses can be used to form public keys for users, with the corresponding private keys being created by a Trusted Authority (TA) who is in possession of a system-wide master secret. Then a party Alice who wishes, for example, to encrypt to a party Bob need only know Bob's identifier and the system-wide public parameters. This approach eliminates certificates and the associated processing and management overheads from public key cryptography. The first efficient and secure constructions for IBE were not forthcoming till the work of Cocks [12], and the pairing-based solutions of Sakai, Ohigishi and Kasahara [22] and Boneh and Franklin [6]. Boneh and Franklin [6] also proposed

S.D. Galbraith and K.G. Paterson (Eds.): Pairing 2008, LNCS 5209, pp. 354–375, 2008.

the first security models for IBE and gave schemes secure in the random oracle model [5]. Since the publication of these first results, there has been an explosion of interest in IBE and related cryptographic primitives.

1.1 Motivation and Contributions

Historically, anonymous encryption was motivated in the context of mobile communication. In the standard public key setting, an entity B sends a user A ciphertexts of messages encrypted under A's public key (and vice versa), over a wireless network. It is reasonable to assume that A and B will want to keep their identities hidden from an eavesdropper who can see all ciphertexts on the network. This is possible only when ciphertexts do not leak information about the public keys used to create them, a notion formalised as key-privacy in [4].

If an IBE scheme is used instead of a standard public key scheme, the equivalent notion is that of recipient anonymity: the ciphertext should not leak the identity of the (intended) recipient. In this setting, we assume that there is a single global TA issuing keys to all users in the system, and that all ciphertexts are created using the public parameters of that single global TA. With a small number of exceptions (upon which we elaborate in the related work section below), the security models proposed for IBE to date all consider such a single-TA setting.

It is however possible to envisage scenarios as above but with multiple, independent TAs (perhaps sharing some common system parameters). In some applications, each user may only have a single private key issued by one of the TAs, while in others, users could have multiple private keys for the same identity string with the different private keys being issued by different TAs. In both settings, in addition to the usual IBE security notions of indistinguishability and recipient anonymity, the notion of *TA anonymity* arises as being both natural and of fundamental importance. Here, we desire that an adversary should find it difficult to distinguish ciphertexts produced using different TA master public keys, even if the ciphertext is for the same message and identity string. As well as being a natural security notion for the multi-TA setting, TA anonymity may have practical significance. For example, we can imagine a coalition of TAs operating in a wireless setting where all ciphertexts can be captured from the network by an adversary. In such a scenario, if the ciphertext were to somehow leak the identity of the TA, then this would open up avenues for traffic analysis. In a hostile environment, traffic analysis can lead to the leaking of information relating to which entities are communicating and how frequently, which can often reveal important intelligence about the nature of operations.

In this paper we extend the usual indistinguishability and recipient anonymity notions for IBE security to the multi-TA setting, and, in addition, formalize the notion of TA anonymity. We introduce a modified version of the Fujisaki-Okamoto conversion for the multi-TA setting, proving that our modified transformation lifts security and anonymity properties from the CPA to the CCA setting. We then apply these results to study the security and anonymity of the Boneh-Franklin [6] and the Sakai-Kasahara [21] IBE schemes in the multi-TA setting.

As well as formalising the notion of TA anonymity, our work also establishes new results concerning the recipient anonymity of important IBE schemes. For example, to the best of our knowledge, no CCA-secure variant of the Boneh-Franklin IBE scheme was previously known to have recipient anonymity. Moreover, we show that the Sakai-Kasahara scheme (and a CCA-secure variant of it) enjoys recipient anonymity, contradicting a claim of [7].

1.2 Related Work

Anonymity. In the standard public key setting, the notion of key-privacy [4] captures the requirement that an adversary in possession of a ciphertext cannot tell which public key was used to create the ciphertext, i.e the ciphertext should not leak information about the public key. The equivalent notion in the IBE setting is the notion of recipient anonymity, i.e the ciphertext should not leak the identity of the recipient. The systematic study of recipient anonymity was initiated in [1], motivated both by its intrinsic interest in IBE and for its application in constructing PEKS (Public Key Encryption with Keyword Search) schemes from IBE schemes. Since then, recipient anonymity has quickly become a standard security property for IBE schemes. IBE schemes known to offer recipient anonymity include the CPA-secure `BasicIdent` scheme of Boneh and Franklin [6] and the IBE schemes of Gentry [16].

Multi-TA Security for IBE. Holt [18] also considered security of IBE in the multi-TA setting, motivated by earlier work on anonymous credential systems [19,9]. Two notions of key privacy for IBE were outlined in [18]. The first, termed "identity-based indistinguishability of identity under chosen plaintext attack" (ID-II-CPA), is just the standard single-TA recipient anonymity notion. The second is termed "identity-based indistinguishability of key generator under chosen plaintext attack" (ID-IKG-CPA), and is roughly similar to our notion of multi-TA TA anonymity under chosen plaintext attack (m-TAA-CPA). However, the ID-IKG-CPA security model in [18], while allowing corruption of TAs, does not allow the adversary to extract any user private keys at all. Our m-TAA-CPA model is strictly stronger, allowing both corruption of TAs and extraction of private keys (even for the challenge TA)[1]. Moreover, [18] only considers the CPA setting, showing that the `BasicIdent` scheme of [6] has ID-II-CPA and ID-IKG-CPA security. However, even the proofs for these CPA cases are at best informal. In this paper, we consider the CCA setting, use stronger security notions, and give rigorous proofs.

Wang and Cao [24] gave examples of IBE schemes enjoying reduced ciphertext expansion and reduced computation when the sender sends the same message to a single identity using multiple, different master public keys belonging to different TAs, such that the message can be recovered with a private key issued

[1] Holt's work allows the adversary to dynamically instantiate new TAs during its attack but without any adversarial input to the set up process, while we set up all the TAs at the start of the security games. These two approaches are easily seen to have equivalent strength.

for that identity by any one of the TAs. However, the security models presented in [24] are the standard single-TA, indistinguishability-based security models, and no consideration is given to how security may be affected by encrypting the same message using multiple master public keys. In addition, the schemes of [24] reuse randomness to enhance efficiency, and this is not formally addressed in the security analysis. Barbosa and Farshim [3] consider the security of multi-recipient IBE with randomness re-use, but only in the single-TA setting.

Chase [10] has considered Attribute Based Encryption (ABE), a generalisation of IBE, in the setting of multiple authorities. In her work, a user is equipped with private keys corresponding to attributes from different TAs and the user is only able to decrypt a ciphertext if he possesses a threshold of attributes from different TAs. Chase does not seem to consider the issue of TA anonymity.

Anonymity for Hierarchical IBE. Anonymity properties for IBE have already been studied in the hierarchical setting [1,8]. Anonymous Hierarchical IBE (AHIBE) is related to, but different from, our notion of TA anonymity for IBE. In AHIBE, a single root TA generates public parameters and a master secret, using which the master secrets of all sub-TAs are produced. Ciphertexts are then anonymous, in that an adversary cannot distinguish which identity was used when producing a ciphertext, where now identities are comprised of a vector of strings identifying a hierarchy of TAs and a final user. On other hand, in our multi-TA setting, there is no single root authority, but rather a group of independent TAs (who may share some common parameters). The "right" generalisation of our multi-TA IBE concept to the hierarchical setting would then involve multiple, independent root TAs, each being the root of a tree of TAs and users. Thus we would have a forest of trees, and would then wish to study anonymity properties of ciphertexts in this multi-HIBE setting. We leave further development of this line of research to future work.

Fujisaki-Okamoto Conversions. Yang *et al.* [25] and Kitagawa *et al.* [20] considered the adaptation of the Fujisaki-Okamoto conversions of [14] and [13] to the IBE setting, showing that simple modifications of the original Fujisaki-Okamoto approaches can be used to build IBE schemes with IND-CCA security from schemes having only OW-CPA and IND-CPA security, respectively, in the random oracle model. We adapt the Fujisaki-Okamoto technique of [13] to the multi-TA setting, showing how it lifts security and anonymity properties from the CPA to the CCA setting.

2 Background and Definitions

In this section, we provide basic definitions needed for the remainder of the paper.

Definition 1. *A pairing-friendly group generator* `PairingGen` *is a polynomial time algorithm with input* 1^k *and output a tuple* (G, G_T, e, q, P)*. Here* G, G_T *are groups of prime order* q*,* P *generates* G*, and* $e : G \times G \rightarrow G_T$ *is a bilinear,* *non-degenerate and efficiently computable map. By convention, G is an additive group and G_T multiplicative.*

For ease of presentation, we work exclusively in the setting where e is symmetric; our definitions and results can be generalised to the asymmetric setting where $e : G_1 \times G_2 \rightarrow G_T$, with G_1 and G_2 being different groups. Further details concerning the basic choices that are available when using pairings in cryptography can be found in [15].

Definition 2. *A function $\epsilon(k)$ is said to be negligible if, for every c, there exists k_c such that $\epsilon(k) \leq k^{-c}$ for every $k \geq k_c$.*

Definition 3. *We define the advantage of an algorithm $\mathcal{A}$ in solving the Bilinear Diffie-Hellman (BDH) problem in (G, G_T) to be:*

$$\boldsymbol{Adv}_{\mathcal{A}}^{BDH}(k) = \Pr(\mathcal{A}(aP, bP, cP) = e(P, P)^{abc})$$

where $a, b, c \leftarrow \mathbb{Z}_q$. Here, we implicitly assume that parameters (G, G_T, e, q, P) are given to $\mathcal{A}$ as additional inputs. We say that the BDH problem is hard *in (G, G_T) if no polynomial-time algorithm that solves the BDH problem in (G, G_T) has a non-negligible advantage.*

Definition 4. *We define the advantage of an algorithm $\mathcal{A}$ in solving the ℓ-Bilinear Diffie-Hellman Inversion (ℓ-BDHI) problem in (G, G_T) to be:*

$$\boldsymbol{Adv}_{\mathcal{A}}^{\ell\text{-}BDHI}(k) = \Pr(\mathcal{A}(xP, x^2P, \ldots, x^{\ell}P) = e(P, P)^{1/x})$$

where $x \leftarrow \mathbb{Z}_q$. Here, we implicitly assume that parameters (G, G_T, e, q, P) are given to $\mathcal{A}$ as additional inputs. We say that the ℓ-BDHI problem is hard *in (G, G_T) if no polynomial-time algorithm that solves the ℓ-BDHI problem in (G, G_T) has a non-negligible advantage.*

Definition 5. *A (single-TA) IBE scheme is defined in terms of four algorithms:*

- `Setup`: On input 1^k, outputs a master public key *mpk* which includes system parameters *params*, and a master secret key *msk*. We assume that *params* contains descriptions of the message and ciphertext spaces, MsgSp and CtSp, and that $\mathsf{MsgSp} \subset \{0,1\}^*$.
- `KeyDer`: A key derivation algorithm that on input *mpk*, *msk* and identifier $id \in \{0,1\}^*$, returns a private key usk_{id}. This algorithm may or may not be randomized.
- `Enc`: An encryption algorithm that on input *mpk*, identifier $id \in \{0,1\}^*$ and message $m \in \mathsf{MsgSp}$, returns a ciphertext $c \in \mathsf{CtSp}$. This algorithm is usually randomized; in subsequent descriptions, we will write $c = \mathtt{Enc}(mpk, id, m; r)$ when we wish to emphasize that randomness r (drawn from some space RSp) is used when performing an encryption.
- `Dec`: A decryption algorithm that on input *mpk*, a private key usk_{id} and a ciphertext $c \in \mathsf{CtSp}$, returns either a message $m \in \mathsf{MsgSp}$ or a failure symbol $\perp$.

These algorithms must satisfy the standard consistency requirement that decryption undoes encryption, i.e. $\forall m \in \mathsf{MsgSp}, \mathtt{Dec}(mpk, usk_{id}, c) = m$ where $c = \mathtt{Enc}(mpk, id, m)$.

3 Multi-TA Security

We formalize IBE in the multi-TA setting and the associated notions of security. A multi-TA IBE scheme is defined in terms of five algorithms:

- `CommonSetup`: On input 1^k, outputs *params*, a set of system parameters shared by all TAs; $\mathcal{TA} = \{ta_i : 1 \leq i \leq n\}$ will represent the set of (labels of) TAs, where $n = n(k) \in \mathbb{N}$.
- `TASetup`: On input *params*, outputs a master public key *mpk* (which includes *params*), and a master secret key *msk*. This algorithm is randomized and executed independently for each TA in $\mathcal{TA}$.
- `KeyDer`, `Enc`, `Dec`: These are all as per a normal IBE scheme.

Note that we explicitly include a `CommonSetup` algorithm which outputs *params*, a set of system parameters shared by all TAs. The different TAs will of course have different master public keys and master secret keys. Our model is capable of handling situations where no such common system parameters are used, simply by setting *params* to be the security parameter 1^k. Nevertheless, it is not unreasonable to assume that the different TAs may share some common system parameters (e.g. the output of a pairing parameter generator in the Boneh-Franklin IBE scheme), since cryptographic schemes and related parameters are often standardised by bodies like ISO, NIST or IEEE P1363, and then used in multiple domains by different authorities. Indeed, the IEEE P1363.3 working group aims to produce a set of standards specific to identity based cryptography and we may expect specific recommendations for cryptographic parameters to be produced by this group in due course. For the concrete schemes considered in this paper, common parameters are needed in order to achieve our notion of TA anonymity; doing so without having some (non-trivial) common parameters is an interesting open problem.

We also need a standard consistency requirement on such a scheme. In addition, in applications, we may require a robustness condition – decrypting a ciphertext created using an identity and the master public key of one TA should fail to decrypt using a private key for that (or any other) identity issued by another TA. We return to this issue in Section 5.

In the security games defined below, *TASet* represents the set of TAs that have been compromised, i.e queried for their master secret keys, $IDSet_{ta}$ represents the set of identities queried for private keys for each $ta \in \mathcal{TA}$, while $CSet_{ta}$ represents the set of identity/ciphertext pairs on which decryption queries have been performed for each $ta \in \mathcal{TA}$. In these games, $MPK = \{mpk_{ta} : ta \in \mathcal{TA}\}$ and $MSK = \{msk_{ta} : ta \in \mathcal{TA}\}$ represent the set of all master public keys and all master secret keys, respectively. For each experiment defined below, we associate to an adversary $\mathcal{A}$ and a bit $b \in \{0,1\}$, the advantage of the adversary for a given "notion-attack" combination, which is defined as:

$$\mathbf{Adv}_{\mathcal{A}}^{\text{notion-atk}}(k) = \left|\Pr[\mathbf{Exp}_{\mathcal{A}}^{\text{notion-atk-1}}(k) = 1] - \Pr[\mathbf{Exp}_{\mathcal{A}}^{\text{notion-atk-0}}(k) = 1]\right|$$

A scheme is said to be "notion-atk"-secure if the advantage of all PPT adversaries is negligible as a function of the security parameter k.

We focus below on Chosen Ciphertext Attacks (CCA) for three different security notions: indistinguishability, recipient anonymity and TA anonymity. Removing adversarial access to decryption oracles gives the same notions of security against a Chosen Plaintext Attack (CPA).

In each of the first two cases (namely, indistinguishability of messages and recipient anonymity), setting $n = 1$ and removing access to the `Corrupt` oracle gives us a security notion that coincides with a known (single-TA) IBE security notion. Formally, to obtain a (single-TA) IBE scheme, we need to combine the `CommonSetup` and `TASetup` algorithms of the multi-TA scheme into a single `Setup` algorithm. In what follows, we will refer to this scheme as being the *corresponding single-TA IBE scheme.* In the third case, TA anonymity, the security notion is inappropriate for the single-TA setting.

3.1 m-IND-CCA Security

We first define the m-IND-CCA security notion that captures indistinguishability of messages under chosen ciphertext attacks in the multi-TA setting.

Experiment $\mathbf{Exp}_{\mathcal{A}}^{\text{m-IND-CCA-b}}(k)$ $params \leftarrow \texttt{CommonSetup}(1^k)$ $TASet \leftarrow \emptyset$ $\forall ta \in \mathcal{TA}$, $(mpk_{ta}, msk_{ta}) \leftarrow \texttt{TASetup}(params)$ $IDSet_{ta} \leftarrow \emptyset$, $CSet_{ta} \leftarrow \emptyset$ $(ta, id, m_0, m_1, state) \leftarrow$ $\mathcal{A}^{\texttt{Corrupt},\texttt{KeyDer},\texttt{Dec}}(\texttt{find}, MPK)$ $c^* \leftarrow \texttt{Enc}(mpk_{ta}, id, m_b)$ $b' \leftarrow \mathcal{A}^{\texttt{Corrupt},\texttt{KeyDer},\texttt{Dec}}(\texttt{guess}, c^*, state)$ If $\{m_0, m_1\} \not\subseteq \mathsf{MsgSp}$ or $\lvert m_0\rvert \neq \lvert m_1\rvert$ or $m_0 = m_1$ then return 0 If $ta \notin TASet$, $id \notin IDSet_{ta}$ and $(id, c^*) \notin CSet_{ta}$ then return b' else return 0	Oracle $\texttt{Corrupt}(ta)$ $TASet \leftarrow TASet \cup \{ta\}$ Return msk_{ta} Oracle $\texttt{KeyDer}(ta, id)$ $IDSet_{ta} \leftarrow IDSet_{ta} \cup \{id\}$ $usk_{id,ta} \leftarrow \texttt{KeyDer}(msk_{ta}, id)$ Return $usk_{id,ta}$ Oracle $\texttt{Dec}(ta, id, c)$ $CSet_{ta} \leftarrow CSet_{ta} \cup (id, c)$ $usk_{id,ta} \leftarrow \texttt{KeyDer}(msk_{ta}, id)$ $m \leftarrow \texttt{Dec}(mpk_{ta}, usk_{id,ta}, c)$ Return m

The following theorem relates the m-IND-CCA security of a multi-TA IBE scheme to the IND-CCA security of the corresponding single-TA IBE scheme.

Theorem 1. *Let $atk \in \{CPA, CCA\}$. Then for any m-IND-atk adversary $\mathcal{A}$ against a multi-TA IBE scheme with n TAs having advantage ε and running in time t, there exists an IND-atk adversary $\mathcal{B}$ against the corresponding single-TA IBE scheme with advantage $\frac{\varepsilon}{n}$ and running in time $O(time(\mathcal{A}))$.*

Proof. Suppose there is an m-IND-atk adversary $\mathcal{A}$ against a multi-TA IBE scheme having advantage ε and running in time t. We show how to construct an algorithm $\mathcal{B}$ that uses $\mathcal{A}$ to break the IND-atk security of the corresponding single-TA IBE scheme.

$\mathcal{B}$'s input from its challenger is the public key *mpk* of the single-TA scheme which, by our definitions, includes some public parameters *params* that are output by the `CommonSetup` part of the `Setup` algorithm of the single-TA scheme. $\mathcal{B}$'s task is to break the IND-atk property of the scheme and it does this by acting as a challenger for $\mathcal{A}$.

$\mathcal{B}$ first sets up a multi-TA IBE scheme. It does this by first taking *params* from the public key of the single-TA scheme. If n is the number of TAs in the multi-TA setting, it first picks $i \stackrel{\$}{\leftarrow} \{1, \dots n\}$ and sets $mpk_{ta_i} = mpk$ (note it does not know the corresponding master secret key for this TA). For the remaining $n-1$ TAs it generates the master public keys and master secret keys itself using the `TASetup` algorithm. $\mathcal{B}$ now gives the set of n master public keys to $\mathcal{A}$.

$\mathcal{A}$ then makes a series of TA corrupt queries, extraction queries (and decryption queries in the CCA setting) which $\mathcal{B}$ answers using either its knowledge of the relevant master secret key or by relaying queries to its own challenger. If $\mathcal{A}$ makes a corrupt query on ta_i then $\mathcal{B}$ aborts the simulation.

$\mathcal{A}$ also makes a single query in the challenge phase; if $\mathcal{A}$'s selected TA in this phase is not ta_i, then $\mathcal{B}$ aborts, otherwise $\mathcal{B}$ again uses its own challenger to answer the query. When $\mathcal{A}$ terminates by outputting a bit b', $\mathcal{B}$ simply relays this bit to its challenger.

This completes our description of $\mathcal{B}$'s simulation. Note that $\mathcal{A}$'a view of the simulation is identical to its view in a real attack, unless $\mathcal{B}$ aborts. Moreover $\mathcal{B}$'s output b' is correct if $\mathcal{A}$'s is. It is easy to see that $\mathcal{B}$ aborts with probability $1/n$ and that $\mathcal{B}$ runs in time $O(time(\mathcal{A}))$. The result follows.

3.2 m-RA-CCA Security

Our m-RA-CCA security notion captures the notion of recipient anonymity in the presence of chosen ciphertext attackers, in the multi-TA setting. The single-TA version of the m-RA-CPA security notion was studied in detail in [1], where it was named IBE-ANO-CPA security.

Halevi [17] provides a simple sufficient condition for an IND-CPA public key encryption scheme to have key-privacy: given public keys pk_0 and pk_1 and the encryption of a random message under pk_b for a bit b chosen at random, even a computationally unbounded adversary should have negligible advantage in determining which public key was used. Abdalla *et al.* [1] extended this condition to study recipient anonymity of IND-CPA-secure IBE schemes. We further extend these ideas to study multi-TA IBE schemes in the following sections.

Here, as throughout, we suppress "IBE", since all of our work is in the ID-based setting. We use "RA' in place of "ANO" because we wish to study two forms of anonymity, *viz* recipient anonymity (RA) and TA anonymity (TAA).

Experiment $\mathbf{Exp}_{\mathcal{A}}^{\text{m-RA-CCA-b}}(k)$
$params \leftarrow \texttt{CommonSetup}(1^k)$
$TASet \leftarrow \emptyset$
$\forall ta \in \mathcal{TA}$, $(mpk_{ta}, msk_{ta}) \leftarrow \texttt{TASetup}(params)$, $IDSet_{ta} \leftarrow \emptyset$ and $CSet_{ta} \leftarrow \emptyset$
$(ta, id_0, id_1, m, state) \leftarrow \mathcal{A}^{\texttt{Corrupt},\texttt{KeyDer},\texttt{Dec}}(\texttt{find}, MPK)$
$c^* \leftarrow \texttt{Enc}(mpk_{ta}, id_b, m)$
$b' \leftarrow \mathcal{A}^{\texttt{Corrupt},\texttt{KeyDer},\texttt{Dec}}(\texttt{guess}, c^*, state)$
If $m \notin \mathsf{MsgSp}$ or $id_0 = id_1$ then return 0
If $ta \notin TASet$, $id_0 \notin IDSet_{ta}$, $id_1 \notin IDSet_{ta}$, $(id_0, c^*) \notin CSet_{ta}$ and $(id_1, c^*) \notin CSet_{ta}$ then return b' else return 0

Oracle $\texttt{Corrupt}(ta)$
$TASet \leftarrow TASet \cup \{ta\}$
Return msk_{ta}

Oracle $\texttt{KeyDer}(ta, id)$
$IDSet_{ta} \leftarrow IDSet_{ta} \cup \{id\}$
$usk_{id,ta} \leftarrow \texttt{KeyDer}(msk_{ta}, id)$
Return $usk_{id,ta}$

Oracle $\texttt{Dec}(ta, id, c)$
$CSet_{ta} \leftarrow CSet_{ta} \cup (id, c)$
$usk_{id,ta} \leftarrow \texttt{KeyDer}(msk_{ta}, id)$
$m \leftarrow \texttt{Dec}(mpk_{ta}, usk_{id,ta}, c)$
Return m

Theorem 2. *Let atk $\in$ {CPA, CCA}. Then for any m-RA-atk adversary $\mathcal{A}$ against a multi-TA IBE scheme with n TAs having advantage ε and running in time t, there exists an RA-atk adversary $\mathcal{B}$ against the corresponding single-TA IBE scheme with advantage $\frac{\varepsilon}{n}$ and running in time $O(time(\mathcal{A}))$.*

The proof is similar to that of Theorem 1 and is omitted.

3.3 m-RA-RE-CCA Security

In order to establish the m-RA-CPA/m-RA-CCA security of concrete schemes, it is helpful to work with a related notion, m-RA-RE-CPA/m-RA-RE-CCA security. Our treatment here follows that of [1], with appropriate modifications for the multi-TA setting.

In handling the challenge phase in the following game, the challenger encrypts a random message m' in place of the adversary's choice of message m, hence the choice "RE" in m-RA-RE-CCA to signify "randomized encryption".

Experiment $\mathbf{Exp}_{\mathcal{A}}^{\text{m-RA-RE-CCA-b}}(k)$
$params \leftarrow \texttt{CommonSetup}(1^k)$
$TASet \leftarrow \emptyset$
$\forall ta \in \mathcal{TA}$, $(mpk_{ta}, msk_{ta}) \leftarrow \texttt{TASetup}(params)$, $IDSet_{ta} \leftarrow \emptyset$ and $CSet_{ta} \leftarrow \emptyset$
$(ta, id_0, id_1, m, state) \leftarrow \mathcal{A}^{\texttt{Corrupt},\texttt{KeyDer},\texttt{Dec}}(\texttt{find}, MPK)$
$m' \xleftarrow{\$} \mathsf{MsgSp}$ with $|m'| = |m|$;
$c^* \leftarrow \texttt{Enc}(mpk_{ta}, id_b, m')$
$b' \leftarrow \mathcal{A}^{\texttt{Corrupt},\texttt{KeyDer},\texttt{Dec}}(\texttt{guess}, c^*, state)$
If $m \notin \mathsf{MsgSp}$ or $id_0 = id_1$ then return 0
If $ta \notin TASet$, $id_0 \notin IDSet_{ta}$, $id_1 \notin IDSet_{ta}$, $(id_0, c^*) \notin CSet_{ta}$ and $(id_1, c^*) \notin CSet_{ta}$ then return b' else return 0

Oracle $\texttt{Corrupt}(ta)$
$TASet \leftarrow TASet \cup \{ta\}$
Return msk_{ta}

Oracle $\texttt{KeyDer}(ta, id)$
$IDSet_{ta} \leftarrow IDSet_{ta} \cup \{id\}$
$usk_{id,ta} \leftarrow \texttt{KeyDer}(msk_{ta}, id)$
Return $usk_{id,ta}$

Oracle $\texttt{Dec}(ta, id, c)$
$CSet_{ta} \leftarrow CSet_{ta} \cup (id, c)$
$usk_{id,ta} \leftarrow \texttt{KeyDer}(msk_{ta}, id)$
$m \leftarrow \texttt{Dec}(mpk_{ta}, usk_{id,ta}, c)$
Return m

The following result relates the notions of m-RA-atk security and m-RA-RE-atk security; a single-TA version of this result for atk = CPA was given in [1].

Lemma 1. *Let* m-IBE *be a multi-TA IBE scheme that is m-IND-atk-secure and m-RA-RE-atk-secure. Then* m-IBE *is also m-RA-atk-secure. Here atk* $\in \{CPA, CCA\}$.

Proof. Let $\mathcal{A}$ be a poly-time algorithm (PTA) attacking the m-RA-atk security of a scheme m-IBE. It is easy to construct PTAs $\mathcal{A}_1$, $\mathcal{A}_3$ attacking the m-IND-atk security of m-IBE, and a PTA $\mathcal{A}_2$ attacking m-RA-RE-atk security of m-IBE such that:

$$\begin{aligned}
&\mathbf{Adv}_{\mathcal{A}}^{\text{m-RA-atk}}(k) \\
= &\ |\Pr[\mathbf{Exp}_{\mathcal{A}}^{\text{m-RA-atk-1}}(k) = 1] - \Pr[\mathbf{Exp}_{\mathcal{A}}^{\text{m-RA-atk-0}}(k) = 1]| \\
= &\ |\Pr[\mathbf{Exp}_{\mathcal{A}}^{\text{m-RA-atk-1}}(k) = 1] - \Pr[\mathbf{Exp}_{\mathcal{A}}^{\text{m-RA-RE-atk-1}}(k) = 1] \\
&+ \Pr[\mathbf{Exp}_{\mathcal{A}}^{\text{m-RA-RE-atk-1}}(k) = 1] - \Pr[\mathbf{Exp}_{\mathcal{A}}^{\text{m-RA-RE-atk-0}}(k) = 1] \\
&+ \Pr[\mathbf{Exp}_{\mathcal{A}}^{\text{m-RA-RE-atk-0}}(k) = 1] - \Pr[\mathbf{Exp}_{\mathcal{A}}^{\text{m-RA-atk-0}}(k) = 1]| \\
\leq &\ |\Pr[\mathbf{Exp}_{\mathcal{A}_1}^{\text{m-RA-atk-1}}(k) = 1] - \Pr[\mathbf{Exp}_{\mathcal{A}_1}^{\text{m-RA-RE-atk-1}}(k) = 1]| \\
&+ |\Pr[\mathbf{Exp}_{\mathcal{A}_2}^{\text{m-RA-RE-atk-1}}(k) = 1] - \Pr[\mathbf{Exp}_{\mathcal{A}_2}^{\text{m-RA-RE-atk-0}}(k) = 1]| \\
&+ |\Pr[\mathbf{Exp}_{\mathcal{A}_3}^{\text{m-RA-RE-atk-0}}(k) = 1] - \Pr[\mathbf{Exp}_{\mathcal{A}_3}^{\text{m-RA-atk-0}}(k) = 1]| \\
\leq &\ \mathbf{Adv}_{\mathcal{A}_1}^{\text{m-IND-atk}}(k) + \mathbf{Adv}_{\mathcal{A}_2}^{\text{m-RA-RE-atk}}(k) + \mathbf{Adv}_{\mathcal{A}_3}^{\text{m-IND-atk}}(k)
\end{aligned}$$

3.4 m-TAA-CCA Security

The m-TAA-CCA security notion formalizes TA anonymity: a ciphertext should not leak which TA's master public key was used to compute the ciphertext. We work with chosen ciphertext adversaries in the multi-TA setting. As explained above, TA anonymity is a necessary condition to achieve fully private communication thwarting adversarial activity like traffic analysis in the multi-TA setting.

Experiment $\mathbf{Exp}_{\mathcal{A}}^{\text{m-TAA-CCA-b}}(k)$
$params \leftarrow \texttt{CommonSetup}(1^k)$
$TASet \leftarrow \emptyset$
$\forall ta \in \mathcal{TA}$, $(mpk_{ta}, msk_{ta}) \leftarrow \texttt{TASetup}(params)$,
$IDSet_{ta} \leftarrow \emptyset$ and $CSet_{ta} \leftarrow \emptyset$
$(ta_0, ta_1, id, m, state) \leftarrow \mathcal{A}^{\texttt{Corrupt},\texttt{KeyDer},\texttt{Dec}}(\texttt{find}, MPK)$
$c^* \leftarrow \texttt{Enc}(mpk_{ta_b}, id, m)$
$b' \leftarrow \mathcal{A}^{\texttt{Corrupt},\texttt{KeyDer},\texttt{Dec}}(\texttt{guess}, c^*, state)$
If $m \notin \mathsf{MsgSp}$ or $ta_0 = ta_1$ then return 0
If $ta_0 \notin TASet$, $ta_1 \notin TASet$, $id \notin IDSet_{ta_0}$, $id \notin IDSet_{ta_1}$, $(id, c^*) \notin CSet_{ta_0}$ and $(id, c^*) \notin CSet_{ta_1}$ then return b' else return 0

Oracle $\texttt{Corrupt}(ta)$
$TASet \leftarrow TASet \cup \{ta\}$
Return msk_{ta}

Oracle $\texttt{KeyDer}(ta, id)$
$IDSet_{ta} \leftarrow IDSet_{ta} \cup \{id\}$
$usk_{id,ta} \leftarrow \texttt{KeyDer}(msk_{ta}, id)$
Return $usk_{id,ta}$

Oracle $\texttt{Dec}(ta, id, c)$
$CSet_{ta} \leftarrow CSet_{ta} \cup (id, c)$
$usk_{id,ta} \leftarrow \texttt{KeyDer}(msk_{ta}, id)$
$m \leftarrow \texttt{Dec}(mpk_{ta}, usk_{id,ta}, c)$
Return m

3.5 m-TAA-RE-CCA Security

Again, when proving m-TAA-RE-CCA security for a concrete scheme it is sometimes easier to work with the related m-TAA-RE-CCA security notion, which we define next.

Experiment $\mathbf{Exp}_{\mathcal{A}}^{\text{m-TAA-CCA-b}}(k)$
$params \leftarrow \texttt{CommonSetup}(1^k)$
$TASet \leftarrow \emptyset$
$\forall ta \in \mathcal{TA}$, $(mpk_{ta}, msk_{ta}) \leftarrow \texttt{TASetup}(params)$,
$IDSet_{ta} \leftarrow \emptyset$ and $CSet_{ta} \leftarrow \emptyset$
$(ta_0, ta_1, id, m, state) \leftarrow$
$\mathcal{A}^{\texttt{Corrupt},\texttt{KeyDer},\texttt{Dec}}(\texttt{find}, MPK)$
$m' \xleftarrow{\$} \mathsf{MsgSp}$ with $|m'| = |m|$;
$c^* \leftarrow \texttt{Enc}(mpk_{ta_b}, id, m')$
$b' \leftarrow \mathcal{A}^{\texttt{Corrupt},\texttt{KeyDer},\texttt{Dec}}(\texttt{guess}, c^*, state)$
If $m \notin \mathsf{MsgSp}$ or $ta_0 = ta_1$ then return 0
If $ta_0 \notin TASet$, $ta_1 \notin TASet$, $id \notin IDSet_{ta_0}$, $id \notin IDSet_{ta_1}$, $(id, c^*) \notin CSet_{ta_0}$ and $(id, c^*) \notin CSet_{ta_1}$ then return b' else return 0.

Oracle $\texttt{Corrupt}(ta)$
$TASet \leftarrow TASet \cup \{ta\}$
Return msk_{ta}

Oracle $\texttt{KeyDer}(ta, id)$
$IDSet_{ta} \leftarrow IDSet_{ta} \cup \{id\}$
$usk_{id,ta} \leftarrow \texttt{KeyDer}(msk_{ta}, id)$
Return $usk_{id,ta}$

Oracle $\texttt{Dec}(ta, id, c)$
$CSet_{ta} \leftarrow CSet_{ta} \cup (id, c)$
$usk_{id,ta} \leftarrow \texttt{KeyDer}(msk_{ta}, id)$
$m \leftarrow \texttt{Dec}(mpk_{ta}, usk_{id,ta}, c)$
Return m

Lemma 2. *Let* m-IBE *be a multi-TA IBE scheme that is m-IND-atk-secure and m-TAA-RE-atk-secure. Then* m-IBE *is also m-TAA-atk-secure. Here atk* $\in \{CPA, CCA\}$.

The proof is similar to that of Lemma 1 and is omitted.

3.6 A Combined Security Notion: m-IND-TAA-RA-CCA Security

Finally, we define an m-IND-RA-TAA-CCA experiment that simultaneously captures message indistinguishability, recipient anonymity, and TA anonymity in the multi-TA setting for chosen ciphertext adversaries.

Experiment $\mathbf{Exp}_{\mathcal{A}}^{\text{m-IND-RA-TAA-CCA-b}}(k)$
$params \leftarrow \texttt{CommonSetup}(1^k)$
$TASet \leftarrow \emptyset$
$\forall ta \in \mathcal{TA}$, $(mpk_{ta}, msk_{ta}) \leftarrow \texttt{TASetup}(params)$,
$IDSet_{ta} \leftarrow \emptyset$ and $CSet_{ta} \leftarrow \emptyset$
$(ta_0, ta_1, id_0, id_1, m_0, m_1, state) \leftarrow$
$\mathcal{A}^{\texttt{Corrupt},\texttt{KeyDer},\texttt{Dec}}(\texttt{find}, MPK)$
$c^* \leftarrow \texttt{Enc}(mpk_{ta_b}, id_b, m_b)$
$b' \leftarrow \mathcal{A}^{\texttt{Corrupt},\texttt{KeyDer},\texttt{Dec}}(\texttt{guess}, c^*, state)$
If $\{m_0, m_1\} \not\subseteq \mathsf{MsgSp}$ or $|m_0| \neq |m_1|$
then return 0
If $(ta_0 = ta_1$ and $id_0 = id_1$ and $m_0 = m_1)$
then return 0
If $ta_0 \notin TASet$, $ta_1 \notin TASet$, $id_0 \notin IDSet_{ta_0}$, $id_1 \notin IDSet_{ta_1}$, $(id_0, c^*) \notin CSet_{ta_0}$ and $(id_1, c^*) \notin CSet_{ta_1}$ then return b' else return 0.

Oracle $\texttt{Corrupt}(ta)$
$TASet \leftarrow TASet \cup \{ta\}$
Return msk_{ta}

Oracle $\texttt{KeyDer}(ta, id)$
$IDSet_{ta} \leftarrow IDSet_{ta} \cup \{id\}$
$usk_{id,ta} \leftarrow \texttt{KeyDer}(msk_{ta}, id)$
Return $usk_{id,ta}$

Oracle $\texttt{Dec}(ta, id, c)$
$CSet_{ta} \leftarrow CSet_{ta} \cup (id, c)$
$usk_{id,ta} \leftarrow \texttt{KeyDer}(msk_{ta}, id)$
$m \leftarrow \texttt{Dec}(mpk_{ta}, usk_{id,ta}, c)$
Return m

Lemma 3. *Let* m-IBE *be a multi-TA IBE scheme that is m-IND-atk-secure, m-RA-atk-secure and m-TAA-atk-secure. Then* m-IBE *is also m-IND-RA-TAA-atk-secure. Here atk* $\in \{CPA, CCA\}$.

Proof. The proof (informally) follows by noting that if m-IBE is m-TAA-atk-secure, then the challenger may replace the triple (ta_0, id_0, m_0) with (ta_1, id_0, m_0) in its response to the challenge query without the adversary being able to detect the change. Likewise, using m-RA-atk security, the challenger may then replace (ta_1, id_0, m_0) with (ta_1, id_1, m_0). Finally, using m-IND-atk security, the challenger can replace (ta_1, id_1, m_0) with (ta_1, id_1, m_1), again, without the adversary being able to detect the change. This informal argument can be made rigorous using a sequence of games.

A combined m-IND-RA-CCA security notion can also be defined and it is easy to show that m-IND-RA-CCA security holds for a scheme that has both m-IND-CCA and m-RA-CCA security, using a similar strategy as above. In the single-TA setting, we obtain IND-RA-CCA and IND-RA-CPA security notions. The latter security notion for IBE was used to prove the security of PEKS schemes in [1]. Similarly, we define combined m-IND-TAA-CPA and m-IND-TAA-CCA security notions.

4 Extending the Fujisaki-Okamoto Conversion to Multi-TA IBE Schemes

In two separate but related strands of work, Fujisaki and Okamoto studied the problem of building Public Key Encryption (PKE) schemes which are secure in a very strong sense (IND-CCA) from PKE schemes which are secure in a weaker sense.

In [14], Fujisaki and Okamoto gave a generic conversion that takes any OW-CPA-secure PKE scheme satisfying a mild technical condition (γ-uniformity) and outputs a PKE scheme that is IND-CCA-secure in the Random Oracle Model. Yang *et al.* [25] investigated how to adapt this particular Fujisaki-Okamoto (FO) technique to the ID-based setting.

Similarly, in [13], Fujisaki and Okamoto gave a generic conversion that takes any IND-CPA-secure PKE scheme and outputs a PKE scheme that is IND-CCA-secure in the Random Oracle Model. The security analysis in [13] is significantly simpler than that of [14]. Kitagawa *et al.* [20] investigated how to modify this particular FO technique for the ID-based setting.

We now describe a modified FO conversion for IBE in the multi-TA setting. We are able to show that in the multi-TA setting, we can apply this modified conversion to build an IBE scheme that has m-IND-RA-TAA-CCA security from an IBE scheme that is m-IND-RA-TAA-CPA-secure and γ-uniform. We extend the ideas of [13,20]. In particular, we include additional parameters in the input to the hash function used in the scheme. This allows us to efficiently respond to hash queries, simplifies book-keeping in the proof, and yields a simulation that has a reduced running time in comparison to an application of the unmodified Fujisaki-Okamoto transformation.

We begin by defining a suitable notion of γ-uniformity for the multi-TA setting.

Definition 6. *Let Π be a multi-TA IBE scheme with space of randomness RSp. Then Π is said to be γ-uniform if, for any fixed choice of $c \in \mathsf{CtSp}$, $m \in \mathsf{MsgSp}$, $id \in \{0,1\}^*$ and $ta \in \mathcal{TA}$, we have:*

$$\Pr\left[c = \mathsf{Enc}(mpk_{ta}, id, m; r) : r \xleftarrow{\$} \mathsf{RSp}\right] \leq \gamma.$$

Now let $\Pi = \{\texttt{CommonSetup}, \texttt{TASetup}, \texttt{KeyDer}, \texttt{Enc}, \texttt{Dec}\}$ be a multi-TA IBE scheme. Then $\Pi' = \{\texttt{CommonSetup}', \texttt{TASetup}', \texttt{KeyDer}', \texttt{Enc}', \texttt{Dec}'\}$ denotes a new multi-TA IBE scheme with algorithms defined as follows.

Let $l_0 + l_1$ be the bit length of messages in Π, where l_0 will be the bit length of messages in Π', and let RSp be the space of randomness used by $\texttt{Enc}$.

- $\texttt{CommonSetup}'$: As in $\texttt{CommonSetup}$, but in addition, we pick a hash function $H : \{0,1\}^* \times \{0,1\}^* \times \{0,1\}^{l_0} \times \{0,1\}^{l_1} \rightarrow \mathsf{RSp}$.
- $\texttt{TASetup}'$:As in $\texttt{TASetup}$.
- $\texttt{KeyDer}'$: As in $\texttt{KeyDer}$.
- $\texttt{Enc}'$: This algorithm takes as input mpk_{ta} for $ta \in \mathcal{TA}$, $id \in \{0,1\}^*$, and a message $m \in \{0,1\}^{l_0}$. Its output is:

$$\texttt{Enc}'(mpk_{ta}, id, m) = \texttt{Enc}(mpk_{ta}, id, m||\sigma; H(mpk_{ta}, id, m, \sigma))$$

  where $\sigma \xleftarrow{\$} \{0,1\}^{l_1}$. So Π' has randomness space $\{0,1\}^{l_1}$.
- $\texttt{Dec}'$: Let c denote a ciphertext to be decrypted using a private key $usk_{id,ta}$ issued by TA ta with master public key mpk_{ta} for identity id. This algorithm works as follows:
  1. Compute $m' = \texttt{Dec}(mpk_{ta}, usk_{id,ta}, c)$.
  2. Let $m = [m']^{l_0}$ and $\sigma = [m']_{l_1}$ where $[a]^b$ and $[a]_b$ denote the first and last b bits of a string a respectively.
  3. Test if $\texttt{Enc}(mpk_{ta}, id, m||\sigma; H(mpk_{ta}, id, m, \sigma)) = c$. If not output $\perp$; otherwise output m as the decryption of c.

Theorem 3. *Modelling H as a random oracle, if Π is a multi-TA IBE scheme that is m-IND-RA-TAA-CPA-secure and γ-uniform for some negligible γ, then Π' is m-IND-RA-TAA-CCA-secure.*

In more detail, suppose Π is a γ-uniform IBE encryption scheme. Let $\mathcal{A}$ be an m-IND-RA-TAA-CCA adversary which has advantage $\epsilon(k)$ against Π' and which runs in time $t(k)$. Suppose $\mathcal{A}$ makes at most q_H queries to H, at most q_E extraction queries, and at most q_D decryption queries. Suppose further that executing $\texttt{Enc}$ *once needs at most time τ. Then there is an m-IND-RA-TAA-CPA adversary $\mathcal{B}$ which has advantage at least $\epsilon'(k)$ against Π, with running time $t'(k)$, such that*

$$\epsilon'(k) = 2(\frac{\epsilon + 1}{2} - \frac{q_h}{2^{l_1} - 1})(1 - \gamma)^{q_d} - 1$$

and

$$t'(k) = O(t(k) + q_h \tau).$$

Proof. Suppose there is an m-IND-RA-TAA-CCA adversary $\mathcal{A}$ against Π' with advantage $\epsilon(k)$ and running in time $t(k)$. We show how to construct an adversary $\mathcal{B}$ that uses $\mathcal{A}$ to break the m-IND-RA-TAA-CPA-security of Π

$\mathcal{B}$'s input is the set of all master public keys MPK. $\mathcal{B}$ gives $\mathcal{A}$ the set MPK. $\mathcal{A}$ also has access to random oracle H that is controlled by $\mathcal{B}$. $\mathcal{A}$ then makes a series of queries which $\mathcal{B}$ answers as follows.

- **H-queries:** $\mathcal{B}$ maintains a list of tuples $\langle mpk_i, id_i, m_i, \sigma_i, g_i, c_i\rangle$. We refer to this list as the H^{list}. The list is initially empty.
  When $\mathcal{A}$ makes a H query on (mpk, id, m, σ), $\mathcal{B}$ responds as follows:
  - If the query (mpk, id, m, σ) already appears in a tuple $\langle mpk_i, id_i, m_i, \sigma_i, g_i, c_i\rangle$ then $\mathcal{B}$ responds with $H(mpk, id, m, \sigma) = g_i$.
  - Otherwise $\mathcal{B}$ picks $g \xleftarrow{\$} \mathsf{RSp}$, generates $c = \mathtt{Enc}(mpk_{ta}, id, m||\sigma; g)$, adds the tuple $\langle mpk, id, m, \sigma, g, c\rangle$ to the H^{list} and responds to $\mathcal{A}$ with $H(mpk, id, m, \sigma) = g$.
- **Corrupt Queries:** If $\mathcal{A}$ issues a TA corrupt query on $ta \in \mathcal{TA}$, then $\mathcal{B}$ simply relays this query to its challenger, which responds with the corresponding master secret key msk_{ta}. $\mathcal{B}$ then passes the resulting key to $\mathcal{A}$.
- **Extraction Queries:** If $\mathcal{A}$ issues an extraction query on (ta, id), then $\mathcal{B}$ forwards (ta, id) to its challenger, which responds with the private key $usk_{id,ta}$. $\mathcal{B}$ forwards this key to $\mathcal{A}$.
- **Decryption Queries:** If $\mathcal{A}$ issues a decryption query on (ta, id, c), $\mathcal{A}$ responds as follows:
  - Searches for a tuple $\langle mpk_i, id_i, m_i, \sigma_i, g_i, c_i\rangle$ from the H^{list} such that $mpk_{ta} = mpk_i$, $id = id_i$ and $c = c_i$.
  - If such a tuple exists, then outputs m, else outputs $\perp$.
- **Challenge:** $\mathcal{A}$ outputs data $(ta_0, ta_1, id_0, id_1, m_0, m_1)$ on which it wishes to be challenged. This data is subject to the usual restrictions (see Section 3.6). $\mathcal{B}$ chooses two l_1 bit strings σ_0 and σ_1 uniformly at random, subject to the condition that they be distinct, and sends $(ta_0, ta_1, id_0, id_1, m_0||\sigma_0, m_1||\sigma_1)$ to its challenger. $\mathcal{B}$'s challenger picks a random bit b and sets

$$c^* = \mathtt{Enc}(mpk_{ta_b}, id_b, m_b||\sigma_b; r)$$

  where $r \in \mathsf{RSp}$. $\mathcal{B}$ forwards c^* to $\mathcal{A}$.

After the Challenge query has been issued, if the adversary $\mathcal{A}$ makes a hash oracle query on either $(ta_0, id_0, m_0, \sigma_0)$ or $(ta_1, id_1, m_1, \sigma_1)$ then the adversary $\mathcal{B}$ simply outputs $b' = 0$ or $b' = 1$, respectively, as its guess for the value of the bit b. If neither hash query is made then, at the end of $\mathcal{A}$'s attack, $\mathcal{B}$ simply outputs the same bit b' that $\mathcal{A}$ outputs. $\mathcal{B}$ wins if $b' = b$. This completes our description of the simulation.

Our analysis now follows closely the analysis in [13]. We define the following events and probabilities.

Let $\Pr[\mathsf{Succ}\mathcal{A}]$ be the probability that adversary $\mathcal{A}$ outputs a bit $b' = b$. Similarly, let $\Pr[\mathsf{Succ}\mathcal{B}]$ be the probability that adversary $\mathcal{B}$ outputs a bit $b' = b$. For notational convenience, we let ϵ denote $\mathcal{A}$'s advantage in the simulation.

Let $\texttt{Ask}_b$ be the event that $\mathcal{A}$ asks a hash query that coincides with$(mpk_{ta_b}, id_b, m_b, \sigma_b)$ and $\texttt{Ask}_{\bar{b}}$ be the event that $\mathcal{A}$ asks a hash query that coincides with $(mpk_{ta_{\bar{b}}}, id_{\bar{b}}, m_{\bar{b}}, \sigma_{\bar{b}})$. Notice that these two queries are distinct because $\sigma_0 \neq \sigma_1$.

We define $\mathcal{F}$ to be the event that $\mathcal{B}$ fails to answer a decryption query correctly at some point during the game so that $\Pr[\neg\mathcal{F}]$ is the probability that $\mathcal{B}$ answers all decryption queries correctly during the simulation. Now,

$$\begin{aligned}\Pr[\texttt{Succ}\mathcal{A}] = &\Pr[\texttt{Succ}\mathcal{A}|\texttt{Ask}_b]\cdot\Pr[\texttt{Ask}_b] \\ &+\Pr[\texttt{Succ}\mathcal{A}|(\neg\texttt{Ask}_b)\wedge\texttt{Ask}_{\bar{b}}]\cdot\Pr[(\neg\texttt{Ask}_b)\wedge\texttt{Ask}_{\bar{b}})] \\ &+\Pr[\texttt{Succ}\mathcal{A}|(\neg\texttt{Ask}_b)\wedge(\neg\texttt{Ask}_{\bar{b}})]\cdot\Pr[(\neg\texttt{Ask}_b)\wedge(\neg\texttt{Ask}_{\bar{b}})].\end{aligned}$$

Similarly,

$$\begin{aligned}\Pr[\texttt{Succ}\mathcal{B}] = &\Pr[\texttt{Succ}\mathcal{B}|\texttt{Ask}_b]\cdot\Pr[\texttt{Ask}_b] \\ &+\Pr[\texttt{Succ}\mathcal{B}|(\neg\texttt{Ask}_b)\wedge\texttt{Ask}_{\bar{b}}]\cdot\Pr[(\neg\texttt{Ask}_b)\wedge\texttt{Ask}_{\bar{b}})] \\ &+\Pr[\texttt{Succ}\mathcal{B}|(\neg\texttt{Ask}_b)\wedge(\neg\texttt{Ask}_{\bar{b}})]\cdot\Pr[(\neg\texttt{Ask}_b)\wedge(\neg\texttt{Ask}_{\bar{b}})].\end{aligned}$$

From the conditions of the simulation, we have the following:

$$\begin{aligned}\Pr[\texttt{Succ}\mathcal{B}|\texttt{Ask}_b] &= 1, \\ \Pr[\texttt{Succ}\mathcal{B}|(\neg\texttt{Ask}_b)\wedge\texttt{Ask}_{\bar{b}}] &= 0, \\ \Pr[\texttt{Succ}\mathcal{A}|(\neg\texttt{Ask}_b)\wedge(\neg\texttt{Ask}_{\bar{b}})] &= \Pr[\texttt{Succ}\mathcal{B}|(\neg\texttt{Ask}_b)\wedge(\neg\texttt{Ask}_{\bar{b}})].\end{aligned}$$

Therefore,

$$\begin{aligned}\Pr[\texttt{Succ}\mathcal{B}]-\Pr[\texttt{Succ}\mathcal{A}] &= \Pr[\texttt{Ask}_b](1-\Pr[\texttt{Succ}\mathcal{A}|\texttt{Ask}_b]) \\ &\quad+\Pr[(\neg\texttt{Ask}_b)\wedge\texttt{Ask}_{\bar{b}}](0-\Pr[\texttt{Succ}\mathcal{A}|(\neg\texttt{Ask}_b\wedge\texttt{Ask}_{\bar{b}})]) \\ &\geq -\Pr[(\neg\texttt{Ask}_b)\wedge\texttt{Ask}_{\bar{b}}].\end{aligned}$$

Since even a computationally unbounded adversary has no information about what the string $\sigma_{\bar{b}}$ is (except that it is distinct from σ_b and so is uniformly distributed on a set of size $2^{l_1}-1$), and our adversary makes at most q_h queries to the oracle H, we infer that $\Pr[(\neg\texttt{Ask}_b)\wedge\texttt{Ask}_{\bar{b}}] \leq \frac{q_h}{2^{l_1}-1}$. Hence,

$$\begin{aligned}\Pr[\texttt{Succ}\mathcal{B}] &\geq \Pr[\texttt{Succ}\mathcal{A}]-\Pr[(\neg\texttt{Ask}_b)\wedge\texttt{Ask}_{\bar{b}}] \\ &\geq \frac{\epsilon+1}{2}-\frac{q_h}{2^{l_1}-1}.\end{aligned}$$

The event $\mathcal{F}$ occurs only when $\mathcal{A}$ submits a decryption query (ta, id, c) such that

$$c = \texttt{Enc}(mpk_{ta}, id, m||\sigma; H(mpk_{ta}, id, m, \sigma))$$

without first querying H on input $(mpk_{ta}, id, m, \sigma)$. Now observe that, given values ta, id, c, there is at most one possible message $m' = m||\sigma$ that could result from decrypting ciphertext c under the private key $usk_{id,ta}$, namely $m' = \texttt{Dec}(mpk_{ta}, usk_{id,ta}, c)$. Applying the definition of γ-uniformity, and noting that the randomness r that would be used to form c for the scheme Π' is still uniformly distributed whenever the relevant hash query has not been made, we see that $\mathcal{B}$ fails to properly answer each decryption query with probability at most γ. Therefore $\Pr[\neg\mathcal{F}] \leq (1-\gamma)^{q_d}$.

Hence, we have

$$\mathbf{Adv}_{\mathcal{B}}(k) = 2\Pr[\mathtt{Succ}\mathcal{B}] \cdot \Pr[\neg\mathcal{F}] - 1 \geq 2(\frac{\epsilon+1}{2} - \frac{q_h}{2^{l_1}-1})(1-\gamma)^{q_d} - 1.$$

For the running time analysis, note that in addition to the running time of $\mathcal{A}$, the adversary $\mathcal{B}$ has to run the encryption algorithm **Enc** at most q_h times. Therefore $t'(k) = O(t(k) + q_h\tau)$. □

Notice that the above theorem as stated requires the initial scheme Π to have all three security properties (IND, RA and TAA) in order to convert from CPA-security to CCA-security. In fact, it is easy to prove versions of Theorem 3 that convert IND-RA-CPA security to IND-RA-CCA security and IND-TAA-CPA security to IND-TAA-CCA security. However, the proof technique does not allow us to prove that the conversion preserves either of our anonymity properties in isolation – we need the base scheme Π to also be IND-secure.

We leave as an open problem to find a modified version of the "other" FO conversion (from [14]) that preserves anonymity properties in the multi-TA setting.

4.1 Applying the Modified FO Conversion to `BasicIdent`

We describe and analyse a multi-TA scheme `m-BasicIdent` that is based on the scheme `BasicIdent` from [6]. This scheme is defined as follows:

$\mathtt{CommonSetup}(1^k)$:

- $(G, G_T, e, q, P) \leftarrow \mathtt{PairingGen}(1^k)$.
- Output $params = (G, G_T, e, q, P, H_1, H_2, n)$ where $H_1 : \{0,1\}^* \rightarrow G$, $H_2 : G_T \rightarrow \{0,1\}^n$ for some $n = n(k)$.
- $\mathsf{MsgSp} = \{0,1\}^n$, $\mathsf{CtSp} = G_1 \times \{0,1\}^n$, $\mathsf{RSp} = \mathbb{Z}_q$.

$\mathtt{TASetup}(params)$

- Set $s \xleftarrow{\$} \mathbb{Z}_q, Q = sP$.
- Set $mpk = (params, Q)$.
- Set $msk = s$.
- Output (mpk, msk).

$\mathtt{KeyDer}^{H_1}(ta, id)$:

- Set $usk_{id,ta} = msk_{ta} \cdot H_1(id)$.
- Output $usk_{id,ta}$.

$\mathtt{Enc}^{H_1,H_2}(ta, id, m)$:

- Parse mpk_{ta} as $(params, Q_{ta})$.
- Set $r \xleftarrow{\$} \mathbb{Z}_q$.
- Set $T = e(H_1(id), Q_{ta})^r$.
- Output $c = (rP, m \oplus H_2(T))$.

$\mathtt{Dec}^{H_2}(ta, usk_{id,ta}, c)$:

- Parse c as (U, V).
- Set $T = e(usk_{id,ta}, U)$.
- Output $m = V \oplus H_2(T)$.

The scheme `m-BasicIdent`.

We next show the scheme that results from applying the modified Fujisaki-Okamoto tranformation to the `m-BasicIdent` scheme above.

$\texttt{CommonSetup}'(1^k)$:

- $(G, G_T, e, q, P) \leftarrow \texttt{PairingGen}(1^k)$.
- Output $params = (G, G_T, e, q, P, H_1, H_2, H_3, l_0, l_1, n)$ where $H_1 : \{0,1\}^* \rightarrow G$, $H_2 : G_T \rightarrow \{0,1\}^n$ for some $n = n(k)$, $l_0 + l_1 = n$, and $H_3 : \{0,1\}^* \times \{0,1\}^* \times \{0,1\}^{l_0} \times \{0,1\}^{l_1} \rightarrow \mathbb{Z}_q$.
- $\mathsf{MsgSp} = \{0,1\}^{l_0}$, $\mathsf{CtSp} = G_1 \times \{0,1\}^n$, $\mathsf{RSp} = \{0,1\}^{l_1}$.

$\texttt{TASetup}'$: As in $\texttt{TASetup}$

$\texttt{KeyDer}'$: As in $\texttt{KeyDer}$

$\texttt{Enc}^{H_1,H_2,H_3}(ta, id, m)$:

- Parse mpk_{ta} as $(params, Q_{ta})$.
- Set $\sigma \xleftarrow{\$} \{0,1\}^{l_1}$.
- Set $r = H_3(mpk_{ta}, id, m, \sigma)$.
- Set $T = e(H_1(id), Q_{ta})^r$.
- Output $c = (rP, (m||\sigma) \oplus H_2(T))$.

$\texttt{Dec}^{H_2,H_3}(ta, usk_{id,ta}, c)$:

- Parse c as (U, V).
- Set $T = e(usk_{id,ta}, U)$.
- Set $m' = V \oplus H_2(T)$.
- Set $m = [m']^{l_0}$ and $\sigma = [m']_{l_1}$.
- Test if $r = H_3(mpk_{ta}, id, m, \sigma)$. If not, output $\perp$; otherwise output m as the decryption of c.

The scheme `FO-m-BasicIdent`.

Lemma 4. *The multi-TA scheme* `m-BasicIdent` *is m-IND-CPA-secure, assuming the hardness of the BDH problem in groups output by* `PairingGen`.

Proof. The single-TA scheme corresponding to `m-BasicIdent` is nothing other than the Boneh-Franklin `BasicIdent` scheme, whose IND-CPA security is known to rest on the hardness of the BDH problem in groups output by `PairingGen` [6]. Now apply Theorem 1.

The following result is an extension of a result from [1] that showed that the `BasicIdent` scheme has recipient anonymity against CPA attackers.

Lemma 5. *The multi-TA scheme* `m-BasicIdent` *is m-RA-CPA-secure and m-TAA-CPA-secure, assuming the hardness of the BDH problem in groups output by* `PairingGen`.

Proof. Ciphertexts c in the `m-BasicIdent` scheme have two parts, namely $U = rP$ and $V = m \oplus H_2(T)$. The value U is chosen uniformly at random from G. If the message m is chosen uniformly at random from $\{0,1\}^n$ then V is also distributed uniformly in $\{0,1\}^n$ and is independent of $H_2(T)$. Thus, in both 0 and 1 worlds of the m-RA-RE-CPA and m-TAA-RE-CPA games, the ciphertext c has exactly the same distribution and any adversary in these RE games will have zero advantage. By Lemma 4, `m-BasicIdent` is m-IND-CPA-secure. Applying Lemmas 1 and 2 yields m-RA-CPA and m-TAA-CPA security for `m-BasicIdent`, assuming the hardness of the BDH problem in groups output by `PairingGen`.

Lemma 6. *The* m-BasicIdent *scheme is γ-uniform for $\gamma = 1/q$.*

Proof. In the m-BasicIdent scheme, the first component of the ciphertext is $U = rP$ where $r \xleftarrow{\$} \mathbb{Z}_q$. It is them immediate that m-BasicIdent is γ uniform with $\gamma = 1/q$.

Theorem 4. *The scheme* FO-m-BasicIdent *obtained by applying the modified FO conversion to the scheme* m-BasicIdent *is m-IND-RA-TAA-CCA-secure, assuming the hardness of the BDH problem in groups output by* PairingGen.

Proof. We obtain the above result by combining Lemmas 4, 5 with Lemmas 3, 6 and Theorem 3.

Thus we have obtained an efficient multi-TA IBE scheme enjoying indistinguishability, recipient anonymity and TA anonymity for the CCA setting, in the random oracle model. We note as a corollary of our analysis that the single-TA version of our scheme offers recipient anonymity. To the best of our knowledge, this is the first such result for a CCA-secure variant of BasicIdent.

4.2 Applying the Modified FO Conversion to the Sakai-Kasahara IBE Scheme

The Sakai-Kasahara IBE scheme [21] has an alternative (and attractive) private key extraction algorithm compared to the Boneh-Franklin scheme. We define m-BasicSK, a multi-TA version of this scheme using symmetric pairings, immediately below, and then provide a sketch security analysis.

CommonSetup(1^k):

- $(G, G_T, e, q, P) \leftarrow$ PairingGen(1^k).
- Output *params* $= (G, G_T, e, q, P, Z, H_1, H_2, n)$ where $Z = e(P,P) \in G_T$, $H_1 : \{0,1\}^* \rightarrow \mathbb{Z}_q$, $H_2 : G_T \rightarrow \{0,1\}^n$ for some $n = n(k)$.
- $\mathsf{MsgSp} = \{0,1\}^n$, $\mathsf{CtSp} = G_1 \times \{0,1\}^n$, $\mathsf{RSp} = \mathbb{Z}_q$.

TASetup$(params)$

- Set $s \xleftarrow{\$} \mathbb{Z}_q, Q = sP$.
- Set $mpk = (params, Q)$.
- Set $msk = s$.
- Output (mpk, msk).

KeyDer$^{H_1}(ta, id)$:

- Output $usk_{id,ta} = \frac{1}{msk_{ta} + H_1(id)} \cdot P$.

Enc$^{H_1,H_2}(ta, id, m)$:

- Parse mpk_{ta} as $(params, Q_{ta})$.
- Set $r \xleftarrow{\$} \mathbb{Z}_q$.
- Set $U = rQ_{ta} + rH_1(id)P$.
- Output $c = (U, m \oplus H_2(Z^r))$.

Dec$^{H_2}(ta, usk_{id,ta}, c)$:

- Parse c as (U, V).
- Set $T = e(usk_{id,ta}, U)$.
- Output $m = V \oplus H_2(T)$.

The scheme m-BasicSK.

The IND-CPA security of the single-TA scheme corresponding to m-BasicSK can be proved by making small modifications to the proof of [11, Theorem 2], which established the OW-CPA security of a closely related scheme based on the hardness of the ℓ-BDHI problem in groups output by PairingGen (for some value ℓ related to the number of queries made by the adversary). Using Theorem 1, we can deduce that m-BasicSK is m-IND-CPA-secure under the same assumption. It is then easy to establish that m-BasicSK is m-RA-CPA-secure and m-TAA-CPA-secure; this requires a similar analysis as in Lemma 5. Moreover, m-BasicSK is γ-uniform for $\gamma = 1/q$. We may now apply Theorem 3 to deduce that the scheme FO-m-BasicSK that is obtained by applying our modified FO conversion to m-BasicSK is m-IND-RA-TAA-CCA-secure, assuming the hardness of the ℓ-BDHI problem in groups output by PairingGen.

Thus we have obtained a second efficient multi-TA IBE scheme enjoying indistinguishability, recipient anonymity and TA anonymity for the CCA setting, in the random oracle model. Our CCA-secure scheme has roughly the same performance as the KEM-DEM-derived scheme of [11], but offers stronger proven anonymity guarantees. We also note that even the recipient anonymity of the single-TA version of m-BasicSK was not previously known – indeed this is explicitly claimed *not* to hold in [7].

5 Conclusion and Future Work

We have given a formal analysis of various security and anonymity notions for multi-TA IBE schemes and the relationships between them. We have also investigated a modified Fujisaki-Okamoto transformation for IBE and shown that this transformation preserves our security and anonymity notions when building a CCA-secure scheme from a CPA-secure one. We investigated the application of this transformation to the Boneh-Franklin BasicIdent scheme and to the Sakai-Kasahara scheme.

In future work, we will investigate further specific IBE schemes and see if they meet the multi-TA security notions introduced in this paper. In particular, it will be interesting to examine the IND-RA-atk-secure IBE schemes of Gentry [16] and see if they can also be proven to be TAA-atk-secure We raised the possibility of adapting the "other" FO conversion of [14] so as to preserve our multi-TA security notions. Another open problem suggested by this work is its generalization to the hierarchical IBE (HIBE) setting, where the anonymity properties of ciphertexts generated using different root TA master public keys could be studied.

Finally, the subject of robustness of IBE in the single-TA and multi-TA settings requires further investigation: when using an IND-RA-TA-CCA-secure scheme in practice in a fully anonymous communications system, users will need to be able to decide whether or not a ciphertext is intended for their consumption. Seemingly the only way for a user to do this is to attempt a trial decryption using his private key, relying on the decryption algorithm to reject the ciphertext if the wrong private key has been used. However, there is nothing intrinsic to our formal definitions or security models that guarantees that decryption will always output "$\perp$" when the wrong private key is used, though such a robustness property is clearly desirable (as it would prevent attacks where an adversary fooled a user into decrypting a ciphertext intended for another party to obtain a meaningful message upon which the decrypting party might then act). Robustness in this sense for standard public key encryption and IBE schemes is the subject of a recent paper of Abdalla *et al.* [2]. It would be interesting to attempt to extend their results to the multi-TA setting, but it should be noted that the authors of [2] have already established that the FO conversion of [14] does not preserve robustness in general.

Acknowledgements

This research was sponsored in part by the US Army Research Laboratory and the UK Ministry of Defence and was accomplished under Agreement Number W911NF-06-3-0001. The views and conclusions contained in this document are those of the authors and should not be interpreted as representing the official policies, either expressed or implied, of the US Army Research Laboratory, the US Government, the UK Ministry of Defense, or the UK Government. The US and UK Governments are authorized to reproduce and distribute reprints for Government purposes notwithstanding any copyright notation hereon.

The second author is supported by a Dorothy Hodgkin Postgraduate Award, funded by EPSRC and Vodafone and administered by Royal Holloway, University of London.

We are grateful to Nigel Smart for initiating a discussion concerning the anonymity properties of the Sakai-Kasahara IBE scheme and to the anonymous referees for many valuable comments and references.

References

1. Abdalla, M., Bellare, M., Catalano, D., Kiltz, E., Kohno, T., Lange, T., Malone-Lee, J., Neven, G., Paillier, P., Shi, H.: Searchable encryption revisited: Consistency properties, relation to anonymous IBE, and extensions. In: Shoup, V. (ed.) CRYPTO 2005. LNCS, vol. 3621, pp. 205–222. Springer, Heidelberg (2005)
2. Abdalla, M., Bellare, M., Namprempre, C., Neven, G.: Robust public-key and identity-based encryption (unpublished manuscript, 2008)
3. Barbosa, M., Farshim, P.: Efficient identity-based key encapsulation to multiple parties. In: Smart, N. (ed.) Cryptography and Coding 2005. LNCS, vol. 3796, pp. 428–441. Springer, Heidelberg (2005)
4. Bellare, M., Boldyreva, A., Desai, A., Pointcheval, D.: Key-privacy in public-key encryption. In: Boyd, C. (ed.) ASIACRYPT 2001. LNCS, vol. 2248, pp. 566–582. Springer, Heidelberg (2001)
5. Bellare, M., Rogaway, P.: Random oracles are practical: A paradigm for designing efficient protocols. In: ACM Conference on Computer and Communications Security, pp. 62–73 (1993)

6. Boneh, D., Franklin, M.K.: Identity-based encryption from the Weil pairing. In: Kilian, J. (ed.) CRYPTO 2001. LNCS, vol. 2139, pp. 213–229. Springer, Heidelberg (2001)
7. Boyen, X.: The $\mathtt{BB}_1$ identity-based cryptosystem: A standard for encryption and key encapsulation. IEEE P1363.3 (submission, 2006), `http://grouper.ieee.org/groups/1363/IBC/submissions/Boyen-bb1_ieee.pdf`
8. Boyen, X., Waters, B.: Anonymous hierarchical identity-based encryption (without random oracles). In: Dwork, C. (ed.) CRYPTO 2006. LNCS, vol. 4117, pp. 290–307. Springer, Heidelberg (2006)
9. Bradshaw, R.W., Holt, J.E., Seamons, K.E.: Concealing complex policies with hidden credentials. In: Atluri, V., Pfitzmann, B., McDaniel, P.D. (eds.) ACM Conference on Computer and Communications Security, pp. 146–157. ACM, New York (2004)
10. Chase, M.: Multi-authority attribute based encryption. In: Vadhan, S.P. (ed.) TCC 2007. LNCS, vol. 4392, pp. 515–534. Springer, Heidelberg (2007)
11. Chen, L., Cheng, Z., Malone-Lee, J., Smart, N.P.: An efficient ID-KEM based on the Sakai-Kasahara key construction. Cryptology ePrint Archive, Report 2005/224 (2005), `http://eprint.iacr.org/`
12. Cocks, C.: An identity based encryption scheme based on quadratic residues. In: Honary, B. (ed.) Cryptography and Coding 2001. LNCS, vol. 2260, pp. 360–363. Springer, Heidelberg (2001)
13. Fujisaki, E., Okamoto, T.: How to enhance the security of public-key encryption at minimum cost. In: Imai, H., Zheng, Y. (eds.) PKC 1999. LNCS, vol. 1560, pp. 53–68. Springer, Heidelberg (1999)
14. Fujisaki, E., Okamoto, T.: Secure integration of asymmetric and symmetric encryption schemes. In: Wiener, M. (ed.) CRYPTO 1999. LNCS, vol. 1666, pp. 537–554. Springer, Heidelberg (1999)
15. Galbraith, S.D., Paterson, K.G., Smart, N.P.: Pairings for cryptographers. Cryptology ePrint Archive, Report 2006/165 (2006), `http://eprint.iacr.org/`
16. Gentry, C.: Practical identity-based encryption without random oracles. In: Vaudenay, S. (ed.) EUROCRYPT 2006. LNCS, vol. 4004, pp. 445–464. Springer, Heidelberg (2006)
17. Halevi, S.: A sufficient condition for key-privacy. Cryptology ePrint Archive, Report 2005/005 (2005), `http://eprint.iacr.org/`
18. Holt, J.E.: Key privacy for identity based encryption. Cryptology ePrint Archive, Report 2006/120 (2006), `http://eprint.iacr.org/`
19. Holt, J.E., Bradshaw, R.W., Seamons, K.E., Orman, H.K.: Hidden credentials. In: Jajodia, S., Samarati, P., Syverson, P.F. (eds.) WPES, pp. 1–8. ACM, New York (2003)
20. Kitagawa, T., Yang, P., Hanaoka, G., Zhang, R., Watanabe, H., Matsuura, K., Imai, H.: Generic transforms to acquire CCA-security for identity based encryption: The cases of FOpkc and REACT. In: Batten, L.M., Safavi-Naini, R. (eds.) ACISP 2006. LNCS, vol. 4058, pp. 348–359. Springer, Heidelberg (2006)
21. Sakai, R., Kasahara, M.: ID based cryptosystems with pairing on elliptic curve. Cryptology ePrint Archive, Report 2003/054 (2003), `http://eprint.iacr.org/`
22. Sakai, R., Ohgishi, K., Kasahara, M.: Cryptosystems based on pairing. In: The 2000 Sympoium on Cryptography and Information Security, Okinawa, Japan, January, pp. 26–28 (2000)
23. Shamir, A.: Identity-based cryptosystems and signature schemes. In: Blakely, G.R., Chaum, D. (eds.) CRYPTO 1984. LNCS, vol. 196, pp. 47–53. Springer, Heidelberg (1985)

24. Wang, S., Cao, Z.: Practical identity-based encryption (IBE) in multiple PKG environments and its applications. Cryptology ePrint Archive, Report 2007/100 (2007), http://eprint.iacr.org/
25. Yang, P., Kitagawa, T., Hanaoka, G., Zhang, R., Matsuura, K., Imai, H.: Applying Fujisaki-Okamoto to identity-based encryption. In: Fossorier, M.P.C., Imai, H., Lin, S., Poli, A. (eds.) AAECC 2006. LNCS, vol. 3857, pp. 183–192. Springer, Heidelberg (2006)

Author Index

Lecture Notes in Computer Science

Sublibrary 4: Security and Cryptology

For information about Vols. 1–4004
please contact your bookseller or Springer

Vol. 5209: S.D. Galbraith, K.G. Paterson (Eds.), Pairing-Based Cryptography – Pairing 2008. XI, 377 pages. 2008.

Vol. 5157: D. Wagner (Ed.), Advances in Cryptology – CRYPTO 2008. XIV, 594 pages. 2008.

Vol. 5155: R. Safavi-Naini (Ed.), Information Theoretic Security. XI, 249 pages. 2008.

Vol. 5154: E. Oswald, P. Rohatgi (Eds.), Cryptographic Hardware and Embedded Systems – CHES 2008. XIII, 445 pages. 2008.

Vol. 5143: G. Tsudik (Ed.), Financial Cryptography and Data Security. XIII, 326 pages. 2008.

Vol. 5137: D. Zamboni (Ed.), Detection of Intrusions and Malware, and Vulnerability Assessment. X, 279 pages. 2008.

Vol. 5134: N. Borisov, I. Goldberg (Eds.), Privacy Enhancing Technologies. X, 237 pages. 2008.

Vol. 5107: Y. Mu, W. Susilo, J. Seberry (Eds.), Information Security and Privacy. XIII, 480 pages. 2008.

Vol. 5086: K. Nyberg (Ed.), Fast Software Encryption. XI, 489 pages. 2008.

Vol. 5057: S.F. Mjølsnes, S. Mauw, S.K. Katsikas (Eds.), Public Key Infrastructure. X, 239 pages. 2008.

Vol. 5037: S.M. Bellovin, R. Gennaro, A.D. Keromytis, M. Yung (Eds.), Applied Cryptography and Network Security. XI, 508 pages. 2008.

Vol. 5023: S. Vaudenay (Ed.), Progress in Cryptology – AFRICACRYPT 2008. XI, 415 pages. 2008.

Vol. 5019: J.A. Onieva, D. Sauveron, S. Chaumette, D. Gollmann, K. Markantonakis (Eds.), Information Security Theory and Practices. XII, 151 pages. 2008.

Vol. 4991: L. Chen, Y. Mu, W. Susilo (Eds.), Information Security Practice and Experience. XIII, 420 pages. 2008.

Vol. 4990: D. Pei, M. Yung, D. Lin, C. Wu (Eds.), Information Security and Cryptology. XII, 534 pages. 2008.

Vol. 4986: M.J.B. Robshaw, O. Billet (Eds.), New Stream Cipher Designs. VIII, 295 pages. 2008.

Vol. 4968: P. Lipp, A.-R. Sadeghi, K.-M. Koch (Eds.), Trusted Computing - Challenges and Applications. X, 191 pages. 2008.

Vol. 4965: N. Smart (Ed.), Advances in Cryptology – EUROCRYPT 2008. XIII, 564 pages. 2008.

Vol. 4964: T. Malkin (Ed.), Topics in Cryptology – CT-RSA 2008. XI, 437 pages. 2008.

Vol. 4948: R. Canetti (Ed.), Theory of Cryptography. XII, 645 pages. 2008.

Vol. 4939: R. Cramer (Ed.), Public Key Cryptography – PKC 2008. XIII, 397 pages. 2008.

Vol. 4920: Y.Q. Shi (Ed.), Transactions on Data Hiding and Multimedia Security III. IX, 91 pages. 2008.

Vol. 4896: A. Alkassar, M. Volkamer (Eds.), E-Voting and Identity. XII, 189 pages. 2007.

Vol. 4893: S.W. Golomb, G. Gong, T. Helleseth, H.-Y. Song (Eds.), Sequences, Subsequences, and Consequences. X, 219 pages. 2007.

Vol. 4890: F. Bonchi, E. Ferrari, B. Malin, Y. Saygin (Eds.), Privacy, Security, and Trust in KDD. IX, 173 pages. 2008.

Vol. 4887: S.D. Galbraith (Ed.), Cryptography and Coding. XI, 423 pages. 2007.

Vol. 4886: S. Dietrich, R. Dhamija (Eds.), Financial Cryptography and Data Security. XII, 390 pages. 2007.

Vol. 4876: C. Adams, A. Miri, M. Wiener (Eds.), Selected Areas in Cryptography. X, 409 pages. 2007.

Vol. 4867: S. Kim, M. Yung, H.-W. Lee (Eds.), Information Security Applications. XIII, 388 pages. 2008.

Vol. 4861: S. Qing, H. Imai, G. Wang (Eds.), Information and Communications Security. XIV, 508 pages. 2007.

Vol. 4859: K. Srinathan, C.P. Rangan, M. Yung (Eds.), Progress in Cryptology – INDOCRYPT 2007. XI, 426 pages. 2007.

Vol. 4856: F. Bao, S. Ling, T. Okamoto, H. Wang, C. Xing (Eds.), Cryptology and Network Security. XII, 283 pages. 2007.

Vol. 4833: K. Kurosawa (Ed.), Advances in Cryptology – ASIACRYPT 2007. XIV, 583 pages. 2007.

Vol. 4817: K.-H. Nam, G. Rhee (Eds.), Information Security and Cryptology - ICISC 2007. XIII, 367 pages. 2007.

Vol. 4812: P. McDaniel, S.K. Gupta (Eds.), Information Systems Security. XIII, 322 pages. 2007.

Vol. 4784: W. Susilo, J.K. Liu, Y. Mu (Eds.), Provable Security. X, 237 pages. 2007.

Vol. 4779: J.A. Garay, A.K. Lenstra, M. Mambo, R. Peralta (Eds.), Information Security. XIII, 437 pages. 2007.

Vol. 4776: N. Borisov, P. Golle (Eds.), Privacy Enhancing Technologies. X, 273 pages. 2007.

Vol. 4752: A. Miyaji, H. Kikuchi, K. Rannenberg (Eds.), Advances in Information and Computer Security. XIII, 460 pages. 2007.

Vol. 4734: J. Biskup, J. López (Eds.), Computer Security – ESORICS 2007. XIV, 628 pages. 2007.

Vol. 4727: P. Paillier, I. Verbauwhede (Eds.), Cryptographic Hardware and Embedded Systems - CHES 2007. XIV, 468 pages. 2007.

Vol. 4691: T. Dimitrakos, F. Martinelli, P.Y.A. Ryan, S. Schneider (Eds.), Formal Aspects in Security and Trust. VIII, 285 pages. 2007.

Vol. 4677: A. Aldini, R. Gorrieri (Eds.), Foundations of Security Analysis and Design IV. VII, 325 pages. 2007.

Vol. 4657: C. Lambrinoudakis, G. Pernul, A.M. Tjoa (Eds.), Trust, Privacy and Security in Digital Business. XIII, 291 pages. 2007.

Vol. 4637: C. Kruegel, R. Lippmann, A. Clark (Eds.), Recent Advances in Intrusion Detection. XII, 337 pages. 2007.

Vol. 4631: B. Christianson, B. Crispo, J.A. Malcolm, M. Roe (Eds.), Security Protocols. IX, 347 pages. 2007.

Vol. 4622: A. Menezes (Ed.), Advances in Cryptology - CRYPTO 2007. XIV, 631 pages. 2007.

Vol. 4593: A. Biryukov (Ed.), Fast Software Encryption. XI, 467 pages. 2007.

Vol. 4586: J. Pieprzyk, H. Ghodosi, E. Dawson (Eds.), Information Security and Privacy. XIV, 476 pages. 2007.

Vol. 4582: J. López, P. Samarati, J.L. Ferrer (Eds.), Public Key Infrastructure. XI, 375 pages. 2007.

Vol. 4579: B.M. Hämmerli, R. Sommer (Eds.), Detection of Intrusions and Malware, and Vulnerability Assessment. X, 251 pages. 2007.

Vol. 4575: T. Takagi, T. Okamoto, E. Okamoto, T. Okamoto (Eds.), Pairing-Based Cryptography – Pairing 2007. XI, 408 pages. 2007.

Vol. 4567: T. Furon, F. Cayre, G. Doërr, P. Bas (Eds.), Information Hiding. XI, 393 pages. 2008.

Vol. 4521: J. Katz, M. Yung (Eds.), Applied Cryptography and Network Security. XIII, 498 pages. 2007.

Vol. 4515: M. Naor (Ed.), Advances in Cryptology - EUROCRYPT 2007. XIII, 591 pages. 2007.

Vol. 4499: Y.Q. Shi (Ed.), Transactions on Data Hiding and Multimedia Security II. IX, 117 pages. 2007.

Vol. 4464: E. Dawson, D.S. Wong (Eds.), Information Security Practice and Experience. XIII, 361 pages. 2007.

Vol. 4462: D. Sauveron, K. Markantonakis, A. Bilas, J.-J. Quisquater (Eds.), Information Security Theory and Practices. XII, 255 pages. 2007.

Vol. 4450: T. Okamoto, X. Wang (Eds.), Public Key Cryptography – PKC 2007. XIII, 491 pages. 2007.

Vol. 4437: J.L. Camenisch, C.S. Collberg, N.F. Johnson, P. Sallee (Eds.), Information Hiding. VIII, 389 pages. 2007.

Vol. 4392: S.P. Vadhan (Ed.), Theory of Cryptography. XI, 595 pages. 2007.

Vol. 4377: M. Abe (Ed.), Topics in Cryptology – CT-RSA 2007. XI, 403 pages. 2006.

Vol. 4356: E. Biham, A.M. Youssef (Eds.), Selected Areas in Cryptography. XI, 395 pages. 2007.

Vol. 4341: P.Q. Nguyên (Ed.), Progress in Cryptology - VIETCRYPT 2006. XI, 385 pages. 2006.

Vol. 4332: A. Bagchi, V. Atluri (Eds.), Information Systems Security. XV, 382 pages. 2006.

Vol. 4329: R. Barua, T. Lange (Eds.), Progress in Cryptology - INDOCRYPT 2006. X, 454 pages. 2006.

Vol. 4318: H. Lipmaa, M. Yung, D. Lin (Eds.), Information Security and Cryptology. XI, 305 pages. 2006.

Vol. 4307: P. Ning, S. Qing, N. Li (Eds.), Information and Communications Security. XIV, 558 pages. 2006.

Vol. 4301: D. Pointcheval, Y. Mu, K. Chen (Eds.), Cryptology and Network Security. XIII, 381 pages. 2006.

Vol. 4300: Y.Q. Shi (Ed.), Transactions on Data Hiding and Multimedia Security I. IX, 139 pages. 2006.

Vol. 4298: J.K. Lee, O. Yi, M. Yung (Eds.), Information Security Applications. XIV, 406 pages. 2007.

Vol. 4296: M.S. Rhee, B. Lee (Eds.), Information Security and Cryptology – ICISC 2006. XIII, 358 pages. 2006.

Vol. 4284: X. Lai, K. Chen (Eds.), Advances in Cryptology – ASIACRYPT 2006. XIV, 468 pages. 2006.

Vol. 4283: Y.Q. Shi, B. Jeon (Eds.), Digital Watermarking. XII, 474 pages. 2006.

Vol. 4266: H. Yoshiura, K. Sakurai, K. Rannenberg, Y. Murayama, S.-i. Kawamura (Eds.), Advances in Information and Computer Security. XIII, 438 pages. 2006.

Vol. 4258: G. Danezis, P. Golle (Eds.), Privacy Enhancing Technologies. VIII, 431 pages. 2006.

Vol. 4249: L. Goubin, M. Matsui (Eds.), Cryptographic Hardware and Embedded Systems - CHES 2006. XII, 462 pages. 2006.

Vol. 4237: H. Leitold, E.P. Markatos (Eds.), Communications and Multimedia Security. XII, 253 pages. 2006.

Vol. 4236: L. Breveglieri, I. Koren, D. Naccache, J.-P. Seifert (Eds.), Fault Diagnosis and Tolerance in Cryptography. XIII, 253 pages. 2006.

Vol. 4219: D. Zamboni, C. Krügel (Eds.), Recent Advances in Intrusion Detection. XII, 331 pages. 2006.

Vol. 4189: D. Gollmann, J. Meier, A. Sabelfeld (Eds.), Computer Security – ESORICS 2006. XI, 548 pages. 2006.

Vol. 4176: S.K. Katsikas, J. López, M. Backes, S. Gritzalis, B. Preneel (Eds.), Information Security. XIV, 548 pages. 2006.

Vol. 4117: C. Dwork (Ed.), Advances in Cryptology - CRYPTO 2006. XIII, 621 pages. 2006.

Vol. 4116: R. De Prisco, M. Yung (Eds.), Security and Cryptography for Networks. XI, 366 pages. 2006.

Vol. 4107: G. Di Crescenzo, A. Rubin (Eds.), Financial Cryptography and Data Security. XI, 327 pages. 2006.

Vol. 4083: S. Fischer-Hübner, S. Furnell, C. Lambrinoudakis (Eds.), Trust and Privacy in Digital Business. XIII, 243 pages. 2006.

Vol. 4064: R. Büschkes, P. Laskov (Eds.), Detection of Intrusions and Malware & Vulnerability Assessment. X, 195 pages. 2006.

Vol. 4058: L.M. Batten, R. Safavi-Naini (Eds.), Information Security and Privacy. XII, 446 pages. 2006.

Vol. 4047: M.J.B. Robshaw (Ed.), Fast Software Encryption. XI, 434 pages. 2006.

Vol. 4043: A.S. Atzeni, A. Lioy (Eds.), Public Key Infrastructure. XI, 261 pages. 2006.